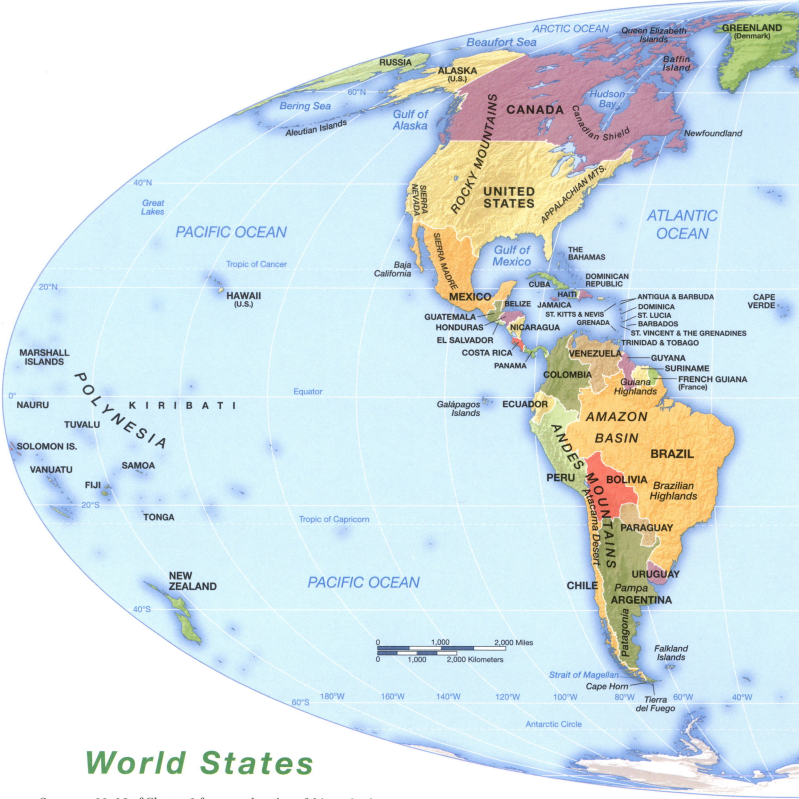

World States

See pages 11–13 of Chapter 1 for an explanation of this projection.

HUMAN GEOGRAPHY
Places and Regions in Global Context

Paul L. Knox
Virginia Tech

Sallie A. Marston
University of Arizona

HUMAN GEOGRAPHY
Places and Regions in Global Context

SIXTH EDITION

PEARSON

Boston   Columbus   Indianapolis   New York   San Francisco   Upper Saddle River   Amsterdam
Cape Town   Dubai   London   Madrid   Milan   Munich   Paris   Montréal   Toronto   Delhi
Mexico City   São Paulo   Sydney   Hong Kong   Seoul   Singapore   Taipei   Tokyo

Geography Editor: Christian Botting
Marketing Manager: Maureen McLaughlin
Project Editor: Anton Yakovlev
VP/Executive Director, Development: Carol Trueheart
Development Editor: Erin Mulligan
Senior Media Producer: Angela Bernhardt
Media Producer: Tim Hainley
Assistant Editor: Sean Hale
Editorial Assistant: Bethany Sexton
Marketing Assistant: Nicola Houston
Managing Editor, Geosciences and Chemistry: Gina M. Cheselka
Project Manager, Production: Wendy Perez
Full Service/Composition: PreMediaGlobal
Full Service Project Manager: Jenna Gray
Senior Art Specialist: Connie Long
Illustrations: Kevin Lear, Spatial Graphics
Design Manager: Mark Ong
Interior and Cover Design: tani hasegawa
Photo Manager: Maya Melenchuk
Photo Researcher: Lily Ferguson, Bill Smith Group
Text Permissions Manager: Beth Wollar
Text Permissions Researcher: Sheri Gilbert
Operations Specialist: Michael Penne

Cover Photo Credit: © Yann Arthus-Bertrand/Altitude

Credits and acknowledgments of content borrowed from other sources and reproduced, with permission, in this textbook appear on the appropriate page beginning on p. 423 for photographs and within the figure captions for all maps and line art.

Library of Congress Cataloging-in-Publication Data

Knox, Paul L.
  Human geography: places and regions in global context / Paul L. Knox, Sallie A.
    Marston. — 6th ed.
    p. cm.
  Rev. ed. of: Places and regions in global context : Prentice Hall, c2010.
  ISBN-13: 978-0-321-76966-4
  1. Human geography. I. Marston, Sallie A. II. Knox, Paul L. Places and regions
    in global context. III. Title.
  GF41.K56 2013
  304.2—dc23
                                        2011039494

1 2 3 4 5 6 7 8 9 10—CRK—15 14 13 12 11

ISBN-10: 0-321-76966-X; ISBN-13: 978-0-321-76966-4 (Student Edition)
ISBN-10: 0-321-81933-0; ISBN-13: 978-0321-81933-8 (Instructor's Review Copy)

Brief Contents

Contents

Chapter 2   THE CHANGING GLOBAL CONTEXT   34

Chapter 3   POPULATION GEOGRAPHY   66

Chapter 6   INTERPRETING PLACES AND LANDSCAPES   184

Chapter 10   URBANIZATION   350

List of Maps

Preface

A nation, like a person, has a mind—a mind that must be kept informed and alert, that must know itself, that understands the hopes and needs of its neighbors—all the other nations that live within the narrowing circle of the world.

> *Franklin Roosevelt, Third Inaugural Address,*
> *Monday, January 20, 1941*

Most people have an understanding of what their own lives are like and know a good deal about their own neighborhood and perhaps even something of the larger city and state in which they live. Yet, even as the countries and regions of the world become more interconnected, most of us still know very little about the lives of people on the other side of our country, or in other societies, or about the ways the lives of those people connect to our own. In order to change the world, to make it a better place for all people, we need to understand not just our little corner of it, but the whole of it—the broad sweep of human geography that constitutes the larger world of which our small corners are just a part.

This book provides an introduction to human geography that will help young men and women to understand critically the world in which they live. To study human geography, to put it simply, is to study the dynamic and complex relationships between peoples and the worlds they inhabit. Our book gives students the basic geographical tools and concepts needed to understand the complexity of places and regions, and to appreciate the interconnections between their own lives and those of people in different parts of the world—to make the world a better place.

NEW TO THE SIXTH EDITION

The sixth edition of *Places and Regions in Global Context* represents a thorough revision. Every part of the book was examined carefully with the dual goals of keeping topics and data current and improving the clarity of the text and the graphics. We have also sought to enhance the utility of the book for both instructors and students.

- Chapter-opening anecdotes introduce students to the subject matter with interesting and varied topics.
- A list of Learning Outcomes in each chapter opener offers students a structured learning path to guide them through the main learning goals for the chapter. These Learning Outcomes are revisited at the end of each chapter with summaries of chapter content correlated to the Learning Outcomes, for the final step of the structured learning path.
- Throughout every chapter, conceptual Apply Your Knowledge questions are integrated within the text, giving students a chance to pause and apply their understanding for a more active learning approach.

- We have increased the focus on basic introductory human geography content in Chapter 1, allowing the text to be more accessible to a wide range of introductory courses and students.
- Material from the final chapter of the fifth edition, on Future Geographies, has been updated and integrated into each of the thematic chapters, placing the "futures" content in context.
- The boxed feature essays on Visualizing Geography have been fundamentally redesigned to incorporate edgy, modern applications and visualizations of geography data.
- The sixth edition also incorporates a comprehensive updating of all of the data, maps, photographs, and illustrative examples.
- We have added or expanded upon quite a few topics, including the global financial crisis; the credit crunch and mortgage foreclosures; climate change and issues of sustainability; global interdependence and food supplies; water supply problems; oil and energy; geopolitics; political ecology; religion; biotechnology and agricultural systems; fast food and slow food; virtual social networks; current events like the Arab Spring and the Japanese earthquake and tsunami; global film and music; big-box retailing and global commodity chains; the "experience economy"; and the landscapes of the polycentric metropolis. These changes are designed to ensure that we offer the most up-to-date coverage in the field of human geography.
- The new MasteringGeography™ platform is linked to the Learning Outcomes and contains a wide range of resources and activities designed to reinforce basic concepts in human geography, including MapMaster™ interactive maps, Google Earth™ activities, geography videos, and more.

OBJECTIVE AND APPROACH

The objective of the book is to introduce the study of human geography by providing not only a body of knowledge about the creation of places and regions, but also an understanding of the interdependence of places and regions in a globalizing world. The approach is aimed at establishing an intellectual foundation that will enable a lifelong and life-sustaining geographical imagination: an essential tool for today's students in order to confront tomorrow's global, national, regional, and local challenges.

The book takes a fresh approach to human geography, reflecting the major changes that have recently been impressed on global, regional, and local landscapes. These changes include the globalization of industry and the related rapid rise of China and India as economic powerhouses, the upwelling of ethnic regionalisms on the heels of decolonization and the formation of new states, the movement of peoples around the world in search of better lives, the physical restructuring of cities, the transformation of traditional agricultural practices throughout much of the world, global environmental change and the movement for sustainability, the eruptions of war and the struggles for peace, and the emerging trend toward transnational political and economic organizations. The approach used in *Places and Regions in Global Context* provides access not only to the new ideas, concepts, and theories that address these changes, but also to the fundamentals of human geography: the principles, concepts, theoretical frameworks, and basic knowledge that are necessary to more specialized studies.

The most distinctive feature of this approach is that it emphasizes the interdependence of both places and processes in different parts of the globe. In overall terms, this approach is designed to provide an understanding of relationships between global processes and the local places in which they unfold. It follows that one of the chief organizing principles is how globalization frames the social and cultural construction of particular places and regions. This approach has several advantages. For example:

- It captures aspects of human geography that are among the most compelling in the contemporary world—the geographical bases of cultural diversity and their impacts on everyday life.

- It encompasses the salient aspects of new emphases in academic human geography—the new emphasis on sustainability and its role in the social construction of spaces and places.

- It makes for an easier marriage between topical and regional material by emphasizing how processes link them—technological innovation and the varying ways technology is adopted and modified by people in particular places.

- It facilitates meaningful comparisons between places in different parts of the world—how the core-generated industrialization of agriculture shapes gender relations in households both in the core and the periphery.

In short, the textbook is designed to focus on geographical processes and to provide an understanding of the interdependence among places and regions without losing sight of their individuality and uniqueness.

Several important themes are woven into each chapter, integrating them into the overall approach:

- the relationship between global processes and their local manifestations;

- the interdependence of people and places, especially the interactive relationships between core regions and peripheral regions;

- the continuing transformation of the political economy of the world system, and of nations, regions, cities, and localities;

- the social and cultural differences that are embedded in human geographies (especially the differences that relate to race, ethnicity, gender, age, and class).

CHAPTER ORGANIZATION

The organization of the book is innovative in several ways. First, the chapters are organized so that the conceptual framework—why geography matters in a globalizing world—is laid out in Chapters 1 and 2 and then deployed in thematic chapters (Chapters 3 through 11). Second, the conceptual framework of the book requires the inclusion of two introductory chapters rather than the usual one. The first describes the basics of a geographic perspective; the second explains the value of the globalization approach.

Third, the distinctive chapter ordering within the book follows the logic of moving from less complex to more complex systems of human social and economic organization, always highlighting the interaction between people and the world around them. The first thematic chapter (Chapter 3) focuses on human population. Its early placement in the book reflects the central importance of people in understanding geography. Chapter 4 deals with the relationship between people and the environment as it is mediated by technology. This chapter explores environmental problems and establishes a central theme: that all human geographical issues are about how people negotiate their environment—whether the natural or the built environment.

The chapter on nature, society, and technology is followed by Chapter 5 on cultural geography. The intention in positioning the cultural chapter here is to signal that culture is the primary medium through which people operate and understand their place in the world. In Chapter 6, the impact of cultural processes on the landscape is explored, together with the ways in which landscape shapes cultural processes.

In Chapter 7, the book begins the move toward more complex concepts and systems of human organization by concentrating on economic development. The focus of Chapter 8 is agriculture. The placement of agriculture after economic development reflects the overall emphasis on globalization. This chapter shows how processes of globalization and economic development have led to the industrialization of agriculture at the expense of more traditional agricultural systems and practices.

The final three thematic chapters cover political geography (Chapter 9), urbanization (Chapter 10), and city structure (Chapter 11). Devoting two chapters to urban geography, rather than a more conventional single chapter, is an important indication of how globalization increasingly leads to urbanization of the world's people and places.

Features

The hallmark feature of our book is the global framework that promotes a strong connection between topical and regional material by emphasizing how their processes are linked (e.g., technological innovation and the varying ways technology is adopted and modified by people and places). This makes for a contemporary approach to human geography, reflecting many trends in the discipline, such as the globalization of industry, the upwelling of ethnic regionalisms on the heels of decolonization and new state formation, and the trend toward transnational political and economic organizations. The global framework also facilitates meaningful comparisons between people and places in different parts of the world, such as how the core-generated industrialization of agriculture shapes gender relations in households, both in the core and in the periphery. It allows us to present the relevant aspects of different emphases in academic human geography (e.g., geopolitics and its role in the social construction of spaces and places). At the same time, a focus on fundamentals gives students access not only to the new ideas, concepts, and theories that address the changes mentioned above but also to the fundamentals of human geography: the principles, concepts, theoretical frameworks, and basic knowledge. The book's thematic structure weaves several important themes into every chapter: the interdependence of people and places, especially the interactive relationships between the core and the periphery; social and cultural differences that are embedded in human geographies, especially the differences that relate to race, ethnicity, gender, age, and class; the relationship between global processes and their local manifestations; and the continuing transformation of the political economy of the world system and nations, regions, cities, and localities.

To signal the freshness of the approach, the book features a superior cartographic program, consisting of rich, diverse, and fully updated maps that help professors better teach their students the important spatial elements inherent to human geography. The cartography program features numerous compound figures that combine maps with photographs and line drawings. The pedagogy of the book employs three different boxed features—"Geography Matters," "Visualizing Geography," and "Window on the World."

Geography Matters Geography Matters boxes examine one of the key concepts of the chapter, providing an extended example of its meaning and implications through both visual illustration and text. The Geography Matters features demonstrate to students that the focus of human geography is on real-world problems.

Visualizing Geography Visualizing Geography boxes treat key concepts of the chapter through modern applications and visualizations of data. These boxes help students to "see" the ways that geography shapes their daily lives.

Window on the World Window on the World boxes take a key concept and explore its application in a particular location. This feature allows students to appreciate the relevance of geographic concepts to world events and brings far-flung places closer to their comprehension.

CONCLUSION

The idea for this book evolved from conversations between the authors and colleagues about how to teach human geography in colleges and universities. Our intent was to find a way not only to capture the exciting changes that are rewriting the world's landscapes and reorganizing the spatial relationships between people but also to demonstrate convincingly why the study of geography matters. Our aim was to show why a geographical imagination is important, how it can lead to an understanding of the world and its constituent places and regions, and how it has practical relevance in many spheres of life.

The first edition of this book was written at the culmination of a significant period of reform in geographic education. One important outcome of this reform was the inclusion of geography as a core subject in Goals 2000: Educate America Act (Public Law 103–227). Another was the publication of a set of national geography standards for K–12 education (*Geography for Life,* published by National Geographic Research and Education for the American Geographical Society, the Association of American Geographers, the National Council for Geographic Education, and the National Geographic Society, 1994). This sixth edition builds on these reforms, offering a fresh and compelling approach to college-level geography.

Acknowledgments

We are indebted to many people for their assistance, advice, and constructive criticism in the course of preparing this book. Among those who provided comments on drafts of the various editions of this book are the following professors:

Christopher A. Airriess (*Ball State University*)
Stuart Aitken (*University of California at San Diego*)
Kevin Archer (*University of South Florida*)
Sarah Bednarz (*Texas A&M University*)
Brian J. L. Berry (*University of Texas at Dallas*)
Brian W. Blouet (*College of William and Mary*)
George O. Brown, Jr. (*Boston College*)
Michael P. Brown (*University of Washington*)
Henry W. Bullamore (*Frostburg State University*)
Edmunds V. Bunske (*University of Delaware*)
Craig Campbell (*Youngstown State University*)
Dylan Clark (*University of Colorado*)
David B. Cole (*University of Northern Colorado*)
Mario Cora (*University of Phoenix*)
Jerry Crampton (*George Mason University*)
Christine Dando (*University of Nebraska, Omaha*)
Fiona M. Davidson (*University of Arkansas*)
Jeff DeGrave (*University of Wisconsin at Eau Claire*)
Benjamin Dixon (*SUNY College at Oneonta*)
Vernon Domingo (*Bridgewater State College*)
Patricia Ehrkamp (*Miami University*)
Nancy Ettlinger (*The Ohio State University*)
Paul B. Frederic (*University of Maine*)
Kurtis G. Fuelhart (*Shippensburg University*)
Gary Fuller (*University of Hawaii at Manoa*)
Wilbert Gesler (*University of North Carolina*)
Melissa Gilbert (*Temple University*)
Jeffrey Allman Gritzner (*University of Montana*)
David Gwynn (*Michigan State University*)
Douglas Heffington (*Middle Tennessee State University*)
Andrew Herod (*University of Georgia*)
Peter Hugill (*Texas A&M University*)
David Icenogle (*Auburn University*)
Mary Jacob (*Mount Holyoke College*)
Wendy Jepson (*Texas A&M University*)
Douglas L. Johnson (*Clark University*)
Jin-Kyu Jung (*University of North Dakota*)
Colleen E. Keen (*Minnesota State University*)
Paul Kelley (*University of Nebraska, Lincoln*)
Thomas Klak (*Miami University*)
James Kus (*California State University, Fresno*)
David Lanegran (*Macalester College*)
James Lindberg (*University of Iowa*)
Max Lu (*Kansas State University*)
John C. Lowe (*George Washington University*)
Donald Lyons (*University of North Texas*)
Brian McCabe (*University of New Mexico at Valencia*)
James McCarthy (*Penn State University*)

John Milbauer *(Northeastern State University)*
Byron Miller *(University of Cincinnati)*
Roger Miller *(University of Minnesota)*
Don Mitchell *(Syracuse University)*
Wendy Mitteager *(SUNY College at Oneonta)*
Woodrow W. Nichols, Jr. *(North Carolina Central University)*
Richard Pillsbury *(Georgia State University)*
James Proctor *(University of California at Santa Barbara)*
Mark Purcell *(University of Washington)*
Jeffrey Richetto *(University of Alabama)*
Andrew Schoolmaster *(University of North Texas)*
David Schul *(The Ohio State University, Marion)*
Alex Standish *(Rutgers University)*
Debra Straussfogel *(University of New Hampshire)*
Johnathan Walker *(James Madison University)*
Gerald R. Webster *(University of Alabama)*
Lisa Westwood *(Ruidoso Branch Community College)*
Joseph S. Wood *(George Mason University)*
Wilbur Zelinsky *(Penn State University)*

Special thanks go to our very capable project manager, Anton Yakovlev, and Erin Mulligan, our smart and assiduous development editor, as well as to Jeff Wilson for his swift and insightful research assistance. We thank as well our Pearson team, including Christian Botting, Connie Long, Wendy Perez, Bethany Sexton, and Jenna Gray at PreMediaGlobal. For photo research we thank Lily Ferguson of Bill Smith Group, and for the graphics program we thank Kevin Lear of Spatial Graphics.

Finally, a number of colleagues gave generously of their time and expertise in guiding our thoughts, making valuable suggestions, and providing materials: Alejandro A. Alonso (University of Southern California), Martin Cadwallader (University of Wisconsin), Audra el Vilaly (University of Arizona), Banu Gokariksel (University of North Carolina, Chapel Hill), John Paul Jones, III (University of Arizona), Cindi Katz (City University of New York), Diana Liverman (University of Arizona), Brian Marks (University of Arizona), Sarah Moore (University of Arizona), Sara Smith (University of North Carolina, Chapel Hill), Paul Robbins (University of Arizona), Ian Shaw (University of Glasgow), Keith Woodward (University of Wisconsin).

Paul L. Knox
Sallie A. Marston

About the Authors

Paul L. Knox

Paul Knox received his PhD in Geography from the University of Sheffield, England. After teaching in the United Kingdom for several years, he moved to the United States to take up a position as professor of urban affairs and planning at Virginia Tech. His teaching centers on urban and regional development, with an emphasis on comparative study. He has received the university's award for teaching excellence. He has written several books on aspects of economic geography, social geography, and urbanization; serves on the editorial board of several scientific journals; and is co-editor of a series of books on world cities. In 2008, Professor Knox received the Distinguished Scholarship Award from the Association of American Geographers. He is currently a University Distinguished Professor in the College of Architecture and Urban Studies at Virginia Tech.

Sallie A. Marston

Sallie Marston received her PhD in Geography from the University of Colorado, Boulder. She is a full professor in the School of Geography and Development at the University of Arizona. Her undergraduate teaching focuses on political and cultural geography through innovative forms of pedagogy. She is the recipient of the College of Social and Behavioral Sciences' Outstanding Teaching Award as well as the Graduate College's Graduate Mentor Award. She is the co-editor of five books and author or co-author of over 60 journal articles and book chapters and currently the Principal Investigator of a National Science Foundation grant on art and science collaborations. She serves on the editorial board of several scientific journals.

About Our Sustainability Initiatives

Pearson recognizes the environmental challenges facing this planet, as well as acknowledges our responsibility in making a difference. This book is carefully crafted to minimize environmental impact. The binding, cover, and paper come from facilities that minimize waste, energy consumption, and the use of harmful chemicals. Pearson closes the loop by recycling every out-of-date text returned to our warehouse.

Along with developing and exploring digital solutions to our market's needs, Pearson has a strong commitment to achieving carbon-neutrality. As of 2009, Pearson became the first carbon- and climate-neutral publishing company. Since then, Pearson remains strongly committed to measuring, reducing, and offsetting our carbon footprint.

The future holds great promise for reducing our impact on Earth's environment, and Pearson is proud to be leading the way. We strive to publish the best books with the most up-to-date and accurate content, and to do so in ways that minimize our impact on Earth. To learn more about our initiatives, please visit **www.pearson.com/responsibility.**

PEARSON

The Teaching and Learning Package

The sixth edition provides a complete human geography program for teachers and students.

FOR INSTRUCTORS

MasteringGeography™ for Human Geography: Places and Regions in Global Context

The Mastering platform is the most effective and widely used tutorial, homework, and assessment system for the sciences. It helps instructors maximize class time with customizable, easy-to-assign, and automatically graded assessments that motivate students to learn outside of class and arrive prepared for lecture. These assessments can easily be customized and personalized for an instructor's individual teaching style. The powerful gradebook provides unique insight into student and class performance even before the first test. As a result, instructors can spend class time where students need it most. The Mastering system empowers students to take charge of their learning through activities aimed at different learning styles and engages them in learning science through practice and step-by-step guidance—at their convenience, 24/7. MasteringGeography offers:

- Assignable activities that include MapMaster™ interactive maps, Encounter Human Geography Google Earth™ Explorations, geography videos, Thinking Spatially and Data Analysis activities, end-of-chapter questions, reading quizzes, Test Bank questions, maps, and more.
- Student study area with MapMaster interactive maps, geography videos, glossary flash cards, "In the News" RSS feeds, reference maps, an optional Pearson eText, and more.

Each chapter of *Places and Regions in Global Context* ends with a section describing the online content available for that chapter. **www.masteringgeography.com**

MasteringGeography with Pearson eText gives students access to the text whenever and wherever they can access the Internet. The eText pages look exactly like the printed text and include powerful interactive and customization functions, including links to the multimedia.

Instructor Resource Manual (Download Only) for Human Geography: Places and Regions in Global Context by Wendy Mitteager (0321811666)

For download only, the *Instructor Resource Manual* is intended as a resource for both new and experienced instructors. It includes a variety of lecture outlines, additional source materials, teaching tips, advice about how to integrate visual supplements (including the Web-based resources), and various other ideas for the classroom. **www.pearsonhighered.com/irc**

TestGen® Computerized Test Bank (Download Only) for Human Geography: Places and Regions in Global Context by Paul Kaldjian (0321811674)

TestGen® is a computerized test generator that lets instructors view and edit Test Bank questions, transfer questions to tests, and print the test in a variety of customized formats. This Test Bank includes over 2,000 multiple-choice, true/false, and short-answer/essay questions. Questions are correlated to the revised U.S. National Geography Standards and Bloom's Taxonomy to help instructors better map the assessments against both broad and specific teaching and learning objectives. The Test Bank is also available in Microsoft Word®, and is importable into Blackboard. **www.pearsonhighered.com/irc**

Instructor Resource DVD for Human Geography: Places and Regions in Global Context (0321811135)

The *Instructor Resource Center on DVD* provides high-quality electronic versions of photos and illustrations from the book, as well as customizable PowerPoint™ lecture presentations by Wendy Mitteager, Classroom Response System questions in PowerPoint by Fiona Davidson, and the *Instructor Resource Manual* and *Test Bank* in Microsoft Word® and TestGen formats. The DVD includes all of the illustrations and photos from the text in presentation-ready JPEG files. For easy reference and identification, all resources are organized by chapter. **www.pearsonhighered.com/irc**

Teaching College Geography: A Practical Guide for Graduate Students and Early Career Faculty (0136054471)

This two-part resource provides a starting point for becoming an effective geography teacher from the very first day of class. Part One addresses "nuts-and-bolts" teaching issues. Part Two explores being an effective teacher in the field, supporting critical thinking with GIS and mapping technologies, engaging learners in large geography classes, and promoting awareness of international perspectives and geographic issues.

Aspiring Academics: A Resource Book for Graduate Students and Early Career Faculty (0136048919)

Drawing on several years of research, this set of essays is designed to help graduate students and early career faculty start their careers in geography and related social and environmental sciences. *Aspiring Academics* stresses the interdependence of teaching, research, and service—and the importance of achieving a healthy balance of professional and personal life—while doing faculty work. Each chapter provides accessible, forward-looking advice on topics that often cause the most stress in the first years of a college or university appointment.

Practicing Geography: Careers for Enhancing Society and the Environment (0-321-81115-1)

This book examines career opportunities for geographers and geospatial professionals in business, government, nonprofit, and educational sectors. A diverse group of academic and industry professionals share insights on career planning, networking, transitioning between employment sectors, and balancing work and home life. The book illustrates the value of geographic expertise and technologies through engaging profiles and case studies of geographers at work.

AAG Community Portal This website is intended to support community-based professional development in geography and related disciplines. Here you will find activities providing extended treatment of the topics covered in both books. The activities can be used in workshops, graduate seminars, brown bags, and mentoring programs offered on campus or within an academic department. You can also use the discussion boards and contributions tool to share advice and materials with others. **www.pearsonhighered.com/aag/**

FOR STUDENTS

MasteringGeography™ for Human Geography: Places and Regions in Global Context The Mastering platform is the most effective and widely used tutorial, homework, and assessment system for the sciences. The Mastering system empowers students to take charge of their learning through activities aimed at different learning styles, and engages them in learning science through practice and step-by-step guidance—at their convenience, 24/7. MasteringGeography offers:

- Assignable activities that include MapMaster™ interactive maps, *Encounter Human Geography* Google Earth Explorations, geography videos, Thinking Spatially and Data Analysis activities, end-of-chapter questions, reading quizzes, Test Bank questions, and more.
- Student study area with MapMaster™ interactive maps, geography videos, glossary flash cards, "In the News" RSS feeds, reference maps, an optional Pearson eText, and more. **www.masteringgeography.com**

MasteringGeography with Pearson eText gives students access to the text whenever and wherever they can access the Internet. The eText pages look exactly like the printed text and include powerful interactive and customization functions, including links to the multimedia.

Study Guide for Human Geography: Places and Regions in Global Context by Michael A. Pretes (0321811682) This study guide includes chapter notes, key terms, 8–10 review questions, and 10–15 activities per chapter and Web references throughout.

Goode's World Atlas 22nd edition (0321652002) *Goode's World Atlas* has been the world's premiere educational atlas since 1923, and for good reason. It features over 250 pages of maps, from definitive physical and political maps to important thematic maps that illustrate the spatial aspects of many important topics. The 22nd edition includes 160 pages of new, digitally produced reference maps, as well as new thematic maps on global climate change, sea level rise, CO_2 emissions, polar ice fluctuations, deforestation, extreme weather events, infectious diseases, water resources, and energy production.

Television for the Environment *Life* Human Geography Videos on DVD (0132416565) This three-DVD set is designed to enhance any human geography course. It contains 14 complete video programs (average length 25 minutes) covering a wide array of issues affecting people and places in the contemporary world, including international immigration, urbanization, global trade, poverty,

and environmental destruction. The videos included on these DVDs are offered at the highest quality to allow for full-screen viewing on a computer and projection in large lecture classrooms.

Television for the Environment *Earth Report* Geography Videos on DVD (0321662989) This three-DVD set is designed to help students visualize how human decisions and behavior have affected the environment and how individuals are taking steps toward recovery. With topics ranging from the poor land management promoting the devastation of river systems in Central America to the struggles for electricity in China and Africa, these 13 videos from Television for the Environment's global *Earth Report* series recognize the efforts of individuals around the world to unite and protect the planet.

Television for the Environment *Life* World Regional Geography Videos on DVD (013159348X) From the Television for the Environment's global *Life* series this two-DVD set brings globalization and the developing world to the attention of any world regional geography course. These 10 full-length video programs highlight matters such as the growing number of homeless children in Russia, the lives of immigrants living in the United States trying to aid family still living in their native countries, and the European conflict between commercial interests and environmental concerns.

Dire Predictions: Understanding Global Warming by Michael Mann, Lee R. Kump (0136044352) Periodic reports from the Intergovernmental Panel on Climate Change (IPCC) evaluate the risk of climate change brought on by humans. But the sheer volume of scientific data remains inscrutable to the general public, particularly to those who may still question the validity of climate change. In just over 200 pages, this practical text presents and expands upon the essential findings in a visually stunning and undeniably powerful way to the lay reader. Scientific findings that provide validity to the implications of climate change are presented in clear-cut graphic elements, striking images, and understandable analogies.

***Encounter Human Geography* Workbook and Web site** by Jess C. Porter (0321682203) *Encounter Human Geography* provides rich, interactive explorations of human geography concepts through Google Earth™. Students explore the globe through themes such as population, sexuality and gender, political geography, ethnicity, urban geography, migration, human health, and language. All chapter explorations are available in print format as well as online quizzes, accommodating different classroom needs. All worksheets are accompanied with corresponding Google Earth™ media files, available for download as part of MasteringGeography, and, for those who do not use MasteringGeography, also from **www.mygeoscienceplace.com.**

***Encounter World Regional Geography* Workbook & Website** by Jess C. Porter *(0321681754)* This provides rich, interactive explorations of world regional geography concepts through Google Earth™. Students explore the globe through the themes of Environment, Population, Culture, Geopolitics, and Economy and Development, answering multiple-choice and short-answer questions. All chapter explorations are available in print format as well as online quizzes, accommodating different classroom needs. All worksheets are accompanied with corresponding Google Earth™ media files, available for download as part of Mastering-Geography, and, for those who do not use MasteringGeography, also from **www.mygeoscienceplace.com.**

A structured learning path organizes each chapter.

1 GEOGRAPHY MATTERS

Learning Outcomes

- Explain how the study of geography has become essential for understanding a world that is more complex, interdependent, and changing faster than ever before.
- Identify four examples of how places influence inhabitants' lives.
- Distinguish the differences among major map projections and describe their relative strengths and weaknesses.
- Explain how geographers use geographic information systems (GIS) to merge and analyze data.
- Summarize the five concepts that are key to spatial analysis and describe how they help geographers to analyze relationships between peoples and places.
- Describe the importance of distance in shaping human activity.
- Summarize the three concepts that are key to regional analysis and explain how they help geographers analyze relationships between peoples and places.

For three days in September 2010 the streets of Maputo, the capital of Mozambique, were filled with thousands of striking workers protesting rising food prices. Mozambique, a former Portuguese colony, is one of the most impoverished nations in the world, ranked 175th out of 179 countries on the United Nations Human Development Index. Seventy percent of the population of 23 million survives beneath the poverty line and an estimated 54 percent are unemployed. The statutory minimum wage is US$37 a month; annual per capita income is only $807. The riots and strikes, organized primarily by text messaging, were prompted by a 30 percent increase in the price of bread. In the course of the riots, 13 people were killed, including a 6-year-old girl and a 12-year-old boy, and more than 400 wounded when police opened fire on demonstrators.

The problem of food shortages and rising food prices in Mozambique and in many other places is a reflection of the increasing geographic interdependence of the world. The situation is partly the result of increasing food consumption in other parts of the world, especially in booming China and India, where many have stopped growing their own food and have the cash to buy a lot more of it. Increasing meat consumption has helped drive up demand for feed grain, and this in turn has driven up the price of grain everywhere. Speculators in international commodity markets have joined the fray, further accelerating price rises. Another key linkage concerns energy prices. High oil prices push up fertilizer [...] of transporting food from farm to market. The popularity of [...] droughtness is straining food supplies, especially in the United [...] subsidies for ethanol have lured farmers away from growing c[...] this is climate change. Harvests in many countries have been [...] weather. In 2009–2010, there were prolonged droughts in [...] region of Africa; catastrophic floods in India, Indonesia, Pak[...] and a record-breaking heat wave that affected China, North [...] States along with parts of Canada, Russia, South Korea, and [...]

Geographers learn by finding out where things are and why [...] phy is about recognizing and understanding the interdepende[...] without losing sight of the uniqueness of each specific place[...] settings with distinctive physical, social, and cultural attribu[...] encompass many places, all or most of which share attributes[...] places elsewhere. Maps are also important tools for introduci[...] way that places and regions are made and altered.

Learning Outcomes

- Explain how the study of geography has become essential for understanding a world that is more complex, interdependent, and changing faster than ever before.
- Identify four examples of how places influence inhabitants' lives.
- Distinguish the differences among major map projections and describe their relative strengths and weaknesses.
- Explain how geographers use geographic information systems (GIS) to merge and analyze data.
- Summarize the five concepts that are key to spatial analysis and describe how they help geographers to analyze relationships between peoples and places.
- Describe the importance of distance in shaping human activity.
- Summarize the three concepts that are key to regional analysis and explain how they help geographers analyze relationships between peoples and places.

▼ NEW! Apply Your Knowledge

Conceptual Apply Your Knowledge questions and tasks are integrated throughout the chapter sections, giving students a chance to stop and check their understanding of the reading and key concepts.

APPLY YOUR KNOWLEDGE Referring to spatial analysis concepts, discuss a national or international environmental issue. Provide examples of how complementarity, transferability, intervening opportunities, and diffusion each relate to the issue you have chosen. ∎

NEW! Learning Outcomes ▶

Learning Outcomes in each chapter opener offer students a structured learning path to guide them through the main learning goals for the chapter. The assignable content in MasteringGeography™ is also correlated to these and other Learning Outcomes.

NEW! Learning ▶ Outcomes Revisited

End-of-chapter Learning Outcomes Revisited are found at the end of each chapter and summarize chapter content correlated to the Learning Outcomes stated in each chapter's opener.

Learning Outcomes Revisited

- Explain how the study of geography has become essential for understanding a world that is more complex, interdependent, and changing faster than ever before.

  Geography matters because it enables us to understand where we are both literally and figuratively. Geography provides an understanding of the interdependency of people and places and an appreciation of how and why certain places are distinctive or unique. From this knowledge, we can begin to understand the implications of future spatial patterns.

- Identify four examples of how places influence inhabitants' lives.

  Specific places provide the settings for people's daily lives. Places are settings for social interaction that, among other things, structure the daily routines of people's economic and social lives; provide both opportunities for—and constraints on—people's long-term social well-being; establish a context in which everyday commonsense knowledge and experience are gathered; provide a setting for processes of socialization; and provide an arena for contesting social norms.

- Distinguish the differences among major map projections and describe their relative strengths and weaknesses.

  The choice of map projection depends largely on the purpose of the map. Equidistant projections allow distance to be represented as accurately as possible but in only one direction (usually north-south). Conformal projections render compass directions accurately but tend to exaggerate the size of northern continents. Equivalent projections portray areas on Earth's surface in their true proportions but result in world maps on which many locations appear squashed and have unsatisfactory outlines.

Future Geographies

In 2011, the world contained nearly 7 billion people. The population division of the UN Department of Social and Economic Affairs projects that the world's population will continue to increase by 1.2 percent annually to mid-century, resulting in 9.3 billion people by 2050. The distribution of this projected population growth is noteworthy: Over the next half-century, population growth is predicted to occur overwhelmingly in regions least able to support it. Just six countries will account for half the increase in the world's population, including Bangladesh, China, India, Indonesia, Nigeria, and Pakistan (**Figure 3.31**). Meanwhile, Europe and North America will experience very low and in some cases zero population growth. Collectively, the periphery will grow 58 percent, as opposed to 2 percent for the core regions through 2050. The periphery will account for 99 percent of the expected increment in world population in this period.

Nevertheless, it is increasingly clear that human population will not continue to grow indefinitely, even in countries with already high populations. Demographic transition theory has prompted some population analysts to suggest that many of the economic, political, social, and technological transformations associated with continued urbanization and industrialization across the globe will shift the world's population from a period of high growth to one of low growth, or even population decline. This will be due to falling fertility. Current total fertility of 3.11 children per woman will fall to 2.04 by mid-century—just below replacement level but still above the current rate in core regions.

FIGURE 3.31 Shanghai, China Considered to be one of the world's largest city by the UN, Shanghai has nearly 14 million people within the core city and its inner suburbs. The populations of cities like Shanghai are expected to continue to grow through the first half of the twenty-first century, though urban planners raise serious concerns about how urban systems like water, sewer, transportation, and energy supply—to name a few—will be able to keep up.

◄ NEW! Future Geographies

The Future Geographies section at the end of each chapter emphasizes the unique, modern, forward-looking appeal of this text.

NEW! Review and Discussion questions ►

Review and Discussion questions at the end of each chapter can be easily assigned for group discussions or student essays.

REVIEW AND DISCUSSION

1. Discuss in your group whether the United States can still be considered a core country (see definitions on p. 46). Develop three reasons in support and against your arguments. What statistical evidence supports your argument? Also consider which countries might be moving into the core or into the semiperiphery or periphery? What evidence could you find to support such transitions?

2. Two major events have had significant impacts on places and regions, namely the droughts in Russia in the summer of 2010 and the Egyptian revolution of February 2011. Discuss in your group how these events are related to globalization. Provide two examples for each case.

3. Consider the core, the semiperiphery, and the periphery, not from the perspective of states, but from the viewpoint of transnational corporations (see p. 50). Research two corporations and find out where their corporate headquarters are located, develop a list of what they produce, and, if possible, find out where their products come from. Take into consideration any raw materials and manufacturing that are involved. For example, your group might want to consider cell phone companies. Where are the companies located? Are they in the core, semiperiphery, or periphery? Where do they get the material to make cell phones, within the core, semiperiphery, or periphery? And finally where are the cells phones manufactured? Once you have compiled this information, develop a world map displaying your information and the ways the transnational corporations' products move between the core, semiperiphery, and periphery.

UNPLUGGED

1. The present-day core regions of the world-system, shown in Figure 2.14 (p. 50), are those that dominate trade, control the most advanced technologies, have high levels of productivity within diversified economies, and enjoy relatively high per capita incomes. What statistical evidence can you find in your local library to support this characterization? Consider and list what core countries produce. (*Hint:* Look for data in annual compilations of statistics published in annual reviews, such as the Encyclopaedia Britannica's *Yearbook* and in the annual reports of organizations, such as the United Nations Development Programme [UNDP], the World Bank, and the World Resources Institute, and consider per capita incomes and how they differ between countries). How could differences in trade be understood from these publications?

2. The idea of an international division of labor is based on the observation that different countries tend to specialize in the production or manufacture of particular commodities, goods, or services. In what product or products do the following countries specialize: Bolivia, Ghana, Guinea, Libya, Namibia, Peru, and Zambia? You will find the data you need in a good statistical yearbook, such as the Encyclopaedia Britannica's *Yearbook*, or in a world reference atlas or economic atlas. Take one product that you found and use in your everyday life and reflect on the process it underwent to come into your possession.

3. Consider the division of labor in your place of work (either as a student or outside the university). How are tasks divided and who is responsible for what types of work? Is there an apparent division of labor (see p. 46)? If so, list three examples of how work is divided in your workplace. Think of who does what. And consider why certain tasks are divided differently.

◄ REVISED! Unplugged questions

End-of-chapter Unplugged questions challenge students to go out into the field and apply the geographic knowledge acquired in the chapter to real-world situations.

The superior cartography found in *Human Geography: Places and Regions in Global Context* comprises scores of rich, diverse, and fully updated maps that help professors better teach their students the important spatial elements inherent to human geography.

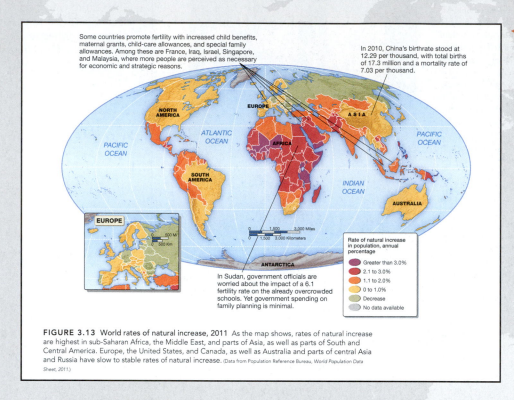

Some countries promote fertility with increased child benefits, maternal grants, child-care allowances, and special family allowances. Among these are France, Iraq, Israel, Singapore, and Malaysia, where more people are perceived as necessary for economic and strategic reasons.

In 2010, China's birthrate stood at 12.29 per thousand, with total births of 17.3 million and a mortality rate of 7.03 per thousand.

In Sudan, government officials are worried about the impact of a 6.1 fertility rate on the already overcrowded schools. Yet government spending on family planning is minimal.

Rate of natural increase in population, annual percentage
- Greater than 3.0%
- 2.1 to 3.0%
- 1.1 to 2.0%
- 0 to 1.0%
- Decrease
- No data available

FIGURE 3.13 World rates of natural increase, 2011 As the map shows, rates of natural increase are highest in sub-Saharan Africa, the Middle East, and parts of Asia, as well as parts of South and Central America. Europe, the United States, and Canada, as well as Australia and parts of central Asia and Russia have slow to stable rates of natural increase. (Data from Population Reference Bureau, *World Population Data Sheet, 2011*.)

◀ Current data

Up-to-date information gives readers access to the most current demographic statistics and data. This example is from the Population Reference Bureau's 2011 World Population Data Sheet.

HIGH MASS CONSUMPTION

Exploitation of comparative advantages in international trade

DRIVE TO MATURITY
Development of wider industrial and commercial base

Investment in manufacturing exceeds 10 percent of national income; development of modern social, economic, and political institutions

TAKE-OFF
Development of a manufacturing sector

Installation of physical infrastructure (roads, railways, etc.) and emergence of social/political elite

PRECONDITIONS FOR TAKE-OFF
Commercial exploitation of agriculture and extractive industry

Transition triggered by external influence, interests, or markets

TRADITIONAL SOCIETY
Limited technology; static society

FIGURE 7.15 Stages of economic development This diagram illustrates a model based on the idea of successive stages of economic development. Each stage is seen as leading to the next, though some regions or countries may take longer than others to make the transition from one stage to the next. According to this view, now regarded as overly simplistic, places and regions follow parallel courses within a world that is steadily modernizing. Late starters eventually make progress, but at speeds determined by their resource endowments, their productivity, and the wisdom of their people's policies and decisions.

◀ Everyday Life and Mental maps

These maps depict people's perceptions of their surroundings, highlighting the ways in which maps are useful for everyday life.

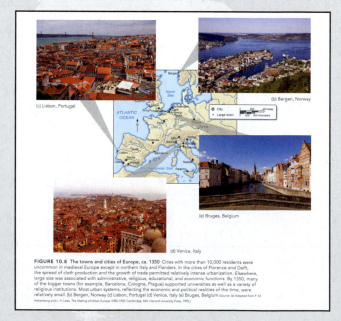

(c) Lisbon, Portugal
(b) Bergen, Norway
(e) Bruges, Belgium
(d) Venice, Italy

FIGURE 10.8 The towns and cities of Europe, ca. 1350 Cities with more than 10,000 residents were uncommon in medieval Europe except in northern Italy and Flanders. In the cities of Florence and Delft, the spread of cloth production and the growth of trade permitted relatively intense urbanization. Elsewhere, large size was associated with administrative, religious, educational, and economic functions. By 1350, many of the bigger towns (for example, Barcelona, Cologne, Prague) supported universities as well as a variety of religious institutions. Most urban systems, reflecting the economic and political realities of the time, were relatively small. (b) Bergen, Norway (c) Lisbon, Portugal (d) Venice, Italy (e) Bruges, Belgium (Source (a) Adapted from P. M. Hohenberg and L. H. Lees, *The Making of Urban Europe 1000–1950*. Cambridge, MA: Harvard University Press, 1995.)

▲ Compound figures

The book features many compound figures that combine maps with photographs and/or illustrations. These figures capture student interest by integrating spatial, visual, and conceptual information.

ENGAGING, RELEVANT APPLICATIONS

Knox and Marston's provocative applications increase student interest, fostering awareness of current issues and developing trends that impact the world and their lives.

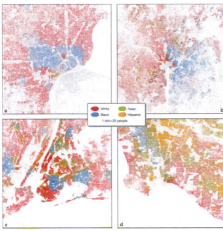

11.1 VISUALIZING GEOGRAPHY
Racial Segregation

Usually, ethnic neighborhoods on maps are sharply bounded blocks of color at the scale of census tracts that contain hundreds, sometimes thousands, of people. On these maps, one dot equals 25 people, and the dots are color-coded based on race: White is pink, black is blue, Hispanic is orange, and Asian is green. The results, based on U.S. Census data from 2000, show that just as every city is different, every city is segregated (or integrated) in different ways.

Detroit, for example, is highly segregated. Its Eight Mile beltway serves as a precise boundary between the city's black

and white populations. In Washington, D.C., there is a sharp east-west divide between white and black. In New York, there are areas of extreme racial concentration, but the sheer number of people in those areas means that boundary regions become intensely rich areas of cross-cultural ferment. Long Beach, California, meanwhile, is almost the opposite: Because no part of the city is particularly dense, there are a number of blended neighborhoods, some of which are more extensive than the racially homogeneous ones.

FIGURE 11.A (a) Detroit (b) Washington, D.C. (c) New York (d) Long Beach

Based on infographics by Eric Fischer. See http://www.flickr.com/photos/walkingsf/sets/72157624812674967/with/4981417821/

◀ **REVISED! Visualizing Geography boxed essays**

Reimagined Visualizing Geography feature essays incorporate edgy, modern applications and visualizations of geography data. These interesting and challenging visualizations are unique, and set apart Knox/Marston's visual program.

6.1 GEOGRAPHY MATTERS
Jerusalem, the Holy City

Although many cities in the world have been the object of struggle and conflict over the centuries, none has been as endlessly beset as Jerusalem. Visitors, writers, and residents believe Jerusalem to be the most beautiful city in the world. If there are other serious competitors to that coveted beauty title, Jerusalem certainly has few rivals for the title of the most sacred city in the world, possessing as it does an unmatched Christian, Jewish, and Islamic history. Jerusalem began as a small settlement on the slopes of Mount Moriah. In 997 B.C.E., it was captured by David, king of the Israelites, who made Jerusalem the capital of Israel. Solomon, David's son and successor, built the Great (First) Temple on Mount Moriah to commemorate the place where Abraham offered to sacrifice his son. Though the temple was destroyed centuries ago, the site is central to the Jewish faith.

The history and the map of the city reflect the histories of the various empires that dominated and were succeeded by new and yet more powerful empires (**Figure 6.A**). Nebuchadnezzar, king of Babylon, destroyed the Great Temple in 586 B.C.E. and banished the Jews. But the Babylonian control of Jerusalem eventually gave way to the Persians, under whose rule the Jews were allowed to return and rebuild their temple, known as the Second Temple. The Persian occupation of Jerusalem was swept aside by Alexander the Great (356–323 B.C.E.), the king of Macedonia and one of

world's greatest military leaders, who helped spread the Greek Empire from the southern shores of the Caspian Sea into central Asia. The Romans entered the scene around 63 B.C.E., and Herod the Great was eventually installed to command the Roman Kingdom of Judea from Jerusalem.

During the early Christian period in Jerusalem, the Jews revolted openly against the Roman occupiers. In 132 C.E., the Romans responded by destroying the Second Temple and banishing all Jews from Jerusalem and Palestine. As a result, the Jews scattered north into Babylon and later into Europe and North Africa. The popular myth is that this ancient Jewish diaspora remained in exile until 1948, when the state of Israel was created.

The major Christian influence on Jerusalem, the "Holy City," began when Constantine I (285–337 C.E.), the emperor of the Eastern Roman Empire, converted to Christianity in 313 C.E. This event led to the construction of churches and other buildings dedicated to celebrating the life of Jesus Christ. But Christian influence over the city ceased when Jerusalem eventually succumbed to Islam. In 638 C.E., Jerusalem was designated a holy city of Islam because it was believed that Muhammad's spirit had once made a visit to heaven while he was in the city.

Although for several centuries Jews, Christians, and Muslims were all allowed access to the city of Jerusalem, by the tenth

7.2 WINDOW ON THE WORLD
China's Economic Development

Under the leadership of Deng Xiaoping (1978–1997), China embarked on a thorough reorientation of its economy, dismantling Communist-style central planning in favor of private entrepreneurship and market mechanisms and integrating itself into the world economy. Saying that he did not care whether the cat was black or white as long as it caught mice, Deng Xiaoping established a program of "Four Modernizations" (industry, agriculture, science, and defense) and an "open-door policy" that allowed China to be plugged in to the interdependence circuits of the global economy. As a result, China has completely reorganized and revitalized its economy. Agriculture has been decollectivized, with Communist collective farms modified to allow a degree of private profit-taking. State-owned industries have been closed or privatized, and centralized state planning has been dismantled in order to foster private entrepreneurship.

In the 1980s and early 1990s, when the world economy was sluggish, China's manufacturing sector grew by almost 15 percent each year. For example, almost all of the shoes once made in South Korea or Taiwan are now made in China. More than 60 percent of the toys in the world, accounting for more than $10 billion in trade, are also made in China. Since 1992, China has extended its open-door policy, permitted foreign investment aimed at Chinese domestic markets, and normalized trading relationships with the United States and the European Union. In 2001, China was admitted to the World Trade Organization, allowing China to trade more freely than ever before with the rest of the world.

China's increased participation in world trade has created an entirely new situation within the world economy. Chinese manufacturers, operating with low wages, have imposed a deflationary

trend on world prices for manufacturers. The Chinese economy's size makes it a major producer, and its labor costs stay flat year after year because there is an endless supply of people who will work for 60 cents an hour. Meanwhile, the rapid expansion of consumer demand in China has begun to drive up commodity prices in the world market. Overall, China's economy is already the third largest in the world after those of the United States and Japan.

Nowhere have China's "open-door" policies had more impact than in South China, where the Chinese government has deliberately built upon the prosperity of Hong Kong, the former British colony that was returned to China in 1997. The coastline of South China provides many protected bays suitable for harbors and a series of ports, including Quanzhou, Shantou, Xiamen, and, on either side of the mouth of the Zhu Jiang (Pearl River), Macâo and Hong Kong. These ports made possible South China's emergence as a core manufacturing region by providing an interface with the world economy. The established trade and manufacturing of Macâo—a Portuguese colony that was returned to China in 1999—and Hong Kong provided another precondition for success.

When Deng Xiaoping established his "open-door" policy, a third factor kicked in: capital investment from Hong Kong, Taiwan, and the Chinese diaspora. By 1993, more than 15,000 manufacturers from Hong Kong alone had set up businesses in neighboring Guangdong Province, and a similar number established subcontracting relationships, contracting out processing work to Chinese companies. Today, the cities and special economic zones of South China's "Gold Coast" provide a thriving export-processing platform that has driven double-digit annual economic growth for much of the past two decades. The population of Shenzhen (**Figure 7.B**)

FIGURE 7.B Shenzhen The city of Shenzhen, just across the border from the Special Administrative Region of Hong Kong.

◀ **Window on the World**

Window on the World feature essays take a key concept and explore its application in a particular location. This feature allows students to appreciate the relevance of geographic concepts to world events and brings some far-flung places closer to their comprehension.

▲ **Geography Matters**

Geography Matters feature essays each examine one of the key concepts of the chapter, providing an extended example of its meaning and implications through both visual illustration and text. The Geography Matters features demonstrate to students that the focus of human geography is on real-world problems.

MasteringGeography™

This online homework and tutoring system delivers self-paced activities that provide individualized coaching, focus on course objectives, and are responsive to each student's progress. The Mastering system helps instructors maximize class time with customizable, easy-to-assign, and automatically graded assessments that motivate students to learn outside of class and arrive prepared for lecture. **www.masteringgeography.com**

Engaging Experiences

MasteringGeography™ provides a personalized, dynamic, and engaging experience for each student that strengthens active learning. Survey data show that the immediate feedback and tutorial assistance in MasteringGeography motivate students to do more homework. The result is that students learn more and improve their test scores.

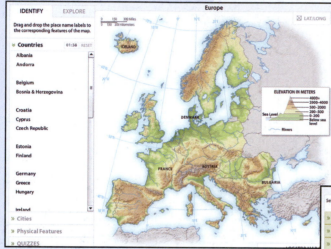

MapMaster™ Place Name Interactive Map Activities

Place Name Interactive Map Activities have students identify place names of political and physical features at regional and global scales, as well as explore select country data from the CIA World Factbook. Multiple-choice quizzes for the place name labeling activities and country data sets offer instructors flexible opportunities for summative assessment and pre- and post-testing.

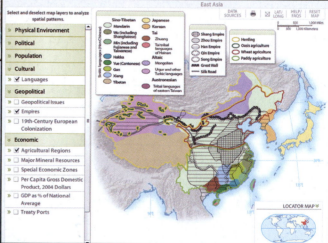

MapMaster™ Layered Thematic Interactive Map Activities

Layered Thematic Interactive Map Activities act as a mini-GIS tool, allowing students to layer various thematic maps to analyze spatial patterns and data at regional and global scales. Multiple-choice and short-answer quizzes are organized around themes of Physical Environment, Population, Culture, Geopolitics, and Economy, giving instructors flexible, modular options for assessing student learning.

Encounter Human Geography Activities

Pearson's **Encounter Activities** provide rich, interactive explorations of geography concepts through GoogleEarth™. *Encounter Human Geography* uses this exciting online geobrowser technology to explore the themes of human and cultural geography. All explorations include corresponding Google Earth KMZ media files, and questions include hints and specific wrong-answer feedback to help coach students toward mastery of the concepts.

Geography Videos

A variety of short geography videos help students get a sense of place, visualize, and explore a range of places and topics. Covering issues of economy, development, globalization, climate and climate change, culture, and more, there are 10-20 multiple-choice and short-answer questions for each of the 34 episodes. These video quizzes allow instructors to test students' understanding and application of concepts, and offer hints and specific wrong-answer feedback to guide students toward mastery of the concepts.

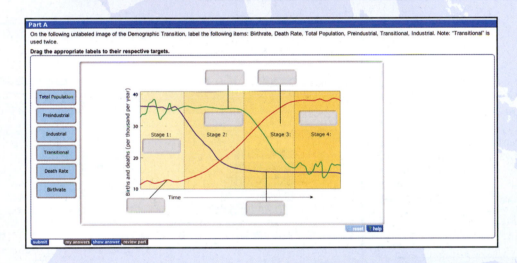

Thinking Spatially and Data Analysis

These activities help students develop spatial reasoning and critical thinking skills by identifying and labeling features from maps, illustrations, photos, graphs, and charts. Students then examine related data sets, answering multiple-choice and increasingly higher order conceptual short-answer questions.

Student Resources in MasteringGeography

- MapMaster™ interactive maps
- Practice chapter quizzes
- Geography videos
- "In the News" RSS feeds
- Glossary flash cards
- Optional Pearson eText and more

Callouts to MasteringGeography appear at the end of each chapter to direct students to extend their learning beyond the textbook.

Proven Results

The Mastering platform is the only online homework system with research showing that it improves student learning. A wide variety of published papers based on NSF-sponsored research and tests illustrate the benefits of the Mastering program. Results documented in scientifically valid efficacy papers are available at **www.masteringgeography.com/site/results**

MasteringGeography™

A Trusted Partner

The Mastering platform was developed by scientists for science students and instructors, and has a proven history with over 10 years of student use. Mastering currently has more than 1.5 million active registrations with active users in 50 states and in 41 countries.

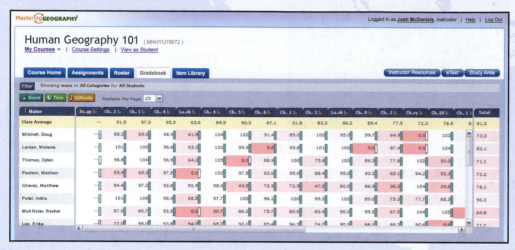

◄ Gradebook

Every assignment is automatically graded. Shades of red highlight struggling students and challenging assignments.

Gradebook Diagnostics ►

Gradebook Diagnostics provide unique insight into class and student performance. With a single click, charts summarize the most difficult problems, struggling students, grade distribution, and even score improvement over the duration of the course.

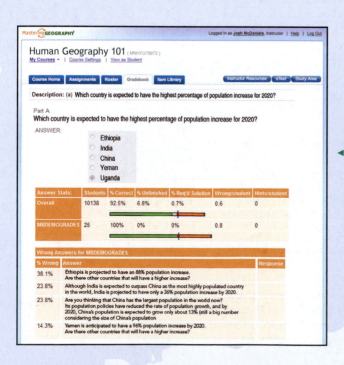

◄ Student Performance Data

MasteringGeography provides at-a-glance statistics on each class as well as national results. Wrong-answer summaries give unique insight into your students' misconceptions and facilitate just-in-time teaching adjustments.

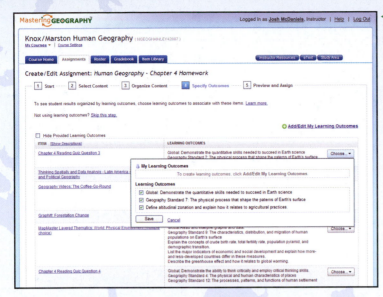

NEW! Learning Outcomes

MasteringGeography tracks student performance against each instructor's learning outcomes. Instructors can

- Add their own or use the publisher-provided learning outcomes to track student performance and report it to their administrations.

- View class performance against the specified learning outcomes.

- Export results to a spreadsheet that can be customized further and/or shared with the chair, dean, administrator, or accreditation board.

Mastering offers a data-supported measure to quantify students' learning gains and to share those results quickly and easily.

Annotation Function
Allows students to take notes.

Google®-Based Search function.

Highlight Function
Lets sudents highlight what they want to remember.

Zoom
Lets students zoom in and out for better viewing.

Pearson eText ▶

Pearson eText gives students access to the text whenever and wherever they can access the Internet. The eText pages look exactly like the printed text and include powerful interactive and customization features.

Interactive Glossary
Provides pop-up definitions and terms.

Instructor Notes
Instructors might also share his or her notes and highlights with the class.

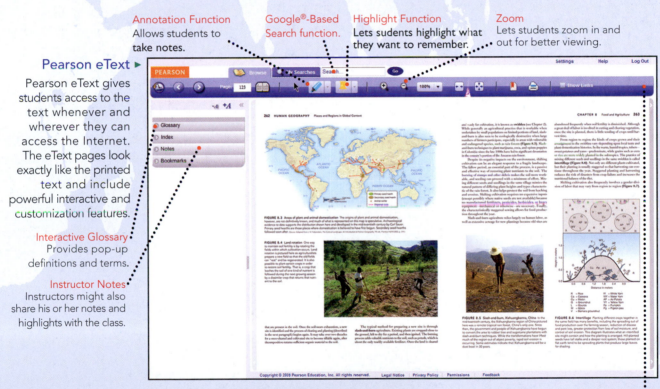

Hyperlinks
Links to quizzes, tests, activities, and animations.

Instructor Resources in MasteringGeography

Assignable activities, with options for automatically graded assessments, include

- MapMaster™ interactive maps with automatically graded assessments

- Videos
- Encounter Human Geography Google Earth™ Explorations
- Reading quizzes
- Test Bank questions
- Thinking Spatially and Data Analysis activities
- End-of-chapter questions and more

HUMAN GEOGRAPHY
Places and Regions in Global Context

1

GEOGRAPHY MATTERS

Learning Outcomes

- Explain how the study of geography has become essential for understanding a world that is more complex, interdependent, and changing faster than ever before.

- Identify four examples of how places influence inhabitants' lives.

- Distinguish the differences among major map projections and describe their relative strengths and weaknesses.

- Explain how geographers use geographic information systems (GIS) to merge and analyze data.

- Summarize the five concepts that are key to spatial analysis and describe how they help geographers to analyze relationships between peoples and places.

- Describe the importance of distance in shaping human activity.

- Summarize the three concepts that are key to regional analysis and explain how they help geographers analyze relationships between peoples and places.

For three days in September 2010 the streets of Maputo, the capital of Mozambique, were filled with thousands of striking workers protesting rising food prices. Mozambique, a former Portuguese colony, is one of the most impoverished nations in the world, ranked 175th out of 179 countries on the United Nations Human Development Index. Seventy percent of the population of 23 million survives beneath the poverty line and an estimated 54 percent are unemployed. The statutory minimum wage is US$37 a month; annual per capita income is only $807. The riots and strikes, organized primarily by text messaging, were prompted by a 30 percent increase in the price of bread. In the course of the riots, 13 people were killed, including a 6-year-old girl and a 12-year-old boy, and more than 400 wounded when police opened fire on demonstrators.

The problem of food shortages and rising food prices in Mozambique and in many other places is a reflection of the increasing geographic interdependence of the world. The situation is partly the result of increasing food consumption in other parts of the world, especially in booming China and India, where many have stopped growing their own food and have the cash to buy a lot more of it. Increasing meat consumption has helped drive up demand for feed grain, and this in turn has driven up the price of grain everywhere. Speculators in international commodity markets have joined the fray, further accelerating price rises. Another key

Two men attempt to brace the front gate as hundreds of mothers of malnourished children try to push into a therapeutic feeding center in Maradi, Niger.

linkage concerns energy prices: High oil prices push up fertilizer prices, as well as the cost of transporting food from farm to market. The popularity of biofuels as an alternative to hydrocarbons is straining food supplies, especially in the United States, where generous federal subsidies for ethanol have lured farmers away from growing crops for food. Compounding all this is climate change. Harvests in many countries have been seriously disrupted by extreme weather. In 2009–2010, there were prolonged droughts in Argentina, China, and the Sahel region of Africa; catastrophic floods in India, Indonesia, Pakistan, Sri Lanka, and Thailand; and a record-breaking heat wave that affected China, North Africa, Europe, and the United States along with parts of Canada, Russia, South Korea, and Japan.

Geographers learn by finding out where things are and why they are there. Human geography is about recognizing and understanding the interdependence among places and regions, without losing sight of the uniqueness of each specific place. **Places** are specific geographic settings with distinctive physical, social, and cultural attributes. **Regions** are territories that encompass many places, all or most of which share attributes different from the attributes of places elsewhere. Maps are also important tools for introducing geographers' ideas about the way that places and regions are made and altered.

WHY GEOGRAPHY MATTERS

The importance of geography as a subject of study is becoming more widely recognized as people everywhere struggle to understand a world that is increasingly characterized by instant global communications, rapidly changing international relationships, unexpected local changes, and growing evidence of environmental degradation. Many more schools now require courses in geography than just a decade ago, and the College Board has added the subject to its Advanced Placement program. Meanwhile, many employers are coming to realize the value of employees with expertise in geographical analysis and an understanding of the uniqueness, influence, and interdependence of places. Through an appreciation of the diversity and variety of the world's peoples and places, geography provides real opportunities not only to contribute to local, national, and global development but also to understand and promote multicultural, international, and feminist perspectives on the world.

Most people want to understand the intrinsic nature of the world in which we live. Geography enables us to understand where we are both literally and figuratively. Geography provides knowledge of Earth's physical and human systems and of the interdependency of living things and physical environments. That knowledge, in turn, provides a basis for people to cooperate in the best interests of our planet. Geography also captures the imagination: It stimulates curiosity about the world and the world's diverse inhabitants and places. By obtaining a better understanding of the world, people can overcome closed-mindedness, prejudice, and discrimination.

APPLY YOUR KNOWLEDGE List three reasons why a corporate employer would feel it is important for prospective employees to have some knowledge of geography. ■

WHY PLACES MATTER

An appreciation of the diversity and variety of peoples and places is a theme that runs through all of *human geography,* the study of the spatial organization of human activity and of people's relationships with their environments. This theme is inherently interesting to nearly all of us. *National Geographic* magazine has become a venerable institution by bringing us monthly updates of the seemingly endless variety of landscapes and communities around the world. More than 5 million households, representing about 19 million regular readers, subscribe to this magazine for its intriguing descriptions and striking photographs. Millions more read it occasionally in offices, lobbies, waiting rooms, or online.

Yet many Americans often seem content to confine their interest in geography to the pages of glossy magazines, to television documentaries, or to one-week packaged vacations. It has become part of the conventional wisdom—both in the United States and around the world—that many Americans have little real appreciation or understanding of people and places beyond their own daily routines. This is perhaps putting it too mildly. Surveys have revealed widespread ignorance among a high proportion of Americans, not only of the fundamentals of the world's geography but also of the

FIGURE 1.1 Flood victims A resident of Onagawa, Japan stands next to a building destroyed by a tsunami on March 26, 2011. Although major disasters are covered widely in the media, only a minority of Americans are able to understand them in geographic context.

diversity and variety within the United States itself. One Gallup poll found that 70 percent of young adults in the United States could not find their own country on a map of the world. In surveys of young adults in Canada, France, Germany, Great Britain, Italy, Japan, Mexico, Sweden, and the United States, Americans come in next to last in terms of geographic literacy.[1] Neither wars nor natural disasters appear to have compelled majorities of young Americans to absorb knowledge about international places in the news (**Figure 1.1**).

Also striking is Americans' ignorance of how the United States fits into the wider world. Majorities overestimate the total size of the U.S. population and fail to understand how much larger the population of China is. Three-quarters of young adults in the United States believe English is the most commonly spoken native language in the world, rather than Mandarin Chinese. This lack of geographical knowledge carries over to people's perceptions of the United States and its role in the world. A poll in 2010 found, for example, that most Americans believe that the United States spends more than 15 percent of its budget on aid to poor countries; the actual figure has consistently been less than 0.2 percent of the country's gross national income, putting it last among rich industrial nations. So although most people in the United States are fascinated by different places, relatively few have a systematic knowledge of them. Fewer still understand how different places came to be the way they are or why places matter in the broader scheme of things. This lack of understanding is important because geographic knowledge can take us far beyond a simple glimpse of the inherently interesting variety of peoples and places.

[1]Roper ASW, *2002 Global Geographic Literacy Survey.* Washington, DC: National Geographic–Roper ASW, 2002; Roper Public Affairs, *2006 Global Geographic Literacy Survey.* Washington, DC: National Geographic Education Foundation–Roper Public Affairs, 2006.

FIGURE 1.2 Quality of Life A petrochemical plant adjacent to a housing development in Wilmington, California.

The Influence of Places

Places are dynamic, with changing properties and fluid boundaries that are the product of the interplay of a wide variety of environmental and human factors. This dynamism and complexity is what makes places so fascinating for readers of *National Geographic*. It is also what makes places so important in shaping people's lives and in influencing the pace and direction of change. Places provide the settings for people's daily lives and their social relations (patterns of interaction among family members, at work, in social life, in leisure activities, and in political activity). It is in these settings that people learn who and what they are, how they are expected to think and behave, and what life is likely to hold for them.

Places exert a strong influence, for better or worse, on people's physical well-being, opportunities, and lifestyle choices. Living in a small town dominated by petrochemical industries, for example, means a higher probability of being exposed to air and water pollution, having a limited range of job opportunities, and having a relatively narrow range of lifestyle options because of a lack of amenities such as theatres, specialized stores and restaurants, and recreational facilities (**Figure 1.2**). Living in a central neighborhood of a large metropolitan area, on the other hand, usually means having a wider range of job opportunities and a greater choice of lifestyle options because of the variety of amenities accessible within a short distance (**Figure 1.3**). But it also means, among other things, living with a relatively high exposure to crime.

The Meaning of Places

Places also contribute to people's collective memory and become powerful emotional and cultural symbols. Consider the evocative power for most Americans of places like Times Square in New York; the Mall in Washington, D.C.; Hollywood Boulevard in Los Angeles; and Graceland in Memphis. And for many people, ordinary places have special meaning: a childhood neighborhood, a college campus, a baseball stadium, or a family vacation spot. This layering of meanings reflects the way that places are *socially*

FIGURE 1.3 Central city neighborhood Newbury Street shopping street in Boston.

FIGURE 1.4 Tahrir Square, Cairo, Egypt
Egyptian anti-government demonstrators flood
Cairo's landmark Tahrir Square on February 11,
2011, the 18th day of protests against President
Hosni Mubarak, who subsequently resigned.

constructed—given different meanings by different groups for different purposes. Places exist and are constructed by their inhabitants from a subjective point of view.

The meanings given to a place may be so strong that they become a central part of the identity of the people experiencing them. Your **identity** is the sense that you make of yourself through your subjective feelings based on your everyday experiences and social relations. Your own neighborhood, for example, is probably heavily laden with personal meaning and sentiment for you. But your neighborhood may well be viewed very differently, perhaps unsympathetically, by outsiders. This distinction is useful in considering the importance of understanding spaces and places from the viewpoint of the insider—the person who normally lives in and uses a particular place—as well as from the viewpoint of outsiders (including geographers).

Finally, places are the sites of innovation and change, of resistance and conflict (**Figure 1.4**). The unique characteristics of specific places can provide the preconditions for new agricultural practices (such as the development of seed agriculture and the use of plow and draft animals that sparked the first agricultural revolution in the Middle East in prehistoric times—see Chapter 8); new modes of economic organization (such as the high-tech revolution that began in Silicon Valley in the late twentieth century); new cultural practices (the punk movement that began in disadvantaged British housing projects, for instance); and new lifestyles (for example, the hippie lifestyle that began in San Francisco in the late 1960s). It is in specific locales that important events happen, and it is from them that significant changes spread.

Nevertheless, the influence of places is by no means limited to the occasional innovative change. Because of their distinctive characteristics, places always modify and sometimes resist the imprint of even the broadest economic, cultural, and political trends. Consider, for example, the way that a global cultural trend—rock 'n' roll—was modified in Jamaica to produce reggae. And how in Iran and North Korea rock 'n' roll has been resisted by the authorities, with the result that it has acquired an altogether different

kind of value and meaning for the citizens of those countries. Similarly, Indian communities in London developed Bhangra—a "world beat" composite of traditional Punjabi music, Mumbai (Bombay) movie scores, and Western disco. Cross-fertilization with local music cultures in New York and Los Angeles has produced Bhangra rap.

To consider a different illustration, think of the ways some communities have declared themselves "nuclear-free" zones: places where nuclear weapons and nuclear reactors are unwelcome or even banned by local laws. By establishing such zones, individual communities are seeking to challenge trends toward using nuclear energy and maintaining nuclear arms. They are, to borrow a phrase, "thinking globally and acting locally." Similarly, some communities have established "GM-free" zones, taking a stance against genetically modified crops and food. In adopting such strategies, they hope to influence thinking in other communities so that eventually their challenge could result in a reversal of established trends.

In summary, places are settings for social interaction that, among other things,

- structure the daily routines of people's economic and social life;

- provide both opportunities and constraints in terms of people's long-term social well-being;

- provide a context in which everyday, commonsense knowledge and experience are gathered;

- provide a setting for processes of socialization; and

- provide an arena for contesting social norms.

APPLY YOUR KNOWLEDGE Explain how and why a particular place has mattered to you. How might others' experience or perception of that same place differ from yours? How does your place influence your health or job prospects? ∎

STUDYING HUMAN GEOGRAPHY

The study of geography involves the study of Earth as created by natural forces and modified by human action. This, of course, covers an enormous amount of subject matter. There are two main branches of geography: physical and human. **Physical geography** deals with Earth's natural processes and their outcomes. It is concerned, for example, with climate, weather patterns, landforms, soil formation, and plant and animal ecology. **Human geography** deals with the spatial organization of human activities and with people's relationships to their environments. This involves looking at natural physical environments insofar as they influence, and are influenced by, human activity. To that end, the study of human geography must cover a wide variety of phenomena. These include, for example, agricultural production and food security, population change, the ecology of human diseases, resource management, environmental pollution, regional planning, and the symbolism of places and landscapes.

Regional geography combines elements of both physical and human geography. Regional geography is concerned with the way that unique combinations of environmental and human factors produce territories with distinctive landscapes and cultural attributes. The concept of region is used by geographers to apply to larger-sized territories that encompass many places, all or most of which have similar attributes distinct from the attributes of other places.

What is distinctive about the study of human geography is not so much the phenomena that are studied as the way they are approached. The contribution of human geography is to reveal *how and why geographical relationships are important*, in relation to a wide spectrum of natural, social, economic, political, and cultural phenomena. Thus, for example, human geographers are interested not only in patterns of agricultural production but also in the geographical relationships and interdependencies that are both causes and effects of such patterns. To put it in concrete terms, geographers are interested not only in what specialized agricultural subregions (for example, the dairy farming area of Jutland, Denmark) are like but also in the role of subregions such as Jutland in national and international agro-food systems (their interdependence with producers, distributors, and consumers in other places and regions—see Chapter 8).

Geography is very much an applied discipline as well as a means of understanding the world. Geographers employed in business, industry, and government are able to use geographic theories and techniques to understand and solve a wide variety of specific problems. A great deal of the research undertaken by geography professors has an applied focus (see **Box 1.1**, "Geography Matters: Geographers at Work").

THE BASIC TOOLS AND METHODS OF HUMAN GEOGRAPHERS

In general terms, the basic tools employed in geography are similar to those in other disciplines. Like other social scientists, human geographers usually begin with observation. Information must be collected and data recorded. This can involve many different methods and tools. Fieldwork (surveying, asking

questions, using scientific instruments to measure and record things), laboratory experiments, and archival searches all are used by human geographers to gather information about geographical relationships. Geographers also use **remote sensing**, the collection of information about parts of Earth's surface by means of aerial photography or satellite imagery designed to record data on visible, infrared, and microwave sensor systems (**Figure 1.5** see page 9). For example, agricultural productivity can be monitored by remotely sensed images of crops, and energy efficiency can be monitored by remotely sensed levels of heat loss from buildings.

Once data have been obtained through some form of observation, the next important step is to portray and describe them through *visualization* or *representation*. This can involve a variety of tools, including written descriptions, charts, diagrams, tables, mathematical formulas, and maps. Visualization and representation are important activities because they allow large amounts of information to be explored, summarized, and presented to others. They are nearly always a first step in the analysis of geographical relationships, and they are important in conveying the findings and conclusions of geographic research.

At the heart of geographic research, as with other kinds of research, is the *analysis* of data. The objective of analysis, whether of quantitative or qualitative data, is to discover patterns and establish relationships so that hypotheses can be established and models can be built. Models, in this sense, are abstractions of reality that help explain the real world. They require tools that allow us to generalize about things. Once again, we find that geographers are like other social scientists in that they utilize a wide range of analytical tools, including conceptual and linguistic devices, maps, charts, and mathematical equations.

In many ways, therefore, the tools and methods of human geographers are parallel to those used in other sciences, especially the social sciences. In addition, geographers increasingly use some of the tools and methods of the humanities—interpretive analysis and inductive reasoning, for example—together with ethnographic research (the systematic recording of human cultures) and textual analysis. The most distinctive tools in the geographer's kit bag are maps and geographic information systems (GIS).

Maps

Maps not only describe data but also serve as important sources of data and tools for analysis. Because of their central importance to geographers, they can also be objects of study in their own right. Geographic information systems involve an organized set of computer hardware, software, and spatially coded data that is designed to capture, store, update, manipulate, and display geographically referenced information.

Maps are representations of the world. They are usually two-dimensional graphic representations that use lines and symbols to convey information or ideas about spatial relationships. Maps express particular interpretations of the world, and they affect how we understand the world and how we see ourselves in relation to others. As such, all maps are "social products." In general, maps reflect the power of the people who draw them up. Just

Career options for geography majors are diverse, challenging, and exciting. Most geography graduates are able to find careers in which they have the opportunity to make a positive contribution to the world through their analytical skills. These careers include **cartography**, geographic information systems (GIS), laboratory analysis, private consulting, urban and regional planning, international development, teaching, and management in private industry. Because of the broad nature of the field, these careers engage every aspect of human activity on every scale from the local to the global. The following examples reflect this diversity:

- *International Affairs.* Knowledge and understanding of regional histories and geographies, along with their ability to analyze the interdependence of places and regions, enables geographers to effectively contribute to discussions of international policy. Work within governmental agencies, corporations, and nonprofit institutions in shaping international strategies is especially important in the context of accelerating globalization.

- *Location of Public Facilities.* Geographers use specialized techniques to analyze the location patterns of particular population groups, to analyze transportation networks, and to analyze patterns of geographic accessibility to sites. Such analysis enables geographers to determine the most effective locations for new public facilities, such as clinics, emergency rooms, social centers for the elderly, and shelters for the homeless.

- *Marketing and Location of Industry.* Similar techniques are used to determine the most efficient, or most profitable, locations for new factories, stores, and offices. Geographical research is also used to analyze the changing geography of supply and demand, allowing industry to determine whether, and where, to relocate. Basic techniques of geographic analysis are also used in geodemographic research, an important aspect of marketing.

- *Geography and the Law.* Geographical analysis is increasingly involved in resolving complex social and environmental issues. One important example is the issue of property development and the implications of environmental hazards such as flooding, coastal erosion, toxic-waste dumps, and earthquake fault zones for policies, codes, and regulations. Another is the issue of drawing up geographical boundaries for political units in ways that ensure equal representation by population size and racial and political composition—even as populations are changing between one election and another.

- *Disease Ecology.* By analyzing social and environmental aspects of human diseases, geographers are able to shed light on the causes of disease, to predict the spread of particular outbreaks, and to suggest ways in which the incidence of disease might be controlled.

- *Urban and Regional Planning.* Urban and regional planning adopts a systematic, creative approach to address and resolve physical, social, and economic concerns of neighborhoods, cities, suburbs, metropolitan areas, and larger regions. Planners work directly on preserving and enhancing the quality of life in communities, protecting the environment, promoting equitable economic opportunity, and managing growth and change of all kinds. Planning has roots in engineering, law, architecture, social welfare, and government, but geography, because of its focus on the interdependence of peoples and places, offers the best preparation for specialized professional training in urban and regional planning.

- *Economic Development.* An ability to understand the interdependence of places and to analyze the unique economic, environmental, cultural, and political attributes of specific regions enables geographers to contribute effectively to strategies and policies aimed at economic development. Geographers are involved in applied research and policy formulation concerning economic development all over the globe, addressing the problems not only of individual places and regions but also of the entire world economy.

- *Security.* Geographers' knowledge and understanding of geopolitics, political geography, demographics, medical geography, and cultural geography, together with an appreciation of the interdependent relationships among local, regional, and global systems, provide a sound basis for work in many areas of security. Knowing how places and regions "work" also means knowing about their vulnerability to potential security risks.

FIGURE 1.A Urban and Regional Planning Planning professionals examine a model of Scharnhauser Park, a new town near Stuttgart, Germany.

Landsat satellite images are digital images captured from spectral bands both visible and invisible to the human eye. Different kinds of vegetation cover, soils, and built environments are reflected by different colors in the processed image. This Landsat image is of the Washington, DC and Baltimore, Maryland Area.

Aerial photographs can be helpful in explaining what would otherwise require expensive surveys and detailed cartography. They are especially useful in working with multidisciplinary teams. This example shows the lower Connecticut river near the town of Old Lyme. The photograph was taken during the Connecticut River Marsh Restoration Project.

Using a collection of satellite-based observations, scientists and visualizers stitched together months of observations of the land surface, oceans, sea ice, and clouds into a seamless, true-color mosaic of every square kilometer (0.386 square mile) of the planet. Much of the information contained in this image came from a single remote-sensing device—NASA's Moderate Resolution Imaging Spectroradiometer, or MODIS—flying more than 700 km above the Earth on board the Terra satellite.

FIGURE 1.5 Remotely sensed images Remotely sensed images can provide new ways of seeing the world, as well as unique sources of data on all sorts of environmental conditions. Such images can help explain problems and processes. Examples of such applications include studies of the deforestation of the Amazon rain forest, urban encroachment onto farmland, water pollution, and bottlenecks in highway systems. (*Source:* Top left photo courtesy of Spaceimaging.com; top right photo courtesy Joel Sartore www.joelsartore.com.)

including things on a map—literally "putting something on the map"—can be empowering. The design of maps—what they include, what they omit, and how their content is portrayed—inevitably reflects the experiences, priorities, interpretations, and intentions of their authors. The most widely understood and accepted maps—"normal" maps—reflect the view of the world that is dominant in universities and government agencies.

Maps that are designed to represent the *form* of Earth's surface and to show permanent (or at least long-standing) features such as buildings, highways, field boundaries, and political boundaries

FIGURE 1.6 Topographic maps Topographic maps represent the Earth's surface in both horizontal and vertical dimensions. This extract is from a Swiss map of Lugano at the scale of 1:25,000. The height of landforms is represented by contours (lines that connect points of equal vertical distance above sea level), which on this map are drawn every 20 meters. Features such as roads, power lines, built-up areas, and so on are shown by stylized symbols. Note how the closely spaced contours of the hill slopes represent the shape and form of the land. (*Source:* Extract from Carta Nazionale Della Svizzera, Sheet 1353, 1:25,000 series, Ufficio Federale di Topografia, 3084 Wabern, Switzerland. Edizione 1998.)

are called *topographic maps* (**Figure 1.6**). The usual device for representing the form of Earth's surface is the *contour*, a line that connects points of equal vertical distance above or below a zero data point, usually sea level.

Maps that are designed to represent the spatial dimensions of particular conditions, processes, or events are called *thematic maps*. These can be based on any one of a number of devices that allow cartographers or map makers to portray spatial variations or spatial relationships. One of these is the *isoline*, a line (similar to a contour) that connects places of equal data value (for example precipitation, as in **Figure 1.7**). Maps based on isolines are known as *isopleth maps*. Another common device used in thematic maps is the *proportional symbol*. Circles, squares, spheres, cubes, or some other shape can be drawn in proportion to the frequency of occurrence of some particular phenomenon or event at a given location. **Figure 1.8** shows an example of a map featuring proportional circles. Symbols such as arrows or lines can also be drawn proportionally in order to portray flows of things between particular places. *Dot maps*, in which a single dot or other symbol represents a specified number of

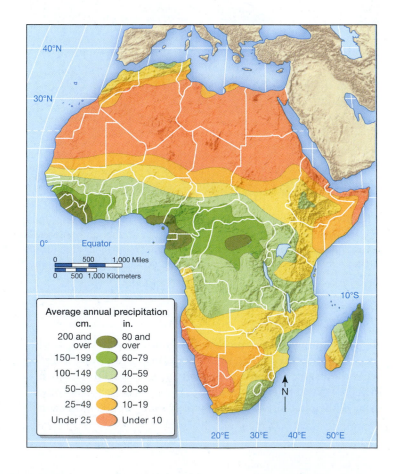

FIGURE 1.7 Isoline maps Isoline maps portray spatial information by connecting points of equal data value. Contours on topographic maps are isolines. This map shows average annual precipitation for the continent of Africa. (*Source:* Reprinted with permission of Pearson, from D. Hess and D. Tasa, *McKnight's Physical Geography: A Landscape Appreciation*, 10th edition, 2011, p. 36.)

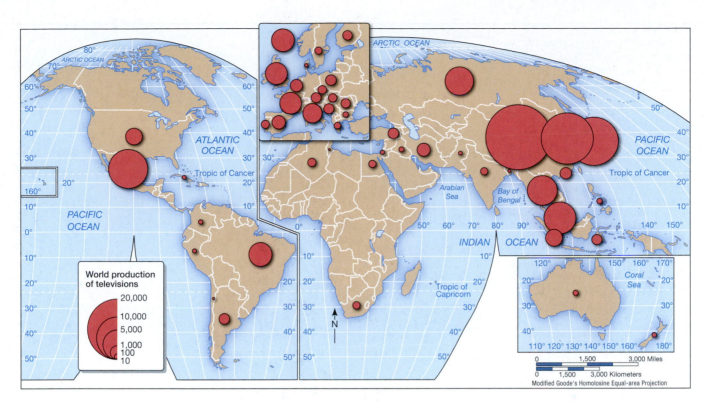

FIGURE 1.8 An example of proportional symbols in thematic mapping Proportional circles showing worldwide television receiver production. East Asia has become the leading area in the manufacture of this commodity. (Adapted from F. P. Stutz and B. Warf, *The World Economy: Geography, Business, Development,* 6th edition, Pearson, 2012, p. 204.)

occurrences of some particular phenomenon or event, can be effectively portray simple distributions. Another device is the *choropleth map,* in which tonal shadings are graduated to reflect variations in numbers, frequencies, or densities (see, for example, **Figures 7.1**, p. 217, and **10.1**, p. 352). Finally, thematic maps can be based on *located charts,* in which graphs or charts are located by place or region (**Figure 1.9**).

Map Scales

A *map scale* is the ratio between linear distance on a map and linear distance on Earth's surface. It is usually expressed in terms of corresponding lengths, as in "one centimeter equals one kilometer," or as a *representative fraction* (in this case, 1/100,000) or ratio (1:100,000). *Small-scale* maps are maps based on small representative fractions (for example, 1/1,000,000 or 1/10,000,000). A map drawn on this page to the scale of 1:10,000,000 would cover about half of the United States; a map drawn to the scale of 1:16,000,000 would easily cover the whole of Europe. *Large-scale* maps are maps based on larger representative fractions (e.g., 1/25,000 or 1/10,000). A map drawn on this page to the scale 1:10,000 would cover a typical suburban subdivision; a map drawn to the scale of 1:1,000 would cover just a block or two of it.

Map Projections

A **map projection** is a systematic rendering on a flat surface of the geographic coordinates of the features found on Earth's surface. Because Earth's surface is curved and it is not a perfect sphere, it is impossible to represent on a flat plane, sheet of paper, or monitor screen without some distortion. Cartographers have devised a number of different techniques for projecting latitude and

longitude (see **Figure 1.15**) onto a flat surface, and the resulting representations of Earth each have advantages and disadvantages. None of them can represent distance correctly in all directions, though many can represent compass bearings or area without distortion.

The choice of map projection depends largely on the purpose of the map.

Projections that allow distance to be represented as accurately as possible are **equidistant projections**. These projections can represent distance accurately in only one direction (usually north-south), although they usually provide accurate scale in the perpendicular direction (which in most cases is the equator). Equidistant projections are often more aesthetically pleasing for representing Earth as a whole, or large portions of it. An example is the polyconic projection (**Figure 1.10**).

Projections on which compass directions are rendered accurately are known as **conformal projections**. On the Mercator projection (**Figure 1.10**), for example, a compass bearing between any two points is plotted as a straight line. As a result, the Mercator projection has been widely used in navigation for hundreds of years.

Some projections are designed such that compass directions are correct only from one central point. These are known as *azimuthal projections*. They can be equidistant, as in the Azimuthal Equidistant projection (**Figure 1.10**), which is sometimes used to show air-route distances from a specific location, or equal-area, as in the Lambert Azimuthal Equal-Area projection.

Projections that portray areas on Earth's surface in their true proportions are known as **equal-area or equivalent projections**. Such

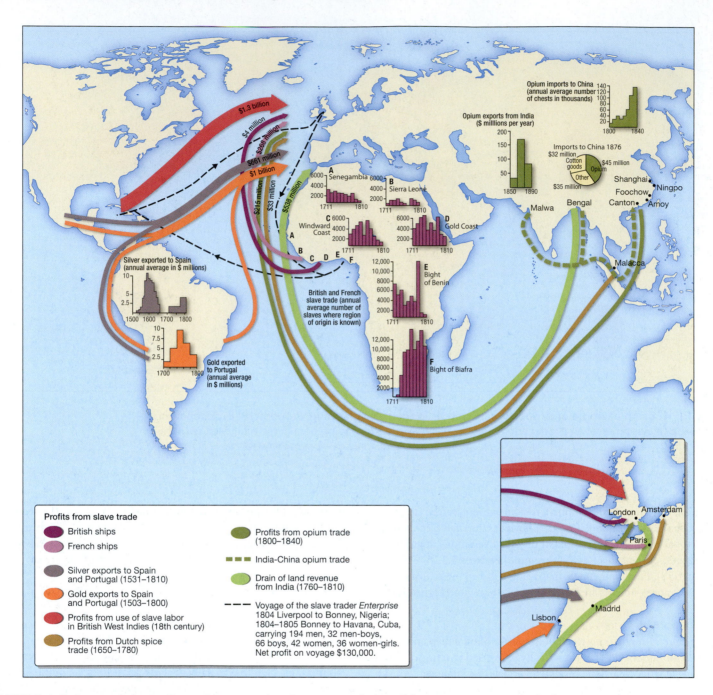

FIGURE 1.9 Located charts By combining graphs, charts, or symbols with base maps, a great deal of information can be conveyed in a single figure. This example illustrates the profits generated through European plunder of global minerals, spices, and human beings over a 300-year period. (Adapted from B. Crow and A. Thomas, *Third World Atlas*. Milton Keynes: Open University Press, 1982, p. 27.)

projections are used where the cartographer wishes to compare and contrast distributions on the Earth's surface: the relative area of different types of land use, for example. Examples of equal-area projections include the Eckert IV projection, Bartholomew's Nordic projection and the Mollweide projection **(Figure 1.10)**. Equal-area projections such as the Mollweide projection are especially useful for thematic maps showing economic, demographic, or cultural data. Unfortunately, preserving accuracy in terms of area tends to result in world maps on which many locations appear squashed and have unsatisfactory outlines.

For some applications, aesthetic appearance is more important than conformality, equivalence, or equidistance, so cartographers have devised a number of other projections. Examples include the Times projection, which is used in many world atlases, and the Robinson projection, which is used by the National Geographic Society in many of its publications.

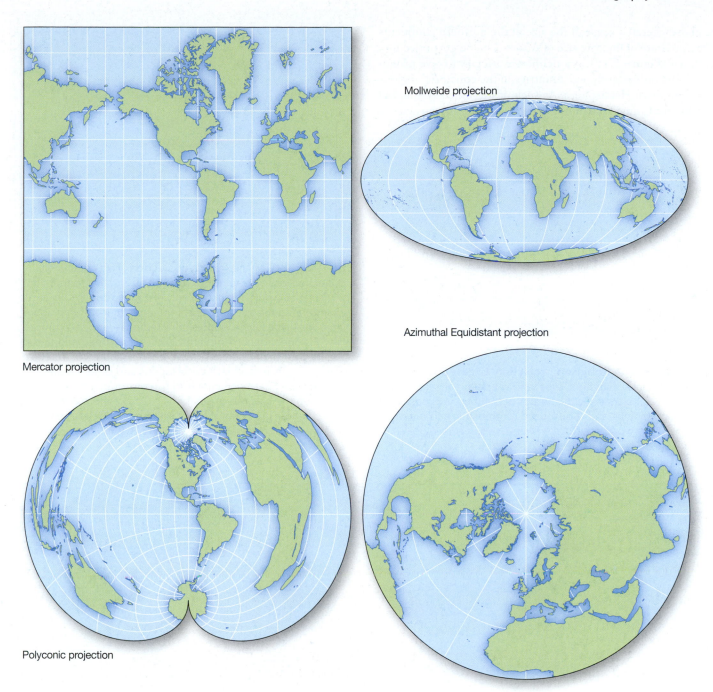

Mollweide projection

Mercator projection

Azimuthal Equidistant projection

Polyconic projection

FIGURE 1.10 Comparison of map projections Different map projections have different properties. The polyconic projection is true to scale along each east-west parallel and along the central north-south meridian. It is neither conformal nor equal-area, and it is free of distortion only along the central meridian. On the Mercator projection, compass directions between any two points are true, and the shapes of landmasses are true but their relative size is distorted. On the Azimuthal Equidistant projection, distances measured from the center of the map are true, but direction, area, and shape are increasingly distorted as the distance from the center point increases. On the Mollweide projection, relative sizes are true, but shapes are distorted.

The Robinson projection (**Figure 1.11**) is a compromise projection that distorts both area and directional relationships but provides a general-purpose world map.

There are also political considerations to take into account when considering different projections. Countries may appear larger and so more "important" on one projection rather than another. The Mercator projection (see **Figure 1.10**) was widely used as the standard classroom wall map of the world for many years, and its image of the world is deeply ingrained into general consciousness. As a result, many Europeans and North Americans have an exaggerated sense of the size of the northern continents and are unaware of the true size of Africa. The Peters projection, in contrast (**Figure 1.12**), is a deliberate attempt to give prominence to the underdeveloped countries of the equatorial regions and the Southern Hemisphere. As such, it was officially adopted by the World Council of Churches and by numerous agencies of

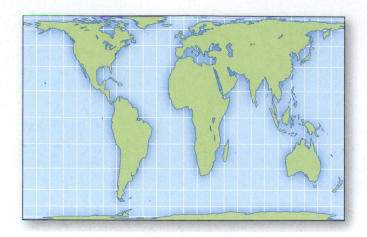

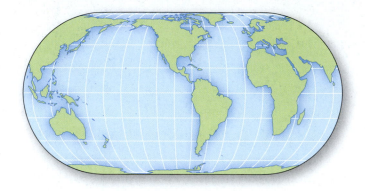

FIGURE 1.11 **The Robinson projection** On the Robinson projection, distance, direction, area, and shape are all distorted in an attempt to balance the properties of the map. It is designed purely for appearance and is best used for thematic and reference maps at the world scale. (*Source:* After E. F. Bergman, *Human Geography: Cultures, Connections, and Landscapes,* © 1995 by Pearson, p. 12.)

FIGURE 1.12 **The Peters projection** This equal-area projection was an attempt to offer an alternative to traditional projections, which, filmmaker and historian Arno Peters argued, exaggerated the size and apparent importance of the higher latitudes—that is, the world's core regions—and so promoted the "Europeanization" of Earth. While it has been adopted by the World Council of Churches, the Lutheran Church of America, and various agencies of the United Nations and other international institutions, it has been criticized by cartographers in the United States on the grounds of aesthetics: One of the consequences of equal-area projections is that they distort the shape of landmasses. (*Source:* Adapted from E. F. Bergman, *Human Geography: Cultures, Connections, and Landscapes,* © 1995 by Pearson, p. 13.)

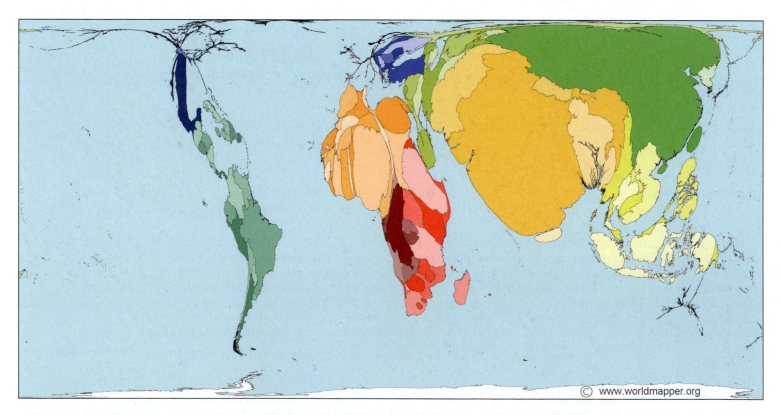

FIGURE 1.13 **Example of a cartogram** In a cartogram, space is distorted to emphasize a particular attribute of places or regions. This example shows the relative size of countries based on **the proportion of all people living on US$10 or less a day**; the cartographers have maintained the shape of each country as closely as possible to make the map easier to read. As you can see, population-based cartograms are very effective in demonstrating spatial inequality. *Source:* © Copyright SASI Group (University of Sheffield) and Mark Newman (University of Michigan).

the United Nations and other international institutions. Its unusual shapes give it a shock value that gets people's attention. For some, however, its unusual shapes are unattractive; it has been likened to laundry hung out to dry.

One particular kind of map projection that is sometimes used in small-scale thematic maps is the *cartogram*. In this kind of projection, space is transformed according to statistical factors, with the largest mapping units representing the greatest statistical values. **Figure 1.13** shows a cartogram of the world in which the relative size of countries is based not on area but on the proportion of people with extremely low incomes. The deliberate distortion of the shapes of the continents in this sort of projection dramatically emphasizes spatial variations.

APPLY YOUR KNOWLEDGE Describe four different scenarios that illustrate situations in which the use of (a) equidistant, (b) conformal, (c) azimuthal, and (d) equal-area projections would be appropriate. ∎

Geographic Information Systems (GIS)

The combination of high-performance computing and computerized record keeping has led to an unprecedented increase in the volume and popularity of geographic data. **Geographic information systems (GIS)** have rapidly grown to become one of the most important methods of geographic analysis, particularly in the military and commercial worlds. The software in GIS incorporates programs to store and access spatial data, to manipulate those data, and to draw maps.

The primary requirement for data to be used in GIS is that the locations for the variables—the characteristics under consideration—are known. Location may be annotated by x, y, and z coordinates of longitude, latitude, and elevation, or by such systems as ZIP codes or highway mile markers. Any variable that can be located spatially can be fed into a GIS. Data capture—putting the information into the system—is the most time-consuming component of GIS work. Different sources of data, using different systems of measurement, scales, and systems of representation, must be integrated with one another; changes must be tracked and updated. Many GIS operations in the United States, Europe, Japan, and Australia have begun to contract out such work to firms in countries where labor is cheaper. India has emerged as a major data-conversion center for GIS.

The most important aspect of GIS, from an analytical point of view, is that they allow data from several different sources, on different topics and at different scales, to be merged. This allows analysts to emphasize the spatial relationships among the objects being mapped. A geographic information system makes it possible to link, or integrate, information that is difficult to associate through any other means.

GIS technology can render visible many aspects of geography that were previously unseen. GIS can, for example, produce incredibly detailed maps based on millions of pieces of information—maps that could never have been drawn by human hands. One example of such a map is the satellite image reconstruction of the vegetation cover of the United States shown in **Figure 1.14**. At the other extreme of spatial scale, GIS can put places under the microscope, creating detailed new insights using huge databases and effortlessly browsable media.

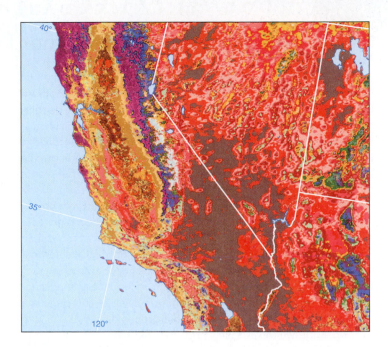

FIGURE 1.14 Map of land cover This extract is from a map of land cover in the United States that was compiled from several data sets using GIS technology. These included 1-kilometer resolution, Advanced Very High Resolution Radiometer (AVHRR) satellite imagery, and digital data sets on elevation, climate, water bodies, and political boundaries. Each of the 159 colors on the U.S. map represents a specific vegetation region. The purples and blues represent various subregions of western coniferous forests, the yellows are grasslands, and the reds are shrublands. The gray-brown region is the barren area of the Mojave Desert. (*Source:* United States Geological Survey, Map of Seasonal Land Cover Regions, 1993; see T. Loveland, J. W. Merchant, J. F. Brown, D. O. Ohlen, B. C. Reed, P. Olson, and J. Hutchinson, "Seasonal Land Cover Regions of the United States," *Annals, Association of American Geographers, 85,* 1995, 339–355.)

Many advances in GIS have come from military applications. GIS allows infantry commanders to calculate line of sight from tanks and defensive emplacements, allows cruise missiles to fly below enemy radar, and provides a comprehensive basis for military intelligence. Beyond the military, GIS technology allows an enormous range of problems to be addressed. For instance, it can be used to decide how to manage farmland, to monitor the spread of infectious diseases, to monitor tree cover in metropolitan areas, to assess changes in ecosystems, to analyze the impact of proposed changes in the boundaries of legislative districts, to identify the location of potential business customers, to identify the location of potential criminals, and to provide a basis for urban and regional planning.

Some of the most influential applications of GIS have resulted from geodemographic research (see **Box 1.2**, "Visualizing Geography: Geography and Consumer Preferences"). **Geodemographic research** uses census data and commercial data (such as sales data and property records) about the populations of small districts in creating profiles of those populations for market research. The digital media used by GIS make such applications very flexible. With GIS it is possible to zoom in and out, evaluating spatial relationships at different spatial scales. Similarly, it is possible to vary the appearance and presentation of maps, using different colors and rendering techniques.

Geography and Consumer Preferences

Market research consultants, using statistical methods similar to those used by urban social geographers, are able to identify the geography of different types of households according to their distinctive consumption patterns and preferences as well as their socioeconomic and demographic attributes and their typical residential settings. The best-known and most comprehensive classification of households in the United States is the Nielsen PRIZM segmentation system. Block-level data from the U.S. Census of population is merged with household demographic and lifestyle data from list-based records for over 120 million households, along with detailed profiles of more than 890,000 households from sources such as R. L. Polk & Co's new-car buyers and Experian Simmons National Consumer Study which provides measures of consumer buying behavior and lifestyle descriptors. The PRIZM® segmentation system from Nielsen crunches data into fourteen major socio-geographic groups: Urban Uptown, Midtown Mix, Urban Cores, Elite Suburbs, The Affluentials, Middleburbs, Inner Suburbs, Second City Society, City Centers, Micro-City Blues, Landed Gentry, Country Comfort, Middle America, and Rustic Living. Each of these, in turn, is subdivided, giving a total of 66 distinctive market segments, based on characteristics such as income, education, occupation, and home value.

The following examples illustrate the kind of market segmentation that is possible. One of the most affluent segments is Young Digerati. This relatively wealthy segment is part of the Urban Uptown social group and enjoyed an estimated median household income of $91,104 in 2010. Young Digerati are tech-savvy and tend to live in fashionable neighborhoods on the urban fringe. Highly educated and ethnically mixed, these communities are typically filled with trendy apartments and condos, fitness clubs and clothing boutiques, casual restaurants, and all types of bars—from juice to coffee to microbrew. Their distinctive consumer preferences include water skiing, Audi A3 cars, the *Economist* magazine, and the Independent Film Channel.

American Dreams is also part of the Urban Uptown social group, however this segment demonstrates how ethnically diverse the United States has become: Just under half the residents are Hispanic, Asian, or African American, and one in ten speaks a language other than English. These middle-aged immigrants and their children live in upper-middle-class comfort. The estimated median household income of the American Dreams neighborhoods is $58,517 and residents tend to buy Lexus IS vehicles, shop at Old Navy, listen to motivational tapes, read *Black Enterprise* magazine, and watch TeleFutura.

Bedrock America belongs to the Rustic Living social group and is made up of economically challenged families (an estimated median household income $28,978) in small, isolated towns located throughout America's heartland. With modest educations, sprawling families,

and service jobs, many of these residents struggle to make ends meet. One quarter live in mobile homes. One in three have not finished high school. Dodge Ram Flex Fuel is the favorite choice for those purchasing new vehicles; fishing and hunting are preferred pastimes in Bedrock America, and residents tend to order from Avon, read *Parents Magazine*, and watch *The Young and the Restless*.

Figures 1.B and 1.C show the location of these three PRIZM segments in the New York City, New York, and Seattle, Washington, metro area. All sorts of marketers and market researchers use geodemographic data such as these (often at a much more detailed level) not only to identify and find potential consumers of houses and apartments, cars, consumer goods, and magazines but also to help make location decisions about stores, restaurants, gas stations, and supermarkets.

Source: Nielsen MyBestSegments 2010 (www.mybestsegments.com)

TABLE 1.A   A Socio-spatial Classification of Consumer Groups

Source: Nielsen PRIZM 2010. PRIZM is a trademark or registered trademark of The Nielsen Company (US), LLC.

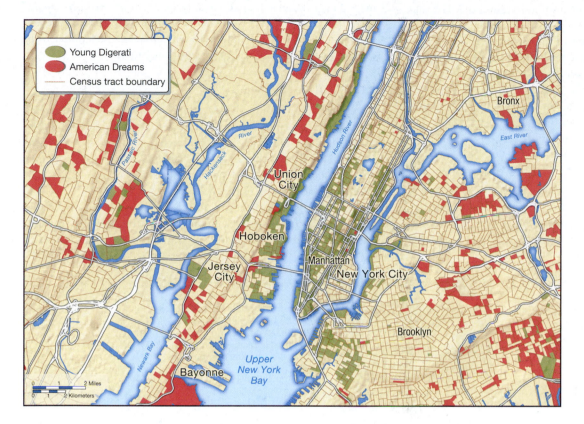

FIGURE 1.B The Geography of Consumer Groups The location of selected market segments in and around New York City.

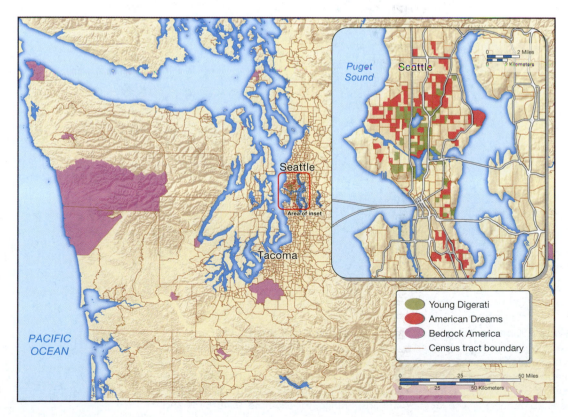

FIGURE 1.C The Geography of Consumer Groups The location of selected market segments in and around Seattle.

Within the past 5 years, applications of GIS have resulted in the creation of more maps than were created in all previous human history. One result is that as maps have become more commonplace, more people and more businesses have become more spatially aware. Nevertheless, some critics have argued that GIS has been exploited by those who already possess power and control to increase the level of surveillance of the population. The fear is that GIS may be helping to create a world in which people are not treated and judged by who they are and what they do, but more by where they live. People's credit ratings, ability to buy insurance, and ability to secure a mortgage, for example, are all routinely judged in part by GIS-based analyses that take into account the attributes and characteristics of their neighbors.

APPLY YOUR KNOWLEDGE Choose the geographic representations and tools that you would use to illustrate alternative views of a local issue that is currently the subject of public discussion in your community. Explain your choice. ■

SPATIAL ANALYSIS

The study of many geographic phenomena can be approached in terms of their arrangement as points, lines, areas, or surfaces on a map. This is known as **spatial analysis**. *Location, distance, space, accessibility,* and *spatial interaction* are five concepts that are key to spatial analysis. Although these concepts may be familiar from everyday language, they require some elaboration.

Location

Location is often nominal; that is, it is expressed solely in terms of the names given to regions and places. We speak, for example, of Washington, D.C., or of Georgetown, a location within Washington, D.C. Location can also be used as an absolute concept, whereby locations are fixed mathematically through coordinates of latitude and longitude (**Figure 1.15**).

Latitude refers to the angular distance of a point on Earth's surface, measured in degrees, minutes, and seconds north or south from the equator. The equator is assigned a value of 0°. Lines of

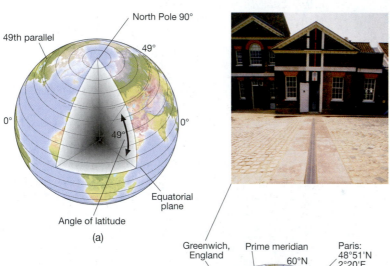

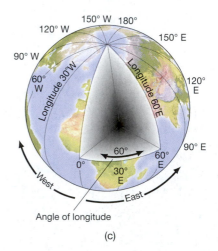

The prime meridian at the Royal Observatory in Greenwich, England. The observatory was founded by Charles II in 1675 with the task of setting standards for time, distance, latitude, and longitude—the key components of navigation.

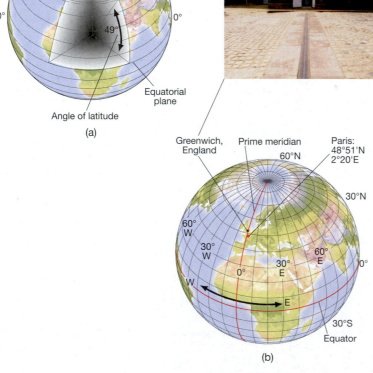

FIGURE 1.15 Latitude and longitude Lines of latitude and longitude provide a grid that covers Earth, allowing any point on Earth's surface to be accurately referenced. Latitude is measured in angular distance (i.e., degrees and minutes) north or south of the equator, as shown in (a). Longitude is measured in the same way, but east and west from the prime meridian, a line around Earth's surface that passes through both poles (North and South) and the Royal Observatory in Greenwich, just to the east of central London, in England. Locations are always stated with latitudinal measurements first. The location of Paris, France, for example, is 48°51′N and 2°20′E, as shown in (b). (a) and (c), adapted from R. W. Christopherson, *Geosystems: An Introduction to Physical Geography*, 2nd ed., © 1994, pp. 13 and 15. (b), adapted from E. F. Bergman, *Human Geography: Cultures, Connections, and Landscapes*, © 1995, Figs. 1–10 and 1–13.

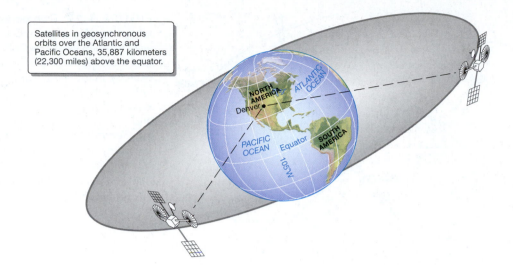

Satellites in geosynchronous orbits over the Atlantic and Pacific Oceans, 35,887 kilometers (22,300 miles) above the equator.

FIGURE 1.16 The importance of site and situation The location of telecommunications activities in Denver, Colorado, provides a good example of the significance of the geographic concepts of site (the physical attributes of a location) and situation (the location of a place relative to other places and human activities). Because of its site and situation, Denver is a major center for cable television and associated specialized support companies. Denver's site, 1.6 kilometers (1 mile) above sea level, is important because it gives commercial transmitters and receivers a better "view" of communications satellites. Its situation, on the 105th meridian and equidistant between the telecommunications satellites that are in geostationary orbit over the Pacific and Atlantic oceans, allows it to send cable programming directly not just to the whole of the Americas but also to Europe, the Middle East, India, Japan, and Australia—to every continent, in fact, except Antarctica. This precludes "double-hop" transmission (in which a signal goes up to a satellite, then down, then up and down again), which increases costs and decreases picture quality. Before the location of telecommunications facilities in Denver, places east or west of the 105th meridian had to double-hop some of their transmissions because satellite dishes would not have a clear "view" of both the Pacific and Atlantic telecommunications satellites.

latitude around the globe run parallel to the equator, which is why they are sometimes referred to as *parallels*. **Longitude** refers to the angular distance of a point on Earth's surface, measured in degrees, minutes, and seconds east or west from the *prime meridian*. The prime meridian is the line that passes through both poles and through Greenwich, England, which is assigned a value of 0°. Lines of longitude, called *meridians,* run from the North Pole (latitude 90° north) to the South Pole (latitude 90° south). Georgetown's coordinates are precisely 38°55′N, 77°00′E.

Thanks to the **Global Positioning System (GPS)**, it is very easy to determine the latitude and longitude of any given point. The Global Positioning System consists of 21 satellites (plus 3 spares) that orbit Earth on precisely predictable paths, broadcasting highly accurate time and locational information. The GPS is owned by the U.S. government, but the information transmitted by the satellites is freely available to everyone around the world. Basic GPS receivers can relay latitude, longitude, and height to within 100 meters day or night, in all weather conditions, in any part of the world. The most precise GPS receivers, costing thousands of dollars, are accurate to within a centimeter. The GPS has dramatically increased the accuracy and efficiency of collecting spatial data. In combination with GIS and remote sensing, GPS has revolutionized mapmaking and spatial analysis.

Location can also be *relative*, fixed in terms of site or situation. **Site** refers to the physical attributes of a location: its terrain, its soil, vegetation, and water sources, for example. **Situation** refers to the location of a place relative to other places and human activities: its accessibility to routeways, for example, or its nearness

to population centers (**Figure 1.16**). Washington, D.C., has a low-lying riverbank site and is situated at the head of navigation of the Potomac River, on the eastern seaboard of the United States.

Finally, location also has a *cognitive* dimension, in that people have cognitive images of places and regions, compiled from their own knowledge, experiences, and impressions. **Cognitive images** (sometimes referred to as *mental maps*) are psychological representations of locations that spring from people's individual ideas and impressions of these locations. These representations can be based on direct experiences, on written or visual representations of actual locations, on hearsay, on imaginations, or on a combination of these sources. Location in these cognitive images is fluid, depending on a given individual's changing information and perceptions of the principal landmarks in their environment.

Some things may not be located in a person's cognitive image at all. **Figure 1.17** shows a cognitive image of Washington, D.C. Georgetown is shown on this mental map, even though it is some distance from the residence of the person who sketched her image of the city. Less well-known and less distinctive places do not appear on this particular image.

Distance

Distance is also useful as an *absolute* physical measure, whose units we may count in kilometers or miles. Distance can also be a *relative* measure, expressed in terms of time, effort, or cost. It can take more or less time, for example, to travel 10 kilometers from point A to point B than it does to travel 10 kilometers from point A to point C.

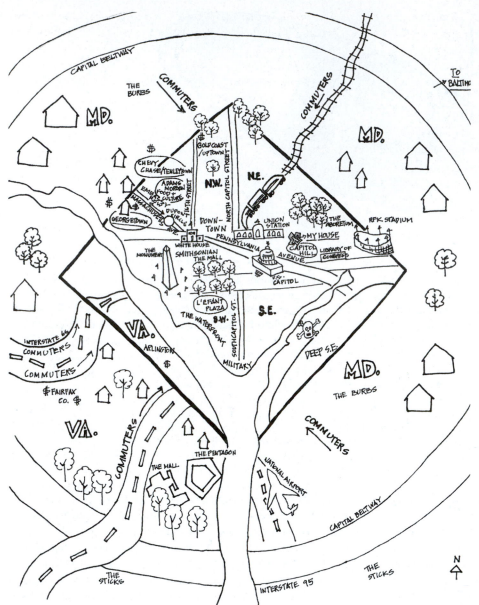

FIGURE 1.17 One person's cognitive image of Washington, D.C. This sketch was drawn by Rasheda DuPree, an urban affairs major at Virginia Tech, as part of a class exercise in recalling locations within students' hometowns. Rasheda has included many of Washington's most prominent landmarks and some of its distinctive districts, including Georgetown. In contrast, there are no recorded locations in the city's southeastern quarter (marked by a skull and crossbones in her sketch) or in the eastern outskirts (marked as "the burbs").

Similarly, it can cost more or less. Geographers have to recognize that distance can sometimes be in the eye of the beholder. It can seem longer or shorter, more or less pleasant, to travel from A to B as compared to traveling from A to C. This is **cognitive distance**, the distance that people perceive as existing in a given situation. Cognitive distance is based on people's personal judgments about the degree of spatial separation between points.

Distance is a fundamental factor in determining real-world relationships and this is a central theme in geography. It was once described as the "first law" of geography: "Everything is related to everything else, but near things are more related than distant things." Waldo Tobler, the geographer who put it this way, is one of many who have investigated the friction of distance, the deterrent or inhibiting effect of distance on human activity. The **friction of distance** is a reflection of the time and cost of overcoming distance.

What these geographers have established is that these effects are not uniform—that is, they are not directly proportional to distance itself. This is true whether distance is measured in absolute

terms (i.e., kilometers) or in relative terms (i.e., time- or cost-based measures). The deterrent effects of extra distance tend to lessen as greater distances are involved. Thus, for example, while there is a big deterrent effect in having to travel 2 kilometers instead of 1 to get to a grocery store, the deterrent effect of the same extra distance (1 kilometer) after already traveling 10 kilometers is relatively small.

This sort of relationship creates what geographers call a distance-decay function. A **distance-decay function** describes the rate at which a particular activity or phenomenon diminishes with increasing distance. Typically, the farther people have to travel, the less likely they are to do so. Distance-decay functions reflect people's behavioral response to opportunities and constraints in time and space. As such, they reflect the **utility** of particular locations. The utility of a specific place or location is its usefulness to a particular person or group. In practice, utility is thought of in different ways by different people in different situations. The emphasis may be on cost, on profitability, on prestige, on security, or on ease of mobility, for example, or more likely on some combination of attributes. However place utility is determined, people in most circumstances tend to *seek to maximize the net utility of location*. Because we seek to maximize the net utility of location, a great deal of human activity is influenced by what University of Washington geographer Richard Morrill once called the "nearness principle." According to this principle—a more explicit version of Tobler's first law—people will seek to

- maximize the overall utility of places at minimum effort;
- maximize connections between places at minimum cost; and
- locate related activities as close together as possible.

The result is that patterns of behavior between people and places come to take on fairly predictable, organized patterns.

> **APPLY YOUR KNOWLEDGE** Provide three examples of the inhibiting effect distance has on human activity. ■

TABLE 1.1 Different Kinds of Spaces Analyzed by Human Geographers

Absolute Space: Mathematical Space	Relative Space: Socioeconomic Space	Relative Space: Experiential/Cultural Space	Cognitive Space: Behavioral Space
Points	Sites	Places	Landmarks
Lines	Situations	Ways	Paths
Areas	Routes	Territories	Districts
Planes	Regions	Domains	Environments
Configurations	Distributions	Worlds	Spatial Layouts

Source: Based on H. Couclelis, "Location, Place, Region and Space," in R. Abler et al., *Geography's Inner Worlds.* New Brunswick, NJ: Rutgers University Press, 1992, Table 10.1, p. 231.

Space

Like distance, space can be measured in absolute, relative, and cognitive terms. **Table 1.1** lists the concepts human geographers use in talking about space in these various ways. Absolute space is a mathematical space described through points, lines, areas, planes, and configurations whose relationships can be fixed precisely through mathematical reasoning. Several ways of analyzing space mathematically are of use to geographers. The conventional way is to view space as a container, defined by rectangular coordinates and measured in absolute units of distance (kilometers or miles, for example). **Topological space**, another mathematical conception of space that geographers find useful, is defined as the connections between, or connectivity of, particular points in space (**Figure 1.18**). Topological space is measured not in terms of conventional measures of distance, but by the nature and degree of connectivity between locations.

Relative measurements of space can take the form of socioeconomic space or of experiential or cultural space (see **Table 1.1**). Socioeconomic space can be described in terms of sites and situations, routes, regions, and distribution patterns. In these terms, spatial relationships are fixed through measures of time, cost, profit, and production, as well as through physical distance. Experiential or cultural space is the space of people with common ties, described through the places, territories, and settings whose attributes carry special meaning for particular groups. Finally, **cognitive space** is defined and measured in terms of people's values, feelings, beliefs, and perceptions about locations, districts, and regions. Cognitive space can be described, therefore, in terms of behavioral space—landmarks, paths, environments, and spatial layouts.

FIGURE 1.18 Topological space Some dimensions of space and aspects of spatial organization do not lend themselves to description simply in terms of distance. The connectivity of people and places is often important: whether they are linked, how they are linked, and so on. These attributes of connectivity define topological space. The map of the Metro system in Milan, Italy, is a topological map, showing how specific points are joined within a particular network. The most important aspects of networks of any kind, from the geographer's viewpoint, are their connectivity attributes. These attributes determine the flows of people and things (goods, information) and the centrality of places. As most Milanese know, the Metro system gives Duomo a very high degree of connectivity because trains on both the M1 and M3 lines stop there. Duomo is therefore relatively central within the "space of flows" of passenger traffic. Missori—nearby in absolute terms—is much less central, however, and much less the focus of passenger flows.

Accessibility

Because it is a fundamental influence on the utility of locations, distance is an important influence on people's behavior. **Accessibility** is generally defined by geographers in terms of relative location: the opportunity for contact or interaction from a given point or location in relation to other locations. It implies proximity, or nearness, to something. Distance is one aspect of accessibility, but it is by no means the only important aspect.

Connectivity is also an important aspect of accessibility because contact and interaction are dependent on channels of communication and transportation: streets, highways, telephone lines, and wave bands, for example. Effective accessibility is a function not only of distance but also of the configuration of networks of communication and transportation. Commercial airline networks provide many striking examples of this. Cities that operate as airline hubs are much more accessible than cities that are served by fewer flights and fewer airlines. Charlotte, N.C., for example (a U.S. Airways hub), is more accessible from Albany, N.Y., than from Richmond, V.A., even though Richmond is 400 kilometers (248 miles) closer to Albany than Charlotte. To get to Richmond from Albany, airline passengers must travel to Charlotte or another hub and change—a journey that takes longer and often costs more.

Accessibility is often a function of economic, cultural, and social factors. In other words, relative concepts and measures of distance are often as important as absolute distance in determining accessibility. A nearby facility, such as a health care clinic, is accessible to us only if we can actually afford the cost of getting there, in other words, only if it seems close according to our own standards of distance, if we can afford to use the facility, if we feel that it is socially and culturally acceptable for us to use it, and so on. To take another example, a day care center may be located just a few blocks from a single-parent family, but the center is not truly accessible if it opens after the parent has to be at work, if the cost of care is too high, or if the parent feels that the staff, children, or other parents at the center are from an incompatible social or cultural group.

Spatial Interaction

Interdependence between places and regions can be sustained only through movement and flows. Geographers use the term **spatial interaction** as shorthand for all kinds of movement and flows involving human activity. Freight shipments, commuting, shopping trips, telecommunications, electronic cash transfers, migration, and vacation travel are all examples of spatial interaction. The fundamental principles of spatial interaction can be reduced to four basic concepts: complementarity, transferability, intervening opportunities, and diffusion.

Complementarity

For any kind of spatial interaction to occur between two places, there must be a demand in one place and a supply that matches, or complements, it in the other. This complementarity can be the result of several factors. One important factor is the variation in physical environments and resource endowments from place to place. For example, a heavy flow of vacation travel from Swedish cities to Mediterranean resorts is largely a function of climatic complementarity. To take another example, the flow of crude oil from Saudi Arabia (with vast oil reserves) to Japan (with none) is a function of complementarity in natural resource endowments.

A second factor contributing to complementarity is the international division of labor that derives from the evolution of the world's economic systems. The more developed countries of the world have sought to establish overseas suppliers for their food, raw materials, and exotic produce, allowing these more developed countries to specialize in more profitable manufacturing and knowledge-based industries (see Chapter 2). Through a combination of colonialism, imperialism, and sheer economic dominance on the part of the more developed countries, less powerful countries have found themselves with economies that directly complement the needs of the more developed countries. Among the many flows resulting from this complementarity are shipments of sugar from Barbados to the United Kingdom, bananas from Costa Rica and Honduras to the United States, palm oil from Cameroon to France, automobiles from France to Algeria, school textbooks from the United Kingdom to Kenya, and investment capital from the United States to less developed countries.

A third contributory factor to complementarity is specialization and **economies of scale**. Places, regions, and countries can derive economic advantages from the efficiencies created through specialization, which allows for larger-scale operations. Economies of scale are cost advantages to manufacturers in high-volume production; the average cost of production falls with increasing output. Among other things, fixed costs (for example, the cost of renting or buying factory space, which is the same—fixed—whatever the level of output from the factory) can be spread over higher levels of output so that the average cost per unit of production falls. Economic specialization results in complementarities, which in turn contribute to patterns of spatial interaction. For example, Israeli farmers specialize in high-value fruit and vegetable crops for export to the European Union, which in return exports grains and root crops to Israel.

Transferability

Another precondition for interdependence between places is *transferability*, which depends on the frictional (or deterrent) effects of distance. Transferability is a function of two things: the costs of moving a particular item, measured in real money and/or time, and the ability of the item to bear these costs. If, for example, the costs of moving a product from one place to another make it too expensive to sell successfully at its destination, then that product does not have transferability between those places.

Transferability varies between places, between kinds of items, and between modes of transportation and communication. The transferability of coal, for example, is much greater between places that are connected by rail or by navigable waterways than between places connected only by highways. This is because it is much cheaper to move heavy, bulky materials by rail, barge, or ship. The transferability of fruit and salad crops, on the other hand, depends more on the speed of transportation and the availability of specialized refrigerated vehicles so the fruits and vegetables stay fresh. While the transferability of money capital is much greater by telecommunications than it is by surface transportation, it is also higher between places where banks are equipped to deal routinely with electronic transfers. Computer microchips have high transferability because they are easy to handle, and transport costs are a small proportion of their value. Computer monitors have lower transferability because of their fragility and their relatively lower value by weight and volume.

Transferability also varies over time, with successive innovations in transport and communications technologies and successive waves of **infrastructure** development (canals, railways, harbor installations, roads, bridges, and so on). New technologies and new or extended infrastructures alter the geography of transport costs and the transferability of particular things between particular places. *As a result, the spatial organization of many different activities is continually changing and readjusting.* The consequent tendency toward a shrinking world gives rise to **time-space convergence**, the rate at which places move closer together in travel or communication time or costs. Time-space convergence results from a decrease in the friction of distance as space-adjusting technologies have, in general, brought places closer together over time

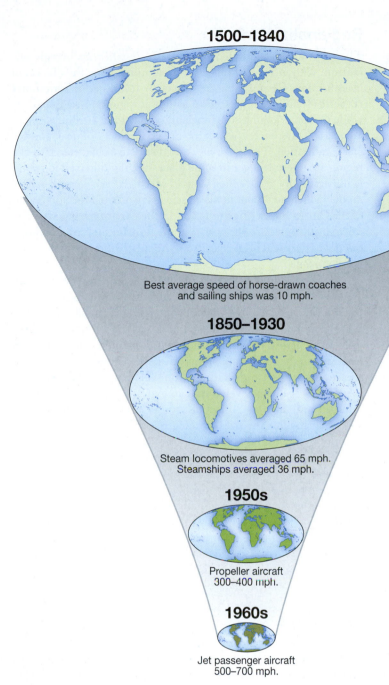

1500–1840

Best average speed of horse-drawn coaches and sailing ships was 10 mph.

1850–1930

Steam locomotives averaged 65 mph. Steamships averaged 36 mph.

1950s

Propeller aircraft 300–400 mph.

1960s

Jet passenger aircraft 500–700 mph.

FIGURE 1.19 Time-space convergence The effects of changing transportation technologies in "shrinking" the world in terms of travel time.

extent to which people in different places are "plugged in" to new technologies. The shrinking of space has important implications for people's everyday conceptions of space and distance and for their level of knowledge about other places.

Intervening Opportunity

While complementarity and transferability are preconditions for spatial interaction, intervening opportunities are more important in determining the *volume* and *pattern* of movements and flows. Intervening opportunities are alternative origins and/or destinations. Such opportunities are not necessarily situated directly between two points or even along a route between them. Thus, to take one of our previous examples, for Swedish families considering a Mediterranean vacation in Greece, resorts in Spain, southern France, and Italy are all likely to be intervening opportunities because they can probably be reached more quickly and cheaply than resorts in Greece.

The size and relative importance of alternative destinations are important aspects of the concept of intervening opportunity. For our Swedish families, Spanish resorts probably offer the greatest intervening opportunity because they contain the largest aggregate number of hotel rooms and vacation apartments. We can therefore state the principle of intervening opportunity as follows: Spatial interaction between an origin and a destination will be proportional to the number of opportunities at that destination and inversely proportional to the number of opportunities at alternative destinations.

Spatial Diffusion

Disease outbreaks, technological innovations, political movements, and new musical fads all originate in specific places and subsequently spread to other places and regions. The way that things spread through space and over time—**spatial diffusion**—is one of the most important aspects of spatial interaction and is crucial to an understanding of geographic change.

Diffusion seldom occurs in an apparently random way, jumping unpredictably all over the map. Rather, it occurs as a function of statistical probability, which is often based on fundamental geographic principles of distance and movement. The diffusion of a contagious disease, for example, is a function of the probability of physical contact, modified by variations in individual resistance to the disease. The result is typically a "wave" of diffusion that describes an S-curve, with a slow buildup, rapid spread, and final leveling off (**Figure 1.20**).

It is possible to recognize several spatial tendencies in patterns of diffusion. In *expansion diffusion* (also called *contagion diffusion*), a phenomenon spreads because of the proximity of carriers, or agents of change, who are fixed in their location. An

(**Figure 1.19**). Overland travel between New York and Boston, for example, has been reduced from 3.5 days (in 1800) to 5 hours (in the 2000s) as the railroad displaced stagecoaches and was in turn displaced by interstate automobile travel. Other important space-adjusting innovations include air travel and air cargo; telegraphic, telephonic, and satellite communications systems; national postal services; package delivery services; and modems, fiber-optic networks, and electronic-mail software.

What is most significant about the latest developments in transport and communication is that they are not only global in scope but also are able to penetrate to local scales. As this penetration occurs, some places that are distant in kilometers are becoming closer together, while some that are close in terms of absolute space are becoming more distant in terms of their ability to reach one another electronically. Much depends on the mode of communication—the

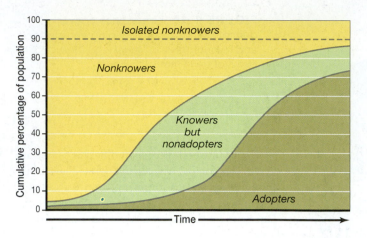

FIGURE 1.20 Spatial diffusion The spatial diffusion of many phenomena tends to follow an S-curve of slow buildup, rapid spread, and leveling off. In the case of the diffusion of an innovation, for example, it usually takes a while for enough potential adopters to find out about the innovation, and even longer for a critical mass of them to adopt it. After that the innovation spreads quite rapidly, until most of the potential adopters have been exposed. (Adapted from D. J. Walmsley and G. J. Lewis, *Human Geography: Behavioral Approaches.* London: Longman, 1984, Fig. 5.3, p. 52.)

example would be the diffusion of an agricultural innovation, such as the use of hybrid seed stock, among members of a local farming community.

With *hierarchical diffusion* (also known as cascade diffusion), a phenomenon can be diffused from one location to another without necessarily spreading to people or places in between. An example would be the spread of a fashion trend from large metropolitan areas to successively smaller cities, towns, and rural settlements.

Many patterns and processes of diffusion reflect both expansion and hierarchical diffusion, as different aspects of human interaction come into play in different geographic settings. The diffusion of outbreaks of communicable diseases, for example, usually involves a combination of expansion and hierarchical diffusion.

APPLY YOUR KNOWLEDGE Referring to spatial analysis concepts, discuss a national or international environmental issue. Provide examples of how complementarity, transferability, intervening opportunities, and diffusion each relate to the issue you have chosen. ▪

REGIONAL ANALYSIS

Not all geographic phenomena are most effectively understood through spatial analysis. Geographers also seek to understand the complex relationships between peoples and places in terms of the similarities and differences among and between them and the identities and qualities associated with them. Here the key concepts are regionalization, landscape, and sense of place.

Regionalization

The geographer's equivalent of *scientific classification* is **regionalization**, with individual places or areal units being the objects of classification. Geographers are especially interested in **functional regions** (sometimes referred to as nodal regions)—regions that, while they may exhibit some variability in certain attributes, share an overall coherence in structure and economic, political, and social organization. The coherence and distinctive characteristics of a region are often stronger in some places than in others. This point is illustrated by geographer Donald Meinig's *core-domain-sphere* model, which he set out in his classic essay on the Mormon region of the United States (**Figure 1.21**). In the core of a region the distinctive attributes are very clear; in the domain they are dominant but not to the point of exclusivity; in the sphere they are present but not dominant.

People's own conceptions of place, region, and identity may generate strong feelings of regionalism and sectionalism that feed back into the processes of place-making and regional differentiation. **Regionalism** is a term used to describe situations in which different religious or ethnic groups with distinctive identities coexist within the same state boundaries, often concentrated within a particular region and sharing strong feelings of collective identity. If such feelings develop into an extreme devotion to regional interests and customs, the condition is known as **sectionalism**. Regionalism often involves ethnic groups whose aims include autonomy from a national state and the development of their own political power (see Chapter 9). In certain cases, enclaves of ethnic minorities are claimed by the government of a country other than the one in which they reside. Such was the case, for example, of Serbian enclaves in Croatia, claimed by nationalist Serbs. **Irredentism** is the assertion by the government of a country that a minority living outside its formal borders belongs to it historically and culturally. In some circumstances, as with Serbia's claims on Serbian enclaves in Croatia in the early 1990s, irredentism can lead to war.

Landscape

Geographers think of landscape as a comprehensive product of human action such that every landscape is a complex repository of society. It is a collection of evidence about our character and experience, our struggles and triumphs as humans. To understand better the meaning of landscape, geographers have developed different categories of landscape types based on the elements contained within them.

Ordinary landscapes (or vernacular landscapes, as they are sometimes called) are the everyday landscapes that people create in the course of their lives together. From crowded city centers to leafy suburbs and quiet rural villages, landscapes are lived in and changed and influence and change the perceptions, values, and behaviors of the people who live and work in them.

Symbolic landscapes, by contrast, represent particular values or aspirations that builders and financiers want to impart to a larger public. For example, the neoclassical architecture of the buildings of the federal government in Washington, D.C., along with the streets, parks, and monuments of the capital, constitute a symbolic landscape intended to communicate a sense of power but also of democracy in its imitation of the Greek city-state.

FIGURE 1.21 The Mormon culture region Cultural attributes often gradually shade from one region to another, rather than having a single, clear-cut boundary. Geographer Donald Meinig's work on the Mormon culture region of the United States identified a "core" region, which exhibits all the attributes of Mormon culture; a "domain," where not all these attributes may be present (or may be less intense); and a "sphere," where some attributes of Mormon culture are present but often as a minority. (Adapted from D. Meinig, "The Mormon Culture Region: Strategies and Patterns in the Geography of the American West," *Annals, Association of American Geographers*, 55, 1965, pp. 191–220.)

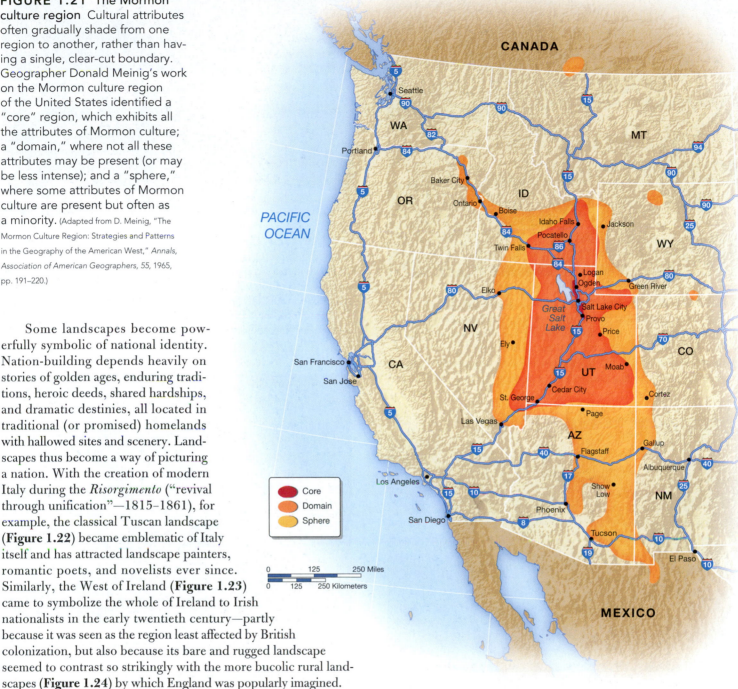

Some landscapes become powerfully symbolic of national identity. Nation-building depends heavily on stories of golden ages, enduring traditions, heroic deeds, shared hardships, and dramatic destinies, all located in traditional (or promised) homelands with hallowed sites and scenery. Landscapes thus become a way of picturing a nation. With the creation of modern Italy during the *Risorgimento* ("revival through unification"—1815–1861), for example, the classical Tuscan landscape (**Figure 1.22**) became emblematic of Italy itself and has attracted landscape painters, romantic poets, and novelists ever since. Similarly, the West of Ireland (**Figure 1.23**) came to symbolize the whole of Ireland to Irish nationalists in the early twentieth century—partly because it was seen as the region least affected by British colonization, but also because its bare and rugged landscape seemed to contrast so strikingly with the more bucolic rural landscapes (**Figure 1.24**) by which England was popularly imagined.

Geographers now recognize that there are many layers of meaning embedded in the landscape. Landscapes reflect people's dreams and ideas as well as their material lives. The messages embedded in landscapes can be read as signs about values, beliefs, and practices, though not every reader will take the same message from a particular landscape (just as people may differ in their interpretation of a passage from a book). In short, landscapes both produce and communicate meaning, and one of our tasks as geographers is to interpret those meanings.

Sense of Place

Everyday routines experienced in familiar settings allow people to derive a pool of shared meanings. Often this carries over into people's attitudes and feelings about themselves and their locality.

When this happens, the result is a self-conscious sense of place. A **sense of place** refers to the feelings evoked among people as a result of the experiences and memories they associate with a place and to the symbolism they attach to that place. It can also refer to the character of a place as seen by outsiders: its distinctive physical characteristics and/or its inhabitants.

For *insiders,* this sense of place develops through shared dress codes, speech patterns, public comportment, and so on. A crucial concept here is that of the **lifeworld**, the taken-for-granted pattern and context for everyday living through which people conduct their day-to-day lives without conscious attention. People become familiar with one another's vocabulary, speech

FIGURE 1.22 The power of place In some countries, particular landscapes have become powerfully symbolic of national identity. In Italy it is the classical landscape of Tuscany, with its scattered farms and villas, elegant cypress trees, silvery-green olive trees, and rolling fields with a rich mixture of cereals, vegetables, fruit trees, and vines, as in this photograph of Val d'Orcia.

FIGURE 1.24 Castle Combe, Wiltshire, England The well-ordered and picturesque landscape of the southern parts of rural England have long been taken to be emblematic of England as a whole and of the values and ideals of its people—even though urban and industrial development, together with modern agricultural practices, have brought about significant changes to both landscapes and society.

FIGURE 1.23 The national landscape of Ireland The rugged landscapes of the West of Ireland have come to symbolize the whole country to many people, both within Ireland and beyond. This photograph shows part of rural County Kerry.

FIGURE 1.25 Intersubjectivity Routine encounters such as this, in Chiavenna, Italy, help to develop a sense of community and a sense of place among residents.

patterns, dress codes, gestures, and humor as a result of routine encounters and shared experiences in bars and pubs, cafés and restaurants, shops and street markets, and parks. This is known as **intersubjectivity**: shared meanings that are derived from everyday practice. Elements of daily rhythms (such as mid-morning grocery shopping with a stop for coffee, the *aperitivo* en route from work to home, and the after-dinner stroll) are all critical to the intersubjectivity that is the basis for a sense of place within a community (**Figure 1.25**). The same is true of elements of weekly rhythms, such as street markets and farmers' markets, and of seasonal rhythms, such as festivals.

These rhythms, in turn, depend on certain kinds of spaces and places: not only streets, squares, and public open spaces but also "third places" (after home, first, and workplace, second): the sidewalk cafés, pubs, post offices, drugstores, corner stores, and family-run *trattoria* that are the loci of routine activities and sociocultural transactions. Third places accommodate "characters," "regulars," and newcomers, as well as routine patrons and, like public spaces, facilitate casual encounters as well as settings for sustained conversations. The nature and frequency of routine encounters and shared experiences depend a great deal on the attributes of these spaces and places.

FIGURE 1.26 Community art Community art can provide an important element in the creation of a sense of place for members of local communities. This example is from the Mission district in San Francisco.

A sense of place also develops through familiarity with the history and symbolism of particular elements of the physical environment—a mountain or lake, the birthplace of someone notable, the site of some particularly well-known event, or the expression of community identity through art (**Figure 1.26**). Sometimes it is deliberately fostered by the construction of symbolic structures such as monuments and statues. Often it is a natural outcome of people's familiarity with one another and their surroundings. Because of this consequent sense of place, insiders feel at home and "in place."

For *outsiders,* a sense of place can be evoked only if local landmarks, ways of life, and so on are distinctive enough to evoke a significant common meaning for people who have no direct experience of them. Central London, for example, is a setting that carries a strong sense of place to outsiders who have a sense of familiarity with the riverside panoramas, busy streets, and distinctive monuments and historic buildings.

APPLY YOUR KNOWLEDGE What are the most distinctive characteristics of the region in which you live? How would you describe its landscapes? What, from your perspective, gives your community a sense of place? ■

DEVELOPING A GEOGRAPHICAL IMAGINATION

A **geographical imagination** allows us to understand changing patterns, processes, and relationships among people, places, and regions. Developing this capacity is increasingly important as the pace of change around the world increases to unprecedented levels.

It is often useful to think of places and regions as representing the cumulative legacy of successive periods of change. Following this approach, we can look for superimposed layers of development, or evidence of the imprint of different phases of local development (see **Box 1.3**, "Windows on the World: South Beach, Miami Beach"). We can show how some patterns and relationships last, while others are modified or obliterated. We can show how different places bear the imprint of different kinds of change, perhaps in different sequences and with different outcomes. To do so, we must be able to identify the kinds of changes that are most significant.

We can prepare our geographical imagination to deal with an important aspect of spatial change by making a distinction between the *general* and the *unique.* This distinction helps account for geographical diversity and variety because it provides a way of understanding how and why one kind of change can result in a variety of spatial outcomes. Because the *general effects* of a particular change always involve some degree of modification as they are played out in different environments, *unique outcomes* result.

Although we can usually identify some general outcomes of major episodes of change, there are almost always some unique outcomes, too. Let us take two related examples. The Industrial Revolution of nineteenth-century Europe provides a good example of a major period of change. A few of the general spatial outcomes were increased urbanization, regional specialization in production, and increased interregional and international trade. At one level, places could be said to have become increasingly alike: generic coalfield regions, industrial towns, ports, downtowns, worker housing, and suburbs.

It is clear, however, that these general outcomes were mediated by the different physical, economic, cultural, and social attributes of different places. Beneath the dramatic overall changes in the geography of Europe, new layers of diversity and variety also existed. Industrial towns developed their own distinctive character as a result of their manufacturing specialties, their politics, the personalities and objectives of their leaders, and the reactions and responses of their residents. Downtowns were differentiated from one another as the general forces of commerce and land economics played out across different physical sites and within different patterns of land ownership. Various local socioeconomic and political factors gave rise to different expressions of urban design. Meanwhile, some places came to be distinctive because they were almost entirely bypassed by this period of change, their characteristics making them unsuited to the new economic and spatial order (**Figure 1.27**).

The second example of general and unique outcomes of change is the introduction of the railroad, one of the specific changes involved in the Industrial Revolution. In general terms, the railroad contributed to time-space convergence, to the reorganization of industry into larger market areas, to an increase in interregional and international trade, and to the interconnectedness of urban systems. Other unique outcomes, however, have also contributed to distinctive regional geographies. In Britain, the railroad was introduced to an environment that was partially industrialized and densely settled. The increased efficiencies provided by the railroad helped to turn Britain's economy into a highly integrated and intensively urbanized national economy. In Spain, however, the railroad was introduced to an environment that was less urbanized and industrialized and less able to afford the costs of railroad construction. The result was that the relatively few Spanish towns connected by the railroads gained a massive comparative advantage. This situation laid the foundation for a modern Spanish space-economy that was much less integrated than Britain's, with an urban system dominated by just a few towns and cities.

South Beach, Miami Beach

South Beach provides a vibrant illustration of a place that embodies the layered legacy of successive waves of development and change. In just over 100 years, South Beach has gone from a coconut plantation to a speculative development of modest single-family residences; a booming resort and retirement district; a run-down district infiltrated by crime and vice; and a renovated district on the National Register of Historic Places. Its latest incarnation is as an exclusive residential enclave—an international destination district and celebrity haunt. In the process, it has changed from a segregated Anglo American community to a predominantly Jewish community to a multiethnic, cosmopolitan population, the setting for an unabashedly vulgar hedonism. Elements of all of these changes are visible in the landscapes of the district today.

Construction of a small oceanfront residential district began in 1913. A year later, Carl G. Fisher, a successful entrepreneur who made millions selling a business to Union Carbide, began to invest in property development and infrastructure improvements (including a vital road bridge to the mainland). South Beach steadily filled up in the 1920s in a real estate boom supercharged by the palatial homes built by several prominent millionaires. The boom coincided with the blossoming of stylish Modern architecture inspired by the Paris Exhibition of 1925. The new styles reflected a fascination with the streamlined design and comfort features of mass transportation of the era, especially transatlantic ocean liners, luxury automobiles, and passenger trains. Property development and redevelopment continued even during the Great Depression of the 1930s, creating a distinctive district of hotels and residences—about 800 buildings in 1 square mile—in what later became known as Art Deco styles (**Figure 1.D**).

After World War II, South Beach became a glamorous destination district, a seasonal haven for wealthy northerners. The arrival of retirees from the Northeast and Midwest brought a new dimension of ethnic diversity to the city. Jewish retirees were drawn to Miami Beach but had to contend with segregation ordinances that restricted them to the area south of Fifth Street, the southern tip of South Beach. Civil rights activism, combined with the sheer weight of increasing numbers, overcame segregation by the 1960s, and within a few years Jews had come to dominate the entire city of Miami Beach. In South Beach, synagogues, Yiddish theatres, kosher restaurants, and delicatessens proliferated. While there were some upscale hotels and residences, the district was dominated by modest, deteriorating, low-rise buildings populated by retirees on small, fixed incomes.

FIGURE 1.D South Beach Architecture Characteristic elements of buildings in the Art Deco district of South Beach include building forms based on geometric shapes, with sharp angles and/or curved corners; "eyebrows" above windows; jutting towers, ziggurat rooflines, glass-block windows, porthole windows, deck railings, racing stripes, and chevron decorations; the use of the modern materials like chrome, plastic, and aluminium; distinctive signage and lettering with exaggerated or multiline details; and neon lighting.

Meanwhile, as metropolitan Miami absorbed increasing numbers of immigrants from Cuba and Latin America, South Beach was infiltrated by drug dealers and the associated crime and vice. Wealthy households moved out, leaving behind less affluent households, such as the retirees on fixed incomes. For a while, South Beach acquired the nickname of "God's Waiting Room." By the mid-1970s, South Beach was down-at-heel, no longer attracting many new retirees and on the cusp of irretrievable deterioration. It was saved by the activism of conservationists, who secured a square-mile portion of South Beach as the Miami Beach Art Deco District on the National Register of Historic Places. Conservationists promoted the idea of renovating the buildings by applying a palette of pastel hues—flamingo and bubblegum pink, peach, periwinkle, purple, aqua, and lavender blue—on what had previously been whitewashed surfaces. Hotels and restaurants also began to invest in classic Art Deco neon lighting that provides an exotic nightscape. The makeover was enough to attract new investment, new visitors, and new residents. The combination of distinctive architecture and wide stretches of soft, almost-white sandy beaches, together with the subtropical light, made South Beach an attractive location for photo shoots. Many of the large New York–based modeling agencies moved in, bringing an influx of fashion industry professionals. Sidewalk cafés, boutiques, nightclubs, and chic restaurants followed, along with more models, celebrities, and international tourists.

South Beach became a hedonistic setting, a 24-hour playground dominated by the young and affluent. Restored Art Deco and Mediterranean Revival buildings were joined by new contemporary architecture and ultraluxurious condominiums. The southern tip of the district, branded as "SoFi," was redeveloped with exclusive, resortlike condominium towers (**Figure 1.E**). Inevitably, the district's demography changed again. Jewish households had represented almost two-thirds of the total population of Miami Beach in 1980. By 2010, the figure was less than 15 percent. Similarly, whereas elderly households had long dominated South Beach, less than 15 percent of its population was aged 50 or more by 2010. The majority were aged between 20 and 40. The traditional flow of Jewish retirees on modest incomes had been priced out of the district, while the more affluent gravitated to Broward and Palm Beach counties. Today, South Beach has developed a unique cultural mix, which is in many ways like the Mediterranean European coast or parts of Latin America. Spanish is the first language of more than half of the population, while English is the first language of only about a third.

Colombian discos, European club scenes, traditional American bars, Cuban cafés, and restaurants specializing in "Eurasian," "New World," and "Nuevo Latino" cuisine have displaced the kosher delis and seafood shacks. Bikini shows and wet T-shirt contests have replaced the mahjong and canasta games of retirees.

FIGURE 1.E South Beach Lifestyle Luxury condominiums at the southern tip of South Beach.

Lounge music from the 1950s has been displaced by Europop, samba from Brazil, salsa from Colombia, tango from Argentina, reggae from the English-speaking Caribbean, flamenco from Spain, and meringue and bachata from the Dominican Republic. The influence of sun, sea, and the warm weather has fostered a distinctive body-consciousness and standard of public exposure that would be censured in many other American cities.

Source: (Based on "South Beach" in Paul Knox, *Palimpsests: Biographies of 50 City Districts*, Basel: Actar/Birkhauser, 2012.)

FIGURE 1.27 Hersbrück, Germany Hersbrück was once a prosperous regional center on an overland trade route—the "Golden Road"—between Nuremberg and Prague. After 1806, when Napoleon redrew the political map of Europe, the reorganization of the European economy, together with the onset of the Industrial Revolution, left Hersbrück somewhat isolated and economically disadvantaged. Hersbrück was never drawn into the industrial development of Germany and is not well connected to the transportation infrastructure of canals, railways, or major highways.

APPLY YOUR KNOWLEDGE Describe the region in which you live, emphasizing the imprint of different periods of development. Which features of the region can be said to be the result of general spatial effects, and which are unique? ■

Future Geographies

Places and regions are constantly in a state of change. This raises the question of what we may expect to see in the future. An understanding of geographic processes and principles, together with a knowledge of past and present spatial patterns, allows us to make informed judgements about future geographies: an important dimension of applying our geographical imagination.

Whereas much of the world had remained virtually unchanged for decades, even centuries, the Industrial Revolution and long-distance, high-speed transportation and communications brought a rapid series of rearrangements to the countryside and to towns and cities in many parts of the world. Today, with a globalized economy and global telecommunications and transportation networks, places have become much more interdependent, and still more of the world is exposed to increasingly urgent imperatives to change.

Will social networking sites bring about new patterns of human interaction? Will we be able to cope with the environmental stresses of increasing industrialization and rapid population growth? Will the United States retain its position as the world's most powerful and influential nation? What kind of problems will the future bring for local, regional, and international development? What new technologies are likely to have the most impact in

reshaping human geographies? Will globalization undermine regional cultures? These are just a few of the many questions that spring from the key themes in human geography.

As we begin to look to the future, we can appreciate that some dimensions of human geography are more certain than others (**Figure 1.28**). We can only guess, of course, at some aspects of the future. Two of the most speculative realms are those of politics and technology, which are both likely to spring surprises at any time. On the other hand, in some ways the future is already here, embedded in the world's institutional structures and in the dynamics of its populations. We know, for example, a good deal about the demographic trends of the next quarter century, given present populations, birth and death rates, and so on. We also know a good deal about the distribution of environmental resources and constraints, about the characteristics of local and regional economies, and about the legal and political frameworks within which geographic change will probably take place. The tools and concepts of human geography allow us to understand change in terms of local place-making processes that are subject to certain broader principles of spatial organization as well as the overall framework of the global economy. In subsequent chapters, as we look more closely at specific aspects of human geography, we shall be able to see how geographic process and principles can suggest future patterns and pathways of change.

APPLY YOUR KNOWLEDGE Make a list of likely future changes to the geography of the region in which you live. ∎

FIGURE 1.28 Affluent China Some aspects of future geographies are relatively certain. The unprecedented shift in relative wealth and economic power roughly from West to East now under way will continue, though it will inevitably increase social and spatial inequalities within the East, the consequences of which are highly uncertain.

CONCLUSION

Human geography is the systematic study of the location of peoples and human activities across Earth's surface and of their relationships to one another. An understanding of human geography is important both from an intellectual point of view (that is, understanding the world around us) and a practical point of view (for example, contributing to environmental quality, human rights, social justice, business efficiency, political analysis, and government policymaking).

Human geography reveals how and why geographical relationships matter in terms of cause and effect in relation to economic, social, cultural, and political phenomena. Human geographers strive to recognize these wider processes and broad geographical patterns without losing sight of the uniqueness of specific places.

Geography matters because it is in specific places that people learn who and what they are and how they should think and behave. Places are also a strong influence, for better or worse, on people's physical well-being, their opportunities, and their lifestyle choices. Places also contribute to peoples' collective memory and become powerful emotional and cultural symbols. Places are the sites of innovation and change, of resistance and conflict.

To investigate specific places we must be able to frame our studies of them within the compass of the entire globe. This is important for two reasons. First, the world consists of a complex mosaic of places and regions that are interrelated and interdependent in many ways. Second, place-making forces—especially economic, cultural, and political forces that influence the distribution of human activities and the character of places—are increasingly operating at global and international scales. In the next chapter, we describe the changing global context that has shaped places and regions around the world.

Learning Outcomes Revisited

- Explain how the study of geography has become essential for understanding a world that is more complex, interdependent, and changing faster than ever before.

  Geography matters because it enables us to understand where we are both literally and figuratively. Geography provides an understanding of the interdependency of people and places and an appreciation of how and why certain places are distinctive or unique. From this knowledge, we can begin to understand the implications of future spatial patterns.

- Identify four examples of how places influence inhabitants' lives.

  Specific places provide the settings for people's daily lives. Places are settings for social interaction that, among other things, structure the daily routines of people's economic and social lives; provide both opportunities for—and constraints on—people's long-term social well-being; establish a context in which everyday commonsense knowledge and experience are gathered; provide a setting for processes of socialization; and provide an arena for contesting social norms.

- Distinguish the differences among major map projections and describe their relative strengths and weaknesses.

  The choice of map projection depends largely on the purpose of the map. Equidistant projections allow distance to be represented as accurately as possible but in only one direction (usually north-south). Conformal projections render compass directions accurately but tend to exaggerate the size of northern continents. Equivalent projections portray areas on Earth's surface in their true proportions but result in world maps on which many locations appear squashed and have unsatisfactory outlines.

- Explain how geographers use geographic information systems (GIS) to merge and analyze data.

  New technologies combine high-performance computing, global positioning systems (GPS), and computerized record keeping. The most important aspect of these technologies, from an analytical point of view, is that they allow data from several different sources, on different topics and at different scales, to be merged.

- Summarize the five concepts that are key to spatial analysis and describe how they help geographers to analyze relationships between peoples and places.

  The study of many geographic phenomena can be approached in terms of their arrangement as points, lines, areas, or surfaces on a map. This is known as spatial analysis. Location, distance, space, accessibility, and spatial interaction are five concepts that are key to spatial analysis. Each of these concepts is multi-faceted and can be applied to different spatial scales. Together, they provide a powerful set of tools for describing and analyzing places and regions.

- Describe the importance of distance in shaping human activity.

  The first law of geography is that "Everything is related to everything else, but near things are more related than are distant things." Human activity is influenced by the "nearness principle," according to which people tend to seek to maximize the overall utility of places at minimum effort, to maximize connections between places at minimum cost, and to locate related activities as close together as possible.

- Summarize the three concepts that are key to regional analysis and explain how they help geographers analyze relationships between peoples and places.

  The key concepts of regional analysis are regionalization, landscape, and sense of place. Regionalization is the geographer's equivalent of scientific classification; landscapes embody many layers of meaning and reflect the influence of past processes of change, while sense of place derives from everyday routines experienced in familiar settings. Geographers also seek to understand the complex relationships between peoples and places in terms of the similarities and differences among and between them and the identities and qualities associated with them.

KEY TERMS

accessibility *(p. 21)*
cartography *(p. 8)*
cognitive distance *(p. 20)*
cognitive image *(p. 19)*
cognitive space *(p. 21)*
conformal projection *(p. 11)*
distance-decay function *(p. 20)*
economies of scale *(p. 22)*
equal-area (equivalent) projection *(p. 11)*
equidistant projection *(p. 11)*

friction of distance *(p. 20)*
functional region *(p. 24)*
geodemographic research *(p. 15)*
geographical imagination *(p. 27)*
geographic information systems (GIS) *(p. 15)*
Global Positioning System (GPS) *(p. 19)*
human geography *(p. 7)*
identity *(p. 6)*
infrastructure *(p. 22)*
intersubjectivity *(p. 26)*
irredentism *(p. 24)*

latitude *(p. 18)*
lifeworld *(p. 25)*
longitude *(p. 19)*
map projections *(p. 11)*
ordinary landscapes *(p. 24)*
physical geography *(p. 7)*
place *(p. 3)*
region *(p. 3)*
regional geography *(p. 7)*
regionalism *(p. 24)*
regionalization *(p. 24)*
remote sensing *(p. 7)*

sectionalism *(p. 24)*
sense of place *(p. 25)*
site *(p. 19)*
situation *(p. 19)*
spatial analysis *(p. 18)*
spatial diffusion *(p. 23)*
spatial interaction *(p. 22)*
symbolic landscapes *(p. 24)*
time-space convergence *(p. 22)*
topological space *(p. 21)*
utility *(p. 20)*

REVIEW AND DISCUSSION

1. Consider the term *symbolic landscapes*. Identify and list five examples of symbolic landscapes in your town (see **Figures 1.22–1.24**). For each of the landscapes, interpret the values, beliefs, and aspirations that it embodies. If necessary, find a picture of the specific landscape to aid with your observations.

2. Decide on a location that all members of your group are familiar with, such as your downtown, city hall, or student union. As individuals, take 10 minutes to make a cognitive sketch of the agreed-upon location (see **Figure 1.17** for an example of a cognitive sketch). Come

back together as a group and compare your drawings. How did the cognitive sketches in your group differ? How were they similar? List five things that are most striking about your sketches.

3. Consider the features that give your specific location a sense of place by creating a list of at least ten things. Make sure your list is specific; take into consideration thinks like vocabulary, speech patterns, clothing, and in-jokes, as well as physical geography. Once this is complete, consider how an "outsider" would interact with and understand your place.

UNPLUGGED

1. Consider spatial interaction from the point of view of your own life. Take an inventory of food you consume in a day, noting, where possible, the location where each item of food was produced, bought, and consumed. Was the store where you bought the food part of a regional, national, or international chain? Where was the food processed? How far did it travel from point of production to the grocery store to your plate? How would you go about representing the journey your food traveled on a thematic map?

2. Describe, as exactly and concisely as possible, the site (see p. 19) of your campus. Then describe its situation (see p. 19). Think of three reasons why the campus is sited and situated where it is. Would there be a better location in your community or further afield? If so, why?

3. Choose a local landscape, one with which you are familiar, and write a short essay (500 words, or two double-spaced typed pages) about how the landscape has evolved over time. Note especially any evidence that physical environmental conditions have shaped any of the human elements in the landscape, as well any evidence of people having modified the physical landscape.

4. **Figures 1.22–1.24** show examples of landscapes that have acquired a strong symbolic value because of the buildings, events, people, or histories with which they are associated. Find five photographs of landscapes that have strong symbolic value to a large number of citizens of your own region or country, and state in 25 words or less why each setting has acquired such value.

Log in to **www.masteringgeography.com** for MapMaster™ interactive maps, geography videos, RSS feeds, flash cards, Web links, an eText version of *Human Geography: Places and Regions in Global Context,* and self-study quizzes to enhance your study of geography matters.

MapMaster™ presents 13 Place Name and 13 Layered Thematic interactive maps to help students practice and master their geographic literacy, spatial reasoning, and critical thinking skills.

2

THE CHANGING GLOBAL CONTEXT

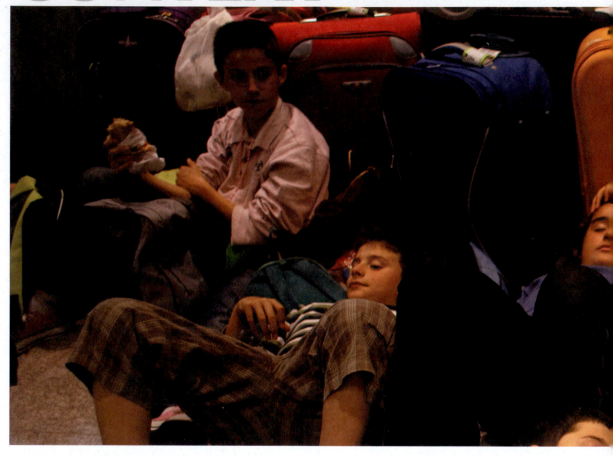

Throughout April and May 2010, after lying dormant for nearly 200 years, the Eyjafjallajökull volcano, one of Iceland's largest, erupted. Ash from the eruption halted air traffic to, from, and within Europe and showed just how interdependent places and regions around the world have become. Thousands of flights were cancelled, stranding millions of passengers in airports from North America to Asia. As the vast, high-altitude plume of volcanic ash spread, travel chaos worsened, forcing aviation authorities to close more airspace and ground more airplanes to forestall damage to jet engines. Suddenly, the vulnerability of the highly interconnected world was exposed. Top military and political leaders were not exempt from the travel disruptions. General Stanley McChrystal, the commander of United States and NATO forces in Afghanistan, had to take a 17-hour bus trip from Paris to Berlin in order to get to an important briefing. Angela Merkel, the German chancellor, was visiting San Francisco; to get back to her office in Berlin she had to fly to Rome via Portugal and then take a bulletproof limousine from Italy to Germany. In Russia, a sudden black market in transportation tickets, not all of them authentic, sprang up.

With most European airports closed, specialized fruit and vegetable farmers and flower growers in Africa, the Middle East, and the Caribbean were cut off from their markets. Kenya, for

Passengers wait for a flight inside El Prat de Llobregat Airport in Barcelona, Spain, after flights between Europe and North America were either delayed or canceled due to a spreading cloud of volcanic ash stretching across much of the northern Atlantic.

example, 5,000 miles away from the Icelandic ash plume, normally ships about 850,000 kilograms (937 tons) of fresh produce to Europe every night. As soon as air traffic to Europe was halted, the flower farms that employ thousands of people began to lay off workers, and roses, lilies, and carnations wilted. Of no use to locals, most flowers were thrown into compost pits. Vegetables, such as baby corn, zucchini, broccoli, green beans, and carrots were also left to rot. Hundreds of thousands of laborers were laid off and the effects of the stalled horticultural industry—Kenya's leading hard currency earner—were felt throughout the country.

Things quickly returned to normal when Eyjafjallajökull's activity subsided at the end of May. But the disruption that it had caused was a sharp reminder of the complexity of the world's geographies. In this chapter we take a long-term, big-picture look at changing human geographies, emphasizing the evolving interdependence among places and regions. We show how geographical divisions of labor have evolved with the growth of a worldwide system of trade and politics and with the changing opportunities provided by successive technology systems. As a result of this evolution, the world is now structured around a series of core regions, semi-peripheral regions, and peripheral regions; and globalization seems to be intensifying, rather then diminishing, many of the differences among places and regions. ∎

THE PREMODERN WORLD

The essential foundation for human geography is an ability to understand places and regions as components of a constantly changing global system. In this sense, all geography is historical geography. Built into every place and each region is the legacy of major changes in world geography. The world is an evolving, competitive, political-economic system that has developed through successive stages of geographic expansion and integration. This evolution has affected the roles of individual places in different ways. It has also affected the nature of the interdependence among places. This explains why places and regions have come to be distinctive and how this distinctiveness has formed the basis of geographic variability. To understand the sequence of major changes in world geography, we need to begin with the hearth areas of the first agricultural revolution.

Hearth Areas

The first agricultural revolution involved a transition from hunter-gatherer groups to agricultural-based **minisystems** that were both more extensive and more stable. A minisystem is a society with a *reciprocal* social economy. That is, each individual specializes in particular tasks (tending animals, cooking, or making pottery, for example) who freely give any excess product to others. The recipients reciprocate in turn by giving up the surplus product of their own specialization. Such societies are found only in subsistence-based economies. Because they do not have (or need) an extensive physical infrastructure, minisystems are limited in geographic scale.

The transition to minisystems began in the Proto-Neolithic (or early Stone Age) period, between 9000 and 7000 B.C.E., and was based on a series of technological preconditions: the use of fire to process food, the use of grindstones to mill grains, and the development of improved tools to prepare and store food. The key breakthrough was the evolution and diffusion of a system of slash-and-burn agriculture (also known as "swidden" cultivation—see Chapter 8). **Slash-and-burn** is a system of cultivation in which plants are harvested close to the ground, the stubble left to dry for a period, and then ignited. The burned stubble provides fertilizer for the soil. Another important breakthrough was the domestication of cattle and sheep, a technique that had become established in a few regions by Neolithic times.

These agricultural breakthroughs could take place only in certain geographic settings: where natural food supplies were plentiful; where the terrain was diversified (thus offering a variety of habitats and species); where soils were rich and relatively easy to till; and where there was no need for large-scale irrigation or drainage. Archaeological evidence suggests that the breakthroughs took place independently in several agricultural hearth areas and that agricultural practices diffused slowly outward from each (**Figures 2.1** and **2.2**). **Hearth areas** are geographic settings where new practices have developed and from which they have spread. The main agricultural hearth areas were situated in four broad regions:

- ***In the Middle East:*** in the so-called Fertile Crescent around the foothills of the Zagros Mountains (parts of present-day Iran and Iraq), along the floodplains of the Tigris and Euphrates rivers, around the Dead Sea Valley (Jordan and Israel), and on the Anatolian Plateau (Turkey).

FIGURE 2.1 Old World hearth areas The first agricultural revolution took place independently in several hearth areas where hunter-gatherer communities began to experiment with locally available plants and animals in ways that eventually led to their domestication. From these hearth areas, improved strains of crops, domesticated animals, and new farming techniques diffused slowly outward. Farming supported larger populations than were possible with hunting and gathering, and the extra labor allowed the development of other specializations, such as pottery making and jewelry.

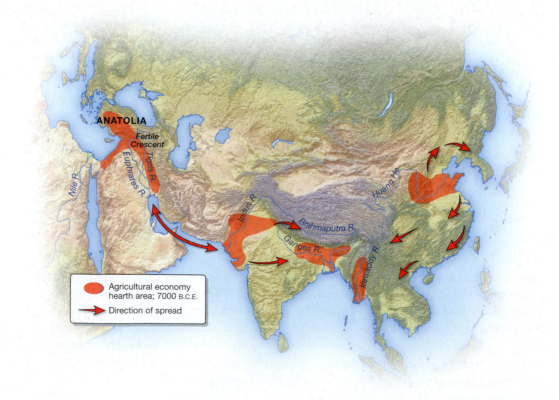

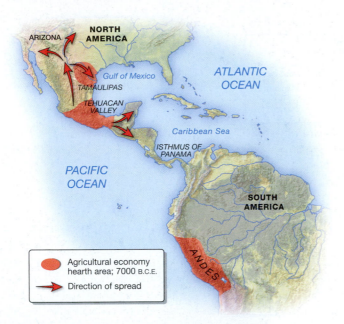

FIGURE 2.2   New World hearth areas The agricultural hearth area in Central America eventually gave rise to the Aztec and Mayan world-empires; the hearth areas of the Andes in South America developed into the Inca world-empire.

FIGURE 2.3   A remnant minisystem An Amazon tribal group, photographed in May 2008. Minisystems are rooted in subsistence-based social economies, organized around reciprocity. Each person specializes in certain tasks and freely gives any surplus to others, who in turn pass on their surplus. Today, few of the world's remnant minisystems are "pure," unaffected by contact with the rest of the world. This group is believed to be related to the Tano and Aruak tribes of the Peru-Brazil border region. Brazil's National Indian Foundation believes there may be as many as 68 "uncontacted" groups around Brazil, although only 24 have been officially confirmed. Anthropologists say almost all of these tribes know about Western civilization and have sporadic contact with prospectors, rubber tappers, and loggers but choose to turn their backs on civilization, usually because they have been attacked.

- *In South Asia:* along the floodplains of the Ganga (Ganges), Brahmaputra, Indus, and Irawaddy rivers (Assam, Bangladesh, Burma, and northern India).

- *In China:* along the floodplain of the Huang He (Yellow) River.

- *In the Americas:* in Mesoamerica (the middle belt of the Americas that extends north to the North American Southwest and south to the Isthmus of Panama) around Tamaulipas and the Tehuacán Valley (Mexico), in Arizona and New Mexico, and along the western slopes of the Andes in South America.

The transition to food-producing minisystems had several important implications for the long-term evolution of the world's geographies:

1. It allowed much higher population densities and encouraged the proliferation of settled villages.

2. It brought about a change in social organization, from loose communal systems to systems that were more highly organized on the basis of kinship. Kin groups provided a natural way of assigning rights over land and resources and of organizing patterns of land use.

3. It allowed some specialization in nonagricultural crafts, such as pottery, woven textiles, jewelry, and weaponry.

4. Specialization led to a fourth development: the beginnings of barter and trade between communities, sometimes over substantial distances.

Most minisystems vanished a long time ago, although some remnants have survived to provide material for Discovery Channel and National Geographic "Life in a Time Warp" specials (**Figure 2.3**). Examples of these residual and fast-disappearing minisystems are the bushmen of the Kalahari, the hill tribes of Papua New Guinea, and the tribes of the Amazon rain forest. They contribute powerfully to regional differentiation and sense of place in a few enclaves around the world, but their most important contribution to contemporary human geographies is that they provide a stark counterpoint to the landscapes and practices of the rest of the contemporary world.

APPLY YOUR KNOWLEDGE List and describe three examples of traditional crafts that were originally developed in the agricultural hearth areas of Arizona and New Mexico. ■

The Growth of Early Empires

The higher population densities, changes in social organization, craft production, and trade brought about by the first agricultural revolution provided the preconditions for the emergence of several "world-empires." A **world-empire** is a group of minisystems that have been absorbed into a common political system. In world-empires, wealth flows from producer classes to an elite class in the form of taxes or tribute. This redistribution of wealth is most often achieved through military coercion, religious persuasion, or a combination of the two. The best-known world-empires were the largest and longest lasting of the ancient

FIGURE 2.4 **The Premodern World** (a) Greek colonies and the extent of the Roman empire. This map shows the distribution of the Greek poleis (city-states) and Carthaginian colonies and the spread of the Roman empire from 218 B.C.E. to 117 C.E. (Source: Adapted from R. King et al., The Mediterranean. London: Arnold, 1977, pp. 59 and 64.)

(b) The Roman world-empire relied on a highly developed infrastructure of roads, settlements, and utilities. Many of the roads that were built by Romans became established as major routes throughout Europe. Wherever they could, Roman surveyors laid out roads in straight lines. This photo shows part of a Roman road that crosses the moorlands across Blackstone Edge, to the east of Littleborough, Greater Manchester, in Northern England.

(c) The Pont du Gard aqueduct near Remoulins, France. Completed in 19 B.C.E., the whole project is a testament to the surveying and engineering skills of the Romans.

civilizations—Egypt, Greece, China, Byzantium, and Rome (**Figure 2.4a**). These world-empires brought important new elements to the evolution of the world's geographies. One was the emergence of *colonization*; the other was *urbanization* (see Chapter 10).

Colonization

Colonization—the physical settlement in a new territory of people from a colonizing state—was in part an indirect consequence of the operation of the **law of diminishing returns**. This law refers to the tendency for productivity to decline after a certain point with the continued addition of capital and/or labor to a given resource base. World-empires could support growing populations only if overall levels of productivity could be increased. While some productivity gains could be achieved through better agricultural practices, harder work, and improvements in farm technology, a fixed resource base meant that as populations grew, overall levels of productivity fell. For each additional person working the land, the gain in production per worker was less.

The usual response of empire builders to these diminishing returns was to enlarge the resource base by colonizing nearby land. This colonization had immediate spatial consequences in terms of establishing dominant/subordinate spatial relationships between world-empires and colonies. Colonization played a role in establishing hierarchies of settlements and creating improved transportation networks as well. The military underpinnings of colonization also meant that new towns and cities came to be carefully sited for strategic and defensive reasons.

The legacy of these important changes is still apparent in today's landscapes. The clearest examples are in Europe, where the Roman world-empire colonized an extensive territory that was controlled through a highly developed system of towns and connecting roads. Most of today's important Western European cities had their origin as Roman settlements. In quite a few it is possible to trace the original street layouts. In some it is possible to glimpse remnants of Roman defensive city walls, paved streets, aqueducts, viaducts, arenas, sewage systems, baths, and public buildings (**Figure 2.4b,c**). In the modern European countryside we can still read the legacy of the Roman world-empire in arrow-straight roads built by Roman engineers and maintained by successive generations.

These early world-empires were also significant in developing a base of geographic knowledge. Greek scholars, for example, developed the idea that places embody fundamental relationships between people and the natural environment and that the study of geography provides the best way of addressing the *interdependencies* between places and between people and nature. The Greeks were also among the first to appreciate the practical importance and utility of geographic knowledge, not least in politics, business, and trade. The word *geography* is in fact derived from the Greek language, the literal translation meaning "earth-writing" or "earth-describing." As Greek civilization developed, descriptive geographical writing came to be an essential tool for recording information about sea and land routes and for preparing colonists and merchants for the challenges and opportunities of faraway places.

Some colonial world-empires were exceptional in that they were based on a particularly strong central state, with totalitarian rulers who were able to organize large-scale communal land-improvement schemes using forced labor. These world-empires were found in China, India, the Middle East, Central America, and the Andean region of South America. Their dependency on large-scale land-improvement schemes (particularly irrigation and drainage schemes) as the basis for agricultural productivity has led some scholars to characterize them as *hydraulic societies*. Today, their legacy can be seen in the landscapes of terraced fields in places like Sikkim, India, and East Java, Indonesia.

Urbanization

Towns and cities became essential as centers of administration for early world-empires. Towns served as military garrisons and as theological centers for the ruling classes, who used a combination of military and theological authority to hold their empires together. While these early world-empires were successful, they gave rise not only to monumental capital cities but also to a whole series of smaller settlements, which acted as intermediate centers in the flow of tribute and taxes from colonized territories.

The most successful world-empires, such as the Greek and Roman, established quite extensive urban systems. In general, the settlements in these urban systems were not very large—typically ranging from a few thousand inhabitants to about 20,000. The seats of empire grew quite large, however. The Mesopotamian city of Ur, in present-day Iraq, for example, has been estimated to have reached a population of around 200,000 by 2100 B.C.E., and Thebes, the capital of Egypt, is thought to have had more than 200,000 inhabitants in 1600 B.C.E. Athens and Corinth, the largest cities of ancient Greece, had populations between 50,000 and 100,000 by 400 B.C.E. Rome at the height of the Roman Empire (around C.E. 200) may have had as many as a million inhabitants. The most impressive thing about these cities, though, was not so much their size as their degree of sophistication: elaborately laid out, with paved streets, piped water, sewage systems, massive monuments, grand public buildings, and impressive city walls.

The Geography of the Premodern World

Figure 2.5 shows the generalized framework of human geographies in the Old World as they existed around 1400 C.E. The following characteristics of this period are important:

1. Harsher environments in continental interiors were still characterized by isolated, subsistence-level, hunting-and-gathering minisystems.

2. The dry belt of steppes and desert margins stretching across the Old World from the western Sahara to Mongolia was a continuous zone of pastoral minisystems (minisystems based on herding animals, usually moving with the animals from one grazing area to another).

3. The principal areas of sedentary agricultural production (with permanently settled farmers) extended in a discontinuous arc from Morocco to China, with two main outliers (not shown on the map in Figure 2.5), in the central Andes and in Mesoamerica.

The dominant centers of global civilization were China, northern India (both of them hydraulic society variants of world-empires), and the Ottoman Empire of the eastern Mediterranean. They were all linked by the Silk Road, a series of overland trade routes between China and Mediterranean Europe (**Figure 2.6**).

By 1400 C.E., other important world-empires had developed in Southeast Asia, in Muslim city-states of coastal North Africa, in the grasslands of West Africa, around the gold and copper mines of East Africa, and in the feudal kingdoms and merchant towns of Europe. Over time, all of these more developed realms were interconnected through trade, which meant that several emerging centers of **capitalism** came into existence. Capitalism is a form of economic and social organization characterized by the profit motive and the control of the means of production, distribution, and exchange of goods by private ownership.

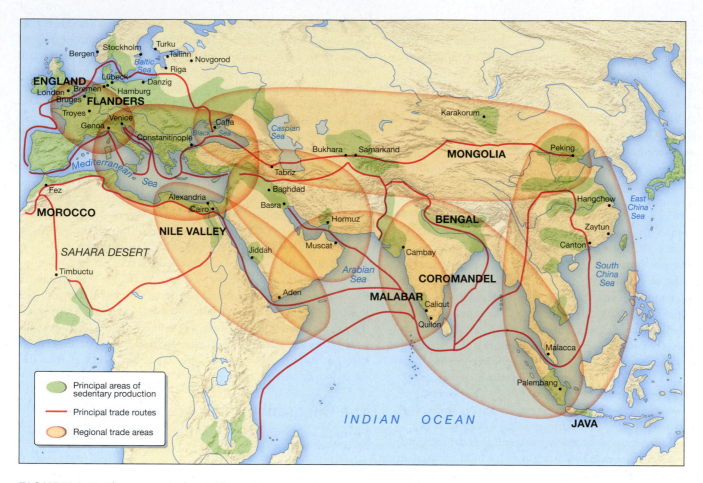

FIGURE 2.5   The precapitalist Old World, circa 1400 C.E. Principal areas of sedentary agricultural production are shaded. Some long-distance trade took place from one region to another, but for the most part it was limited to a series of overlapping regional circuits of trade. (*Source:* Adapted from R. Peet, *Global Capitalism: Theories of Societal Development.* New York: Routledge, 1991; J. Abu-Lughod, *Before European Hegemony: The World-System* A.D. *1200–1350.* New York: Oxford University Press, 1989; and E. R. Wolf, *Europe and the People Without History.* Berkeley: University of California Press, 1983.)

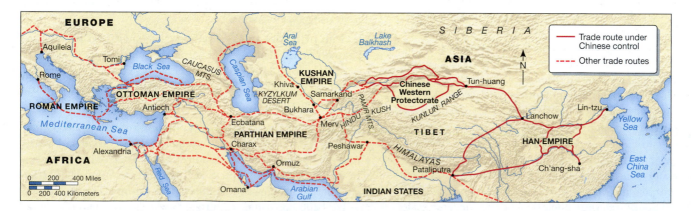

FIGURE 2.6   The Silk Road This map shows the trade routes of the Silk Road as they existed between 112 B.C.E. and 100 C.E. From Roman times until Portuguese navigators found their way around Africa and established seaborne trade routes, the Silk Road provided the main East-West trade route between Europe and China. This shifting trail of caravan tracks facilitated the exchange of silk, spices, and porcelain from the East and gold, precious stones, and Venetian glass from the West. The ancient cities of Samarkand, Bukhara, and Khiva stood along the Silk Road, places of glory and wealth that astonished Western travelers, such as Marco Polo in the thirteenth century. These cities were East-West meeting places for philosophy, knowledge, and religion, and in their prime they were known for producing scholars in mathematics, music, architecture, and astronomy, such as Al Khoresm (780–847), Al Biruni (973–1048), and Ibn Sind (980–1037). The cities' prosperity was marked by impressive feats of Islamic architecture.

Port cities were particularly important to capitalism. Among the leading centers were the city-state of Venice; the Hanseatic League of independent city-states in northwestern Europe (including Bergen, Bremen, Danzig, Hamburg, Lübeck, Riga, Stockholm, and Tallinn and affiliated trading outposts in other cities such as Antwerp, Bruges, London, Turku, and Novgorod); and Cairo, Calicut, Canton, and Malacca in North Africa and Asia. Traders in these port cities began to organize the production of agricultural specialties, textiles, and craft products in their respective hinterlands. The **hinterland** of a town or city is its sphere of economic influence—the area from which it collects products to be exported and through which it distributes imports. By the fifteenth century, several established regions of budding capitalism existed: northern Italy, Flanders, southern England, the Baltic Sea region, the Nile Valley, Malabar, Coromandel, Bengal, northern Java, and southeast coastal China.

Between roughly 500 and 1400 C.E., geographic knowledge was preserved and expanded by Chinese and Islamic scholars. Chinese maps of the world from the same period were more accurate than those of European cartographers because the Chinese were able to draw on information brought back by imperial China's admirals, who successfully navigated much of the Pacific and Indian oceans. Chinese geographers recognized, for example, that Africa was a southward-pointing triangle, whereas on European and Arabic maps of the time it was always represented as pointing eastward.

With the rise of Islamic power in the Middle East and the Mediterranean in the seventh and eighth centuries C.E., centers of scholarship emerged in places such as Baghdad, Damascus, Cairo, and Granada, Spain. Here, surviving Greek and Roman texts were translated into Arabic by scholars such as Al-Battani, Al-Farghani, and Al-Khwarazmi. These Islamic scholars were also able to draw on Chinese geographical writing and cartography brought back by traders along the Silk Road. The requirement that the Islamic religious faithful undertake at least one pilgrimage to Mecca created a demand for travel guidebooks. It also brought scholars from all over the Arab world into contact with one another, stimulating considerable debate over different philosophical views of the world and of people's relationship with nature.

APPLY YOUR KNOWLEDGE Explain how the "law of diminishing returns" contributed to the colonization of new territories by the Roman empire. Describe how such colonization was linked to urbanization. ■

AN INTERDEPENDENT WORLD GEOGRAPHY

When exploration beyond European shores began to be seen as an important way of opening up new opportunities for trade and economic expansion, a modern world-system emerged. A **world-system** is an *interdependent* system of countries linked by political and economic competition. The term *world-system*, which was coined by historian Immanuel Wallerstein, is hyphenated to emphasize the interdependence of places and regions around the world.

By the sixteenth century, new techniques of shipbuilding and navigation had begun to bind more and more places and regions together through trade and political competition. As a result, more and more peoples around the world were exposed to one another's technologies and ideas. Their different resources, social structures, and cultural systems resulted in quite different pathways of development, however. Some societies were incorporated into the new, European-based international economic system faster than others; some resisted incorporation; and some sought alternative systems of economic and political organization. Australia and New Zealand, for example, were discovered by Europeans only in the late eighteenth century and were barely penetrated, if at all, by the European world-system. Regions not yet absorbed into the world-system are called **external arenas**.

With the emergence of this modern world-system at the beginning of the sixteenth century, a whole new geography began to emerge. Although several regions of budding capitalist production existed, and although imperial China could boast of sophisticated achievements in science, technology, and navigation, it was European merchant capitalism that reshaped the world. Several factors motivated European overseas expansion. A relatively high-density population and a limited amount of cultivable land meant that it was a continuous struggle to provide enough food. Meanwhile, the desire for overseas expansion was intensified both by competition among a large number of small monarchies and by inheritance laws that produced large numbers of impoverished aristocrats with little or no land of their own. Many of these landless nobles were eager to set out for adventure and profit.

Added to these motivating factors were the enabling factors of innovations in shipbuilding, navigation, and gunnery. In the mid-1400s, for example, the Portuguese developed a cannon-armed ship—the caravel—that could sail anywhere, defend itself against pirates, pose a threat to those who were initially unwilling to trade, and carry enough goods to be profitable. Naval power enabled the Portuguese and the Spanish to enrich their economies with gold and silver capital plundered from the Americas. The quadrant (1450) and the astrolabe (1480) enabled accurate navigation and mapping of ocean currents, prevailing winds, and trade routes. Europeans embarked on a succession of voyages of discovery (**Figure 2.7**), seeking out new products and new markets.

Equipped with better maps and navigation techniques, Europeans sent adventurers in search of gold and silver and also to commandeer land, decide on its use, and exploit coerced labor to produce high-value crops (such as sugar, cocoa, cotton, and indigo) on **plantations**, large landholdings that usually specialize in the production of one particular crop for market (**Figure 2.8**). Those regions whose populations were resistant to European disease and which also had high population densities, a good resource base, and strong governments were able to keep Europeans at arm's length. For the most part, these regions were in South and East Asia. Their dealings with Europeans were conducted through a series of coastal trading stations. Textiles were an important commodity, as reflected by the origin of certain words in the English language:

The word "satin" comes from the name of an unknown city in China that Arab traders called Zaitun. "Khaki" is the Hindi word for "dusty." The word "calico" comes from India's southwestern coastal city of Calicut; "chintz," from the Hindi name for a printed calico; "cashmere," from Kashmir.

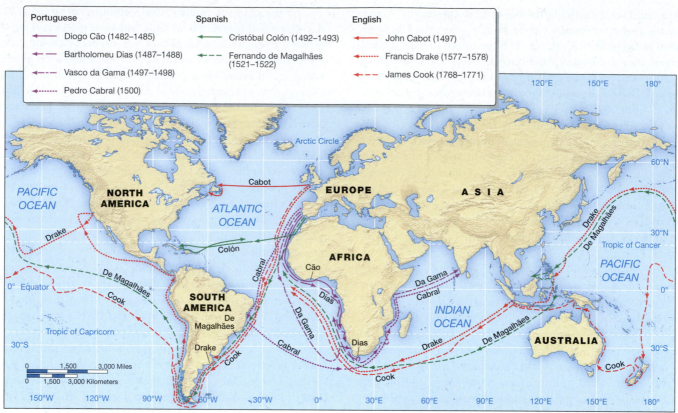

Portuguese	Spanish	English
← Diogo Cão (1482–1485)	← Cristóbal Colón (1492–1493)	← John Cabot (1497)
← Bartholomeu Dias (1487–1488)	--- Fernando de Magalhães (1521–1522)	···· Francis Drake (1577–1578)
← Vasco da Gama (1497–1498)		← — James Cook (1768–1771)
···· Pedro Cabral (1500)		

FIGURE 2.7 The European Age of Discovery The European voyages of discovery can be traced to Portugal's Prince Henry the Navigator (1394–1460), who set up a school of navigation and financed numerous expeditions with the objective of circumnavigating Africa in order to establish a profitable sea route for spices from India. The knowledge of winds, ocean currents, natural harbors, and watering places amassed by Henry's captains was an essential foundation for the subsequent voyages of Cristóbal Colón (Columbus), da Gama, de Magalhães (Magellan), and others. The end of the European Age of Discovery was marked by Captain James Cook's voyages to the Pacific.

"Percale" comes from the Farsi word *pargalah*. Another Farsi derivative is "seersucker," whose bands of alternating smooth and puckered fabric prompted a name that literally means milk and sugar. Still another Farsi borrowing is "taffeta," which comes from the Farsi for "spun." The coarse cloth we call "muslin" is named for Mosul—the town in Iraq—while "damask" is a short form of Damascus. Cotton takes its name from *qutun*, the Arabic name of the fiber.[1]

Within Europe, meanwhile, innovations in business and finance (banking, loan systems, credit transfers, commercial insurance, and courier services, for example) helped to increase savings, investment, and commercial activity. European merchants and manufacturers also became adept at **import substitution**—copying and making goods previously available only by trading. The result was the emergence of Western Europe as the core region of a world-system that had penetrated and incorporated significant portions of the rest of the world.

For Europe, this overseas expansion stimulated still further improvements in technology. These included new developments in nautical mapmaking, naval artillery, shipbuilding, and sailing. The whole experience of overseas expansion also provided a great practical school for entrepreneurship and investment. In this way,

[1]B. Wallach, *Understanding the Cultural Landscape*. New York: Guilford Press, 2005, p. 148.

FIGURE 2.8 Cotton plantation
Plantation agriculture dominated the many territories acquired by European settlers. This lithograph, dated 1884, shows a cotton plantation on the Mississippi River.

the self-propelling growth of merchant capitalism was intensified and consolidated.

For the periphery, European overseas expansion meant dependency (as it has ever since for many of the world's peripheral regions). At worst, territory was forcibly occupied and labor systematically exploited. At best, local traders were displaced by Europeans, who imposed their own terms of economic exchange. Europeans soon destroyed most of the Muslim shipping trade in the Indian Ocean, for example, and went on to capture a large share of the oceangoing trade within Asia, selling Japanese copper to China and India, Persian carpets to India, Indian cotton textiles to Japan, and so on.

As revolutionary as these changes were, however, they were constrained by a technology that rested on wind and water power, on wooden ships and structures, and on wood for fuel. Grain mills, for example, were built of wood and powered by water or wind. They could generate only modest amounts of power and only at sites determined by physical geography, not human choice. Within the relatively small European landmass, wood being grown for structural use and for fuel competed for acreage with food and textile fiber crops.

More important, however, was the inherent limits in the size and strength of timber, which imposed structural limits on the size of buildings, the diameter of waterwheels, the span of bridges, and so on. In particular, it imposed limits on the size and design of ships, which in turn imposed limits on the volume and velocity of world trade. The expense and relative inefficiency of horse- or ox-drawn wagons for overland transportation also meant that for a long time the European world-system could penetrate into continental interiors only along major rivers.

After 300 years of evolution, roughly between 1450 and 1750, the world-system had incorporated only parts of the world. The principal spheres of European influence were Mediterranean North Africa, Portuguese and Spanish colonies in Central and South America, Indian ports and trading colonies, the East Indies, African and Chinese ports, the Greater Caribbean, and British and French territories in North America. The rest of the world functioned more or less as before, with slow-changing geographies based on modified minisystems and world-empires that were only partially and intermittently penetrated by market trading.

APPLY YOUR KNOWLEDGE Search the Internet to find a map of Brazil in the 1600s as well as a current map of Brazil. How would you interpret the historical map in terms of the world-system of the time? How has it changed over time? ■

Core and Periphery in the New World System

With the new production and transportation technologies of the Industrial Revolution (from the late 1700s), capitalism truly became a global system that reached into virtually every part of the inhabited world and into virtually every aspect of people's lives. It is important to recognize that the Industrial Revolution was really an extended transition to new forms of organization and new technologies and that its effects were uneven, reflecting the influence of principles of spatial organization. In Europe, the cradle of the Industrial Revolution, it took the best part of a century for industrialization to work its way across European landscapes, with very different outcomes for different regions (see Box 2.1: "Geography Matters: The Diffusion of Industrialization in Europe").

Human geographies were recast again, this time with a more interdependent dynamic. New production technologies, based on

The Industrial Revolution began in England toward the end of the eighteenth century and eventually resulted not only in the complete reorganization of the geography of the original European core of the world-system but also in an extension of the world-system core to the United States and Japan.

In Europe, three distinctive waves of industrialization occurred. The first, between about 1790 and 1850, was based on the initial cluster of industrial technologies (steam engines, cotton textiles, and ironworking) and was very localized (**Figure 2.A**). It was limited to a few regions in Britain where industrial entrepreneurs and workforces had first exploited key innovations and the availability of key resources (coal, iron ore, water). Although these regions shared the common impetus of certain key innovations, each of them retained its own technological traditions and industrial style. From the start, then, industrialization was a regional-scale phenomenon.

The second wave of industrialization, between about 1850 and 1870, involved the diffusion of industrialization to most of the rest of Britain and to parts of northwest Europe, particularly the coalfields of northern France, Belgium, and Germany (Figure 2A). This second wave also brought a certain amount of reorganization to the first-wave industrial regions as new technologies (steel, machine tools, railroads, steamships) brought new opportunities, new locational requirements, new business structures, and new forms of societal organization. Railroads and steamships made more places accessible, bringing their resources and their markets into the sphere of industrialization. These new activities brought some significant changes to the logic of industrial location.

The importance of railway networks, for example, attracted industry away from smaller towns on the canal systems toward larger towns with good rail connections. The importance of steamships for coastal and international trade attracted industry to larger ports. At the same time, the importance of steel produced concentrations of heavy industry in places that had nearby supplies of coal, iron ore, and limestone. The scale of industry increased as improved technologies and transportation made larger markets accessible to firms. Local, family firms became small companies that were regional in scope. Small companies grew to become powerful firms serving national markets. Specialized business, legal, and financial services emerged within larger cities. The growth of new occupations transformed the structure of social classes, and this transformation in turn was reflected in the politics and landscapes of industrial regions.

FIGURE 2.A The spread of industrialization in Europe
European industrialization began with the emergence of small industrial regions in several parts of Britain, where early industrialization drew on local mineral resources, water power, and nascent industrial technologies. As new rounds of technologies emerged, industrialization spread to other regions with the right locational attributes: access to raw materials and energy sources, good communications, and large labor markets.

more efficient energy sources, helped raise levels of productivity and create new and better products that stimulated demand, increased profits, and created a pool of capital for further investment. New transportation technologies triggered successive phases of geographic expansion, allowing for internal development as well as for external colonization and **imperialism** (the deliberate exercise of military power and economic influence by powerful states in order to advance and secure their national interests—see Chapter 7).

Since the seventeenth century, the world-system has been consolidated, with stronger economic ties between countries. It has also been extended, with all the world's countries eventually becoming involved to some extent in the interdependence of the capitalist system. Although there have been some instances of resistance and adaptation, the overall result is that a highly structured relationship between places and regions has emerged. This relationship is organized around three tiers: *core*, *semiperipheral*, and *peripheral* regions. These broad geographic divisions have evolved—and are still evolving—through a combination of processes of private economic competition and competition among states.

The **core regions** of the world-system at any given time are those that dominate trade, control the most advanced technologies, and have high levels of productivity within diversified economies. As a result, they enjoy relatively high per capita incomes. The first core regions of the world-system were the trading hubs of Holland and England, joined soon afterward by France (**Figure 2.9**). But the continuing success of core regions depends on their dominance and exploitation of other regions. This dominance in turn depends on the participation of these other regions within the world-system. Initially, such participation was achieved by military enforcement, then by European colonialism.

Colonialism involves the establishment and maintenance of political and legal domination by a state over a separate and alien society. This domination usually involves some colonization (that is, the physical settlement of people from the colonizing state) and always results in economic exploitation by the colonizing state. After World War II, the sheer economic and political influence of the core regions was sufficient to maintain their dominance without political and legal control, and colonialism was gradually phased out.

Regions that have remained economically and politically unsuccessful throughout this process of incorporation into the world-system are peripheral. **Peripheral regions** are characterized by dependent and disadvantageous trading relationships, by primitive or obsolescent technologies, and by undeveloped or narrowly specialized economies with low levels of productivity.

Transitional between core regions and peripheral regions are semiperipheral regions. **Semiperipheral regions** are able to exploit peripheral regions but are themselves exploited and dominated by core regions. They consist mostly of countries that were once peripheral. The existence of this semiperipheral category underlines the fact that neither peripheral status nor core status is necessarily permanent. The United States and Japan both achieved core status after having been peripheral; Spain and Portugal, part of the original core in the sixteenth century, became semiperipheral in the nineteenth century but are now once more part of the core. Quite a few countries, including Brazil, India, Mexico, South Korea, and Taiwan, have become semiperipheral after first having been incorporated into the periphery of the world-system and then developing a successful manufacturing sector.

An important determinant of these changes in status is the effectiveness of states in ensuring the international competitiveness of their domestic producers. They can do this in several ways: by manipulating markets (protecting domestic manufacturers by charging taxes on imports, for example); by regulating their economies (enacting laws that help to establish stable labor markets, for example); and by creating physical and social infrastructures (spending public funds on road systems, ports, educational systems, and so on). Because some states are more successful than others in pursuing these strategies, the hierarchy of three geographical tiers is not rigid. Rather, it is fluid, providing a continually changing framework for geographical transformation within individual places and regions.

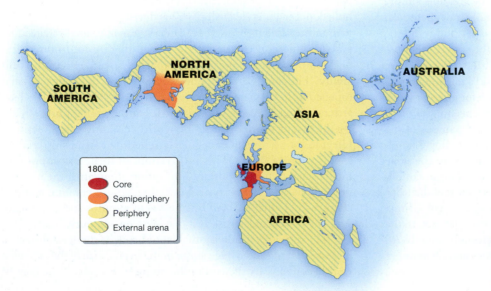

FIGURE 2.9 The world-system core, semiperiphery, and periphery in 1800

(*Source*: The Fuller Projection™ Map design is a trademark of the Buckminster Fuller Institute, © 1938, 1967 & 1992. All rights reserved. www.bfi.org.)

The colonization and imperialism that accompanied the expansion of the world-system was closely tied to the evolution of world leadership cycles. **Leadership cycles** are periods of international power established by individual states through economic, political, and military competition. In the long term, success in the world-system depends on economic strength and competitiveness, which brings political influence and pays for military strength. With a combination of economic, political, and military power, individual states can dominate the world-system, setting the terms for many economic and cultural practices and imposing their particular ideology by virtue of their preeminence. The modern world-system has so far experienced several distinct leadership cycles. In much simplified terms, they have involved dominance by Portugal (for most of the sixteenth century), the Netherlands (for the first three-quarters of the seventeenth century), Great Britain (from the early eighteenth century through the early twentieth century), and the United States (from the 1950s)—see Chapter 9.

This kind of dominance is known as hegemony. **Hegemony** refers to domination over the world economy, exercised—through a combination of economic, military, financial, and cultural means—by one national state in a particular historical epoch. Over the long run, the costs of maintaining this kind of power and influence tend to weaken the hegemon. This phase of the cycle, when the dominant nation is weakened, is known as imperial overstretch. It is followed by another period of competitive struggle, which brings the possibility of a new dominant world power.

APPLY YOUR KNOWLEDGE Look in recent issues of national newspapers or news magazines and select a current international issue involving former colonial territories. Explain how the issue may relate to the legacy of colonialism. (Hint: You might consider oil in the Middle East or copper mining in the Democratic Republic of the Congo.) ■

Organizing the Periphery

The growth and internal development of the core regions simply could not have taken place without the foodstuffs, raw materials, and markets provided by the colonization of the periphery and the incorporation of more and more territory into the sphere of industrial capitalism. Early in the nineteenth century, the industrial core nations embarked on the penetration of the world's inland midcontinental grassland zones in order to exploit them for grain or stock production. This led to the settlement, through the emigration of European peoples, of the temperate prairies and pampas of the Americas, the veld in southern Africa, the Murray-Darling Plain in Australia, and the Canterbury Plain in New Zealand. At the same time, as the demand for tropical plantation products (sugar, cotton, coffee, cocoa, and rubber, for example) increased, most of the tropical world came under the political and economic control—direct or indirect—of one or another of the industrial core nations. In the second half of the nineteenth century, and especially after 1870, there was a vast increase in the number of colonies and the number of people under colonial rule.

The International Division of Labor

The fundamental logic behind all this colonization was economic: the need for an extended arena for trade, an arena that could supply foodstuffs and raw materials in return for the industrial goods of the core. The outcome was an international division of labor, driven by the needs of the core and imposed through its economic and military strength. This **division of labor** involved the specialization of different people, regions, and countries in certain kinds of economic activities. In particular, colonies began to specialize in the production of commodities meeting certain criteria:

- where an established demand existed in the industrial core (for foodstuffs and industrial raw materials, for example);
- where colonies held a **comparative advantage** in specializations that did not duplicate or compete with the domestic suppliers within core countries (tropical agricultural products like cocoa and bananas, for example, simply could not be grown in core countries).

The result was that colonial economies were founded on narrow specializations that were oriented to and dependent upon the needs of core countries. Examples of these specializations were many: bananas in Central America; cotton in India; coffee in Brazil, Java, and Kenya; copper in Chile; cocoa in Ghana; jute in East Pakistan (now Bangladesh); palm oil in West Africa; rubber in Malaya (now Malaysia) and Sumatra; sugar in the Caribbean islands; tea in Ceylon (now Sri Lanka); tin in Bolivia; and bauxite in Guyana and Surinam. Most of these specializations persist today. For example, 45 of the 55 countries in sub-Saharan Africa still depend on just three products—tea, cocoa, and coffee—for more than half of their export earnings.

This new global economic geography took some time to establish, and the details of its pattern and timing were heavily influenced by technological innovations. The incorporation of the temperate grasslands into the commercial orbit of the core countries, for example, involved changes in regional landscapes resulting from critical innovations—such as barbed wire, the railroad, and refrigeration.

The single most important innovation stimulating the international division of labor, however, was the development of metal-hulled, oceangoing steamships. This development was cumulative, with improvements in engines, boilers, transmission systems, fuel systems, and construction materials adding up to produce dramatic improvements in carrying capacity, speed, range, and reliability. The Suez Canal (opened in 1869) and the Panama Canal (opened in 1914) were also critical, providing shorter and less hazardous routes between core countries and colonial ports of call. By the eve of World War I, the world economy was effectively integrated by a system of regularly scheduled steamship trading routes (**Figure 2.10**). This integration, in turn, was supported by the second most important innovation stimulating the international division of labor: a network of telegraph communications (**Figure 2.11**) that enabled businesses to monitor and coordinate supply and demand across vast distances on an hourly basis.

The international division of labor brought about a substantial increase in trade and a huge surge in the overall size of the capitalist world economy. By the end of the nineteenth century, the core of

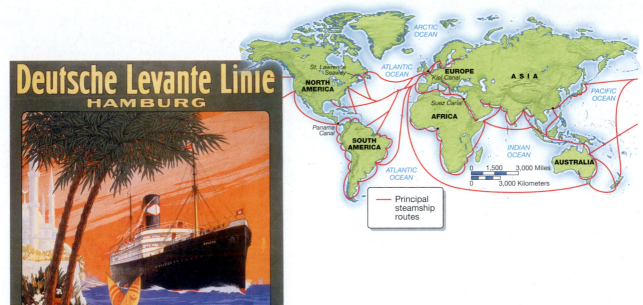

FIGURE 2.10 Major steamship routes in 1920 The shipping routes reflect (1) the transatlantic trade between the core regions of the world-system at the time and (2) the colonial and imperial relations between the world's core economies and the periphery. Transoceanic shipping boomed with the development of steam-turbine engines for merchant vessels and with the construction of shipping canals. When the 82 kilometers (51 miles) of the Panama Canal opened in 1914, shipping could move between the Atlantic and the Pacific without having to go around South America, saving thousands of kilometers of steaming. The poster dates from 1924.

FIGURE 2.11 The international telegraph network in 1900 For Britain, submarine telegraph cables were the nervous system of its empire. Of the global network of 246,000 kilometers (152,860 miles) of submarine cable, Britain had laid 169,000 kilometers (105,015 miles).

the world-system had extended to include the United States and Japan (**Figure 2.12**). The peripheral regions of the world contributed a great deal to this growth. By 1913, Africa and Asia provided more *exports* to the world economy than either North America or the British Isles. Asia alone was *importing* almost as much, by value, as North America. The industrializing countries of the core bought increasing amounts of foodstuffs and raw materials from the periphery, financed by profits from the export of machinery and manufactured goods. Britain, the hegemonic power of the period, drew on a trading empire that was truly global (**Figure 2.13**).

FIGURE 2.12 The world-system core, semiperiphery, and periphery in 1900 (*Source:* The Fuller Projection™ Map design is a trademark of the Buckminster Fuller Institute, © 1938, 1967 & 1992. All rights reserved. www.bfi.org.)

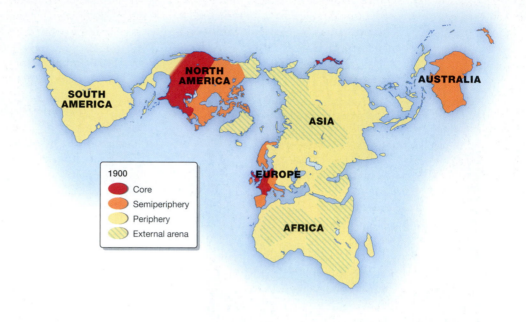

FIGURE 2.13 The British Empire in the late 1800s Protected by the all-powerful Royal Navy, the British merchant navy established a web of commerce that collected food for British industrial workers and raw materials for its industries, much of it from colonies and dependencies appropriated by imperial might and developed by British capital. So successful was the trading empire that Britain also became the hub of trade for other states. (*Source:* Adapted from P. Hugill, *World Trade Since 1431.* Baltimore: Johns Hopkins University Press, 1993, p. 136.)

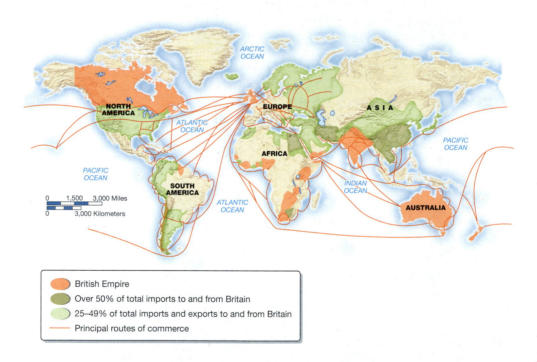

Patterns of international trade and interdependence became increasingly complex. Britain used its capital to invest not only in peripheral regions but also in profitable industries in other core countries, especially the United States. At the same time, these other core countries were able to export cheap manufactured items to Britain. Britain financed the purchase of these goods, together with imports of food from its dominion states (Canada, South Africa, Australia, and New Zealand) and colonies, through the export of its own manufactured goods to peripheral countries. India and China, with large domestic markets, were especially important. A widening circle of exchange and dependence developed, with constantly shifting patterns of trade and investment.

APPLY YOUR KNOWLEDGE Look at some of the clothes and products that you possess and see where they were made. List the materials that go into making your clothing. Can you speculate about where the cotton was grown? Who drove it to the factory? Who produced the cotton? Who sewed the clothing together? How do these questions relate to the principles of division of labor and comparative advantage? ■

Imperialism: Imposing New Geographies on the World

The incorporation of the periphery was by no means entirely motivated by this basic logic of free trade and investment. Although

Britain was the dominant power in the late nineteenth century, several other European countries (notably Germany, France, and the Netherlands), together with the United States—and later Japan—were competing for global influence. This competition developed into a scramble for territorial and commercial domination. The core countries engaged in preemptive geographic expansionism in order to protect their established interests and to limit the opportunities of others. They also wanted to secure as much of the world as possible—through a combination of military oversight, administrative control, and economic regulation—in order to ensure stable and profitable environments for their traders and investors.

This combination of circumstances defined a new era of imperialism in the final quarter of the nineteenth century. Africa, more than any other peripheral region, was given an entirely new geography. It was carved up into a patchwork of European colonies and protectorates in just 34 years, between 1880 and 1914, with little regard for either physical geography or the preexisting minisystems and world-empires. Whereas European interest had previously focused on coastal trading stations and garrison ports, it now extended to the entire continent.

Within just a few years, the whole of Africa became incorporated into the modern world-system, with a geography that consisted of a hierarchy of three kinds of spaces. One consisted of regions and localities organized by European colonial administrators and European investors to produce commodities for the world market. A second consisted of zones of production for local markets, where peasant farmers produced food for consumption by laborers engaged in commercial mining and agriculture. The third consisted of widespread regions of subsistence agriculture whose connection with the world-system was as a source of labor for the commercial regions.

Meanwhile, the major powers also jostled and squabbled over small Pacific islands that had suddenly become valuable as strategic coaling stations for their navies and merchant fleets. Resistance from indigenous peoples was quickly brushed aside by imperial navies with ironclad steamers, high-explosive guns, and troops with rifles and cannons. European weaponry was so superior that Otto von Bismarck, the founder and first chancellor (1871–1890) of the German empire, referred to these conflicts as "sporting wars." Between 1870 and 1900, European countries added almost 22 million square kilometers (8.5 million square miles) and 150 million people to their spheres of control—20 percent of the Earth's land surface and 10 percent of its population.

The imprint of imperialism and colonization on the geographies of the newly incorporated peripheries of the world-system was immediate and profound. The periphery was rendered almost entirely dependent on European and North American capital, shipping, managerial expertise, financial services, and news and communications. Consequently, it also became dependent on European cultural products: language, education, science, religion, architecture, and planning. All of these influences were etched into the landscapes of the periphery in a variety of ways as new places were created, old places were remade, and regions were reorganized.

The discipline of geography played an important role in providing a "scientific" rationale for the domination of peripheral countries by Europeans and North Americans. Prominent geographers argued that civilization and successful economic development are largely the result of "invigorating" temperate climates with marked seasonal variations and varied weather but without prolonged extremes of heat, humidity, or cold. Conversely, tropical climates, they asserted, limit people's vitality. This kind of reasoning reflects an underlying ethnocentrism and environmental determinism. **Ethnocentrism** is the attitude that one's own race and culture are superior to those of others. **Environmental determinism** is a doctrine holding that human activities are shaped and constrained by the environment. Most of the geographic writing in the nineteenth and early twentieth centuries was strongly influenced by this assumption that the physical attributes of geographical settings are the root not only of people's physical differences but also of differences in people's economic vitality, cultural activities, and social structures.

The Third World and Neocolonialism

The imperial world order began to disintegrate shortly after World War II, however. The United States emerged as the new hegemonic power, the dominant state within the world-system core, which came to be called the First World. The Soviet Union and China, opting for alternative, socialist paths of development for themselves and their satellite countries, were seen as a Second World, withdrawn from the capitalist world economy. Their pursuit of alternative political economies was based on radically different values.

By the 1950s, many of the old European colonies began to seek political independence. Some of the early independence struggles were very bloody because the colonial powers were initially reluctant to withdraw from colonies where strategic resources or large numbers of European settlers were involved. In Kenya, in the early 1950s, for example, a militant nationalist movement known as the Mau Mau launched a campaign of terrorism, sabotage, and assassination against British colonists. They killed more than 2,000 white settlers between 1952 and 1956; in return, 11,000 Mau Mau rebels were killed by the colonial army and 20,000 put into detention camps by the colonial administration. By the early 1960s, however, the process of decolonization had become relatively smooth. (In 1962, Jomo Kenyatta, who had been jailed as a Mau Mau leader in 1953, became prime minister of a newly independent Kenya.) The periphery of the world-system now consisted of a "Third World" of politically independent states, some of which adopted a policy of nonalignment; that is, they were not formally aligned with or against either of the major power blocs of the First or Second worlds. They were nevertheless still highly dependent, in economic terms, on the world's core countries.

As newly independent peripheral states struggled from the 1960s onward to be free of their economic dependence through industrialization, modernization, and trade, the capitalist world-system became increasingly integrated and interdependent. The old imperial patterns of international trade broke down and were replaced by more complex patterns. Nevertheless, the newly independent states were still influenced by many of the old colonial links and legacies that remained intact. The result was a neocolonial pattern of international development.

Neocolonialism refers to economic and political strategies by which powerful states in core economies indirectly maintain or extend their influence over other areas or people. Instead of formal, direct rule (colonialism), controls are exerted through such strategies as international financial regulations, commercial relations, and covert intelligence operations. Through neocolonialism, the human geographies of peripheral countries continued to be heavily shaped by the linguistic, cultural, political, and institutional influence of the former colonial powers as well as by their investment and trading activities.

At about the same time, a new form of imperialism was emerging. This was the *commercial imperialism* of giant corporations. These corporations had grown within the core countries through the elimination of smaller firms by mergers and takeovers. By the 1960s, quite a few of them had become so big that they were *transnational* in scope, having established overseas subsidiaries, taken over foreign competitors, or simply bought into profitable foreign businesses.

These **transnational corporations** have investments and activities that span international boundaries, with subsidiary companies, factories, offices, or facilities in several countries. Examples of transnational corporations include Airbus, BP, Halliburton, News Corporation, Siemens, and Virgin Group. By 2007, over 79,000 transnational corporations were operating, 90 percent of which were headquartered in the core states. These corporations control about 790,000 foreign affiliates and account for the equivalent of 11 percent of world **Gross Domestic Product (GDP)** and one-third of world exports.

Transnational corporations have been portrayed as imperialist by some geographers because of their ability and willingness to exercise their considerable power in ways that adversely affect peripheral states. They have certainly been central to a major new phase of geographical restructuring that has been under way for the last 35 years or so. This phase has been distinctive because an unprecedented amount of economic, political, social, and cultural activity has spilled beyond the geographic and institutional boundaries of states. It is a phase of *globalization*, a much fuller integration of the economies of the worldwide system of states and a much greater interdependence of individual places and regions from every part of the world-system (**Figure 2.14**).

APPLY YOUR KNOWLEDGE Provide an example of how neocolonialism reinforces the power and influence of core countries. Please be specific with your example. (Hint: You might want to consider milk production in Jamaica or oil extraction in Nigeria.) What is the role of transnational corporations in neocolonialism? ▪

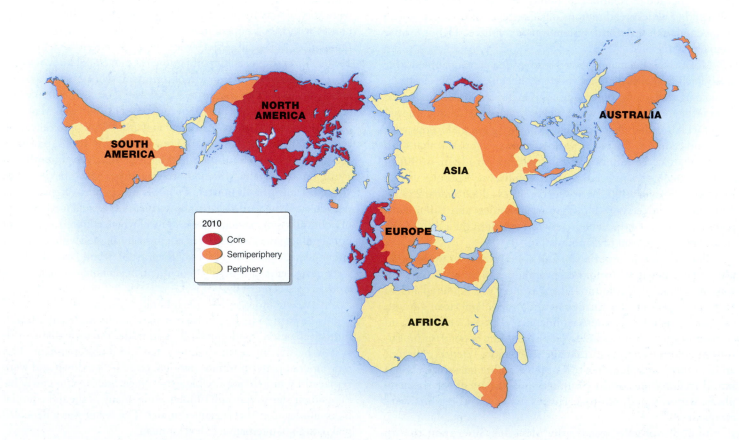

FIGURE 2.14 The world-system core, semiperiphery, and periphery in 2010 (*Source:* The Fuller Projection™ Map design is a trademark of the Buckminster Fuller Institute, © 1938, 1967 & 1992. All rights reserved. www.bfi.org.)

CONTEMPORARY GLOBALIZATION

Globalization is the increasing interconnectedness of different parts of the world through common processes of economic, environmental, political, and cultural change. As we have seen, globalization has been under way since the inception of the modern world-system in the sixteenth century. In the nineteenth century, when the competitive system of states fostered the emergence of international agencies and institutions, global networks of communication, a standardized system of global time, international law, and internationally shared notions of citizenship and human rights, the basic framework of modern globalization came into being. Global connections today, though, differ in at least four important ways from those in the past.

First, they function at much greater *speed* than ever before. Second, globalization operates on a much larger *scale*, leaving few people unaffected and wielding its influence in even the most remote places. Third, the *scope* of global connections is much broader and has *multiple dimensions*: economic, technological, political, legal, social, and cultural, among others. Fourth, the interactions and interdependencies among numerous global actors have created a new level of *complexity* for the relationships between places and regions.

Over the past 35 years, telecommunication technologies, corporate strategies, and institutional frameworks have combined to create a dynamic new geographical framework. Emerging information technologies have helped create a complex and frenetic international financial system, while transnational corporations are now able to transfer their production activities from one part of the world to another in response to changing market conditions and changing transportation and communications technologies (see Chapter 7). Now products, markets, and organizations are both spread and linked across the globe. Governments, in their attempts to adjust to this situation, have sought new ways of dealing with the consequences of globalization, including unprecedented international political and economic alliances such as NAFTA and the European Union (see Chapter 9).

The economic basis of contemporary globalization depends on myriad commodity chains that crisscross global space. **Commodity chains** are networks of labor and production processes that originate in the extraction or production of raw materials and whose end result is the delivery and consumption of a finished commodity (see Box 2.2, "Visualizing Geography: Commodity Chains"). These networks often span countries and continents, linking into vast global assembly lines the production and supply of raw materials, the processing of raw materials, the production of components, the assembly of finished products, and the distribution of finished products. As we shall see in Chapter 7, these global assembly lines are increasingly important in shaping places and regions.

Globalization also has important cultural dimensions (see Chapter 5). One is quite simply the diffusion around the world of all sorts of cultural forms, practices, and artifacts that had previously been confined to specific places or regions. Examples include "ethnic" and regional cuisine, "world" music, Caribbean carnivals, and "charismatic" Christian sects. Another dimension of cultural globalization derives from consumer culture: everything that is sold in international markets, from sneakers, replica soccer shirts, and automobiles to movies and rock concert tours. This had led some observers to believe that globalization is producing a new set of universally shared images, practices, and values—literally, a global culture.

All this adds up to an intensified global connectedness and the beginnings of the world as an interdependent system. Or, to be more precise, this is how it adds up for the 900 million or so of the world's people who are directly tied to global systems of production and consumption and who have access to global networks of communication and knowledge. All of us in this globalizing world are in the middle of a major reorganization of the world economy and a radical change in our relationships to other people and places.

At first glance it might seem that globalization will render geography obsolete—especially in the more developed parts of the world. High-tech communications and the global marketing of standardized products seem as if they might soon wash away the distinctiveness of people and places, permanently diminishing the importance of differences between places. Far from it. The new mobility of money, labor, products, and ideas actually increases the significance of place in some very important ways:

- The more universal the diffusion of material culture and lifestyles, the more valuable regional and ethnic identities become.

- The faster the information highway takes people into cyberspace, the more they feel the need for a subjective setting—a specific place or community—they can call their own.

- The greater the reach of transnational corporations, the more easily they are able to respond to place-to-place variations in labor markets and consumer markets. As a result, economic geography has to be reorganized more frequently and more radically.

- The greater the integration of transnational governments and institutions, the more sensitive people have become to local cleavages of race, ethnicity, and religion.

For some places and regions, globalization is a central reality; for others, it is still a marginal influence. While some places and regions have become more closely interconnected and interdependent as a result of globalization, others have been bypassed or excluded. In short, there is no one experience of globalization. The reality is that globalization is variously embraced, resisted, subverted, and exploited as it makes contact with specific cultures and settings. In the process, places are modified or reconstructed rather than destroyed or homogenized.

Key Issues in a Globalizing World

The integrated global system has increased awareness of a set of common problems—climate change, pollution, disease, crime, poverty, and inequality—that many see as a consequence of globalization. The globalization of the contemporary world—its causes and effects on specific aspects of human geographies at different spatial scales—is a recurring theme through the rest of this book. Here, we note in broad outline the principal issues associated with contemporary globalization.

Environmental Issues

The sheer scale and capacity of the world economy means that humans are now capable of altering the environment at the global scale. The "footprint" of humankind extends to more than

Global commodity chains link the progression of a commodity from design through procurement of raw materials and production to import or export to the point of sale, distribution for sale, marketing, and advertising. They are often entirely internal to the global operations of transnational corporations.

Advances in telecommunications, management techniques, transportation, finance, and other services to industry have made possible the segmentation of corporate production lines, as well as services, across multiple settings. Manufacturing companies now design a product in one country, have it produced by contractors in various countries continents apart, sell the product with its brand name by telephone or Internet almost anywhere in the world, and have other contractors deliver it. The services involved—design, sales, financing, and delivery—can be undertaken without the various actors ever meeting face to face. Advances in technology and management have also permitted the reproduction and standardization of services and products on a global basis. Certain patented services, such as fast-food restaurants (McDonald's, KFC, Burger King), rely on computer-regulated technology to deliver a standard service and product over time and geographic space.

Almost every mass-marketed manufactured product involves a complex commodity chain. Take, for example, the manufacture of Lee Cooper jeans (**Figure 2.B**). Designed in the United States, advertised globally, retailed in stores across Europe, the United States, and the metropolises of semiperipheral and peripheral countries, their manufacture draws on labor and products from around the world. The final stages of the commodity chain of one particular pair of jeans sold in a large discount store of a provincial British city were "in a van that came up the A12 [road] from Lee Cooper's warehouse at Staples Corner, just at the bottom of the M1 [highway] in North London… . Before that they came through the Channel Tunnel in a lorry from similar warehouses in Amiens, France and before that, by boat and train from Tunis in Tunisia … ."[1]

There are three broad types of global commodity chains. The first is *producer-driven*, in which large, often transnational, corporations coordinate production networks. A good example of this is the U.S. pharmaceutical industry. Research and development of drugs are conducted in the United States; materials and components are produced in a global production line; and the drugs are assembled and marketed out of a semiperipheral site (in particular, Puerto Rico and Ireland), where companies enjoy certain competitive advantages not only in the industrial segment of the assembly but in marketing, financial, and management services.

A second type of commodity chain is *consumer-driven*, where large retailers, brand-name merchandisers, and trading companies influence decentralized production networks in a variety of exporting countries, often in the periphery. One good example is the case of Lee Cooper jeans described in this box. Another example is the discount chain-store company Wal-Mart, which contracts directly with producers in low-wage countries (China, in particular) for the bulk of its merchandise (which is sold in the United States to the accompaniment of advertising that invokes community-oriented and even patriotic themes).

The third type, the *marketing-driven* commodity chain, represents a hybrid of the first two types. It involves the production of inexpensive consumer goods—such as colas, beers, breakfast cereals, candies, cigarettes, and infant formula—that are global commodities and carry global brands yet are often manufactured in the periphery and semiperiphery for consumption in those regions. These commodities take their globalized status not only from their recipes and production techniques but, even more important, from their globally contrived cultural identities.

Commodity chains provide varying opportunities for firms and national economies to enter into and improve their position within the global division of labor. Commodity chains are an important dimension of the complex transnationalization of economic space. They are also an important dimension of the complex currents of cultural globalization. In addition to facilitating the standardization of products and services (for those who can afford them) around the world, commodity chains also reflect the inequalities of the global economy. If core-country consumers stop to think about the origins of many of the products they consume, they may recognize what is happening "down the line," where poverty wages and grim working and living conditions are a precondition for the origins of many commodity chains. In contrast, those who work on the farms and plantations and in the workshops and factories at the beginning of commodity chains are acutely aware—thanks to contemporary media—of the dramatically more affluent lifestyles of those who will eventually consume the fruits of their labor.

[1]Abrams, F. and J. Astill, "Story of the Blues," *The Guardian*, 29 May 2001, p. 2; quoted in L. Crewe, "Unravelling Fashion's Commodity Chains." In A. Hughes and S. Reimer (eds.), *Geographies of Commodity Chains*. London: Routledge, 2004, p. 201.

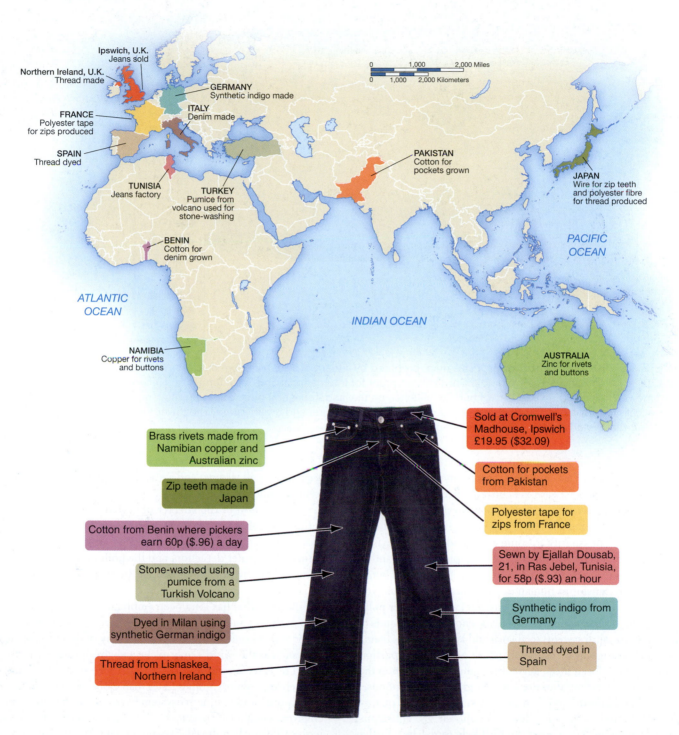

Ipswich, U.K.
Jeans sold

Northern Ireland, U.K.
Thread made

GERMANY
Synthetic indigo made

FRANCE
Polyester tape
for zips produced

ITALY
Denim made

SPAIN
Thread dyed

PAKISTAN
Cotton for
pockets grown

JAPAN
Wire for zip teeth
and polyester fibre
for thread produced

TUNISIA
Jeans factory

TURKEY
Pumice from
volcano used for
stone-washing

PACIFIC
OCEAN

BENIN
Cotton for
denim grown

ATLANTIC
OCEAN

INDIAN OCEAN

NAMIBIA
Copper for rivets
and buttons

AUSTRALIA
Zinc for rivets
and buttons

0   1,000   2,000 Miles
0   1,000   2,000 Kilometers

Sold at Cromwell's
Madhouse, Ipswich
£19.95 ($32.09)

Brass rivets made from
Namibian copper and
Australian zinc

Cotton for pockets
from Pakistan

Zip teeth made in
Japan

Polyester tape for
zips from France

Cotton from Benin where pickers
earn 60p ($.96) a day

Sewn by Ejallah Dousab,
21, in Ras Jebel, Tunisia,
for 58p ($.93) an hour

Stone-washed using
pumice from a
Turkish Volcano

Synthetic indigo from
Germany

Dyed in Milan using
synthetic German indigo

Thread dyed in
Spain

Thread from Lisnaskea,
Northern Ireland

FIGURE 2.B The making of a pair of Lee Cooper jeans (*Source:* Adapted from A. Hughes
and S. Reimer, eds., *Geographics of Commodity Chains.* New York: Routledge, 2004.)

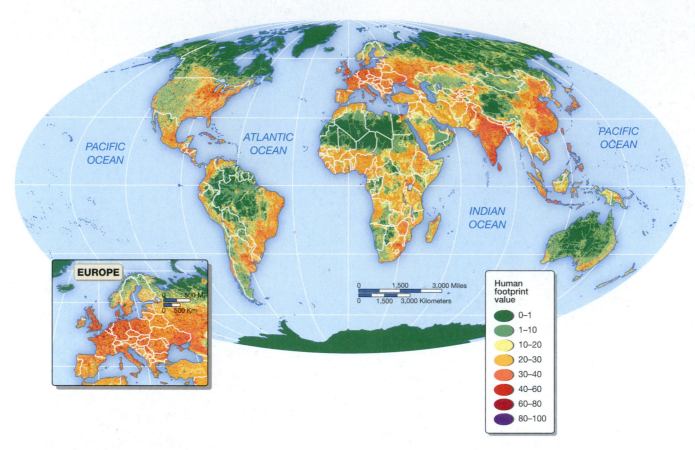

FIGURE 2.15 The human "footprint" This map, prepared by a team of scientists from the New York-based Wildlife Conservation Society and Columbia University's Center for International Science Information Network (CIESIN), shows the extent and intensity of human influence on the land, reflecting population density, agricultural use, access from roads and waterways, electrical power infrastructure, and urbanization. The lower the number, the lesser the overall degree of human influence. (*Source:* www.wcs.org/humanfootprint/)

four-fifths of Earth's surface (**Figure 2.15**). Many of the important issues facing modern society are the consequences—intended and unintended—of human modifications of our physical environment.

Humans have altered the balance of nature in ways that have brought economic prosperity to some areas and created environmental dilemmas and crises in others. For example, clearing land for settlement, mining, and agriculture provides livelihoods and homes for some but also transforms human populations, wildlife, and vegetation. The inevitable by-products—garbage, air and water pollution, hazardous wastes, and so forth—place enormous demands on the capacity of physical systems to absorb and accommodate them.

Climate change as a result of human activity—in particular, our burning of fossil fuels, agriculture, and deforestation that cause emissions of carbon dioxide (CO_2) and other "greenhouse" gases—also has profound implications for environmental quality. Without concerted action to reduce greenhouse gas emissions, the global average surface temperature is likely to rise by a further 1.8–4.0°C this century. Even the lower end of this range would take the temperature increase since preindustrial times above 2°C, the threshold beyond which irreversible and possibly catastrophic changes become far more likely. Projected global warming this century is likely to trigger serious consequences for humanity and other life forms. These consequences may include a rise in sea levels of between 18 and 59 centimeters, which will endanger coastal

areas and small islands, and a greater frequency and severity of extreme weather events.

In addition to the specter of global warming, we are facing serious global environmental degradation through deforestation, desertification, acid rain, loss of genetic diversity, smog, soil erosion, groundwater depletion, and the pollution of rivers, lakes, and oceans. The fate of Lake Baykal, in Russia (**Figure 2.16**), provides a distressing example. It is a place of incredible beauty—"the Pearl of Siberia"—that has long been emblematic of the pristine wilderness of the region. The lake holds 20 percent of the world's freshwater and is home to 2,500 species, many of them found nowhere else, such as the world's only exclusively freshwater seal.

The lake has warmed 1.21°C (2.18°F) since 1946 due to climate change, almost 3 times faster than global air temperatures. The lake's purity and unique ecosystem have also been compromised by environmental mismanagement. The first evidence of this was seen in the 1950s, when the lake's commercial fish populations radically decreased, partly as a result of overfishing and partly as a result of the construction of the Irkutsk dam (which raised the level of the lake and destroyed many of the shallow-water feeding grounds used by the fish). Then, in the 1960s, increasing levels of pollution were carried into the lake by the Selenga River, which supplies about half of the water that flows into the lake. The Selenga originates in mountain ranges to the south but collects human and industrial waste, much

FIGURE 2.16 Lake Baykal Lake Baykal is the world's deepest lake, at 1,615 meters (5,300 feet—over a mile), and contains about 20 percent of all the freshwater on Earth—more than North America's five Great Lakes combined. Lake Baykal is also an unusually ancient lake. Most of the world's large lakes are less than 20,000 years old, but evidence from the 7-kilometer-thick sediment at the bottom of the lake shows it to have been in existence for at least 25 million years, perhaps even 50 million. It has a unique ecology, with over 2,500 recorded plant and animal species, 75 percent of which are found nowhere else. These include the nerpa, Baykal's freshwater seal, separated by more than 3,000 kilometers (1,864 miles) from its nearest relative, the Arctic ringed seal. Ecologists have no understanding of how seals ever got to the lake or how they adapted to freshwater. Although Lake Baykal is now on UNESCO's list of World Natural Heritage sites, pollution from numerous paper mills along the lake shore continues to worsen.

of it untreated, from several large cities before entering Lake Baykal. The Selenga and other rivers also began to carry increasing amounts of agricultural chemicals, such as DDT and PCBs into the lake.

Meanwhile, the purity of the lake's waters caught the attention of Soviet economic planners, who saw the lake as a good location for factories that needed plentiful supplies of pure water. The huge Baikalsk Pulp & Paper Mill was opened in the early 1960s to produce high-quality cellulose for the Russian defense industry. The mill pumps 140,000 tons of waste—including deadly dioxins—into the lake every day, along with 23 tons of pollutants into the atmosphere; over the past 40 years, the mill has spewed over a billion tons of waste into the lake.

When thousands of the lake's freshwater seals began dying in 1997, the lake's fragile ecology came under international scrutiny, and in 1998 the lake was designated a World Heritage Site by UNESCO, the UN cultural agency. In 2007, the Russian government declared the Baikal region a Special Economic Zone, in order to encourage tourism. Nevertheless, it remains to be seen whether Russia can solve its environmental problems at a time when its economy is still in transition.

Environmental issues such as these point to the importance of sustainability. **Sustainability** is about the interdependence of the economy, the environment, and social well-being. This is often couched in terms of the "three Es" of sustainable development, referring to the environment, the economy, and equity in society (**Figure 2.17**). The oft-quoted definition of sustainable development from the Brundtland Report, which examined the issues on the international scale, is "development that meets the needs of the present without compromising the ability of future generations to meet their own needs."[2]

APPLY YOUR KNOWLEDGE Give an example of an environmental concern that affects your home region and suggest how it relates to issues of economic development and social equity. ∎

[2]World Commission on Environment and Development, *Our Common Future* (Brundtland Report), Oxford, UK: Oxford University Press, 1987, p. 40.

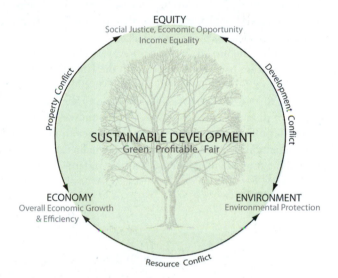

FIGURE 2.17 Sustainability There are three key aspects of sustainability in the long run—the physical environment, equity, and economic efficiency—and there are tensions between each of these.

Health Issues

The increased intensity of international trade and travel has also heightened the risk and speed of the spread of disease. A striking example of the health risks associated with increasing interdependency was the outbreak of H1N1 influenza ("swine flu") in 2009 that quickly spread from Mexico to become a global pandemic that lasted until August 2010. By the time the outbreak had been contained, it had caused widespread panic and serious disruption to business and tourism. The pandemic resulted in more than 14,000 deaths, some as far afield as Australia and Southeast Asia.

Health care professionals are concerned that a new strain of influenza virus is likely to result in an influenza pandemic. A **pandemic** is an epidemic that spreads rapidly around the world with high rates of illness and death. Entirely new flu strains develop several times each century. Because no one has a chance to develop immunity to a new flu strain, it can spread rapidly and widely—and especially so in today's globalized and highly interconnected world. Similarly, there is serious concern about the possibility of epidemics in the human population resulting from zoonotic diseases (diseases originating with other species, e.g., anthrax, avian flu, ebola, West Nile virus). The most significant international health issue so far, however, has been the spread of HIV/AIDS (human immunodeficiency virus/acquired immunodeficiency syndrome).

Over the past 30 years, HIV/AIDS quickly spread around the world from a single hearth area. Medical geographers have concluded that the human immunodeficiency virus (HIV), which causes AIDS, spread in a hierarchical diffusion pattern from a hearth area in Central Africa in the late 1970s (**Figure 2.18**). The virus initially appeared almost simultaneously in the major metropolitan areas of North and South America, the Caribbean, and Europe. These areas then acted as localized diffusion poles for the virus, which next spread to major metropolitan areas in Asia and Oceania and to larger provincial cities in North and South America, the Caribbean, and Europe. Next in this cascading pattern of diffusion were provincial cities in Asia and Oceania and small towns in North and South America, the Caribbean, and Europe.

Today, sub-Saharan Africa is more severely affected by HIV/AIDS than any other part of the world, with the United Nations reporting between 22.5 million people infected in 2009—more than 67 percent of the worldwide total. The infection rate is estimated at 8 percent of all adults compared with a 1 percent world rate, and more than 15 million Africans have lost their lives to HIV/AIDS since it was identified in 1981. It has become the main cause of death in Africa, killing more people than malaria and warfare combined.

Security Issues

As sociologist Ulrich Beck has pointed out, the high degree of interdependence that is now embedded in a globalizing and highly interconnected world has brought about all sorts of security issues. In traditional societies, the risks faced by individuals and groups were associated mostly with hazards generated by nature (disease, flood, famine, and the like), along with socially determined hazards such as invasion and conquest and regressive forms of thought and culture. The industrial societies of the nineteenth and twentieth centuries, with more powerful technologies and weaponry, faced still more hazards, but they were mostly local and regional in nature.

Contemporary society, Beck points out, is characterized by another set of hazards, many of them uncontrollable and with a global reach. Examples include climate change as a result of human activity; the spread of weapons of mass destruction (i.e., nuclear and biological warfare); the risk of accidents involving radiation or contamination by radioactivity from nuclear fuel or nuclear waste; the risk of epidemics in the human population resulting from zoonotic diseases; the risk of epidemic disease in food animals (such as the devastating outbreak of foot-and-mouth disease, which affects cattle and sheep, in parts of northwestern Europe in 2001); and the risk of catastrophic instability in global financial markets (for example, the global financial "meltdown" of 2008: see Chapter 7).

Overall, Beck argues, we are moving toward a **risk society,** in which the significance of wealth distribution is being eclipsed by the distribution of risk and in which politics—both domestic and international—is increasingly about avoiding hazards. As a result, knowledge—especially scientific knowledge—becomes increasingly important as a source of power, while science itself becomes increasingly politicized—as, for example, in the case of global warming.

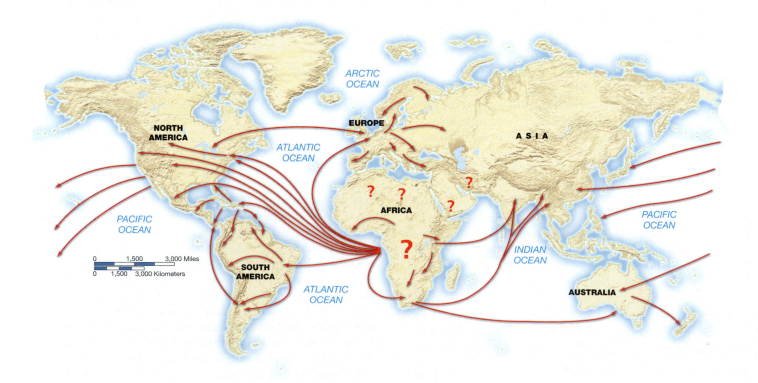

FIGURE 2.18 Diffusion of the HIV virus The probable early diffusion of HIV/AIDS. (Adapted from M. Smallman-Raynor,
A. Cliff, and P. Haggett, London International Atlas of AIDS. Oxford: Blackwell Reference, 1992, Fig. 4.1(c), p. 146.)

FIGURE 2.19 Communications flows between major world regions This diagram shows the flows, in billions of minutes of telecommunications traffic over public telephone networks, between major regions.

(*Source:* Adapted from G. C. Staple, ed., *TeleGeography 1999*. Washington, DC: TeleGeography Inc., 1999, Fig. 4, p. 255.)

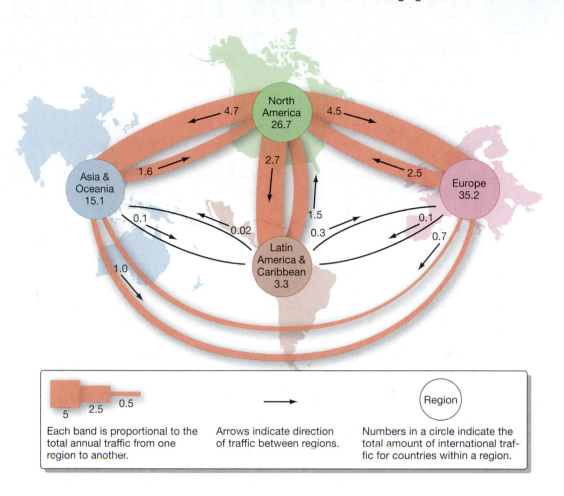

Each band is proportional to the total annual traffic from one region to another.

Arrows indicate direction of traffic between regions.

Numbers in a circle indicate the total amount of international traffic for countries within a region.

International terrorism, another important security issue, can also be attributed in part to globalization. While terrorism has a long history, it is only recently that terrorist attacks have spilled beyond the sites of local conflict. This is largely a result of another set of issues: the cultural and political fallout from the Westernization that is associated with contemporary globalization.

Core-Periphery Disparity Issues

One of the most striking outcomes of contemporary globalization is the consolidation of the core of the world-system. The core is now a close-knit triad of the geographic centers of North America, the European Union, and Japan. These three geographic centers are connected through three main circuits, or flows, of investment, trade, and communication: between Europe and North America, between Europe and the Far East, and among the regions of the Pacific Rim. **Figure 2.19**, for example, shows just how dominant North America has become in accounting for flows of international telephonic communication.

Globalization, although incorporating more of the world more completely into the capitalist world-system, has intensified differences in prosperity between the core and the periphery (see Box 2.3: "Window on the World: Worlds Apart").

According to the United Nations Development Program, the gap between the poorest fifth of the world's population and the wealthiest fifth increased more than threefold between 1965 and 2005. Some parts of the periphery have almost slid off the economic map. In 55 countries, per capita income actually fell during the 1990s. In sub-Saharan Africa,

economic output fell by one-third during the 1980s and stayed low during the 1990s, so that people's standard of living there is now, on average, lower than it was in the early 1960s. In 2010, the fifth of the world's population living in the highest-income countries had:

- 75 percent of world income (the bottom fifth had just 1 percent); and

- 84 percent of world export markets (the bottom fifth had just 1 percent).

While 3 billion people around the world struggle to live on less than US$2.00 a day, the world's billionaires—only 1,000 or so people—were together worth $3.5 trillion (equivalent to more than 5 percent of world GDP). OECD countries (the Organization for Economic Cooperation and Development, an association of 30 industrialized countries), with 19 percent of global population, control more than 75 percent of global trade in goods and services and consume 86 percent of the world's goods.

Such enormous differences lead many people to question the equity of the geographical consequences of globalization. The concept of **spatial justice** is important here because it requires us to consider the distribution of society's benefits and burdens at different spatial scales, taking into account both variations in people's need and in their contribution to the production of wealth and social well-being.

Many people, nations, and ethnic groups around the world feel marginalized, exploited, and neglected as a result of the quickening pace of change. Across much of the peripheral world, the perception

Meet Paul Rust and his family, who live and enjoy life in Zug, Switzerland, the richest canton in the world's richest country. And meet Hussein Sormolo and his family, who live in Addis Ababa, capital of the world's poorest country, Ethiopia.

Hussein Sormolo left the village where he was born for the big city in 1978. He left his eight brothers and seven sisters behind, as the land that the family farmed was being forcibly collectivized by a new regime. Hussein, then 16, traveled 100 miles north to the city in the back of a truck. A kinsman from the same village took him in until he found a job in a bakery. Paul Rust left his village in Switzerland when he was 17 and also ended up in a bakery. The two men are similar in other ways. Both are friendly, hospitable, and generous and love their families. Both work hard. Both like to watch the news. Both are active worshippers, without being religious dogmatists.

Yet their lives are different. Hussein lives with his wife, sons, and daughters in a leaky shack of corrugated asbestos and steel in the Nefas Silk district of Addis Ababa (**Figure 2.C**). Paul lives with his wife in a six-room house (not counting the ground-floor apartment where his son Martin lives with his girlfriend) overlooking the lustrous green waters and steep wooded slopes of Lake Aegeri in Zug (**Figure 2.D**).

The income difference is huge. Hussein supports his wife and three younger children on wages of about $280 a year (more than twice the average income in Ethiopia). Paul and his wife, Hedi, draw roughly $68,000 between them each year from their bakery, though the Rusts are not affluent by Swiss standards (the average income per head in Zug is about $50,000).

It is the rainy season in Addis. Fat raindrops drum against, and often through, the rusting grooves of the corrugated roofs of the houses in Nefas Silk. Nights can be chilly and dank. From Debre Zeit road, the busy street lined with small businesses including the bakery where Hussein works, it's a 10-minute walk to the alley where he lives. Inside the Hussein shack, a single bare light bulb always burns. There is little natural light: There are no glass windows, and the openings punched in the asbestos walls are covered to keep out drafts. Hussein pays his neighbor 18 birr (about $2) a month, almost a tenth of his 200-birr salary, to sublet an electricity supply for the bulb in the shack. The family has no other electrical appliances, apart from a battery-operated radio. Neither Hussein, his wife, Rukia, nor his eldest daughter, Fate, 17, who is lucky enough to be at school, has ever used a computer, taken a photograph, or made a phone call. Hussein and Rukia have a pair of shoes each. They buy new ones every 2 years. They have no savings and the family doesn't take holidays.

Except for feast days, the family eats the same dish every meal—a grey, spongy, bread called injera, spread out like a cloth, and a spicy vegetable stew. Meat, fish, cheese, and eggs are

FIGURE 2.C Hussein Sormolo and his family in Addis Ababa, Ethiopia.

FIGURE 2.D Paul Rust and his family in Zug, Switzerland.

luxuries. They only buy fruit when one of the children is sick. Just under a quarter of the family income is spent on cooking charcoal and cans of water. In a country where only a quarter of the people in the countryside have access to safe drinking water, Hussein's family is lucky. There is a standpipe around the corner with reasonably clean water. That's about where their luck ends. While they used to have a toilet they shared with 26 neighbors, now they have no toilet at all.

The Rust house, not counting the apartment, has three toilets, one each in the bathroom and two shower rooms of the four-story building. On the balconies under its broad, dark, solid eaves are cascades of red flowers. The well-used furnishings inside are not ostentatious, but the building is roomy and comfortable. There is a loft, four bedrooms, two living rooms, a kitchen, an office, a small wine cellar, a workroom, garage parking for three cars (Paul, Hedi, and Martin Rust each have a car) with room for another five on the forecourt. The house has its own elevator.

Paul and Hedi are going on vacation for 2 weeks in Austria this month and usually take another week off at Easter. Each has a mobile phone. The home office has computers and Internet access. They have a TV, a VCR, and a dishwasher. They eat what they want, although their tastes are plain—meat with several vegetables, salad, sometimes a little wine.

Switzerland is a rich country landlocked by other rich countries. Ethiopia is a poor country landlocked by other poor countries.

Unlike other African nations, Ethiopia was not a European colony, but its people have endured regular European military incursions, proxy superpower duels, and local wars that have exacerbated the ravages of famine and disease. Famines in the 1970s, 1980s, and 1990s killed 1.3 million people. Through the 1970s and 1980s, the country was embroiled in ideological and ethnic civil war. Today, almost a million Ethiopians are living with HIV/AIDS and 44 percent of the country's population live below the national poverty line.

Hussein knows little about Switzerland. "I heard about Switzerland on the radio but I don't know. I heard it was a rich country, they help poor countries," he said.

Paul thought he could find Ethiopia on the map. Switzerland is not as aloof from the world as it was, he points out: They joined the boycott of apartheid South Africa. He said his brother helped build a dairy in Nepal 20 years ago. His church has adopted a village in Romania, giving it money for a new church and a school. When the talk turns to immigration, daughter Andrea says, "The really poor people, they can't come to Switzerland, they need money to get here. We work, and have our life, we have our own problems," she concludes. "So we don't think very often of other people's problems. It's a little bit selfish."

Based on an article by James Meek, *The Guardian*, 22 August 2002.

of injustice has been brewing for a long time. Resentment at past colonial and imperial exploitation has been compounded as the more affluent places and regions of the world have become increasingly dependent on the cheap labor and resources of the periphery and as transnational businesses have displaced the traditional economic and social practices of peripheral and semiperipheral regions under the banner of modernization. Thinking about spatial justice is an important aspect of the "geographical imagination" described in Chapter 1 and is a recurring theme in the remainder of this book.

APPLY YOUR KNOWLEDGE Use the Internet to find the degree of international disparity in rates of infant mortality: the number of deaths of infants under 1 year of age, per 1,000 live births (Hint: Good sources are the World Bank, http://data.worldbank.org/, and the United Nations Development Programme, http://hdr.undp.org/en/statistics/). Suggest why some countries still have high rates of infant mortality. ■

Westernization and Cultural Imperialism

Political scientist Benjamin Barber chracterizes the cultural tensions associated with contemporary globalization as "Jihad vs. McWorld." McWorld is shorthand for the pop culture and shallow materialism that is part of Western, capitalist modernization. **"Jihad"** is shorthand for cultural values that are underpinned by religious fundamentalism, traditional tribal allegiances, and opposition to Western materialism. (Note that the term *jihad* properly refers to a struggle waged as a religious duty on behalf of Islam.) Neither "Jihad" nor "McWorld" is conducive to a healthy democracy or civil society, argues Barber, while tensions between the two make for potentially volatile situations.

At the heart of these tensions is a marked disillusionment with the West, especially within traditional Islamic societies. Across much of the world, modernization is now taken to mean Westernization and, more specifically, Americanization (**Figure 2.20**). Most Americans think of modernization as necessary and good but many other people see it as the cause of their exploitation. In most peripheral countries, only a minority can enjoy Western-style consumerism and the impoverished majority is acutely aware of the affluence of the core countries. While the gap between rich and poor countries has actually been widening for several decades, the U.S. aid budget—already low compared to the aid budgets of other developed countries—has been declining substantially. The United States, as a result, tends to be easily portrayed as a swaggering superpower, rigging the world-system to serve its own interests and doing relatively little by way of economic or humanitarian aid.

The current phase of globalization thus involves a distinctive new geopolitical element that has been described as the "new imperialism": the imperialism of the United States, the world's only superpower. Although Americans do not like to think of their country as territorially aggressive or exploitative, the "war on terror" and invasion of Afghanistan and Iraq following the al-Qaeda attacks in 2001 are widely interpreted elsewhere in the world as an exercise in imperialism, motivated in large part by a desire for military control over global oil resources. This interpretation of the United States as the instigator of a new imperialism has been reinforced by military threats against Iran and North Korea, deployment of special forces around the globe, use of "extraordinary rendition" (the apprehension or kidnapping of suspects followed by their transfer to countries known to employ harsh interrogation techniques or torture), and unilateral rejection of international environmental treaties and international aid agreements. What is less widely discussed in the world's newspapers is that this new imperialism is also viewed

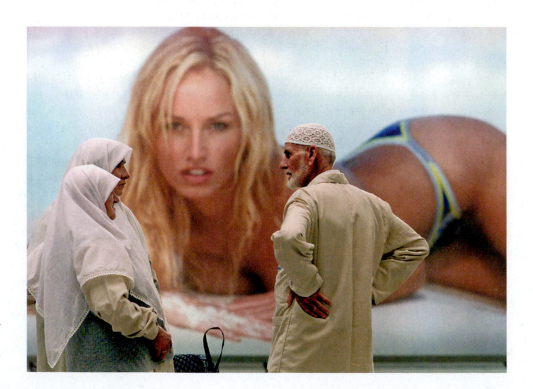

FIGURE 2.20 Westernization Turkish Muslims chat near an illuminated billboard advertising a Turkish Internet company at Istanbul airport.

by some academics as the result of a highly competitive global economic environment in which the United States is no longer able to achieve superiority through innovation, product design, productivity, and marketing and so has had to resort to military intervention.

APPLY YOUR KNOWLEDGE Find a story in a national newspaper that addresses an issue associated with contemporary globalization. Provide three examples from the article that illustrate the increasing interdependence of place and region. ■

Future Geographies

The globalization of the capitalist world-system involves processes that have been occurring for at least 500 years. But since World War II, world integration and transformation have been remarkably accelerated and dramatic. How will the forces of broadening global connectivity—and the popular reactions to them—change the fates and fortunes of world regions whose current coherence owes more to eighteenth- and nineteenth-century European colonialism than to the forces of integration or disintegration in the twenty-first century? To answer this and related questions, we need to understand what the experts think about the processes behind globalization and its future potential. But first we need to understand the very risky issue of predicting the future, how predictions are made, and how useful predictive exercises can be.

There is no shortage of visionary scenarios. Broadly speaking, futurists' projections can be divided into two kinds: optimistic and pessimistic. Optimistic futurists stress the potential for technological innovations to discover and harness new resources, to provide faster and more effective means of transportation and communication, and to make possible new ways of living. This sort of futurism is often characterized by science-fictional cities of mile-high skyscrapers and spaceship-style living pods, by ecological harmony, and by unprecedented social and cultural progress. It projects a world that will be stabilized and homogenized by supranational or even "world" governments. The sort of geography implied by such scenarios is rarely spelled out. Space and place, we are led to believe, will be transcended by technological fixes.

To pessimistic futurists, however, this is just "globaloney." They stress the finite nature of Earth's resources, the fragility of its environment, and population growth rates that exceed the capacity of peripheral regions to sustain them. Such doomsday forecasting scenarios include irretrievable environmental degradation, increasing social and economic polarization, and the breakdown of law and order. The sort of geography associated with these scenarios is also rarely explicit, but it usually involves the probability of a sharp polarization between the haves and have-nots at every geographical scale.

Fortunately, we don't have to choose between the two extreme scenarios of optimism and pessimism. Using our geographical imagination, we can suggest a more grounded outline of future geographies. To do so, we must first glance back at the past. Then, looking at present trends and using what we know about processes of geographic change and principles of spatial organization, we can begin to map out the kinds of geographies that the future most probably holds.

Looking back at the way that the geography of the world-system has unfolded, we can see that a fairly coherent period of economic and geopolitical development occurred between the outbreak of World War I (in 1914) and the collapse of the Soviet Union (in 1989). Some historians refer to this period as the "short twentieth century." It was a period when the modern world-system developed its triadic core of the United States, Western Europe, and Japan. Geopolitics was based on an East-West divide, and geoeconomics was based on a North-South divide. The geographies of specific places and regions within these larger frameworks were shaped by the needs and opportunities of **technology systems** based on the internal combustion engine, oil and plastics, electrical engineering, aerospace industries, and electronics. In this short century, the modern world was established, along with its now familiar landscapes and spatial structures: from the industrial landscapes of the core to the unintended metropolises of the periphery; from the voting blocs of the West to the newly independent nation-states of the South.

Looking around now, much of the established familiarity of the modern world and its geographies seems to be disappearing. We have entered a period of transition, triggered by the end of the Cold War in 1989 and rendered more complex by the geopolitical and cultural repercussions of the terrorist attacks of September 11, 2001, and the global financial "meltdown" of 2008. Obviously, we cannot simply project our future geographies from the landscapes and spatial structures of the past. At the same time, we can only guess at some aspects of the future. But we can draw some conclusions from a combination of existing structures and budding trends. We have to anticipate how the shreds of tradition and the strands of contemporary change might be rewoven into new landscapes and new spatial structures. Among the relative certainties over the next decade or so are the increasing power and influence of China and India, a general shift in relative wealth and economic power from the West to the East, an increased probability of conflict in the Middle East, and an intensification of problems resulting from pressure on food, water, and energy resources. Among the key uncertainties are the speed of climate change, the resolution of the Arab-Israeli conflict, and the effectiveness of alternative energy technologies (Table 2.1). In subsequent chapters, we examine these and related issues in detail.

TABLE 2-1 The 2025 Global Landscape

Relative Certainties	Likely Impact
A global multipolar system is emerging with the rise of China, India, and others. The relative power of nonstate actors—businesses, tribes, religious organizations, and even criminal networks—also will increase.	By 2025 a single "international community" composed of nation-states will no longer exist. Power will be more dispersed with the newer players bringing new rules to the game and the risks will increase that the traditional Western alliances will weaken. Rather than emulating Western models of political and economic development, more countries may be attracted to China's alternative development model.
The unprecedented shift in relative wealth and economic power roughly from West to East now under way will continue.	As some countries become more invested in their economic well-being, incentives toward geopolitical stability could increase. However, the transfer is strengthening states like Russia that want to challenge the Western order.
The United States will remain the single most powerful country but will be less dominant.	Shrinking economic and military capabilities may force the U.S. into a difficult set of tradeoffs between domestic versus foreign policy priorities.
Continued economic growth—coupled with 1.2 billion more people by 2025—will put pressure on energy, food, and water resources.	The pace of technological innovation will be key to outcomes during this period. All current technologies are inadequate for replacing traditional energy architecture on the scale needed.
The number of countries with youthful populations in the "arc of instability"[1] will decrease, but the populations of several youth-bulge states are projected to remain on rapid growth trajectories.	Unless employment conditions change dramatically in youth-bulge states such as Afghanistan, Nigeria, Pakistan, and Yemen, these countries will remain ripe for continued instability and state failure.
The potential for conflict will increase owing to rapid changes in parts of the greater Middle East and the spread of lethal capabilities.	The need for the U.S. to act to balance power in the Middle East will increase, although other outside powers—Russia, China, and India—will play greater roles than today.
Terrorism is unlikely to disappear by 2025, but its appeal could lessen if economic growth continues in the Middle East and youth employment increases. For those terrorists that are active the diffusion of technologies will put dangerous capabilities within their reach.	Opportunities for mass-casualty terrorist attacks using chemical, biological, or less likely, nuclear weapons will increase as technology diffuses and nuclear power (and possibly weapons) programs expand. The practical and psychological consequences of such attacks will intensify in an increasingly globalized world.
Key Uncertainties	Potential Consequences
Will an energy transition away from oil and gas—supported by improved energy storage, biofuels, and clean coal—be completed during the 2025 time frame?	With high oil and gas prices, major exporters such as Russia and Iran will substantially augment their levels of national power, with Russia's GDP potentially approaching that of the U.K. and France. A sustained plunge in prices, perhaps underpinned by a fundamental switch to new energy sources, could trigger a long-term decline for producers as global and regional players.
How quickly will climate change occur and in what locations will its impact be most pronounced?	Climate change is likely to exacerbate resource scarcities, particularly water scarcities.
Will mercantilism stage a comeback and global markets recede?	Descending into a world of resource nationalism increases the risk of great power confrontations.
Will advances toward democracy occur in China and Russia?	Political pluralism seems less likely in Russia in the absence of economic diversification. A growing middle class increases the chances of political liberalization and potentially greater nationalism in China.
Will regional fears about a nuclear-armed Iran trigger an arms race and greater militarization?	Episodes of low-intensity conflict and terrorism taking place under a nuclear umbrella could lead to an unintended escalation and broader conflict.

[1]Countries with youthful age structures and rapidly growing populations mark a crescent or "arc of instability" stretching from the Andean region of Latin America across Sub-Saharan Africa, the Middle East and the Caucasus, and through the northern parts of South Asia.

TABLE 2-1   The 2025 Global Landscape   *(Continued)*

Will the greater Middle East becomes more stable, especially will Iraq stabilize, and will the Arab-Israeli conflict be resolved peacefully?	Turbulence is likely to increase under most scenarios. Revival of economic growth, a more prosperous Iraq, and resolution of the Israeli-Palestinian dispute could engender some stability as the region deals with a strengthening Iran and global transition away from oil and gas.
Will Europe and Japan overcome economic and social challenges caused or compounded by demography?	Successful integration of Muslim minorities in Europe could expand the size of the productive work forces and avert social crisis. Lack of efforts by Europe and Japan to mitigate demographic challenges could lead to long-term declines.
Will global powers work with multilateral institutions to adapt their structure and performance to the transformed geopolitical landscape?	Emerging powers show ambivalence toward global institutions like the UN and IMF, but this could change as these powers become bigger players on the global stage. Asian integration could lead to more powerful regional institutions. NATO faces stiff challenges in meeting growing out-of-area responsibilities with declining European military capabilities. Traditional alliances will weaken.

CONCLUSION

Places and regions everywhere carry the legacy of a sequence of major changes in world geography. The evolution of world geography can be traced from the prehistoric hearths of agricultural development and human settlement, through the trading systems of the precapitalist, preindustrial world, to the modern world. The foundations of the modern world are industrialization, colonization, and the international market economy. Today, these foundations can be seen in the geography of the Information Age, a geography that provides a global context for places and regions.

Today's world is highly integrated. Places and regions have become increasingly interdependent, linked through complex and rapidly changing commodity chains that are orchestrated by transnational corporations. Using new technology systems that allow for instantaneous global telecommunications and flexible patterns of investment and production, these corporations span the world. This integration blurs some national and regional differences as the global marketplace brings about a dispersion of people, tastes, and ideas. The overall result, though, has been an intensification of differences between the core and the periphery. Within this new global context, local differences in resource endowments remain and people's territorial impulses endure. Many local cultures continue to be resilient or adaptive. Fundamental principles of spatial organization also continue to operate.

The emergence of globalization—with its transnational architectural styles, dress codes, retail chains, and popular culture and ubiquitous immigrants, business visitors, and tourists—seems as if it might inevitably impose a sense of placelessness and dislocation, a loss of territorial identity, and an erosion of the distinctive sense of place associated with certain localities. Yet the common experiences associated with globalization are still modified by local geographies. The structures and flows of globalization are variously embraced, resisted, subverted, and exploited as they make contact with specific places and specific communities. In the process, places and regions are reconstructed rather than effaced. Often this involves deliberate attempts by the residents of a particular area to create or re-create territorial identity and a sense of place. Human geographies change, but they don't disappear.

Learning Outcomes Revisited

■ Summarize the distinctive stages of the evolution of the modern world-system.

Premodern geographies were organized around minisystems and regional empires. The modern world-system was established over a long period that began in the late fifteenth century. More and more peoples around the world have become exposed to one another's technologies and ideas since the fifteenth century. Different resources, social structures, and cultural systems resulted in quite different pathways of development, however. Some societies were incorporated into the new, European-based international economic system faster than others; some resisted incorporation; and some sought alternative systems of economic and political organization.

■ Analyze how and why the new technologies of the Industrial Revolution helped to bring about the emergence of a global economic system.

The new technologies of the Industrial Revolution brought about the emergence of a global economic system that reached into almost every part of the world and into virtually every aspect of people's lives. New transportation technologies triggered successive phases of geographic expansion, allowing for an intensive period of external colonization and imperialism. The core of the world-system (Europe) grew to include the United States and Japan, while most of the rest of the world was systematically incorporated into the capitalist world-system as a dependent periphery.

- Examine the changing patterns of interdependence among different world regions.

   Each place and region carries out its own particular role within the competitive world-system. Because of these different roles, places and regions are dependent on one another. The development of each place affects, and is affected by, the development of many other places. Since the seventeenth century, the world-system has been consolidated, with stronger economic ties between countries. It has also been extended, with all the world's countries eventually becoming involved to some extent in the interdependence of the capitalist system and the consequent flows of resources, capital, goods, ideas, and people among places and regions.

- Compare the three tiers that constitute the modern world-system.

   Today, the world-system is highly structured and is characterized by three tiers: core regions, semiperipheral regions, and peripheral regions. The core regions of the world-system are those that dominate trade, control the most advanced technologies, and have high levels of productivity within diversified economies. Peripheral regions are characterized by dependent and disadvantageous trading relationships, by primitive or obsolescent technologies, and by undeveloped or narrowly specialized economies with low levels of productivity. Semiperipheral regions are able to exploit peripheral regions but are themselves exploited and dominated by the core regions. This three-tiered system is fluid, providing a continually changing framework for geographical transformation within individual places and regions.

- Explain how the growth and internal development of the world's core regions could take place only with the foodstuffs, raw materials, and markets provided by the colonization of the periphery.

   Peripheral regions were originally developed and exploited in order to provide the raw materials for industrializing regions and the food supplies for their rapidly growing populations. In the eighteenth and nineteenth centuries, the industrial core nations embarked on the inland penetration of the world's mid-continental grassland zones in order to exploit them for grain or stock production. At the same time, as the demand for tropical plantation products increased, most of the tropical world came under the political and economic control—direct or indirect—of one or another of the industrial core nations. For these peripheral regions, European overseas expansion meant political and economic dependency.

- Identify an example of each of the four key issues caused by globalization—environmental, health, core-periphery disparity, and security issues.

   Many of the important issues facing modern society are the consequences—intended and unintended—of human modifications of our physical environment. In addition, the increased intensity of international trade and travel has also heightened the risk and speed of the spread of disease. Globalization has also intensified differences in prosperity between the core and the periphery, and contemporary society is characterized by new hazards, many of them uncontrollable and with a global reach.

KEY TERMS

capitalism *(p. 39)*
climate change *(p. 54)*
colonialism *(p.45)*
colonization *(p. 38)*
commodity chain *(p. 51)*
comparative advantage *(p. 46)*
core regions *(p. 45)*
division of labor *(p. 46)*
environmental
   determinism *(p. 49)*

ethnocentrism *(p. 49)*
external arena *(p. 41)*
globalization *(p. 51)*
Gross Domestic Product
   (GDP) *(p. 50)*
hearth areas *(p. 36)*
hegemony *(p. 46)*
hinterland *(p. 41)*
imperialism *(p. 45)*
import substitution *(p. 42)*

jihad *(p. 60)*
law of diminishing
   returns *(p. 38)*
leadership cycles *(p. 46)*
minisystem *(p. 36)*
neocolonialism *(p. 49)*
pandemic *(p. 55)*
peripheral regions *(p. 45)*
plantation *(p. 41)*
risk society *(p. 56)*

semiperipheral regions *(p. 45)*
slash-and-burn *(p. 36)*
spatial justice *(p. 57)*
sustainability *(p. 55)*
technology systems *(p. 61)*
transnational
   corporations *(p. 50)*
urbanization *(p. 38)*
world-empire *(p. 37)*
world-system *(p. 41)*

REVIEW AND DISCUSSION

1. Discuss in your group whether the United States can still be considered a core country (see definitions on p. 46). Develop three reasons in support and against your arguments. What statistical evidence supports your argument? Also consider which countries might be moving into the core or into the semiperiphery or periphery? What evidence could you find to support such transitions?

2. Two major events have had significant impacts on places and regions, namely the droughts in Russia in the summer of 2010 and the Egyptian revolution of February 2011. Discuss in your group how these events are related to globalization. Provide two examples for each case.

3. Consider the core, the semiperiphery, and the periphery, not from the perspective of states, but from the viewpoint of transnational

corporations (see p. 50). Research two corporations and find out where their corporate headquarters are located, develop a list of what they produce, and, if possible, find out where their products come from. Take into consideration any raw materials and manufacturing that are involved. For example, your group might want to consider cell phone companies. Where are the companies located? Are they in the core, semiperiphery, or periphery? Where do they get the material to make cell phones, within the core, semiperiphery, or periphery? And finally where are the cells phones manufactured? Once you have compiled this information, develop a world map displaying your information and the ways the transnational corporations' products move between the core, semiperiphery, and periphery.

UNPLUGGED

1. The present-day core regions of the world-system, shown in Figure 2.14 (p. 50), are those that dominate trade, control the most advanced technologies, have high levels of productivity within diversified economies, and enjoy relatively high per capita incomes. What statistical evidence can you find in your local library to support this characterization? Consider and list what core countries produce. (*Hint:* Look for data in annual compilations of statistics published in annual reviews, such as the Encyclopaedia Britannica's *Yearbook* and in the annual reports of organizations, such as the United Nations Development Programme [UNDP], the World Bank, and the World Resources Institute, and consider per capita incomes and how they differ between countries). How could differences in trade be understood from these publications?

2. The idea of an international division of labor is based on the observation that different countries tend to specialize in the production

or manufacture of particular commodities, goods, or services. In what product or products do the following countries specialize: Bolivia, Ghana, Guinea, Libya, Namibia, Peru, and Zambia? You will find the data you need in a good statistical yearbook, such as the Encyclopaedia Britannica's *Yearbook*, or in a world reference atlas or economic atlas. Take one product that you found and use in your everyday life and reflect on the process it underwent to come into your possession.

3. Consider the division of labor in your place of work (either as a student or outside the university). How are tasks divided and who is responsible for what types of work? Is there an apparent division of labor (see p. 46)? If so, list three examples of how work is divided in your workplace. Think of who does what. And consider why certain tasks are divided differently.

Mastering GEOGRAPHY™

Log in to www.masteringgeography.com for MapMaster™ interactive maps, geography videos, RSS feeds, flash cards, Web links, an eText version of *Human Geography: Places and Regions in Global Context*, and self-study quizzes to enhance your study of the changing global context.

MapMaster™ presents 13 Place Name and 13 Layered Thematic interactive maps to help students practice and master their geographic literacy, spatial reasoning, and critical thinking skills.

MapMaster™

3 POPULATION GEOGRAPHY

Learning Outcomes

- Understand the census and other sources of population data and how they are used to describe the geography of population.

- Recognize why populations change, where those changes occur, and what the implications of population change are for the future of different places around the globe.

- Identify the two most important factors in population dynamics, birth and death, and how they shape population characteristics.

- Demonstrate how the movement of population is affected by both push and pull factors, and explain how these factors are key to understanding new settlement patterns.

- Evaluate the challenges of providing for the world's growing population with adequate food and safe drinking water, as well as a sustainable environment.

Iraq has not conducted a national count of its population in over two decades. Most countries assess their population every 10 years. The last population count in Iraq in 1987 occurred during the administration of President Saddam Hussein and was used to conscript a majority of the male population into the army to fight in the Iran-Iraq war. At the time, international observers believed that the census was inaccurate, not only because the bureaucracy needed to conduct a national population count had been seriously dismantled under the dictatorship but also because the recorded numbers were misreported so that some ethnic populations would appear more numerous than others in parts of the country. Since the war against Iraq by the United States and its allies and the deposing of Hussein in 2003, hundreds of thousands of war casualties have occurred and millions of refugees have left the country. Yet the current regime is still reluctant to launch a full national count of the population. Why? Because numbers have a wider impact besides knowing "how many."

A **census**—a straightforward count of the number of people in a country, region, or city—is not only a practical undertaking answering basic demographic questions (how many people there are, where they live, what their ethnic backgrounds are), it is also a political undertaking. For instance, the number of voters in a population routinely determines the number

A worker turns valve on outdated pumping technology at the Shawira oil fields in Kirkuk Province.

of officials that can be elected from that region to represent them in a national government. Population numbers also often decide the amount of aid a region receives.

In Iraq, the current administration is wary of conducting a census because questions important to the future of the oil fields will be answered by the census. And some people who currently have control over oil may lose it and others will gain control. At the center of this dispute is Kirkuk Province, where an unstable ethnic mix of Arabs, Kurds, Turkmen, and others, sits atop 4 percent of the world's oil reserves. The census will statistically establish the majority population in these territories, which the Arab-dominated government in the capital city of Baghdad wants to keep and the Kurds of the northeast want to fold into their semiautonomous enclave. The majority population in this province will have a great deal of control over the future of that oil and be a significant political force not just in Iraq but in the world.

What the Iraq example illustrates is that it's not just the numbers that are important, it's where those numbers are located that matters too. The *where* of population is what drives geographers' interests in demographics, the characteristics of a *human population* including elements such as **gender**, race, age, income, disabilities, educational attainment, and migration patterns among different groups and death rates among others. ∎

THE DEMOGRAPHER'S TOOLBOX

Demography, the study of the characteristics of human populations, is an interdisciplinary undertaking. Geographers study population to understand the areal distribution of Earth's peoples. They are also interested in the reasons for, and the consequences of, the distribution of populations from the international to the local level. While historians study the evolution of demographic patterns, and sociologists study the social dynamics of human populations, geographers focus on the spatial patterns of human populations, the implications of such patterns, and the reasons for them. Using many of the same tools and methods of analysis as other population experts, geographers think of population in terms of the places that populations inhabit. They also consider populations in terms of the way that places are shaped by populations and in turn shape the populations that occupy them.

Censuses and Vital Records

Population experts rely on a wide array of instruments and institutions to carry out their work. Government entities, schools, and hospitals collect information on births, deaths, marriages, migration, and other aspects of population change. The most widely known instrument for assessing the state of the population is the census, a survey developed to obtain information for tax collection.

A census is a straightforward count of the number of people in a country, region, or city. Censuses, however, are not usually so simple. Most are also directed at gathering other information about people, such as previous residences, marital status, and income. Many countries comprehensively assess the characteristics of their national populations every 10 years. In the United States, for example, the Bureau of the Census has surveyed the population every 10 years since 1790. The information gathered is used to apportion seats in the U.S. House of Representatives, as well as to redistribute federal tax funds and other revenues to states, counties, and cities.

In addition to the census, population experts employ other data sources to assess population characteristics. One such source is **vital records**, which report births, deaths, marriages, divorces, and the incidence of certain infectious diseases. These data are collected and recorded by city, county, state, and other levels of government. In combination with the census data, vital records help to show the way the population of an area—whether country, region, or city—is changing. Box 3.1, "Visualizing Geography: A New Sense of Identity," provides just such a view on changing race, ethnicity, and marriage in the United States. Schools, hospitals, police departments, prisons, and other public agencies, such as the U.S. Bureau of Citizenship & Immigration Services and international groups like the World Health Organization, also collect demographic statistics that are useful to population experts.

Limitations of the Census

Censuses are an extremely expensive and labor-intensive undertaking for any governmental jurisdiction, and as a result, they occur rather infrequently. Historically, the United States has undertaken a population census every 10 years, but in 1985 it introduced a quinquennial (every 5 years) census to augment the decennial (every 10 years) census. Most prominent among the reasons for initiating an additional, quinquennial census was that data collected every 10 years quickly became obsolete.

In many peripheral and semiperipheral countries, governments are not always able to finance a decennial census such as the comprehensive surveys undertaken in more developed countries like France or Germany. Cambodia conducted its first complete census in 1962 and did not conduct another until 1998. Liberia conducted its first census in 24 years in 2008. Pop stars were enlisted and billboards about the census were erected by the government to remind villagers to stay home for three days and be counted.

The incompatibility of census dates makes comparisons among, between, and within countries quite difficult. For example, the United States conducted its last decennial census on April 1, 2010. China conducted a comparable decennial census in November 2010, its sixth nationwide census. The difference in collection times between the two censuses makes comparisons very difficult, especially because processes such as international migration can have a significant impact on the actual numbers.

The 2010 U.S. census had been scheduled to undertake the count using handheld computers. But problems with training census takers on the devices led the U.S. Census Bureau to revert to pen-and-paper forms. In contrast, India launched its first **biometric census** in 2011 (**Figure 3.1**). In addition to physically counting the country's over 1 billion people, the government is photographing and fingerprinting every individual over the age of 15, creating a national biometric database enabling the government to issue official identity cards. Over 2.5 million census officials will visit households in more than 7,000 towns and 600,000 villages. The count will also attempt, for the first time, to gather information about the use of the Internet across the country as well as the availability of drinking water and toilets in the home.

FIGURE 3.1 Biometric census taking in India, 2011 India became the first country in the world to collect photographs and fingerprints of all of its population as part of its decennial census exercise. Populations in many western countries have refused to participate in a biometric census count as they are reluctant to have the government possess such information about them.

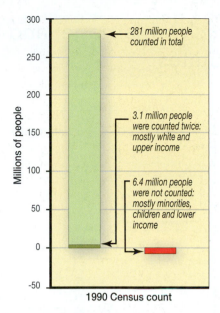

FIGURE 3.2 U.S. Census over- and undercount of population, 1990 The accuracy of the 1990 census was widely criticized by states and cities across the United States. The graph shows the estimated over- and undercount of the population.

(*Source*: Based on illustration from "Are Census Numbers Accurate?" by Allison Plyer, Greater New Orleans Community Data Center. Nov. 26, 2001. Reprinted with permission.)

Despite the massive and costly efforts to count population being undertaken by India and other countries around the world, no census is entirely comprehensive (or comprehensible). All censuses tend to underrepresent nonmainstream kinds of households as well as homeless individuals (**Figure 3.2**). For instance, more than 8 million people were not counted in the U.S. 1990 census, most of whom were Hispanics or African Americans. The city of Inglewood in metropolitan Los Angeles estimated that 13,400 individuals were not found by the census takers, leading to an undercount of 10.9 percent, the largest undercount of any city in the country. Because federal revenue-sharing formulas as well as the apportionment of U.S. House seats are tied to population numbers, urban political officials from many of America's large and medium-size cities complained extensively about the numbers generated from the 1990 census. Billions of dollars in federal grants for education, transportation, and health care are distributed on the basis of population. Undercounted communities stand to lose significant financial support for local schools, roads, and health services, while overcounted communities can reap more grants than warranted.

APPLY YOUR KNOWLEDGE Give an example, other than the Iraq case, of how the census is more than just counting people. ∎

POPULATION DISTRIBUTION AND STRUCTURE

Because human geographers explore the interrelationships and interdependencies between people and places, they are interested in demography. Population geographers bring to demography a special perspective—the spatial perspective—that emphasizes description and explanation of the "where" of population distribution, patterns, and processes.

For instance, the seemingly simple fact that as of early 2011 the world was inhabited by nearly 6.93 billion people is one that geographers like to think about in a more complex way. This number is undeniably phenomenal and increasing with each passing second, but its most important aspect for geographers is its uneven spatial expression from region to region and from place to place. Equally important are the implications and impacts of these differences. Looking at population numbers, geographers ask themselves two questions: Where are these populations concentrated, and what are the causes and consequences of such a population distribution?

Population Distribution

Many geographic reasons exist for the distribution of populations throughout the globe. As the world population density map demonstrates (**Figure 3.3**), some areas of the world are heavily inhabited, others only sparsely. Some areas contain no people whatsoever. Bangladesh and the Netherlands, for example, have high population densities throughout. Egypt, on the other hand, displays a pattern of especially high population concentrations along the coasts and the Nile River but a relatively low population density elsewhere. The sparsest inland population densities in the world occur across the mountains and rolling plateaus of Mongolia. These examples illustrate how environmental and physical factors are important influences on population distributions and concentrations.

Degree of accessibility, topography, soil fertility, climate and weather, water availability and quality, and type and availability of other natural resources are some of the factors that shape population distribution. Other factors are also crucial—first and foremost are a country's political and economic experiences and characteristics. For example, the high population concentrations along Brazil's Atlantic coast date back to the trade patterns set up during Portuguese colonial control in the sixteenth and seventeenth centuries. Another important factor is culture as expressed in religion, tradition, or historical experience. One of the key reasons cities like Medina and Mecca, in the Middle East, comprise important population concentrations is because they are Islamic sacred sites. **Table 3.1** lists population estimates in terms of continental distributions. Asia is far and away the most populous continent. Running a distant second and third are Africa and Europe.

The population clusters that take shape across the globe have a number of physical similarities. Almost all of the world's inhabitants live on 10 percent of the land. Most live near the edges of landmasses, near the oceans or seas, or along rivers with easy access to a navigable waterway. Approximately 90 percent live north of the equator, where the largest proportion of the total land area (63 percent) is located. Finally, most of the world's population lives in temperate, low-lying areas with fertile soils.

Population numbers are significant not only on a global scale. Population concentrations within countries, regions, and even metropolitan areas are also important. For example, much of the population of North Africa is distributed along the coastal areas where most of the large cities are, as well as along the Nile River in Egypt, as noted earlier (**Figure 3.4**).

A New Sense of Identity: Race, Ethnicity, Marriage, and Birth in the Twenty-First Century United States

A 2011 *New York Times* article described a game played by students at the University of Maryland. The object of the game called "What are you?" is to guess the race of a fellow student. A question once considered at best inappropriate and at worst racist is asked with pride by members of the university's Multiracial and Biracial Student Association. What makes the question one that could be asked at universities across the United States is the fact that the current cohort of students moving through college includes the largest group of mixed-race young people in the history of the United States. What's more, this group is just the leading edge of this demographic shift, driven both by immigration and intermarriage.

As the article goes on to state, today one in seven marriages registered in county vital records systems is between spouses of different races or ethnicities with the rates of intermarriage differing significantly by gender, race, and ethnicity. Indeed, multiethnic and multiracial Americans are one of the country's fastest-growing demographic groups. The expectation is that this trend will continue for decades. **Figure 3.A** shows how gender differences in intermarriage are most pronounced among blacks and Asians. Black men and Asian women are twice as likely to marry someone from a different group. Those groups with the highest rate of intermarriage are black Hispanics and Native Americans, with members of the latter group marrying a white person just as often at they marry another native person. Increasing immigration of Asians and Hispanics into the United States in recent decades seems to explain the lower rates of intermarriage among these groups, who have a wider pool of potential partners from which to choose.

Of course, as anyone who watches television, goes to cinemas, or reads the newspaper knows, mixed-race Americans are also increasingly in the public eye as prominent public figures—sports stars, politicians, and business persons. The current U.S. president, Barack Obama, is of mixed race, with a midwestern white North American mother and a Kenyan-born black father. The actor, wrestler, and athlete Dwayne "the Rock" Johnson is Samoan and black Canadian. Tiger Woods, the professional golfer, is of Thai, Chinese, black American, and Native American decent. And the Internet contains hundreds of Web sites aimed at mixed-race audiences that do everything from advertising special products to providing social networking opportunities.

It's unclear what the impact of a growing mixed-race population will have on U.S. society. On the positive side, increasing numbers of mixed-race individuals may be a way toward rejecting race as a social category and thereby reducing bigotry and prejudice (**Figure 3.B**).On the negative side, more multiracial individuals in the population could intensify current social stratification disadvantaging single-race groups such as African Americans. While it is too soon to be able to predict what a predominantly multiracial U.S. population would mean economically, politically, or

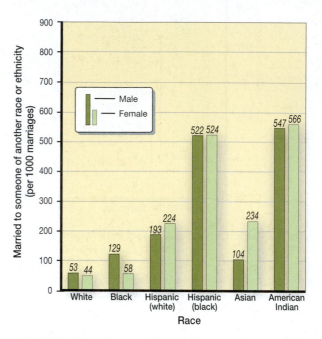

FIGURE 3.A Who is marrying whom This figure displays the number of married people in each group (out of 1,000) who are married to someone of a different race or ethnicity. (*Source:* Based on *The New York Times*, "Black? White? Asian? More Young Americans Choose All of the Above," Sunday, 30 January, 2011, front page and pp. 20–21.)

culturally, or even if there will be any impact in these areas at all, what we do know for sure is that the ethnicity of the U.S. population is changing significantly.

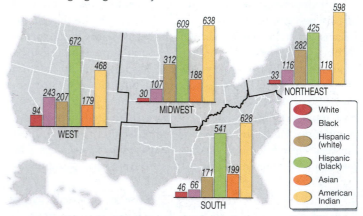

FIGURE 3.B Regional difference in who is marrying whom for both genders The figures on the map indicate how many married individuals out of 1,000 are married to someone of another race or ethnicity. For instance, in the West, whites are more likely to marry outside their racial group than any of the other groups. Looking across all regions and groups, it is possible to tell which groups are marrying outside their own group and where that is happening the most. (*Source:* Based on *The New York Times*, "Black? White? Asian? More Young Americans Choose All of the Above," Sunday, 30 January, 2011. Data sources: Andrew A. Beveridge, Queens College Sociology Department; Census Bureau.)

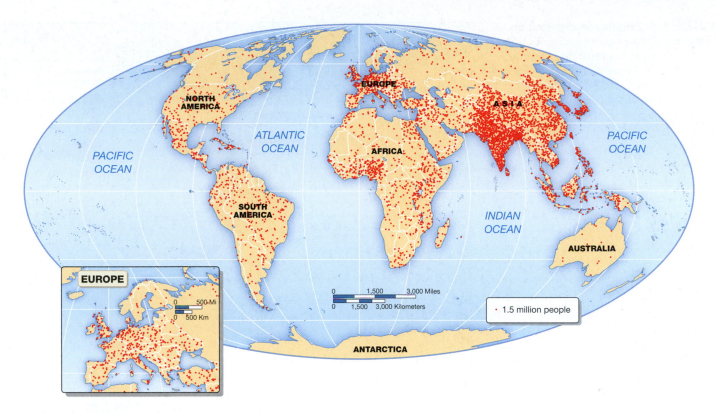

FIGURE 3.3 World population density, 2011 Maps such as this one are useful in understanding the relationships between population distribution and the national contexts within which they occur. China and India are the largest countries in the world with respect to population; the core countries have much smaller populations. (Adapted from World Bank, World Development Indicators, 2007, World Bank: Washington, DC. Updated data from 2011 World Data Sheet: http://www.prb.org/Publications/Datasheets/2010/2010.wpds.aspx)

TABLE 3.1 World Population Estimates by Continents

Continent	Number of Inhabit-ants (in millions)	% of Total Population
Asia	4,216	60
Africa	1,051	15
Europe	740	11
Latin America and Caribbean	596	8.5
North America	346	5
Oceania	37	0.5
Total	6,987	100

Source: Population Reference Bureau Web Site, www.prb.org. World Population Data Sheet, 2011.

APPLY YOUR KNOWLEDGE Name three factors that shape population distribution, and assess how the population in the largest city in your state or province has changed over the last 30 years vis-à-vis these three factors. ∎

Population Density and Composition

Another way to explore population is in terms of **density**, a numerical measure of the relationship between the number of people and some other unit of interest expressed as a ratio. Crude density is probably the most common measurement of population density. **Crude density**, also called **arithmetic density**, is the total number of people divided by the total land area. The metropolitan area of Mexico City, one of the most populous cities in the world, containing over 18 million residents, is a classic high-density urban settlement with a population density of approximately 8,400 persons per square kilometer (about 21,800 persons per square mile; **Figure 3.5**).

The limitation of the crude density ratio—and hence the reason for its "crudeness"—is that it is one-dimensional. It tells us very little about the variations in the relationship between people and land. For that, we need other tools for exploring population density, such as nutritional density or agricultural density. **Nutritional density** is the ratio between the total population and the amount of land under cultivation in a given unit of area. **Agricultural density** is the ratio between the number of agriculturists—people earning their living or subsistence from working the land—per unit of farmable land in a specific

FIGURE 3.4 Population distribution of Egypt Egypt's population distribution is closely linked to the proximity of water. In the north, the population clusters along the Mediterranean, and in the interior, along the banks of the Nile River. (*Source:* Gridded Population of the World [GPW], Version 2. Palisades, NY: CIESIN, Columbia University.)

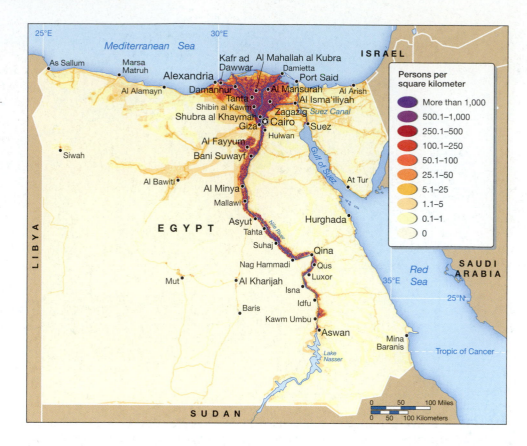

FIGURE 3.5 Population density, Mexico City, Mexico Urban form in Mexico City is the result of many factors. While natural features such as a surrounding lake basin and the presence of active seismic activity are important, social and economic factors include population size and the city's role as a central node in the world system of cities are also key.

area. **Figure 3.6** is a map of the ratio number of physicians to the total population.

In addition to exploring patterns of distribution and density, population geographers also examine population in terms of composition—that is, the subgroups that constitute it. Understanding population composition enables geographers to gather important information about population dynamics. For example, knowing the composition of a population in terms of the total number of males and females, number and proportion of senior citizens and children, and number and proportion of

people active in the workforce provides valuable insights into the ways in which the population behaves now and how it might behave in the future.

For example, countries with populations that contain a high proportion of old people face unique challenges. This is a situation most core countries will soon be facing as their "baby boom" generation ages. The **baby boom** generation includes those individuals born between 1946 and 1964. A considerable amount of a country's resources and energies will be necessary to meet the needs of a large number of people who may no

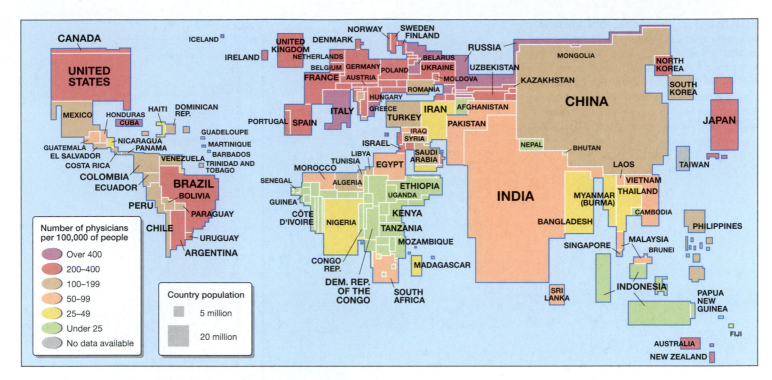

FIGURE 3.6 Health care density Another measure of population density is reflected in this map, which shows the number of people per physician in the total population. Most of the core countries as well as Cuba, the former Soviet Union and some Central Asian and South American countries have the highest ratio of doctors to overall population. Most of the continent of Africa, except South Africa and Egypt, Libya, and Morocco in North Africa, has the lowest ratio, reflecting another dimension of core-periphery inequality. (*Source:* Adapted from H. Veregin (ed.), *Goode's World Atlas*, 22nd ed. Chicago, IL: Rand McNally & Co, 2010, p. 55. Updated data from NationMaster.com, http://www.nationmaster.com/graph/hea_phy_per_1000_peo-physicians-per-1-000-people; World Development Indicators database - http://data.worldbank.org/data-catalog/world-development-indicators.)

longer be contributing in any significant fashion to the creation of the wealth necessary for their maintenance. There might also be a need to import workers to supplement the relatively small working-age population.

Similarly, knowing the number of women of childbearing age in a population, along with other information about their status and opportunities, can provide valuable information about the future growth potential of that population. For example, populations in core countries like Denmark, which has a small number of women of childbearing age relative to the total population size but with high levels of education, socioeconomic security, and wide opportunities for work outside the home, will generally grow very slowly if at all. Peripheral countries like Kenya, on the other hand, where a large number of women of childbearing age have low levels of education and socioeconomic security and relatively few employment opportunities, will continue to experience relatively high rates of population growth, barring unforeseen changes. The variety that exists within a country's population shapes the opportunities and challenges it must confront nationally, regionally, and locally.

Understanding population composition, then, not only tells us about the future demographics of regions but is also quite useful in the present. For example, businesses use population composition data to make marketing decisions and to decide where to locate their operations. For many years, these businesses used laborious computer models to help target their markets. With the development of geographic information systems (GIS), however, this process has been greatly simplified. Assessing the location and composition of particular populations is known as **geodemographic analysis**.

Age-Sex Pyramids

Areal distributions are not the only way that demographers portray population distributions. To display variations within particular subgroups of a population or with respect to certain descriptive aspects, such as births or deaths, demographers also use bar graphs displayed both horizontally and vertically. The most common way for demographers to graphically represent the composition of the population is an **age-sex pyramid**, which is a representation of the population based on its composition according to age and sex. An age-sex pyramid is actually a bar graph displayed horizontally. Ordinarily, males are portrayed on the left side of the vertical axis and females on the right. Age categories are ordered sequentially from the youngest at the bottom of the pyramid to the oldest at the top. By moving up or down the pyramid, one can compare the opposing horizontal bars in order to assess differences in frequencies for each age group.

Age-sex pyramids allow demographers to identify changes in the age and sex composition of populations. For example, an age-sex pyramid depicting Germany's 2000 population clearly revealed the impact of the two world wars, especially with regard to the loss

FIGURE 3.7 Population of Germany, by age and sex, 2000 Germany's population profile is that of a wealthy core country that has passed through the postwar baby boom and currently possesses a low birthrate. It is also the profile of a country whose population has experienced the ravages of two world wars. (*Source:* Adapted from J. McFalls, Jr., "Population: A Lively Introduction," 5th ed., *Population Bulletin, 62*(1), 2007, p. 20.)

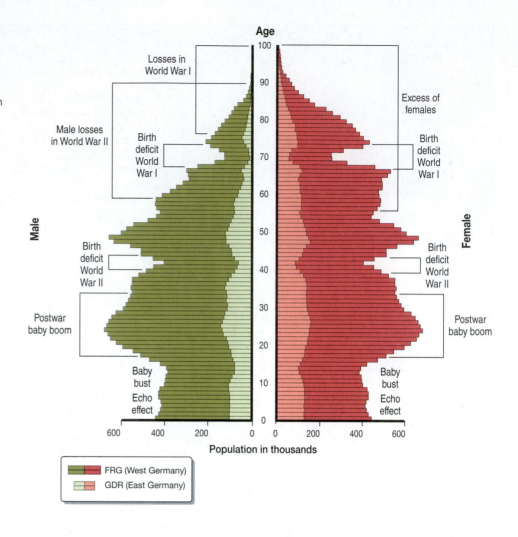

of large numbers of males of military age and the deficit of births during those periods (**Figure 3.7**). Demographers call population groups like these cohorts. A **cohort** is a group of individuals who share a common *temporal* demographic experience. A cohort is not necessarily based only on age, however. Cohorts may be defined based on criteria such as time of marriage or time of graduation.

In addition to revealing the demographic implications of war or other significant events, age-sex pyramids can provide information necessary to assess the potential impacts that growing or declining populations might have. The shape of an age-sex pyramid varies depending on the proportion of people in each age cohort. The pyramid for the peripheral countries, shown in **Figure 3.8**, reveals that many dependent children, ages 0 to 14, exist relative to the rest of the population. The considerable narrowing of the pyramid toward the top indicates that the population has been growing very rapidly in recent years. The shape of this pyramid is typical of peripheral countries with high birthrates and low death rates.

Serious implications are associated with this type of pyramid. First, in the absence of high productivity and wealth, resources are increasingly stretched to their limit to accommodate even elemental schooling, nutrition, and health care for the growing number of children. Furthermore, when these children reach working age, a large number of jobs will have to be created to enable them to support themselves and their families. Also, as they form their own

families, the sheer number of women of childbearing age will almost guarantee that the population explosion will continue. This will be true unless strong measures are taken, such as intensive and well-funded birth-control campaigns, improved education, and outside opportunities for women, as well as modifications of cultural norms that place a high value on large family size.

In contrast, the pyramid for the core countries (**Figure 3.8**) illustrates the typical shape for a country experiencing a slow rate of growth. Most countries in the core are experiencing birthrates that are at or below replacement level. Thus, the pyramid is very columnar, hardly a pyramid at all. People are equally distributed among the cohorts, though the base is perceptibly narrower. Many of the core European countries, such as Denmark, Spain, and Germany, as well as the United States, will face similar demographic challenges, including an extremely large elderly population. In all these countries, however, high levels of production and wealth, combined with low birthrates, translate into a generally greater capacity to provide not only high levels of health, education, and nutrition but also jobs, as those children grow up and join the workforce. Whether these opportunities are equitably distributed among individual members of the population remains to be seen. It is important to note that age-sex pyramids can be constructed at any level from the national to the neighborhood.

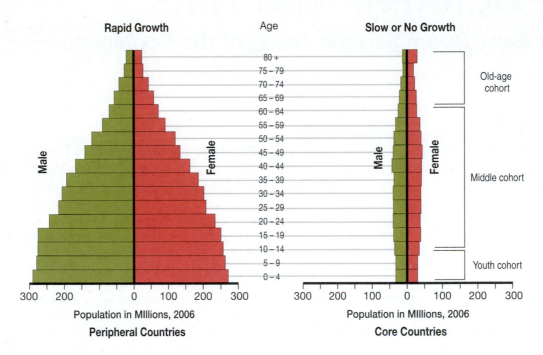

FIGURE 3.8 Population Pyramids of Core and Peripheral Countries Population pyramids vary with the age and sex structure of the population being depicted. We can derive important information about the population growth rates of different countries over time by analyzing changes in the numbers of people in each sex and age category.

TABLE 3.2 Baby Boomer Population Structure

The baby boom demographically dominated the last half of the twentieth century, but its influence will begin to wane in the first half of the new one.

Year	% of Population Who Are Baby Boomers	Age
1990	30	25–44
2000	20	35–59
2020	15	55–79
2040	7	75–85+

Table 3.2 and **Figure 3.9** show the present state and potential future impact of the baby boom cohort, the largest population cohort in U.S. demographic history. Figure 3.9 provides a series of pyramids that illustrate how the configuration changes as the boomers age. The narrower column of younger people rising below the boomer cohort in these pyramids reveals the biggest problem facing this population: a significantly smaller cohort moving into its main productive years having to support a growing cohort of aging and decreasingly productive boomers (**Figure 3.10**). Congressional fights over Social Security, the National Health Care Act, and Medicare funding are only the tip of the iceberg with regard to this problem (see Box 3.2, "Geography Matters: The Baby Boom and the Aging of the Population").

As you might guess based on these discussions, a critical aspect of the population pyramid is the **dependency ratio**, which is a measure of the economic impact of the young and old on the more economically productive members of the population. In order to assess this relation of dependency in a particular population, demographers divide the total population into three age

cohorts, sometimes further dividing those cohorts by sex. The **youth cohort** consists of those members of the population who are less than 15 years of age and generally considered to be too young to be fully active in the labor force. The **middle cohort** consists of those members of the population aged 15 to 64, who are considered economically active and productive. Finally, the **old-age cohort** consists of those members of the population aged 65 and older, who are considered beyond their economically active and productive years. By dividing the population into these three groups, it is possible to obtain a measure of the dependence of the young and old upon the economically active and the impact of the dependent population upon the independent (see Figure 3.8).

APPLY YOUR KNOWLEDGE Why do researchers divide the population of a country into youth, middle, and old-age cohorts? What do these categories indicate about the potential of a country's population? ■

POPULATION DYNAMICS AND PROCESSES

In order to arrive at an understanding of population growth and change, experts look first at two significant factors: fertility and mortality. Birth and death rates are important indicators of a region's level of development and its place within the world economy. To understand population growth overall, however, they must also look at the movement of the population. A simple equation for calculating population growth is $G = B - D + (I - E)$, where G (growth) equals B (births) minus D (deaths)

The Baby Boom and the Aging of the Population

The baby boom generation—those individuals born between 1946 and 1964 in the United States—has been described as "one of the most powerful and enduring demographic influences on this nation." It has certainly been one of the most studied, analyzed, vilified, and sanctified. The case of the baby boom generation provides a useful way of understanding the complex factors that shape demographic change.

Why so many births between 1946 and 1964? Most demographers cannot give a definitive answer to this question. In fact, in the 1940s, population experts were predicting that the American population would stop growing. And although increases in births following a war are expected, the extended surge in births that followed the end of the Second World War came as a surprise to population experts, policymakers, government officials, and most others. Demographic transition theory would have led demographers to anticipate declining births as urbanization and "modernization" accelerated after the war. Sociological theories and predictions based on past trends were not very useful, however, in anticipating the unprecedented phenomenon of the baby boom. To understand it, then, we need to examine a whole host of factors.

Demographic Factors

Demographers insist that the baby boom not be seen as a direct or indirect result of the end of the war. In fact, although marriages (and divorces) did increase dramatically after the war, the accompanying rise in birthrates accounts only for the time from 1946 to 1947. By 1950, births had actually fallen slightly (**Figure 3.C**).

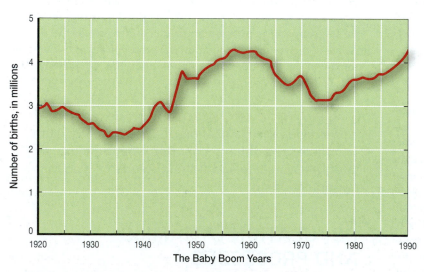

FIGURE 3.C U.S. baby boom crude birthrate The chart very clearly shows the big increase in births following the Second World War. (*Source:* Adapted from L. Bouvier and C. De Vita, "The Baby Boom: Entering Midlife," *Population Bulletin* 46 (3), 1991; and K. Kinsella and V. A. Velkoff, *An Aging World: 2001,* U.S. Census Bureau, Series P95/011, Washington, DC, U.S. Government Printing Office, 2001.)

In 1951, the birthrates once again began to climb and continued to be high until 1964.

Political and Economic Factors

It is significant that the end of the Second World War witnessed a phenomenal expansion in the U.S. economy. Initially stimulated by the war, the expansion was extended by important transformations in transportation and technology. Additionally, government expanded and created programs for education and housing that helped returning veterans start married life with property and the opportunity to improve their economic status by attending college. Transportation—especially the construction of the interstate highway system—helped to fuel **suburbanization**, which meant growth in the building industry, in automobile manufacturing, and in the manufacturing of durable goods for the home. Not surprisingly, demand for labor was high in a growing economy, and young people, most of whom had attained a higher level of education than the previous generation, were able to obtain good jobs with relatively high wages, decent benefits, and reasonable prospects for promotion.

The Aging of the Population

It is important to recognize that something like the baby boom occurred not only in the United States but also in most of the core countries of the world and even in parts of the periphery. The result is that there is a very large cohort of individuals worldwide, currently in their late forties and early to mid-sixties that has had and will continue to have tremendous impacts on the rest of the population, especially as they enter their sixties, seventies, and eighties. In fact, the most fundamental demographic transformation of the twenty-first century is the aging of the population (**Figure 3.D**). In the United States, this is a result not only of the aging of the baby boom cohort but also of longer life expectancies and falling fertility rates, which are also falling all across the globe. In the year 2005, the median age of the world's population was 28, but by 2050 it is expected to increase to 38.

The aging of the population is currently most pronounced in the core countries, where in 1998 the number of older persons (60 and over) actually exceeded the number of children (under 15) for the first time. As a result, for the core countries the median age of the population will rise from 37.5 in 2000 to 45.6 in 2050. In many cases there will be a resulting decrease in national populations. In most cases, populations that are getting both older and smaller are also likely to experience a decrease in the ratio of persons of working age (15–64) to older persons (65 and older); that is, there will be

FIGURE 3.D Active older women in the United States
In core countries, men and women are living longer through life-style changes and improved health care.

fewer working-age people to support the needs of the aging population. What this means is that for countries like Spain, Italy, Japan, and Germany, just to name a few, the prospect of smaller and older populations raises some very serious concerns about employment, economic growth, health care, pensions, and social support services. One possible "solution" to the problem of the aging of the core's population that could have immediate effects is increasing immigration opportunities for young migrants from countries and regions that are experiencing rapid population growth.

And yet, although the aging of the population has been slower in peripheral countries because of high twentieth-century birthrates, the next 50 years will see the median age rise there as well. By 2050, the median age in countries of the periphery is expected to be 36.7 years, up from 24.4 years in 2000. This is actually a bigger jump than in the core. East and Southeast Asia are the regions whose populations are aging most rapidly, with Africa aging the least.

In the United States, the aging of the population is expected to have increasing impacts on the economy, politics, society, and culture. A significant difference of opinion exists among population experts on the future impacts of aging baby boomers. Some see all these highly educated workers as continuing to stimulate the economy in terms of improved productivity and increasing

innovations. Others worry that this cohort will eventually place unfulfillable demands on the nation's social service system. What we do know for certain is that by 2030 the oldest surviving boomers will be age 84 and the youngest will be 65. At that point, a projected 65 million people in the United States will be over the age of 65, compared with roughly 30 million today.

The Impacts on Younger Americans

Demographers, in addition to identifying the existence of the baby boom generation, have also demonstrated the presence of a "baby bust" generation. Two ways exist to determine the extent of this cohort. One is through annual birthrates; if we use this measure, this generation includes those individuals born between 1965 and 1980, when births were below the baby boom average of 4 million births per year. A second way to determine the extent of the baby bust cohort is in terms of the total fertility rate (TFR). Between 1947 and 1964, the TFR ranged between 3.1 and 3.7, well above replacement level. In 1965, however, the TFR dropped below replacement level for the first time in U.S. history, and in 1976 it hit an all-time low of 1.7. Between 1977 and 1987, the TFR remained pretty stable at 1.8, and in 1990 it climbed to 2.09—close to replacement level, marking the end of the baby bust generation.

The term *baby bust* refers to the demographic features of a cohort; cultural terms have also been offered to describe this group as well as groups that have followed it. Those persons born between 1965 and 1975 have also been called Generation X. The cultural label placed on Generation Xers is "slackers," because some of the most famous of this birth cohort, such as Kurt Cobain of the band Nirvana, were associated with grunge music and anti-establishment attitudes.

Those individuals born after the Xers—between 1982 and 1994—are sometimes called Generation Y, the Echo Boomers, or the iGeneration. The Net Generation or Generation Z is the name for people born between the early to mid 1990s and the early 2000s. As their dates make clear, Generations Y and Z are not part of the baby bust cohort. They represent a small rise, or a "boomlet," in birthrates that "echoes" the baby boom. They also represent the first generation born "digital," that is, into a society wired into the Internet.

Generations X, Y, and Z will, of course, be the ones that have eventually to support the boomers as they move toward retirement and beyond. The tremendous size of the baby boom cohort is likely to affect the career and job mobility of members of these generations, as well as labor costs, which will affect taxes, health care, and other benefit costs. Older workers will be more costly for their employers as their salaries and benefits increase in line with their seniority.

The actual impact that the baby boom generation will have into the first half of the twenty-first century remains to be seen. What is clear, however, is that it carries the future of the nation's economic and social security along with it.

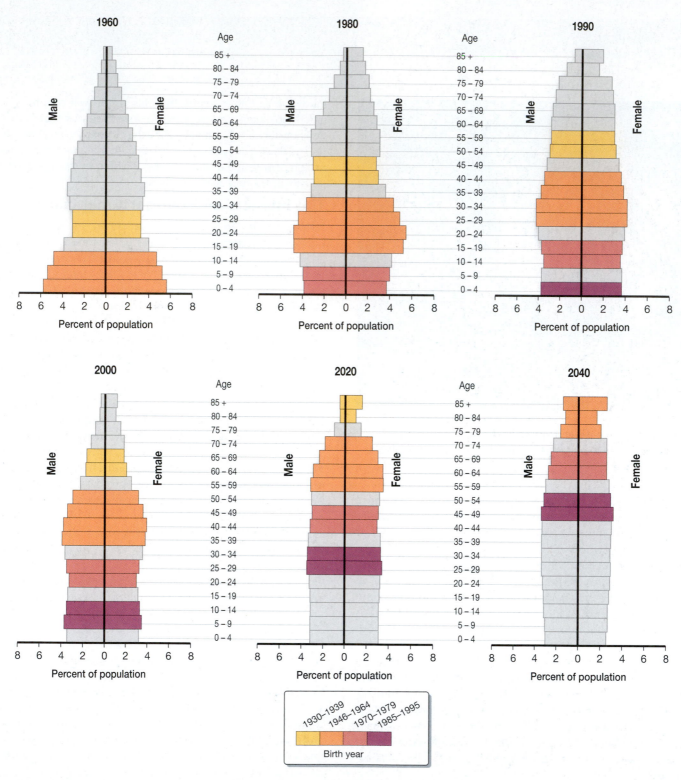

FIGURE 3.9 Population pyramids of U.S. baby boomers, 1960–2040 Population pyramids may be constructed based on estimates of the future. In this series of pyramids, we can clearly see the progression of the baby boomers, shown in orange, up the pyramid as their cohort ages. Note how the "pyramid" becomes a column by the year 2040 as birthrates remain below death rates for each cohort. Note also the significantly higher number of women compared with men in the oldest age group for this same year, reflecting the statistical tendency for women to live longer than men. (Adapted from "Twentieth Century U.S. Generations" *Population Bulletin*, 64(1), 2009, p. 5.)

FIGURE 3.10 The Net Generation Members of the Net Generation, people who are currently in their late teens to late twenties, are faced with the daunting burden of having to help support a huge, aging, baby boom population.

plus I (immigration) minus E (emigration). We look at each of these key population dynamics in turn.

Birth (or Fertility) Rates

The **crude birthrate (CBR)** is the ratio of the number of live births in a single year for every thousand people in the population. The crude birthrate is indeed crude, because it measures the birthrate in terms of the total population and not with respect to a particular age-specific group or cohort. For example, as of 2005, the CBR of the entire U.S. population was 14.0, while the birthrate for Asian American women aged 20 to 24 was 72.0 and for Hispanic women in the same age category 184.6. Clearly, differences exist when we look at specific groups and especially at age and sex cohorts at their reproductive peak.

Although the level of economic development is a very important factor shaping the CBR, other, often equally important, influences also affect it. In particular, it may be heavily affected by the demographic structure of the population, as graphically suggested by age-sex pyramids. In addition, as we mentioned previously in this chapter, an area's CBR is influenced by women's educational achievement, religion, social customs, and diet and health, as well as by politics and civil unrest. Most demographers also believe that the availability of birth-control methods is critically important to a country's or region's birthrate. A world map of the CBR (**Figure 3.11**) shows high levels of fertility in most of the periphery of the world economy and low levels of fertility in the core. The highest birthrates occur in Africa, the poorest region in the world.

The crude birthrate is only one indicator of fertility and in fact is somewhat limited in its usefulness, telling very little about the potential for future fertility levels. Two other indicators formulated by population experts—the total fertility rate and the doubling time—provide more insight into the potential of a population. The **total fertility rate (TFR)** is a measure of the average number of children a woman will have throughout the years that demographers have identified

as her childbearing years, approximately ages 15 through 49 (**Table 3.3**). Whereas the CBR indicates the total number of births per 1,000 people in a given year, the TFR is a more predictive measure that attempts to portray what birthrates will be among a particular cohort of women over time. A population with a TFR of slightly higher than 2 has achieved replacement-level fertility. This means that birthrates and death rates are approximately balanced and there is stability in the population.

Closely related to the TFR is the doubling time of the population. The **doubling time**, as the name suggests, is a measure of how long it will take the population of an area to grow to twice its current size. A country whose population increases at 1.8 percent per year will have doubled in about 40 years. In fact, world population is currently increasing at this rate. By contrast, a country whose population is increasing 3.18 percent annually will double in only 22 years—the doubling time for Kenya. Birthrates and the population dynamics we can project from them, however, tell us only part of the story of the potential of the population for growth. We must also know the death (mortality), rates.

Death (or Mortality) Rates

Countering birthrates and shaping overall population numbers and composition is the **crude death rate (CDR)**, the ratio of the number of deaths in one year to every thousand people in the population. As with crude birthrates, crude death rates often roughly reflect levels of economic development—countries with low birthrates generally have low death rates (**Figure 3.12**).

Although often associated with economic development, CDR is also significantly influenced by other factors. A demographic structure with more men and elderly people, for example, usually means higher death rates. Other important influences on mortality include health care availability, social class, occupation, and even place of residence. Poorer groups in the population have higher death rates than the middle class. In the United States, coal miners have higher death rates than schoolteachers and urban areas often have higher death rates than rural areas. The difference between

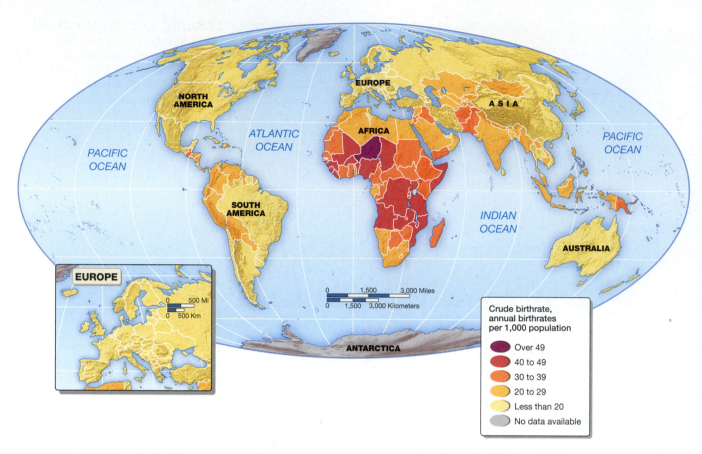

FIGURE 3.11 World crude birthrates, 2011 Crude birthrates and crude death rates are often indicators of the levels of economic development in individual countries. For example, the CBR of Australia offers a stark contrast to that of Ethiopia, a very poor and underdeveloped peripheral country. (Data from Population Reference Bureau, *World Population Data Sheet, 2011.*)

TABLE 3.3 Total Fertility Rates for Selected Countries, 2011

Country	Total Fertility Rate (TFR)
Afghanistan	6.3
Argentina	2.4
China	1.5
India	2.6
Angola	5.7
Mauritius	1.5
Spain	1.4
Russia	1.6
United States	2.0

Source: Population Reference Bureau, *World Population Data Sheet,* 2011.

the CBR and CDR is the rate of **natural increase**—the surplus of births over deaths—or the **natural decrease**—the deficit of births relative to deaths (**Figure 3.13**).

Death rates can be measured for sex and age cohorts; one of the most common measures is the **infant mortality rate**. This figure reflects the annual number of deaths of infants under 1 year of age compared to the total number of live births for that same year. The figure is usually expressed as number of deaths during the first year of life per 1,000 live births. The infant mortality rate has been used by researchers as an important indicator both of the adequacy of a country's health care system and of the general population's access to health care. Global patterns show that infant mortality rates are high in the peripheral countries of Africa and Asia and low in the more developed countries of Europe and North America (**Figure 3.14**). Generally, the core's low rates reflect adequate maternal nutrition and the wider availability of health care resources and personnel.

However, when patterns are examined at the level of countries, regions, and cities, infant mortality rates are not uniform. In the United States, for example, African Americans, as well as other ethnic minorities in urban and rural areas, suffer infant mortality rates that are twice as high as the national average. In east central Europe, Bulgaria has a 9.7 per-thousand infant mortality rate, while the Czech Republic has a rate of 3.3 per thousand. In Israel, the infant mortality rate per thousand is 3.9 and in the neighboring Palestinian Territories it is 25. And when war is introduced into the equation, the infant mortality rate skyrockets. Iraq has 84 infant deaths per 1,000 live births; the rate for surrounding countries such as Saudi Arabia is 18 and in Syria it is 16.

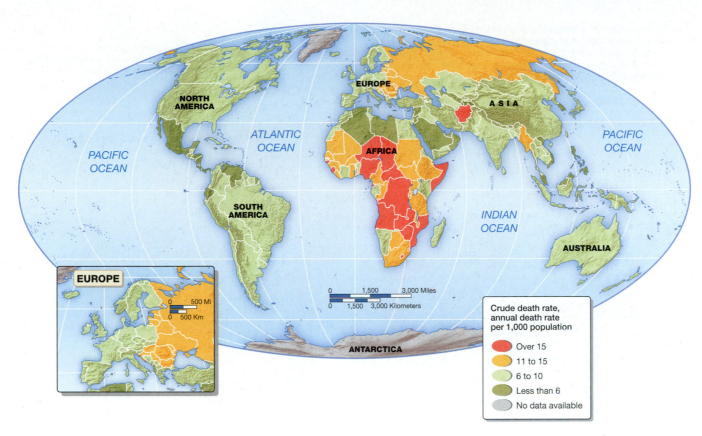

FIGURE 3.12 World crude death rates, 2011 The global pattern of crude death rates varies from crude birthrates. Most apparent is that the difference between highest and lowest crude death rates is relatively smaller than the difference for crude birthrates, reflecting the impact of factors related to the middle phases of the demographic transition. (Data from Population Reference Bureau, *World Population Data Sheet, 2011.*)

Afghanistan's infant mortality rate is 155 per 1,000 live births; its neighbors Iran and Uzbekistan have 29 and 48, respectively. Global patterns often mask regional, local, and even neighborhood variation in mortality rates for both infants and other population cohorts.

Related to infant mortality and the crude death rate is **life expectancy**, the average number of years an infant newborn can expect to live. Not surprisingly, life expectancy varies considerably from country to country, region to region, and even from place to place within cities and among different classes and racial and ethnic groups. In the United States, a child entering the first grade in 2012 can expect to live more than 78 years. If we begin to specify the characteristics of that infant by sex and race, however, variation emerges. An African American male entering the first grade has a life expectancy of 69.5 years, while a 6-year-old Anglo American female can expect to live, on average, 81 years.

Another key factor influencing life expectancy is epidemics, which can quickly and radically alter population numbers and composition. In our times, epidemics can spread rapidly over great distances, largely because people and other disease carriers can travel from one place to another rapidly. Epidemics can have profound effects, from the international to the local level, and reflect the increasing interdependence of a shrinking globe. They may affect different population groups in different ways and,

depending on the quantity and quality of health and nutritional care available, may have a greater or lesser impact on different localities.

One of the most widespread epidemics of modern times is HIV/AIDS (human immunodeficiency virus/acquired immunodeficiency syndrome). The disease is a serious problem in regions ranging from Southeast Asia to sub-Saharan Africa and also affects certain populations in many core countries of Europe as well as the United States. In the United States, for example, HIV/AIDS first arose largely among male homosexuals and intravenous drug users who shared needles. Geographically, early concentrations of AIDS occurred in places with high concentrations of these two subpopulations. It has had perhaps the most severe impact in inner-city areas but has cropped up in every region of the United States, increasingly appearing in the male and female heterosexual population. The rate for blacks is roughly 8 times the rate for whites (67.7 per 100,000 versus 8.2 per 100,000). African American males continue to bear the greatest burden of HIV/AIDS infection.

The pattern of the disease is different, however, in central Africa, where it is overwhelmingly associated with heterosexual, nondrug users and affects both sexes equally. The geographical diffusion of HIV/AIDS in Africa has occurred along roads, rivers, and coastlines, all major transportation routes associated with

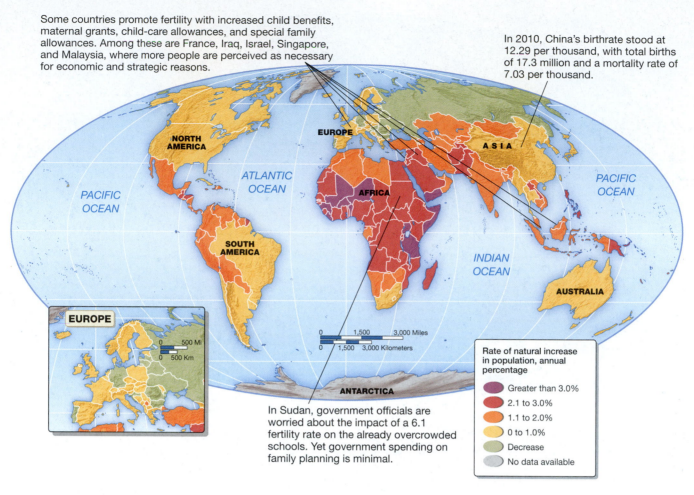

Some countries promote fertility with increased child benefits, maternal grants, child-care allowances, and special family allowances. Among these are France, Iraq, Israel, Singapore, and Malaysia, where more people are perceived as necessary for economic and strategic reasons.

In 2010, China's birthrate stood at 12.29 per thousand, with total births of 17.3 million and a mortality rate of 7.03 per thousand.

In Sudan, government officials are worried about the impact of a 6.1 fertility rate on the already overcrowded schools. Yet government spending on family planning is minimal.

Rate of natural increase in population, annual percentage

- Greater than 3.0%
- 2.1 to 3.0%
- 1.1 to 2.0%
- 0 to 1.0%
- Decrease
- No data available

FIGURE 3.13 World rates of natural increase, 2011   As the map shows, rates of natural increase are highest in sub-Saharan Africa, the Middle East, and parts of Asia, as well as parts of South and Central America. Europe, the United States, and Canada, as well as Australia and parts of central Asia and Russia have slow to stable rates of natural increase. (Data from Population Reference Bureau, *World Population Data Sheet, 2011.*)

regional marketing systems. The central African nations, including Congo (formerly Zaire), Zambia, Uganda, Rwanda, and the Central African Republic, have been hard-hit. In sub-Saharan Africa, however, an estimated 22.5 million people are infected with the virus (**Figure 3.15**). The impact is worst in urban areas, though no area has been immune to the disease's spread.

Medical geographers have made important contributions to the study of the diffusion of HIV/AIDS. **Medical geography** is a subarea of the discipline that specializes in understanding the spatial aspects of health and illness. This spatial perspective includes disease mapping as well as the distribution and diffusion of health and illness. The importance of medical geography lies in the fact that the landscape of disease that we are seeing today is changing. Diseases we thought were eradicated, such as tuberculosis, are coming back; diseases that were previously unknown are emerging, such as ebola; and diseases that have maintained geographic limitations, such as dengue hemorrhagic fever, are spreading.

Although no cure yet exists, some countries such as Finland have successfully slowed the spread of HIV/AIDS through an intensive public relations campaign and the provision of top-notch

health services to all its citizens. In 2000, activists in Durban, South Africa, were effective in forcing several large pharmaceutical companies to slash the price of HIV/AIDS drugs for victims of the disease. The downside of this important success is that as the numbers in treatment expand, the virus will develop resistance to the current drugs and the need for new drugs will grow. World Trade Organization intellectual property rules mean that countries like India that now produce cheap generic versions of the drugs will not be able to legally provide affordable generic versions of the new patented drugs and the high cost of treatment will resume.

APPLY YOUR KNOWLEDGE   How does the level of wealth of a country affect its ability to respond to health issues like HIV/AIDS, infant mortality, or life expectancy in general? ■

Demographic Transition Theory

Many demographers believe that fertility and mortality rates are directly tied to the level of economic development of a country, region, or place. Pointing to the history of demographic change in

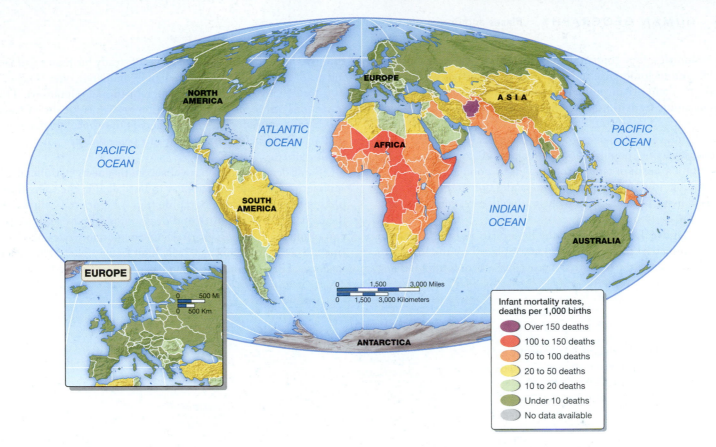

FIGURE 3.14 World infant mortality rates, 2011 The geography of poverty underlies the patterns shown in this map and allows us to analyze the linkages between population variables and social conditions. Infant mortality rates generally parallel crude death rates, with sub-Saharan Africa generally reporting the highest rates. These rates reflect a number of factors, including inadequate or completely absent maternal health care as well as poor nutrition for infants. (Data from Population Reference Bureau, *World Population Data Sheet, 2011.*)

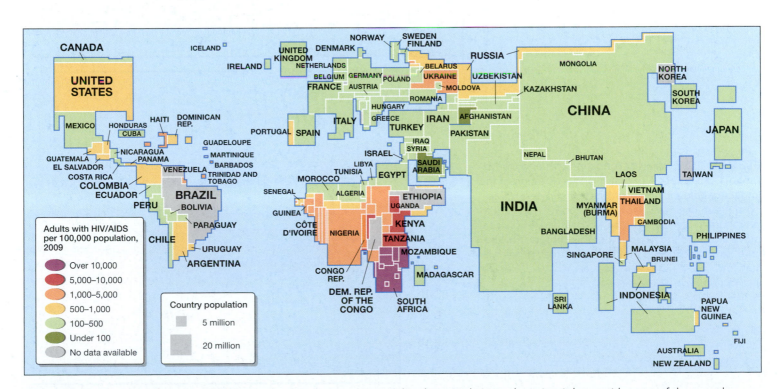

FIGURE 3.15 HIV infection, 2010 HIV infections are concentrated in the periphery and semiperiphery, with most of the people with HIV/AIDS living in Asia and sub-Saharan Africa. Compare the cartograms for Africa with those of North America, Europe, and China. Campaigns in some African countries, especially in Zimbabwe and Uganda, have helped to stem the number of new cases and ultimately the number of deaths. (*Source:* From H. Veregin (ed.), *Goode's World Atlas,* 22nd edition, Chicago, IL: Rand McNally & Co, 2010, p. 54. Updated data from *CIA World Factbook* (percentage of adults (15–49) living with HIV/AIDS. Estimated for year 2009), https://www.cia.gov/library/publications/the-world-factbook/fields/2156.html#ar)

core countries, they contend that many of the economic, political, social, and technological transformations associated with industrialization and urbanization lead to a demographic transition. A **demographic transition** is a model of population change in which high birth and death rates are replaced by low birth and death rates. Once a society moves from a preindustrial economic base to an industrial one, population growth slows. According to the demographic transition model, the slowing of population growth is attributable to improved economic production and higher standards of living brought about by better health care, education, and sanitation.

As **Figure 3.16** illustrates, the high birth and death rates of the preindustrial phase (Phase 1) are replaced by the low birth and death rates of the industrial phase (Phase 4) only after passing through the critical transitional phase of steady birthrates and falling death rates (Phase 2) and then more moderate rates (Phase 3) of natural increase (increase through birth, not migration). This transitional phase of rapid population growth is the direct result of early and steep declines in mortality at the same time that fertility remains at levels characteristic of a place that has not yet been industrialized.

Some demographers have observed that many peripheral and semiperipheral countries appear to be stalled in the transitional phase—caught in a "demographic trap." **Figure 3.17** illustrates the disparity between birth and death rates for core and peripheral countries. Despite a sharp decline in mortality rates, most peripheral countries retain relatively high fertility rates. The reason for this lag in declining fertility rates relative to mortality rates is

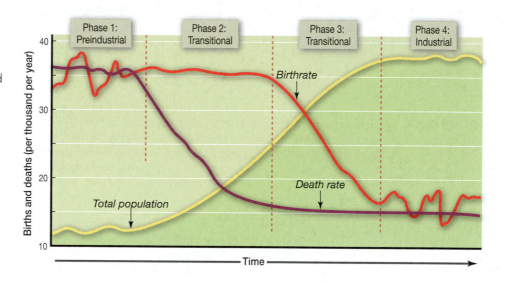

FIGURE 3.16 Demographic transition model The transition from a stable population based on high birth and death rates to one based on low birth and death rates progresses in clearly defined stages, as illustrated by this graph. With basic information about a country's birth and death rates, it is possible to identify that country's position within the demographic transition process. Population experts disagree about the usefulness of the model.

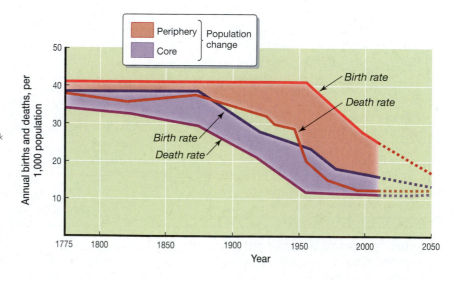

FIGURE 3.17 World trends in birth and death rates, 1775–2050 The graph is an illustration of the impact of affluence on reproductive choices. It also portrays an optimistic view of the future, in which, for peripheral regions birthrates will continue to fall as we move closer to the midpoint of this century. For the core, the projection is that birth and death rates will stay low. (Adapted from T. Allen and A. Thomas, *Poverty and Development in the 1990s.* London: Oxford University Press, 1992; updates from *CIA World Factbook 2011*, https://www.cia.gov/library/publications/the-world-factbook/geos/xx.html.)

TABLE 3.4 Birth and Death Rates for England and Scotland, 1870–1920

The CBRs and CDRs for England and Scotland for the period between 1870 and 1920 illustrate countries moving between Phase 2 and Phase 3 of the demographic transition, in which death rates are lower than birthrates. England and Scotland are clear examples of the way in which the demographic transition has been theorized to operate in the core countries. During the 50-year period covered in the table, both countries were completing their transformation as key industrial regions.

	1870	1880	1890	1900	1910	1920
Crude Birthrate						
England	35.2	35.4	32.5	29.9	27.3	22.7
Scotland	35	34.8	32.3	30.2	28.4	24
Crude Death Rate						
England	22.5	21.4	19.2	18.2	15.4	14.6
Scotland	22.1	21.6	19.2	18.5	16.6	15.3

Source: M. Anderson and D. Morse, "High Fertility, High Emigration, and Low Nuptuality: Adjustment Processes in Scotland's Demographic Experience, 1861–1914, Part I *Population Studies*, 47, 1993, p. 8.

that while societies have developed new and more effective methods for fighting infectious diseases, social attitudes about the desirability of large families are only recently and incrementally changing. **Table 3.4** provides a historical illustration.

Although the demographic transition model is based on actual birth and death statistics, many population geographers and other population experts question its applicability to all places and all times. Although the model adequately describes the history of population change in the core countries, it appears less useful for explaining the demographic history of countries and regions in the periphery. Its significance has therefore been contested. Among other criticisms is that industrialization—which, according to the theory, is central to moving from Phase 2 to Phases 3 and 4—is seldom domestically generated in the peripheral countries. Instead, foreign investment seems to drive peripheral industrialization. As a result, the features of demographic change, such as higher living standards, witnessed in the core countries, where industrialization was largely a result of internal capital investment, have not been as dramatic in many peripheral countries. Other critics of the demographic transition model point to several factors undermining a demographic transition fueled by economic growth in peripheral countries: the shortages of skilled laborers, the absence of advanced educational opportunities for all members of the population (especially women), and limits on technological advances. In other words, while a demographic transition may be a characteristic experience of the core regions of the globe, its applicability to the periphery needs to be premised on a different set of factors.

APPLY YOUR KNOWLEDGE Give an example of why a country might be concerned about its national population being too small. How could this affect its population characteristics? What might a country gain by increasing the birthrate? What might it lose? ■

POPULATION MOVEMENT AND MIGRATION

In addition to the population dynamics of death and reproduction, the third critical influence on population is the movement of people from place to place. Individuals may make far-reaching international or intraregional moves, or they may simply move from one part of a city to another. For the most part, mobility and migration reflect the interdependence of the world-system. For example, global shifts in industrial investment result in local adjustments to those shifts as populations move or remain in place in response to the creation or disappearance of employment opportunities.

Mobility and Migration

One way to describe such movement is by the term **mobility**, the ability to move from one place to another, either permanently or temporarily. Mobility may be used to describe a wide array of human movement, ranging from a journey to work (for example, a daily commute from suburb to city or suburb to suburb) to an ocean-spanning permanent move.

The second way to describe population movement is in terms of **migration**, a long-distance move to a new location. Migration involves a permanent or temporary change of residence from one neighborhood or settlement to another. Moving from a particular location is defined as **emigration**, also known as out-migration. Moving to a particular location is defined as **immigration** or in-migration. For example, a Russian who moves to Israel emigrates from Russia and immigrates to Israel. This type of move, from one country to another, is termed **international migration**. Moves may also occur within a particular country or region, in which case they are called **internal migration**. Both permanent and temporary changes of residence occur for many reasons but most often involve a desire for economic betterment or an escape from adverse political conditions, such as war or oppression.

Governments are concerned about keeping track of migration numbers and rates, as well as the characteristics of the migrant populations, because these factors can have profound consequences for political, economic, and cultural conditions on national, regional, and local levels. For example, a peripheral country that has experienced substantial out-migration of highly trained professionals, such as the Philippines or Kenya, may find it difficult to provide needed services such as health care. Benefiting from low labor costs are countries that have received large numbers of low-skilled in-migrants willing to work for extremely low wages, such as the United States, Germany, and Israel. These countries may also face considerable social stress in times of economic recession, when unemployed citizens begin to blame the immigrants for "stealing" their jobs or receiving welfare benefits.

Demographers have developed several calculations of migration rates. The in-migration and out-migration rates provide the foundation for gross and net migration rates for an area under study. **Gross migration** refers to the total number of migrants moving into and out of a place, region, or country. **Net migration** refers to the gain or loss in the total population of that area as a result of the migration.

Migration rates, however, provide only a small portion of the information needed to understand the dynamics of migration and its effects from the local to the national level. In general terms, migrants make their decisions to move based on push factors and pull factors. **Push factors** are events and conditions that impel an individual to move from a location. They include a wide variety of possible motives, from the idiosyncratic, such as dissatisfaction with the amenities offered at home, to the dramatic, such as war, economic dislocation, or ecological deterioration. **Pull factors** are forces of attraction that influence migrants to move to a particular location.

Usually the decision to migrate is a combination of both push and pull factors, and most migrations are voluntary. In **voluntary migration** an individual chooses to move. Where migration occurs against the individual's will, push factors can produce **forced migration**. Oftentimes, the decision to migrate is a mixed one reflecting both forced and voluntary factors. Forced migration (both internal and international) remains a critical problem in the contemporary world. These migrants may be fleeing a region or country for many reasons, but some of the most common are war, famine (often war-induced), life-threatening environmental degradation or disaster, or governmental coercion or oppression. The terrorist attacks on the World Trade Center and the Pentagon on September 11, 2001, are believed to have caused large refugee movements not only in Central Asia, especially Afghanistan, but also globally. The earthquake in Japan in March 2011 also caused both natives and foreigners to leave the city in large numbers in response to radiation fears and rolling blackouts caused by the damaged Fukushima nuclear plant. Civil strife and general political unrest and state violence also caused refugees to flee unstable situations in Libya, Tunisia, Egypt, and other Middle Eastern countries during 2011. And while **refugees**—individuals who cross national boundaries to seek safety and asylum—are a significant global problem, **internally displaced persons (IDPs)**—the number of individuals who are uprooted within the boundaries of their own country because of conflict or human rights abuse—is also growing globally (see Box 3.3, "Window on the World: Internal Displacement").

APPLY YOUR KNOWLEDGE Identify three push and pull factors that shape the decisions by college students to migrate for employment opportunities. Do you suspect that you will have to migrate after graduation? If so, will you be doing it out of necessity or choice? Please provide specific reasons for both the pull and push factors. ■

International Voluntary Migration

Migration does not always involve force or even a permanent change of residence. Voluntary migration can occur for any number of reasons, such as high wage differentials between places, better experience and job opportunities elsewhere, family links abroad, or local underemployment or unemployment conditions (**Figure 3.18**). Temporary labor migration has long been an indispensable part of the world economic order and has at times been actively encouraged by governments and companies alike. Individuals who migrate temporarily to take up jobs in other countries are known as **guest workers**. Sending workers abroad is an important economic strategy for many peripheral and semiperipheral countries; it lessens local unemployment and also enables workers to send substantial amounts of money to their families at home. This arrangement helps to supplement the workers' family income and supports the dominance of the core in global economic activities. Receiving countries who have developed programs to facilitate the immigration of guest workers gain the many benefits of a workforce they often otherwise would not have.

In the United States, the most controversial form of international voluntary migration is that of **undocumented workers**, those individuals who arrive in the country without official entry visas and are considered by the government to be in the country illegally (**Figure 3.19**). Most of these workers come to the United States from Latin America (the largest share from Mexico). Other migrants who come to the United States with appropriate visas but then fail to return to their country of origin when their visas expire are also considered to be undocumented. Since the United States was founded, the immigration of working-age individuals has been crucial to its growth; it has also been a periodic source of conflict.

The most recent opposition to undocumented migration has been the response to the largest wave of immigration since the 1920s into the United States where for the first time, undocumented immigrants outnumbered documented ones. The number of undocumented immigrants peaked at an estimated 11.9 million in 2008. A 2010 study shows that the figure dropped to about 11.1 million in 2009, the first clear decline in two decades. It is important to keep in mind that 2008 saw the beginning of the so-called "**Great Recession.**" The Great Recession is the late-2000s, the weak economic growth that has followed it, and the related ongoing financial crisis, which was only beginning to show weak signs of abating in early 2011.

Economic historians have shown that anti-immigrant sentiment peaks when times are hard and declines when the economy is booming. The troubled state of the U.S. (and global) economy is likely a strong reason why anti-immigrant feelings are so widespread in the country and worldwide. In April 2010, Arizona passed into state law H.B. 1070, augmenting federal immigration law by requiring certain noncitizens to register with the U.S.

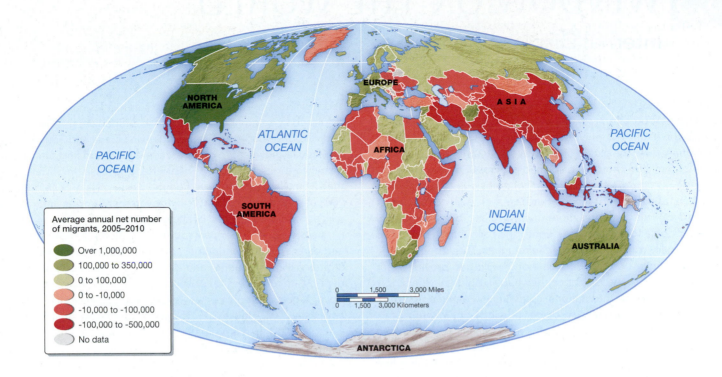

FIGURE 3.18 Global voluntary migration This snapshot map shows those countries that are largely receiving and those that are largely sending migrants. In 2011, 214 million people, over 3 percent of the world's population, lived outside their country of origin. (*Source:* United Nations Dept. of Economic and Social Affairs, Population Division, *International Migration 2009.* www.unmigration.org, www.unpopulation.org.)

FIGURE 3.19 U.S./Mexico border activism The mission of organizations like the Samaritans, shown here, is to end immigration-related death and suffering on the U.S./Mexico border. It is the conviction of this organization that people of conscience must work openly and in community to uphold fundamental human rights.

government and making it illegal for a noncitizen to be in Arizona if he or she is not carrying the required documents. It also restrains state or local government entities from restricting enforcement of federal immigration laws and cracks down on individuals sheltering, hiring, and transporting illegal aliens. Although the law's most controversial provisions were blocked by a federal judge in July 2010, legislative leaders in at least half a dozen other states have said they will propose bills similar to Arizona's.

An increasingly important category of international migrants are **transnational migrants**, so called because they set up homes and/or work in more than one country. Sometimes these migrants are low-paid workers, as in the case of

While the global humanitarian community has paid close attention to refugee populations for decades, the problem of internally displaced persons has only recently begun to draw international attention. Around the globe, at the end of 2010, there were an estimated 27.5 million internally displaced persons (IDPs)—individuals who were uprooted within their own countries due to civil conflict or human rights violations, sometimes by their own governments (**Figure 3.E**). This number is more than twice as high as the global refugee population and it is frequently the case that the plight of IDPs is actually worse than that of the refugees. This is because the IDPs' governments are either unable or unwilling to provide them with the protection or assistance they have a right to expect. It is also the case that the international community is either unaware of them or has not secured the resources needed to help them.

Two of the world's regions with the most difficult IDP challenges at the end of 2010 were Africa and Latin America. Africa possesses nearly 11.1 million IDPs due to rebel activities and intercommunal violence as well as direct abuse by national governments. National

security forces and government-backed militias have deliberately displaced large numbers of people in Zimbabwe, the Democratic Republic of Congo, and Côte d'Ivoire. In Liberia and Somalia, IDPs had virtually nowhere to go to escape attacks and find safe shelter; many were killed or died of hunger and disease. In Sudan, more than 2.5 million people have fled the western state of Darfur since 2003 as a response to attacks by government troops and raids by militias reportedly backed by the government.

In the South American country of Colombia, both guerrillas and paramilitaries continue to target civilian populations through arbitrary killings, looting, and destruction of property in order to depopulate rural areas for political and economic gains and to control or regain strategic territories, displacing some 2 million people as of early 2006 (**Figure 3.F**). The war and the postconflict situation in Iraq has also had a dramatic impact on Iraqi citizens, displacing nearly 1.2 million people internally in addition to the approximately 2.2 million people who have left the country altogether. Other populations of concern to the UNHCR include Azerbaijan, Somalia,

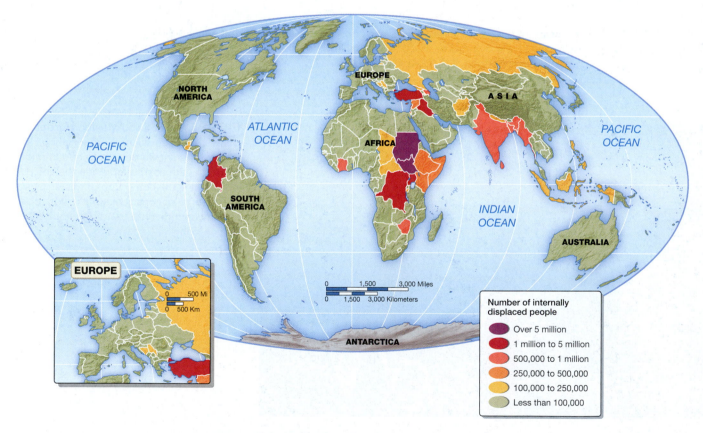

FIGURE 3.E Internally displaced persons worldwide, 2007 This map provides an indication of the geographical extent and numbers of people displaced from their livelihoods and homes but still living within their own countries, oftentimes in temporary camps provided by humanitarian organizations or the UN. As of December 2007, the Internal Displacement Monitoring Center estimated that 27.5 million people had been displaced within their own countries (*Source*: Adapted from Internal Displacement Monitoring Centre, http://www.internal-displacement. org/8025708F004BE3B1/(httpInfoFiles)/9251510E3E5B6FC3C12572BF0029C267/$file/Global_Overview_2006.pdf.)

FIGURE 3.F Resettlement camp, Bogota, Columbia Pictured here is a displaced family camped in a park in the capital city. This family as well as many others also gathered to protest the government's failure to attend to the needs of Columbia's millions of displaced persons.

the country, seeking safe refuge in Tunisia, while few Libyans even attempted to approach the border. With a ban on the entrance of international news organizations into the country, there was no possibility of outside verification of the situation, but the speculation is that fleeing Libyans were being at best turned away and at worst killed at checkpoints throughout the country. Mercenaries and Libyan gunmen loyal to Gaddafi effectively turned citizens into internally displaced persons whose safety was in the hands of hostile forces and who lacked the benefit of outside humanitarian aid.

What is clear is that the ramping up of security globally following the terrorist events of September 11, 2001, is continuing to produce a worldwide displacement crisis by encouraging governments to seek military solutions to conflicts. As a result, international humanitarian and human rights standards, including those relating to the protection of IDPs, have been undermined or ignored altogether in numerous places around the globe (**Table 3.A**). In fact, there is little doubt that the international antiterrorism campaign has enabled some nondemocratic

TABLE 3.A Regional Distribution of IDPs, Refugees, Asylum Seekers, and Stateless Persons, 2009

Region (UN Major area)	Refugees	Asylum seekers	Returned refugees	IDPs protected/ assisted	Returned IDPs	Stateless persons	Various	Total
Africa	2,300,062	436,930	149,480	6,468,788	846,046	100,064	174,197	10,475,567
Asia	5,620,502	67,928	97,584	5,434,532	1,381,234	5,820,357	144,924	18,567,061
Europe	1,628,086	282,714	4,319	420,758	2,260	639,034	92,577	3,069,748
Latin America and Caribbean	367,437	68,785	70	3,303,979	-	118	-	3,740,389
North America	444,895	124,973	-	-	-	-	-	569,868
Oceania	35,558	2,590	-	-	-	-	-	38,148
Total	10,396,540	983,920	251,478	15,628,057	2,229,540	6,559,573	411,698	36,460,806

Source: Data from United Nations High Commission for Refugees, *Statistical Yearbook, 2009.*

Sri Lanka, Serbia and Montenegro, Liberia, Georgia, Bosnia and Herzegovinia, the Russian Federation, and Afghanistan.

At the beginning of 2011, popular uprisings against entrenched government leaders unfolded across North Africa and the Middle East. These have also caused massive population movement and dislocation. Sparked by persistent antigovernment protests in Tunisia, other countries in the region, including Egypt, Algeria, Yemen, Jordan, Syria, Kuwait, Saudi Arabia, and Bahrain, soon followed suit. Although there is massive population dislocation across the region because of this unrest, it is impossible at this time, as events are unfolding both rapidly and unpredictably, to measure the extent of the humanitarian crisis. The worst situation appears to be in Libya, where the de facto chief of state, Colonel Muammar Gaddafi, started a war against his people. Prior to Gaddafi's demise at the hands of Libyan rebel forces in October 2011, thousands of foreign workers from Africa and South Asia fled

governments to characterize armed opposition movements as "terrorists" and to present their counterinsurgency operations as part of an international "war on terror." This posture has earned these governments—many with a long history of instability, military coups, and human rights violations—substantial military support, mainly from the United States. In Indonesia and the Philippines, for example, tens of thousands of people have been displaced because of counterinsurgency operations conducted under the banner of the "war on terror." Importantly, these military campaigns were ongoing before the "war on terror" was launched in 2001. And yet, by invoking the rhetoric of terror, they have, ironically, undermined the protection of civilians, causing them to flee their homes for safe havens elsewhere in the country.

Source: Adapted from *Internal Displacement: A Global Overview of Trends and Developments in 2010,* by the Global IDP Project of the Norwegian Refugee Council, Geneva, Switzerland, 2010.

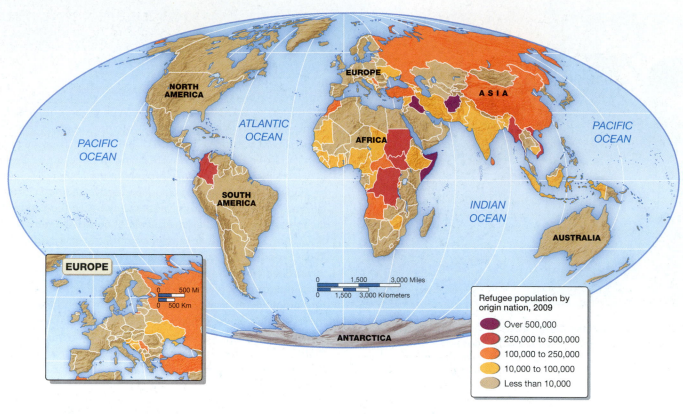

FIGURE 3.20 Refugee-sending countries War is certainly the most compelling factor in forcing refugee migration. Shown are the sending countries, those whose internal situations propelled people to leave. What is perhaps most distressing about this graphic is that refugee populations have increased over the last decade, and almost exclusively so in the periphery. (Data from United Nations High Commission for Refugees, *UNHCR Statistical Yearbook 2009*, page 24. http://www.unhcr.org/4ce5317d9.pdf).

many Mexican immigrant(s) to the United States who take up jobs in the unskilled sectors of manufacturing, agriculture, and the service economy. Other transnational migrants occupy the higher end of the socioeconomic spectrum, such as the Hong Kong Chinese, who have established substantial property and business investments and residences in Canada while maintaining their citizenship status in Hong Kong as well. Transnational migrants seek destinations all over the world, with substantial immigrant(s) labor communities occurring in France, Germany, Saudi Arabia, and South Africa.

International Forced Migration

Forced migration is a worldwide phenomenon (**Figure 3.20**). A recent example of international forced migration is Iraqis fleeing their country during Operation Iraqi Freedom, as **Figure 3.21** shows. Another example is the Lebanese and the Kurds who have also been scattered widely throughout the globe because of war and civil strife. The 1975–1990 civil war and more recent violence in Lebanon have resulted in a particularly large number of emigrations. The failure of the Kurds to establish an autonomous state in the early twentieth century led to their being split among Iran, Iraq, and Turkey, with a small minority in Syria. Many Kurds have moved to different parts of the region or left it altogether because of military aggression, persecution, and the failure of repeated attempts to establish a Kurdish state.

Other prominent examples of international forced migration include the migration of Jews from Germany and Eastern Europe preceding the Second World War and the deportation of Armenians from the Ottoman Empire after the First World War. **Figure 3.22** shows the scattering of Palestinians from their homeland since the establishment of the state of Israel in 1948. The refugees have been fleeing ongoing violence as well as discrimination and land dispossession.

Recently—and especially since September 11, 2001—many European countries, faced with rising numbers of refugees seeking political asylum, have tightened their previously liberal asylum policies. Since September 11, the immigration issue has frequently been reconfigured as a security issue, and national governments across Europe have voted to restrict the conditions under which asylum would be granted. This is despite the fact that many European countries were, until the global recession began in 2008, in need of immigrants to help counter labor shortages in skilled and unskilled jobs. In short, in the last several years, national elections in which xenophobic politicians were either elected or garnered a sizable portion of the vote in normally liberal countries, such as the Netherlands, Switzerland, and France, make it quite clear that Europeans are becoming increasingly resistant to absorbing refugee populations.

In other parts of the world, though, refugee populations are pouring across borders and creating difficult conditions for national governments. The wave of popular uprisings against

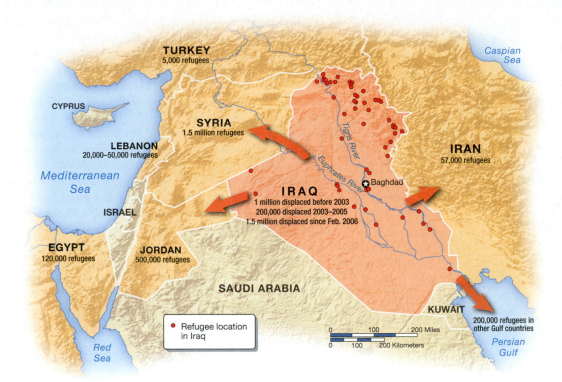

FIGURE 3.21 Iraqi diaspora before 2003 and since Over the last century, increasing numbers of refugees have fled Iraq settling throughout the world. Some left because of persecution, like the Iraqi Jews who fled in 1950–1952. Others left because of wars such a during the First and Second Kurdish Wars in the 1960s and 1970s. The most recent War in Iraq has caused the greatest mass exodus, however, with the United Nations High Commissioner on Refugees estimating the number at 4.7 million (as well as nearly 2 million internally displaced persons) since the 2003 U.S.-led invasion. (*Source:* Data from United Nations High Commission for Refugees. http://www.unhcr.org/487ef7144.pdf.)

decades-old dictatorships that first began in Tunisia in early 2011 has sparked the largest movement of migrants the world has seen since the Second World War. Many of these refugee populations have sought safe passage north to Europe (especially Italy, Spain, France, Greece, and Turkey) as well as south to adjacent counties in Africa south of the Sahel.

Internal Voluntary Migration

One way to understand the geographical patterns of migration is to think in terms of waves of migration. In the United States, for example, three important and overlapping waves of internal migration over the past two centuries altered the population geography. As with a great deal of migration activity, these three major migrations were tied to broad-based political, economic, and social changes.

The first internal migration wave began with colonization and increased steadily through the twentieth century. This wave had two parts. The first was characterized by the large movement of people from the settled eastern seaboard into the interior of the country. This westward expansion, which began spontaneously during the British colonial period—in blatant disregard of British restrictions on such expansion—became official settlement policy after the American Revolution. The federal government encouraged migration over the course of more than a century as part of the country's expansionist strategies (**Figure 3.23**). The second part consisted of massive rural-to-urban migration associated with industrialization, especially occurring during the mid-nineteenth to the early twentieth centuries. **Table 3.5** illustrates how, between 1860 and 2000, the United States was transformed from a rural to an urban society as industrialization created new jobs and as increasingly redundant numbers of agricultural

workers (along with foreign immigrants) moved to urban areas to work in the manufacturing sector.

The second great migration wave, which began early in the 1940s and continued through the 1970s, was the massive and very rapid movement of mostly African Americans out of the rural South to cities in the South, North, and West. Although African Americans formed considerable populations in cities such as Chicago and New York before the onset of this wave, the mechanization of cotton picking pushed additional large numbers of these people out of the rural areas. Tenant cotton picking was a major source of livelihood for African Americans in the Deep South until mechanization reduced the number of jobs available. At the same time, pull factors attracted them to the large cities. In the early 1940s, large numbers of jobs in the defense-oriented manufacturing sector became available as other urban workers joined the war effort. This second wave of migration can be seen as part of a wider pattern of rural-to-urban migration among agricultural workers as industrialization spread globally.

The third internal U.S. migration wave began shortly after World War II and continues into the present. Following the end of the war and directly related to the impact of governmental defense policies and activities on the country's politics and economy, the region of the United States lying below the thirty-seventh parallel, also known as the Sunbelt, emerged as a migration destination. Between 1950 and 1990, this region, which includes 15 states and extends from North Carolina in the east to southern California in the west and Florida and Texas to the south, experienced a 97.9 percent increase in population. Beginning in the late 1970s and early 1980s, the West also began to grow dramatically. At the same time that the West and the South were booming, the Midwest and Northeast, known variously as the Snowbelt, Frostbelt, or Rustbelt, together grew by

only 33.3 percent. **Figure 3.24** is one illustration of the relative decline and disinvestment experienced in the Rustbelt during the 1970s and early 1980s. Since the late 1980s, the Midwest and the Northeast have stabilized and even added population in most cases.

The most compelling explanation for the large-scale population shift characteristic of the third migration wave was the pull of economic opportunity. Rather than being invested in upgrading the aged and obsolescent urban industrial areas of the Northeast and Midwest, venture capital was invested in Sunbelt locations, where cheaper land, lower labor costs, and the absence of labor union organizations made the manufacturing and service-sector activity more profitable. By the 1990 census, there was a decrease in the amount of inmigration to the Sunbelt, but the geography of the U.S. population at the end of the twentieth century was nevertheless almost diametrically opposed to that of 150 years ago. This new population distribution—largely suburban and more evenly spread across the country—illustrates the way in which political and economic transformations play an especially significant role in shaping individual choice and decision making.

Internal Forced Migration

One of the best-known forced migrations in the United States is the "Trail of Tears," a tragic episode in which nearly the entire Cherokee Nation was forced to leave its once treaty-protected Georgia homelands for Oklahoma. **Figure 3.25** illustrates this forced migration as well as the removal of four other eastern tribes to territory west of the Mississippi. Approximately 16,000 Cherokees were forced to march across the continent in the early 1830s, suffering from drought, food scarcity, bitterly cold weather, and sickness along the way. By some estimates at least a quarter of the Cherokees died as a result of the removal.

Placed within the national and international context, the movement of Native American populations during the nineteenth century can be seen as a response to larger political and economic forces. European populations were participating in a massive migration to the United States, and the national economy was on the threshold of an urban-industrial revolution. The eastern Native American populations posed an obstacle to economic expansion, which was dependent upon geographic expansion. Growing Anglo American prosperity, it was believed, had to be secured by taking Indian land.

FIGURE 3.22 Palestinian refugees in the Middle East This map shows the dispersion of Palestinian refugees—in camps and elsewhere—in the states around Israel. One of the biggest sticking points in negotiations between Israelis and Palestinians has been the question of whether refugees will be allowed to return and if so, where they will be allowed to settle, given that most of their land has been occupied by Israeli settlers. (Adapted from *The Guardian*, October 14, 2000, p. 5.)

FIGURE 3.23 Changing demographic center of the United States, 1790–2010 Knowing where the demographic center of a country's population is located at given times allows us to track that population's movement, growth, distribution, and concentration. Here we can clearly perceive the expansion of the American frontier and see that although the east continued to be the most populated part of the country early on, the demographic center moved as more and more of the population dispersed to the West and the population of the West grew through natural increase.

(Reprinted with permission of Prentice Hall, adapted from J. M. Rubenstein, *The Cultural Landscape: An Introduction to Human Geography*, 5th ed., © 1996, p. 122; updated with data from the US 2010 Census, US Census Bureau, http://www.census.gov/newsroom/releases/archives/facts_for_features_special_editions/cb11ff10.html.)

TABLE 3.5 United States Rural to Urban Population Change, 1860–2000

Year	1860	1870	1880	1890	1900	1910	1920	1930	1940	1950	1960	1970	1980	1990	2000
Total U.S. population (in millions)	31.4	38.5	50.1	62.9	76.2	92.2	106	123.2	132.1	151.3	179.3	203.2	226.5	248.7	281.4
Total urban population (in millions)	6.2	9.9	14.1	22.1	30.2	42	54.2	69.1	74.7	96.8	125.2	149.6	167	187	222.3
% Urban	19.8	25.7	28.2	35.1	39.6	45.6	52.2	58.1	58.5	64	69.9	73.6	73.7	75.2	78.9
% Increase in total population	35.6	22.6	30.2	25.5	21	21	15	16.2	7.3	14.5	18.5	13.4	11.4	9.8	13.1
% Increase in urban population	35.6	22.6	30.2	20.5	21	21	15	16.2	7.3	14.5	18.5	13.4	11.4	9.8	18.8

Source: United States Census Bureau.

FIGURE 3.24 U.S. Rustbelt In the mid-twentieth century, the most important automobile manufacturers in the United States located their factories in the Midwest, where skilled labor and raw materials were readily available. By the late 1970s and 1980s, cities like Flint, Michigan, began losing their automobile employment base as companies, such as Ford, moved to suburban locations in the Midwest or out of the region either to the South or to foreign locations, such as Mexico, where labor is far cheaper and environmental laws less stringent.

FIGURE 3.25 Trail of Tears, 1830s
In the Indian Removal Act of 1830, lands held by eastern Native American tribes were exchanged for territory that the federal government had acquired west of the Mississippi, effectively eliminating Indian title to land previously protected by treaties. Without government protection of their rights to their land, members of the Choctaw, Chickasaw, Cherokee, Seminole, Creek, and other eastern tribes had to move west to the lands that had been acquired for them. (Adapted from *Atlas of the North American Indian* by C. Waldman; maps and illustrations by M. Braun. Copyright © 1985 by C. Waldman. Reprinted by permission of Facts on File, Inc.)

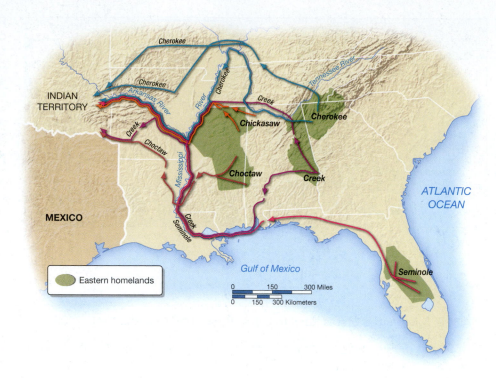

Other, recent examples of internal forced migration are provided by China and South Africa. In the late 1960s and 1970s, as part of the Cultural Revolution the government of China forcibly relocated 10 to 17 million of its citizens to rural communes in order to enforce Chinese Communist dogma and to ease pressures arising from high urban unemployment. The policy has since been disavowed, but the effects on an entire generation of Chinese young people were profound. Another example took place in South Africa between 1960 and 1980, when apartheid policies forced some 3.6 million blacks to relocate to government-created homelands, resulting in much suffering and dislocation. Indeed, civil war, ethnic conflict, famine, deteriorating economic conditions, and political repression have produced an extraordinary series of internal forced migrations in several sub-Saharan countries (**Figure 3.26**).

These forced migrations, both internal and international in scope, become particularly significant in light of changing population. Forecasts predict that 80 percent of the world population increase in the next decade will take place within the poorest countries of the world. Many of these countries have some of the highest rates of forced migration. The combination bodes ill for these countries' prospects for economic and political improvement.

Today, forced **eco-migration**—population movement caused by the degradation of land and essential natural resources—has created a new category of migrants and refugees. In Bangladesh, for example, floodplain settlement that began in the 1960s has led to severe losses of life and property and forced temporary relocation of huge numbers of people whenever severe flooding occurred. It is projected that climate change, particularly the effect of warmer global temperatures on sea level rise, will increasingly force island nations and coastal populations throughout the world to seek more permanent relocation elsewhere.

The International Federation of Red Cross (and Red Crescent) Societies has stated that eco-migrations related to climate change disasters currently cause more population displacement than war and persecution. While the growth of eco-migration has been greatest in sub-Saharan Africa, other areas are also teetering on the brink of environmental disaster. For example, Yemen is rapidly running out of water, China is experiencing dramatic desertification, as are Morocco, Tunisia, and Libya. In Egypt and Turkey, farm lands are being affected, Louisiana and Alaska are losing land to the sea at about 3 m per year, and Tuvalu already has an agreement with New Zealand to accept all 11,600 of its citizens when sea level rise makes the island uninhabitable. Bangladesh, Vietnam, Spain, and Portugal are also concerned about the impact of sea level rise on their coastal cities and populations. And the list of countries likely to be severely affected by climate change extends to central Asia, South America, and North America.

APPLY YOUR KNOWLEDGE Identify an environmental or economic issue in your region. Summarize two ways that this issue might result in push and pull factors that affect migration to and from your community. ■

POPULATION DEBATES AND POLICIES

One big question occupies the agenda of population experts studying world population trends today: How many people can Earth sustain without depleting or critically straining its resource base?

FIGURE 3.26 Acute malnutrition in Ethiopia A casuality of ecological catastrophe was Ethiopia, where late rains, failure of crops, and soaring food prices led to a severe food crisis and dislocation in 2008. Shown here are individuals congregating for international food aid.

The relationship between population and resources, which lies at the heart of this question, has been a point of debate among experts since the early nineteenth century.

Population and Resources

The debate about population and resources originated in the work of an English clergyman named Thomas Robert Malthus (1766–1834), whose theory of population relative to food supply established resources as the critical limiting condition upon population growth. Malthus's theory was published in 1798 in a famous book called *An Essay on the Principle of Population*. In this tract Malthus sets up two important postulates:

- Food is necessary to the existence of human beings.
- The passion between the sexes is necessary and constant.

It is important to put the work of Malthus into the historical context within which it was written. Revolutionary changes—prompted in large part by technological innovations—had occurred in English agriculture and industry and were eliminating traditional forms of employment faster than new ones could be created. This condition led to a fairly widespread belief among wealthy members of English society that a "surplus" of unnecessary workers existed in the population. The displaced agriculturists began to be a heavy burden on charity, and so-called poor laws were introduced to regulate begging and public behavior.

In his treatise, Malthus insisted that "the power of the population is indefinitely greater than the power of the earth to produce subsistence." He also believed that if one accepted this premise, a natural law would follow; that is, the population would inevitably exhaust food supplies. Malthus's response to this imbalance was to advocate for the creation of laws to limit human reproduction, especially among poor people.

Malthus was not without his critics, and influential thinkers such as William Godwin, Karl Marx, and Friedrich Engels disputed his premises and propositions. Godwin argued that "there is no evil under which the human species cannot labor, that man is not competent to cure." Marx and Engels were in general agreement that technological development and an equitable distribution of resources would solve what they saw as a fictitious imbalance between people and food.

The debate about the relationship between population and resources continues to this day, with provocative and compelling arguments for both positions being continually advanced. The geographer David Harvey has explored the population-resources issue in great detail. He has shown that by adopting Malthus's approach, only one outcome is possible—a doomsday conclusion about the limiting effect of resources on population growth. By following Marx's approach, however, quite different perspectives on, and solutions to, the population-resources issue can be generated. These solutions are based on human creativity and socially generated innovation, which allow people to overcome the limitations of their environment.

Neo-Malthusians—people today who share Malthus's perspective—predict a population doomsday. They believe that growing human populations the world over, with their potential to exhaust Earth's resources, pose the most dangerous threat to the environment. Although they acknowledge that the people of the core countries consume the vast majority of resources, they and others argue that only strict demographic controls everywhere will solve the problem, even if they require severely coercive tactics.

A more moderate approach argues that people's behaviors and governmental policies have a much greater impact on the condition of the environment and the state of natural resources than population size in and of itself. Proponents of this approach reject casting the population issue as a biological one in which an ever-growing population will inevitably create ecological catastrophe. They also reject framing it as an economic issue in which technological innovation and the sensitivities of the market will regulate population increases before a catastrophe can occur. Rather, they see the issue as a political one—one that governments have tended to avoid dealing with because they lack the will to redistribute wealth or the resources to reduce poverty, the latter condition being strongly correlated with high fertility.

The question of whether too many people exist for Earth to sustain bedeviled population policymakers and political leaders for most of the second half of the twentieth century. This concern led to the formation of international agencies that monitor and often attempt to influence population change. It also led to the organizing of a series of international conferences that attempted to establish globally applicable population policies. The underlying assumption of much of this policymaking, which has continued into the twenty-first century, is that countries and regions have a better chance of achieving improvement in their level of development if they can keep their population from outstripping the supply of resources and jobs.

Population Policies and Programs

Contemporary concerns about population—especially whether too many people exist for Earth to sustain—have led to the development of international and national policies and programs. A **population policy** is an official government strategy designed to affect any or all of several objectives, including the size, composition, and distribution of population. The implementation of a population policy takes the form of a population program. Whereas a policy identifies goals and objectives, a program is an instrument for meeting those goals and objectives. Most of the international population policies of the last three decades have attempted to reduce the number of births worldwide.

Figure 3.27 provides a picture of the recent history of world population growth by region and a reasonable projection of future growth. By the year 2050, the world is projected

to contain nearly 9 billion people. In comparison, over the course of the entire nineteenth century, fewer than a billion people were added to the population. **Figure 3.28** shows a projection of world population in the year 2020. Over the next century, population growth is predicted to occur almost exclusively in Africa, Asia, and Latin America, while Europe and North America will experience very low, and in some cases zero, population growth. The differing rates of natural increase listed in **Table 3.6** illustrate this point. At the turn of the twentieth century, the core contained 32 countries with zero population growth. By contrast, the periphery contained 39 countries with rates of natural increase of 3.0 or more. A sustained rate of natural increase of 3.0 per year means a population will double in approximately 24 years.

Since 1954, the United Nations has sponsored international conferences at 10-year intervals to develop population policy at the global level. Out of each conference emerged explicit population policies aimed at lowering fertility rates in the periphery and semiperiphery. Significantly, all the world population conferences have recognized that the history, social and cultural practices, development level and goals, and political structures for countries and even regions within countries are highly variable and that one rigid and overarching policy to limit fertility will not work for all. Whereas some programs and approaches will be effective for some countries seeking to cut population numbers, they will be fruitless for others.

For instance, China's family-planning policy of one child per household appears to be effective in driving down the birthrate and damping the country's overall population growth. Because so little data are available on China, however, it is not clear whether the policy is operating in the same fashion throughout the country. For instance, some population experts believe an urban bias exists, and while the policy is adhered to in the cities, it is disregarded in the countryside. Interestingly, in the wake of the May 2008 Sichuan earthquake—with a death toll of approximately

FIGURE 3.27 World population projection by region, 1950–2050 In the projection, population continues to expand in the periphery, though in some regions more than others. Africa is projected to experience the greatest growth, followed by Asia (not including China), where growth is expected to level off by 2150. Less dramatic growth is expected to occur in Latin America, while in the core population, numbers remain constant or drop slightly. Though the total number of people in the world will be dramatically greater by 2050, the forecast indicates a gradual leveling off of world population. (Adapted from I. Hauchler and P. Kennedy (eds.), *Global Trends: The World Almanac of Development and Peace.* New York: Continuum, 1994, p. 109.)

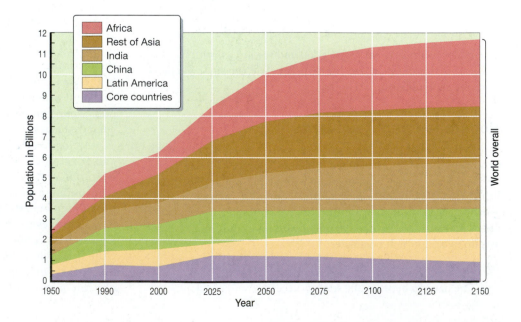

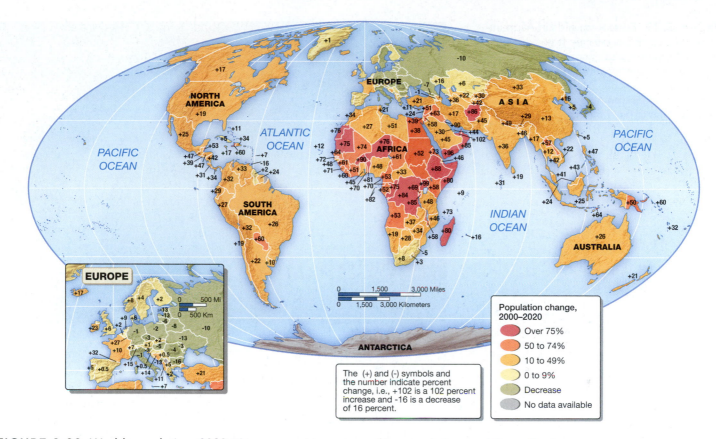

FIGURE 3.28 World population, 2020 This map provides a sense of how much the populations of various countries are expected to change by the year 2020. Although the populations of nearly all countries are expected to increase, it is clear that some populations will grow far more dramatically than others. Notice the substantial growth expected in Saudi Arabia and Afghanistan in contrast to the United States and Europe, where little, if any, population growth will occur. Italy, for example, is expected to lose population, while the Netherlands will grow by only 5 percent.

TABLE 3.6 Global Demographic Indicators, 2011

This table illustrates the substantial population pressures emanating from the global periphery, which contained roughly four times the population of the core with a rate of natural increase also nearly four times as great. While Asian countries possess the largest proportion of the world's population, their rate of natural increase is considered moderate and their level of economic development—measured by their GNI PPP—higher than the average overall for the periphery. Most troubling to demographers, however, are the population dynamics of Africa, with a rate of natural increase of nearly 3 percent per year in many countries and a GNI PPP of only $2,720 per capita.

Region	Population Mid-2011 (millions)	Natural Increase (annual %)	Birthrate (per 1,000 pop.)	Death Rate (per 1,000 pop.)	Life Expectancy at Birth (years)	GNI PPP per Capita ($US 2008)
World	6,987	1.2	20	8	70	$10,240
More Developed	1,242	0.2	11	10	78	$32,470
Less Developed	5,745	1.4	22	8	68	$5,440
Africa	1,051	2.4	36	12	58	$2,720
Asia	4,216	1.1	18	7	72	$6,310
Latin America and Caribbean	596	1.2	18	6	74	$10,130
Europe	740	0.0	11	11	76	$26,390
North America	346	0.5	13	8	78	$44,790

Source: Population Reference Bureau Web site, www.prb.org. *World Population Data Sheet,* 2011.

FIGURE 3.29 Educating girls in Afghanistan Improving the economic status of women is central to the success of controlling population growth. Access to education and employment security are seen as critical factors shaping a woman's decisions about how many children to have and when to have them.

90,000 including children caught in collapsed school buildings—the Chinese government suspended the policy. There is increasing discussion throughout China that the one-child policy should be abandoned for the entire country as the Chinese become increasingly urbanized and the fertility rate is dropping in spite of the policy, not because of it.

Family-planning regulations include setting a legal age for marriage, offering incentives to couples to have only one child, and mandating increasingly severe disincentives for couples who have larger families. In India, family-planning policies offering free contraceptives and family-planning counseling have resulted in lowering the birthrate. Other countries, such as Sri Lanka, Thailand, and Cuba, as well as the Indian state of Kerala, have lowered their birthrates not through regulation of family size, but by increasing access to social resources such as health care and education, particularly for women.

Indeed, as discussed earlier in the chapter, it is now a widely accepted belief among demographers and policymakers that a close relationship exists between women's status and fertility. Women who have access to education and employment tend to have fewer children because they have less of a need for the economic security and social recognition that children are thought to provide (**Figure 3.29**). In Botswana, for instance, women with no formal education have, on average, 5.9 children, while those with 4 to 6 years of schooling have just 3.1 children. In Senegal, women with no education give birth to an average of 7 children. In contrast, the average number of children born to a woman with 10 years of education drops to 3.6. The numbers are comparable for Asia and South America.

More equality between men and women inside and outside the household is also believed to have a significant impact on reducing fertility. Giving both men and women choices about birth control, and educating them about the implications of such choices, appears to be especially successful in small island populations—such as those of Bali, Barbados, and Mauritius—with historically high

population growth. In Mauritius in just 24 years (between 1962 and 1986) the introduction of voluntary constraints lowered the total fertility rate from 5.8 to 1.9. It is hardly any wonder, then, that the 1994 population conference placed such a clear and well-received emphasis on (1) rejecting coercive measures, including government sterilization quotas, that force people to violate their personal moral codes and (2) improving the rights, opportunities, and economic status of girls and women as the most effective way of easing global population growth.

Ultimately, successful family-planning programs rely on the widespread availability of contraceptive commodities and practices. And while 40 years ago contraception was socially and culturally unacceptable to many women in both core and peripheral countries, today the challenge is a different one: There is far more demand for it than there is availability. Ironically, family-planning and reproductive health programs are a victim of their own success. Having effectively increased knowledge about contraceptives over the past four decades, the demand for access to them has increased by over 50 percent.

Today, over 92 percent of all countries support family-planning programs and contraceptives, either directly through government facilities or indirectly through support of nongovernmental activities, such as nonprofit family-planning associations. Despite the pervasiveness of government support for contraceptive methods, however, some 215 million women would like but currently do not have ready access to safe and effective modern means of contraception. The reasons why supply is not keeping up with demand are complex, but they involve several key factors: the increasing cost of contraception commodities; the cost of providing clinicians, facilities, counselors, and education materials; and the uncertain financial support of international donor organizations such as the U.S. Agency for International Development (USAID) (the single largest donor of contraceptives globally) and the UN Population Fund for family-planning programs in poor countries.

FIGURE 3.30 UN World Summit, 2010 The UN Summit on the Millennium development goals met again in September 2010 and concluded with a further commitment to the global action plan to achieve the eight antipoverty goals by their 2015 target date. They also announced a major new initiative for women's and children's health and other initiatives against poverty, hunger, and disease.

TABLE 3.7 Millennium Development Goals (MDGs)

The goals and targets are based on the UN Millennium Declaration, and the UN General Assembly has approved them as part of the Secretary General's road map toward implementing the declaration. UNDP worked with other UN departments, funds, and programs, the World Bank, the International Monetary Fund, and the Organization for Economic Cooperation and Development to identify over 40 quantifiable indicators to assess progress.

Goals and Targets
Goal 1: Eradicate extreme poverty and hunger
Goal 2: Achieve universal primary education
Goal 3: Promote gender equality and empower women
Goal 4: Reduce child mortality
Goal 5: Improve maternal health
Goal 6: Combat HIV/AIDS, malaria, and other diseases
Goal 7: Ensure environmental sustainability
Goal 8: Develop a Global Partnership for Development

Sustainable Development, Gender, and Population Issues

Recognizing the central importance of economic development to improving the lives of women, men, and children throughout the world, international agencies are increasingly turning to sustainable economic development as a way of limiting births and ensuring an improved quality of life through the reduction of poverty. A host of related international conferences have been held to address this theme, the most central one being the UN Millennium Summit. The goals, listed in **Table 3.7**, are aimed at reducing poverty by improving economic development through aid, trade, and debt relief and through the enhancement of democratic governance institutions and careful attention to the impact of development on the environment. The UN Development Program (UNDP) Millennium Development

Goals (MDGs) are based on a partnership between core and peripheral countries. With respect to governance transformations, for instance, UNDP is working with an oil company and Amnesty International in Venezuela to provide the country's judges with a comprehensive understanding of human rights laws, regulations, and issues. With respect to the environment, UNDP is working with farmers in Ethiopia by supporting the planting and marketing of traditional crops and, in the process, strengthening the country's Biodiversity Research Institute and encouraging farmers to create biodiversity banks, while the crops make their incomes more secure.

The eight major MDGs reflect the neoliberal turn in international development. **Neoliberalism** promotes a reduction in the role and budgets of governments, including reduced subsidies and the privatization of formerly publicly owned and operated concerns such as utilities. The goal of neoliberal development policies—such as the one being widely advanced by the UNDP—is to enable peripheral countries to achieve core economic standards of wealth and prosperity while recognizing that preexisting conditions will have to be taken into account to construct a place-specific development path. As the goals imply, enabling more sustainable economic development worldwide is seen as a way of also shaping population growth and the quality of life for populations in the periphery. It is also a way of opening up new markets for core products and services and extending the capitalist world-system.

In 2005, world leaders met again in a follow-up summit meeting to the UN 2000 Millennium Summit. At the "World Summit," representatives (including many leaders) of the 191 (now 192) member states met in New York City for what the UN described as "a once-in-a-generation opportunity to take bold decisions in the areas of development, security, human rights, and reform of the United Nations" (**Figure 3.30**). In addition to reporting on progress and discussing the challenges of meeting the MDGs, the World Summit leaders in 2010 agreed to the formation of a UN Human Rights Council, which was launched in 2006, with the goal of strengthening the UN human rights machinery.

Future Geographies

In 2011, the world contained nearly 7 billion people. The population division of the UN Department of Social and Economic Affairs projects that the world's population will continue to increase by 1.2 percent annually to mid-century, resulting in 9.3 billion people by 2050. The distribution of this projected population growth is noteworthy: Over the next half-century, population growth is predicted to occur overwhelmingly in regions least able to support it. Just six countries will account for half the increase in the world's population, including Bangladesh, China, India, Indonesia, Nigeria, and Pakistan (**Figure 3.31**). Meanwhile, Europe and North America will experience very low and in some cases zero population growth. Collectively, the periphery will grow 58 percent, as opposed to 2 percent for the core regions through 2050. The periphery will account for 99 percent of the expected increment in world population in this period.

Nevertheless, it is increasingly clear that human population will not continue to grow indefinitely, even in countries with already high populations. Demographic transition theory has prompted some population analysts to suggest that many of the economic, political, social, and technological transformations associated with continued urbanization and industrialization across the globe will shift the world's population from a period of high growth to one of low growth, or even population decline. This will be due to falling fertility. Current total fertility of 3.11 children per woman will fall to 2.04 by mid-century—just below replacement level but still above the current rate in core regions.

FIGURE 3.31 Shanghai, China Considered to be one of the world's largest city by the UN, Shanghai has nearly 14 million people within the core city and its inner suburbs. The populations of cities like Shanghai are expected to continue to grow through the first half of the twenty-first century, though urban planners raise serious concerns about how urban systems like water, sewer, transportation, and energy supply—to name a few—will be able to keep up.

CONCLUSION

The geography of population is directly connected to the complex forces that drive globalization. And since the fifteenth century, the distribution of the world's population has changed dramatically as the capitalist economy has expanded, bringing new and different peoples into contact with one another and setting into motion additional patterns of national and regional migrations.

When capitalism emerged in Europe in the fifteenth century, the world's population was experiencing high birthrates, high death rates, and relatively low levels of migration or mobility. Four hundred years later, birth, death, and migration rates vary—sometimes quite dramatically—from region to region, with core countries experiencing low death and birthrates and peripheral and semiperipheral countries generally experiencing high

birthrates and fairly low death rates. Migration rates vary within and outside the core. These variations may be seen as reflections of the level and intensity of political, economic, and cultural connectedness between core and periphery and are difficult to predict.

The example of formerly colonized peoples migrating to their former ruling countries in search of work provides insights into the dynamic nature of the world economy. The same can be said of U.S. migrants who in the 1970s and 1980s steadily left their homes in the Northeast and Midwest to take advantage of the employment opportunities that were emerging in the Sunbelt: the Southeast, Southwest, and West. Both examples show the important role that people play in acting out the dynamics of geographic variety.

Learning Outcomes Revisited

- Understand the census and other sources of population data and how they are used to describe the geography of population.

  Geographers think of population in terms of the places that populations inhabit. They also consider populations in terms of the way that places are shaped by populations and in turn shape the populations that occupy them. Population experts employ the census as well as other data sources such as vital records to assess population characteristics. The data that are collected and the records that are produced are kept by local, county, state, and other levels of government. This information can help to show the way the population of any area—from the national level down to the individual census tract or neighborhood—is changing.

- Recognize why populations change, where those changes occur, and what the implications of population change are for the future of different places around the globe.

  Population geographers bring to demography a special perspective—the spatial perspective—that emphasizes description and explanation of the "where" of population distribution, patterns, and processes. The distribution of population is a result of many factors, such as employment opportunities, culture, water supply, climate, and other physical environment characteristics. Geographers explore these patterns of distribution and density, as well as population composition in order to understand the complex geography of populations. Understanding the reasons for and implications of variation in patterns and composition provides geographers with insight into population change and the potential impacts that growing or declining populations might have.

- Identify the two most important factors in population dynamics, birth and death, and explain how these factors are key to understanding new settlement patterns.

  In order to arrive at a different understanding of population growth and change, experts look first at two significant factors: fertility and mortality. Birth and death rates are simple but central indicators of a place's level of economic development. Fertility and mortality rates provide key insights into how well a country, region, or city is able to provide for its population, especially in terms of income, education, and health care. Population geographers consider life expectancy, immigration, and emigration as well, which also affect population dynamics.

- Demonstrate how the movement of population is affected by both push and pull factors and the way that movement is not always a matter of choice.

  In general terms, migrants make their decisions to move based on push factors and pull factors. Push factors are events and conditions that impel an individual to move from a location. Pull factors are forces of attraction that influence migrants to move to a particular location. Mobility is the capacity to move from one place to another, either permanently or temporarily. Migration, in contrast, is a long-distance move to a new location. Permanent and temporary changes of residence can occur for a variety of reasons. Striving for economic betterment or escaping from adverse political conditions, such as war or oppression, are the most frequent causes. Push factors can produce forced migration, but it is usually the case that the decision to migrate reflects both push and pull factors.

- Evaluate the challenges of providing for the world's growing population with adequate food and safe drinking water, as well as a sustainable environment.

  A moderate response to the question "How can the global economy provide the world's growing population with adequate food and safe drinking water, as well as a sustainable environment?" rejects casting the population issue as a biological one in which an ever-growing population will inevitably create ecological catastrophe. It also rejects framing it as an economic issue in which technological innovation and the sensitivities of the market will regulate population increases before a catastrophe can occur. Importantly, the response to this question is more convincingly understood as a political one. Yet governments across the globe tend to avoid dealing with the population-resource problem because they lack the will to redistribute wealth or the resources to reduce poverty. This leaves the burden on citizens to organize to force government to address the problem and change their own behaviors to lessen its effects.

KEY TERMS

age-sex pyramid *(p. 73)*

agricultural density *(p. 71)*

arithmetic density *(p. 71)*

baby boom *(p. 72)*

biometric census *(p. 68)*

census *(p. 68)*

cohort *(p. 74)*

crude birthrate (CBR) *(p. 79)*

crude death rate (CDR) *(p. 79)*

crude density *(p. 71)*

demographic transition *(p. 84)*

demography *(p. 68)*

density *(p. 71)*

dependency ratio *(p. 75)*

doubling time *(p. 79)*

eco-migration *(p. 94)*

emigration *(p. 85)*

forced migration *(p. 86)*

gender *(p. 67)*

geodemographic
analysis *(p. 73)*

Great Recession *(p. 86)*

gross migration *(p. 86)*

guest workers *(p. 86)*

immigration *(p. 85)*

infant mortality rate *(p. 80)*

internally displaced
persons (IDPs) *(p. 86)*

internal migration *(p. 85)*

international migration *(p. 85)*

life expectancy *(p. 81)*

medical geography *(p. 82)*

middle cohort *(p. 75)*

migration *(p. 85)*

mobility *(p. 85)*

natural decrease *(p. 80)*

natural increase *(p. 80)*

neoliberalism *(p. 99)*

net migration *(p. 86)*

nutritional density *(p. 71)*

old-age cohort *(p. 75)*

population policy *(p. 96)*

pull factors *(p. 86)*

push factors *(p. 86)*

refugees *(p. 86)*

suburbanization *(p. 76)*

total fertility rate (TFR) *(p. 79)*

transnational migrant *(p. 87)*

undocumented workers *(p. 86)*

vital records *(p. 68)*

voluntary migration *(p. 86)*

youth cohort *(p. 75)*

REVIEW AND DISCUSSION

1. Consider the population concentration in your town or city and some of its characteristics. Look up the census and vital records. Pay particular attention to birthrates. Compare 1950 with 2010 census data. What changed with respect to birthrates? What might be one reason for the change or lack of change? Now compare the data with birth data from another city or town. Is there a difference? If so, list two reasons why there might be a difference. If there is not a significant change, why might this be the case?

2. Pick two major cities within the United States and consider them from the perspective of migration. What have been the gross migration patterns of people in and out of the cities? How has migration changed over the last 20 years? List three reasons for these migration patterns.

3. Do an Internet search and find a current example each of refugees and internally displaced people. Compare and contrast these different categorizations. What similarities do these groups share? What are their primary differences? Develop a list of two reasons for each situation. Also, consider how refugees and internally displaced people have changed the population characteristics of their given place. For example, have these factors had an effect on birthrates or infant mortality rates? Please be as specific as possible in your answer by citing data that support your conclusion.

UNPLUGGED

1. The distribution of population is a result of many factors, such as employment opportunities, culture, water supply, climate, and other physical environment characteristics. Look at the distribution of population in your state or province. Is it evenly distributed, or are the majority of people found in only a few cities? List three factors that influence where people live in your state or province.

2. Immigration is an important factor contributing to the increase in the population of the United States. Chances are your great-grandparents, grandparents, parents, or even you immigrated to, or migrated within, your country of residence. Construct your family's immigration or migration history. Identify push and pull factors influencing your family's decision to immigrate to or migrate within your country. Have you had to move within your lifetime? If so, please write a paragraph describing your decision.

3. During the 1990s, an interesting example of return migration occurred. Some areas of the United States that had previously lost large numbers of African Americans began to regain them. Compare data from the 1980 census with data from the 2010 census for the southeastern region of the United States. Which states experienced an increase of African American migrants? What are some of the characteristics of this migrant stream, and why might it have occurred? What impact might this have on electoral issues?

4. Every few years UNESCO publishes a data book on global refugee statistics indicating both sending and receiving countries, among other variables. Locate the most recent data book and identify one country that has been a large sender of refugees and the country that has been the largest receiver of those refugees. The data book provides information not only on the numbers of refugees but on their age, gender, and other variables. Discuss some of the demographic implications for both countries if the refugee population was not to be allowed to return to the sending country. How could this affect population policy in both countries?

Log in to www.masteringgeography.com for MapMaster™ interactive maps, geography videos, RSS feeds, flash cards, web links, an eText version of *Human Geography: Places and Regions in Global Context,* and self-study quizzes to enhance your study of population geography.

MapMaster™ presents 13 Place Name and 13 Layered Thematic interactive maps to help students practice and master their geographic literacy, spatial reasoning, and critical thinking skills.

4 PEOPLE AND NATURE

On March 11, 2011, at 2:46 pm Tokyo time, a massive 9.0 moment magnitude earthquake struck off the eastern coast of Japan. (The moment magnitude scale measures the amount of energy an earthquake releases.) The temblor was the fifth largest in over a century and the most powerful earth-shaking force ever to hit Japan. Near the epicenter, the quake fractured the Japanese landscape, twisting roads, opening up crevasses, and strewing about huge chunks of land. The effect of the temblor was so strong, in fact, that geoscientists believe it may have displaced Earth's rotation, shortening the day by 1.8 microseconds. It also shifted Japan's coastline and changed Earth's balance. Because of the dramatic land displacement, the earthquake set off a huge tsunami with waves as high as 30 feet smashing onto the coast. The waves leveled houses, moved burning buildings into fields, slammed vehicles around, and even perched a large ferry on top of a two-story building. Although the final death toll of the combined earthquake and tsunami was not yet known as of summer 2011, at least eighteen thousand people were estimated to have died and hundreds of thousands were rendered homeless.

Adding insult to injury was the combined effect of the earthquake and tsunami on the Fukushima Daiichi nuclear power plant. Located on the coast, not far from the earthquake's epicenter, the combined force of the earthquake and the tsunami disabled the

Rescue workers search for survivors following the earthquake and tsunami in Natori City.

cooling generators of the plant, resulting in increased heat and pressure buildup and partial meltdown in three of the six reactors. As a result, a number of explosions occurred over the course of several days at Fukushima, sending dangerous levels of radiation into the air and contaminating water, air, land, and foodstuffs within 30 km (about 19 mi) of the plant. The government of Japan rated the nuclear crisis a level seven, the highest rating possible. Drifting radiation fallout contaminated soil, food, and water beyond the evacuation zone as well (12–18 km/7.5–11 mi from the plant). The leakage of radioactive materials appears to have finally been controlled, but the long-term health effects to the terrestrial and marine ecosystems and the people who are part of them will only become known as time passes.

The extraordinary combination of earthquake, tsunami, and nuclear reactor meltdown that occurred in early spring of 2011 reminds us of the vulnerability of even the most sophisticated of human technological achievements in the face of the awesome power of natural systems. This relationship, between humans and their environments, is perhaps the most central of all relationships within the discipline of geography. Indeed, the discipline includes those who study natural systems, those who study human systems, and those who study the connections between them. ▪

NATURE AS A CONCEPT

In this chapter, we explore further the ways that society has used technology to transform and adapt to nature, together with the impact of those technological adaptations on humans and the environment. As discussed briefly in Chapter 2, a simple model of the nature–society relation is that nature, through its grand force and subtle expressions, limits or shapes society. This model is known as environmental determinism. A second model posits that society also shapes and controls nature, largely through technology and social institutions. This second model, explored in this chapter, emphasizes the complexity of nature–society interactions.

Interest in the relationship between nature and society has experienced a resurgence as the scope of environmental problems has widened to include not only those that are locally or regionally contained to those that have implications for the whole planet. The single most dramatic manifestation of this interest occurred in the summer of 1992, when more than 100 world leaders and 30,000 other participants attended the second Earth Summit in Rio de Janeiro (the first Earth Summit was held in Stockholm in 1972) (**Figure 4.1**). The central focus of the 1992 agenda was to ensure a sustainable future for Earth by establishing treaties on global environmental issues, such as climate change and biodiversity.

The signatories to the 1992 Earth Summit conventions created the Commission on Sustainable Development to monitor and report on implementation. A five-year review of Earth Summit progress was undertaken in 1997 through a UN General Assembly meeting in special session. The Rio+20—a celebration of the initial conference held in Rio in 1992—will occur in June 2012. Also known as Earth Summit 2012, its goals are to secure renewed political commitment to sustainable development; to assess progress toward internationally agreed goals on sustainable development; and to address new and emerging challenges. Rio+20 will focus on two specific themes: a green economy in the context of poverty eradication and sustainable development, and an institutional framework for sustainable development.

Some very important changes have occurred since the first summit 40 years ago. One change has been the emergence of international institutions to facilitate and monitor environmental improvements. Another has been real progress on the global phaseout of leaded gasoline. A third has been increasing scientific and popular interest in global environmental issues. Perhaps most significant has been the proliferation of concern for sustainability, especially as it has broadened out from being a conference theme to a set of practices being taken up by governments, corporations, public and private organizations, and individuals. Among these new practices and objectives are environmental restoration, green urban development, alternative energy technologies (such as wind, geothermal, and solar, as well as nuclear), sustainable tourism, "greening" the commodity supply chain, sustainable agriculture, promotion of biomass and biofuels, green real estate development, "green" government practices, the emergence and growth of carbon markets, sustainable buildings, and "green" marketing.

Recent heightened interest in the nature–society relationship stems from the persistence, increasing number, and wider impact of environmental crises. These changes have led to attempts to rethink the relationship between humans and the environment. For example, at the beginning of the twentieth century, Gifford Pinchot, as an influential citizen and later as director of the National Conservation Commission, advocated environmental conservation as he witnessed the nation's forests and wild lands being increasingly given over to development. Aggressive development impinging on open space is still an issue today. Similarly, in 1962, Rachel L. Carson, in her groundbreaking book *Silent Spring*, warned of the dangers of agricultural pesticides to ecosystems. (An **ecosystem** is a community of different species interacting with each other and with the larger physical environment that surrounds it). Yet as the new century moves well into its second decade, the pesticide problem persists in both peripheral and core countries.

In the past, technology was viewed as the apparent solution to most environmental problems, but today's technological

FIGURE 4.1 1972 Earth Summit, Stockholm, Sweden Global concern about environmental issues was formalized in 1972 at the Earth Summit held in Sweden. Pictured here is the Stockholm Royal Opera House where the 1972 Earth Summit was held.

FIGURE 4.3 Vandana Shiva Indian environmental activist who was trained as a physicist and calls herself an ecofeminist and philosopher. She is also a leading figure in the global solidarity movement known as the alter-globalization movement.

FIGURE 4.2 Deformities in frogs A 2005 study published in the *Proceedings of the National Academy of Sciences* found that the combination of pesticide contamination and parasite infection has caused missing legs and extra legs in wood frogs in 43 states in the United States and 5 provinces in Canada. It is believed that exposure to agricultural chemicals may weaken amphibian immune systems, making the frogs more vulnerable to parasitic infection leading to limb deformities. Worldwide, populations of amphibians have been decreasing at an alarming rate and scientists believe that global climate change and other human-induced factors are responsible for this decline.

progress seems often to aggravate rather than to solve such problems (**Figure 4.2**). As a result, researchers and activists have begun to ask different questions and to abandon the assumption that technology is the *only* solution. For instance, philosopher and environmental activist Vandana Shiva has led a broad-based and increasingly global movement advocating changes in agriculture and food production practices (**Figure 4.3**). In 1982, she established the Research Foundation for Science, Technology and Ecology, and in 1991, she founded Navdanya, a social movement with the goal of protecting the diversity and integrity of living resources, especially promoting the use and preservation of native seeds, organic farming, and fair trade. In India, Navdanya has so far successfully conserved more than 5,000 crop varieties, has created awareness of the hazards of genetic engineering, and has defended people whose food rights are under threat from globalization and biopiracy. A controversial practice with many conflicting definitions, *biopiracy* is defined by the *Oxford Dictionaries* (online) as "the practice of commercially exploiting naturally occurring biochemical or genetic material, especially by obtaining patents that restrict its future use, while failing to pay fair compensation to the community from which it originates."

Navdanya and other movements like it have influenced environmental experts, including a number of geographers, to conceptualize nature not as something apart from humans but as inseparable from us. These experts believe that nature and questions about the environment need to be considered in conjunction with society, which shapes our attitudes toward nature and how we identify sources

of and solutions to environmental problems. Such an approach—thinking of nature and society as interactive components of a complex system—enables us to ask new questions and consider new alternatives to our current practices with respect to nature.

In the chapter that follows we examine the nature–society relationship by looking first at different approaches to it. We then examine how changing conceptions of nature have translated into very different uses of and adaptations to it. We conclude the chapter with an examination of sustainable development as a way of addressing global environmental problems and a discussion of the new institutional frameworks and activist organizations that are emerging to promote sustainability.

Nature and Society Defined

The central concepts of this chapter—nature and society—have very specific meanings. Although we discuss the changing conceptions and understandings of nature in some detail, we hold to one basic conception here, that **nature** is a social creation as much as it is the physical universe that includes human beings. Therefore, understandings of nature are the product of different times and different needs. Nature is not only an object, it is a reflection of society in that the philosophies, belief systems, and ideologies people produce shape the way we think about and use nature. The relationship between nature and society is two-way: Society shapes people's understandings and uses of nature at the same time that nature shapes society. The amount of shaping by society is dependent to a large extent on the state of technology and the constraints on its use at any given time.

Society is the sum of the inventions, institutions, and relationships created and reproduced by human beings across particular places and times. Society's relationship with nature varies from place to place and among different social groups. The relationship between society and nature is usually mediated through technology. Knowledge, implements, arts, skills, and the sociocultural context—all are components of technology. If we accept that all of these components are relevant to technology, then we can provide a definition that has three distinguishable, though equally important, aspects. **Technology** is defined as

- physical objects or artifacts (for example, the plow);
- activities or processes (for example, steelmaking);
- knowledge or know-how (for example, biological engineering).

This definition recognizes tools, applications, and understandings equally as critical components of technology. The manifestations and impacts of technology can be measured in terms of concepts, such as level of industrialization and per capita energy consumption.

The definitions provided in this section reflect current thinking on the relationship between society and nature. For centuries, humankind, in response to the constraints of the physical environment, has been as much influenced by prevailing ideas about nature as by its realities. In fact, prevailing ideas about nature have changed over time, as evidence from literature, art, religion, legal systems, and technological innovations makes abundantly clear.

Concern about global environmental changes has given rise to a recent attempt to conceptualize the relationship between social and environmental changes, based on the premise that individual societal changes can be both subtly and dramatically related to environmental changes. A formula now widely used for distinguishing the sources of social impacts on the environment is $I=PAT$; it relates human population pressures on environmental resources to a society's level of affluence and access to technology. More specifically, the formula states that $I=PAT$, where I (impact on Earth's resources) is equal to P (population) times A (affluence, as measured by per capita income) times T (a technology factor). According to this formula, the differential impact on the environment of two households' food consumption in two different places would equal the number of people per household times the per capita income of the household times the type of technology and energy used in producing foodstuffs for that household.

Each of the variables in the formula—population, affluence, and technology—is complex. For example, with regard to population numbers, it is generally believed that fewer people on the planet will result in fewer direct pressures on resources. Some argue, however, that increased world population is quite desirable, since more people means more labor coupled with more potential for the emergence of innovation to solve present and future resource problems. Clearly, there is no simple answer to the question of how many people are too many.

Affluence also cannot simply be assessed in terms of "less is better." Certainly, increasing affluence—a measure of per capita consumption multiplied by the number of consumers and the environmental impacts of their technologies—is a drain on Earth's resources and a burden on Earth's ability to absorb waste. Yet how much affluence is too much is difficult to determine. Furthermore, evidence shows that the core countries, with high levels of affluence, are more effective than the poor countries of the periphery at protecting their environments. Unfortunately, core countries often do so by exporting their noxious industrial processes and waste products to peripheral countries (**Figure 4.4**). By exporting polluting industries and the jobs that go with them, however, core countries may also be contributing to increased affluence in the receiving countries. Given what we know about core countries, such a rise fosters a set of social values that ultimately leads to better protection of the environment in a new place. It is difficult to identify just when environmental consciousness goes from being a

FIGURE 4.4 Electronic waste This map shows the major illegal waste shipment routes for electronic garbage produced in Europe. It also shows which countries are the major waste producers. (*Source:* Adapted from Figure 8.7, Waste Trafficking in Global Environment Outlook 4 (GEO4), 2009. Cartographer/designer Bounford.com and UNEP/GRIDArendal, compiled from multiple sources in UNEP 2006c. http://maps.grida.no/graphic/wastertrafficking.

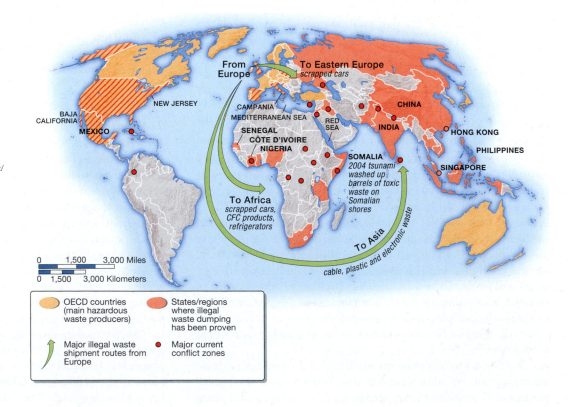

luxury to a necessity. The role of affluence in terms of environmental impacts is, in short, like population, difficult to assess.

Not surprisingly, the technology variable is no less complicated. Technologies affect the environment in three ways, through

- the harvesting of resources;
- the emission of wastes in the manufacture of goods and services;
- the emission of waste in the consumption of goods and services.

A technological innovation can shift demand from an existing resource to a newly discovered, more plentiful one. In addition, technology can sometimes be a solution and sometimes a problem. Both principles can be seen in the case of nuclear energy, widely regarded as cleaner and more efficient than coal or oil as an energy source. Producing this energy creates hazards, however, as the 2011 accident at the Fukushima plant in Japan makes clear.

It is therefore clear that increases in human numbers, in levels of wealth, and in technological capacity are key components of social and economic progress that have had an extremely complex impact on the environment. In the last 100 years, this complexity has come to be seen as a triple-barreled threat to the quality of the natural world and the availability and quality of environmental resources. Before we look more carefully at the specific impacts of populations, affluence, and technology on nature, we need to look first at how differing social attitudes toward nature shape the human behaviors that are a basis for I=PAT.

Nature–Society Interactions

The concept of *adaptation* to the natural environment is part of the geographical subfield of cultural ecology most closely associated with the work of Carl Sauer and his students. **Cultural ecology** is the study of how human society has adapted to environmental challenges like aridity and steep landscapes through technologies such as irrigation and terracing and organizing people to construct and maintain these systems. These adaptations can be seen clearly in the rice terraces of Southeast Asia or the canals and reservoirs of the southwestern United States. More recent adaptations include the use of biotechnology and agricultural chemicals to increase agricultural production and the development of new pharmaceuticals to cope with diseases.

Human adaptation has involved more than simply responding to natural constraints though. In addition, we have produced widespread modifications of environments and landscapes. In some cases, the human use of nature has resulted in environmental degradation or pollution. For example, overcultivation of steep slopes can result in erosion of the soil needed for subsequent agricultural production, and the use of toxic agricultural chemicals has caused the contamination of rivers and lakes. The Industrial Revolution produced a dramatic growth in the emissions of waste material to land, water, and the atmosphere and resulted in serious air pollution and health problems in many areas. These emissions and the associated pollution persist to this day.

The massive transformation of nature by human activity led geographers such as Neil Smith and Margaret Fitzsimmons to claim that there are no more "natural" environments or untouched wildernesses. They used the phrase "social production of nature" to describe the refashioning of landscapes and species by human activity, especially capitalist production and labor processes. Geographers have played a major role in highlighting the

global scope of this transformation through their discussions of the human dimensions of global environmental change and their explorations of the social causes and consequences of changes in global environmental conditions. Of particular concern are global patterns of fossil-fuel use and changes in land use that are producing serious changes in climate and biodiversity through carbon dioxide–induced global warming or deforestation (**Figure 4.5**).

Global climate change is causing sea levels to rise as polar ice caps melt and has increased the strength and severity of violent storms. Warmer oceans surrender greater quantities of water in the form of water vapor. Warmer surface temperatures and more-humid air masses intensify weather systems, resulting in fiercer cyclones and hurricanes. In the summer of 2005, the twin disasters of violent storms and flooding came together in the United States as Hurricane Katrina bore down on a wide swath of the Gulf Coast that extended from Pensacola, Florida, to New Orleans, Louisiana. The hurricane destroyed extensive sections of the built environment and caused the flooding of low-lying areas, especially Greater New Orleans, where thousands died or were injured and more than 1 million people were displaced. The media, politicians, government officials, planners, and engineers labeled Hurricane Katrina the worst natural disaster the United States had ever experienced.

It is now largely accepted that the root causes of the calamity were far from "natural." While there is no doubt that the force of the winds slamming the Gulf Coast were extreme, the fact that districts and parishes in and around New Orleans (where the most dramatic impacts occurred) flooded and so many people (who had not evacuated) died was not because of the hurricane but was the result of avoidable political and social factors. The event is instructive because it provides insight into the ways that global climate change can have catastrophic effects because of the social as well as the environmental vulnerability of populations.

Weak building codes and poor code enforcement in New Orleans made the wood-frame housing stock especially susceptible to damage from high winds. And critically needed improvements to the sinking levees—which hold back the Mississippi River and Lake Ponchartrain and enable various districts of the city of New Orleans to sit nearly 5 feet below sea level—had been repeatedly postponed by the federal government. The strong winds primarily brought down power and communication lines, leaving the region without electricity or phone service but housing fairly intact. But the houses that withstood the rain could not withstand the floodwater. The storm surge that accompanied Katrina breached the levees and caused flooding to occur in over 80 percent of the city. Add to these vulnerabilities the fact that although the city's evacuation plan worked well for many, tens of thousands of people were too poor, disabled, uninformed, or fearful to leave the city before the storm hit and they were put directly in harm's way.

The 2010 earthquake in Haiti is another instance of the impact of a natural disaster being dramatically intensified by poverty and political problems. The quake that hit Haiti on January 12 was a catastrophic 7.0 moment magnitude, with an epicenter near the town of Léogâne, approximately 25 km (16 mi) west of Port-au-Prince, Haiti's capital. It is estimated that 3 million people were affected by the quake with a death toll between 92,000 and 220,000. Around 1.5 million to 1.8 million were made homeless. The government of Haiti has estimated that 250,000 residences and 30,000 commercial buildings collapsed or were severely damaged. What makes the Haiti earthquake so

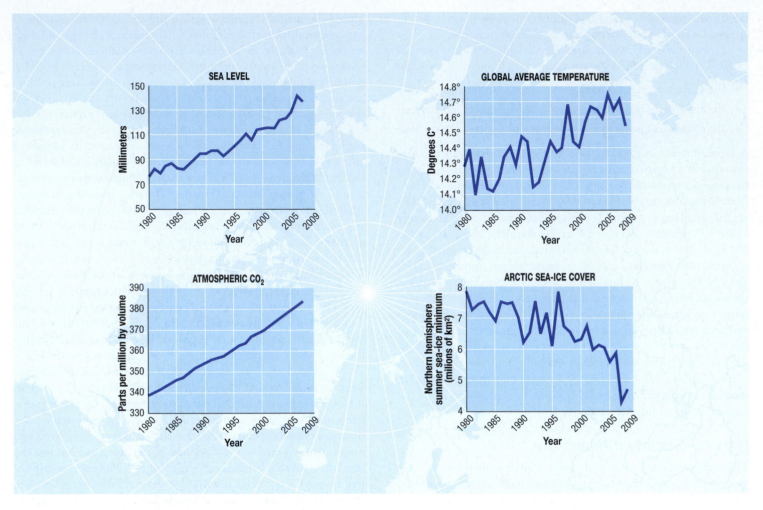

FIGURE 4.5 Composite graphs of climate change The International Geosphere-Biosphere Programme considers the four elements shown here—temperature, carbon dioxide level, sea level, and Artic sea ice—to represent how climate change is affecting the planet. (Adapted with permission from IGBP Climate Change Index: Global-change trends for the public and policy makers. Global Change: International Geosphere-Biosphere Programme (IGBP).)

problematic is that a year later, according to Oxfam International, only 5 percent of the rubble had been cleared as crippling "indecision" continues to stall reconstruction efforts (**Figure 4.6**). Inadequate building codes and poor emergency response also increased the impact of the earthquake significantly. After the quake, government corruption and inefficiency ensure that people's lives will not return to normalcy for a long time.

Atmospheric scientists have been arguing for over two decades that global warming is putting Atlantic and Gulf coastal cities in the United States in a vulnerable position as higher-than-normal ocean temperatures intensify hurricanes and tropical storms. Many believe that Hurricane Katrina is not an example of a "perfect storm," where essential meteorological elements combine to produce an extreme and devastating event, but the perfect example of how global climate change in combination with human practices is transforming Earth's environment in dramatic and devastating ways.

During the twentieth century, global sea level rose by 20 centimeters (7.9 inches), and a recent report by Britain's Meteorological Office warned that flooding will increase more than nine-fold over the twenty-first century, with four-fifths of the increase coming in South and Southeast Asia. A rising sea level would be disastrous for some

countries. About 70 percent of Bangladesh, for example, is at sea level, as is much of Egypt's most fertile land in the Nile delta. On the other hand, farmers in much of Europe and North America would welcome a local rise in mean temperatures, since it would extend their options for the kinds of crops that they could profitably raise.

The causes and consequences of these global climate changes vary considerably by world region. For example, the industrial countries have higher carbon dioxide emissions. Increased carbon dioxide emissions are contributing to rising temperatures through the trapping of heat in Earth's atmosphere. In order to survive in many of the world's peripheral regions, the rural poor are often impelled to degrade and destroy their immediate environment by cutting down forests for fuelwood, leading to the destruction of forests, which help to cool Earth's surface. Thus, both the core and the periphery are contributing to the problem of global change in different, but equally significant, ways (**Figure 4.7**).

Population growth patterns and the changing geography of economic development allow us to predict with some confidence that the air and water pollution generated by low-income countries will more than double in the next 10 to 15 years as they become more industrialized. Thus, environmental problems are becoming

FIGURE 4.6 Haiti, one year after the earthquake Less than 5 percent of debris had been cleared on the one year anniversary of the earthquake, January 11, 2011. The amount of rubble still on the ground would fill dump trucks parked bumper to bumper halfway around the world.

inseparable from processes of demographic change, economic development, and human welfare. In addition, regional environmental problems are becoming increasingly enmeshed in matters of national security and regional conflict. Before the 1980s, the principles of cultural ecology did not include the political dimensions of ecological questions, but since then cultural ecologists have moved away from a strict focus on particular cultural groups' relationship with the environment, placing that relationship within a wider context instead. The result is political ecology, the merging of political economy with cultural ecology. **Political ecology** stresses that human–environment relations can be adequately understood only by relating patterns of resource use to political and economic forces. (See Box 4.1, "Geography Matters: Water Politics.")

APPLY YOUR KNOWLEDGE List three examples of cultural ecology from your own community or campus. (Hint: How specifically has your town or campus adapted to environmental challenges over the last 20 years?) ∎

U.S. ENVIRONMENTAL PHILOSOPHIES AND POLITICAL VIEWS OF NATURE

As we mentioned at the beginning of this chapter, nature is a construct that is very much shaped by social ideas, beliefs, and values. As a result, different societies and different cultures have different views of nature. In the contemporary world, views of nature are dominated by the Western (also known as Judeo-Christian) tradition that

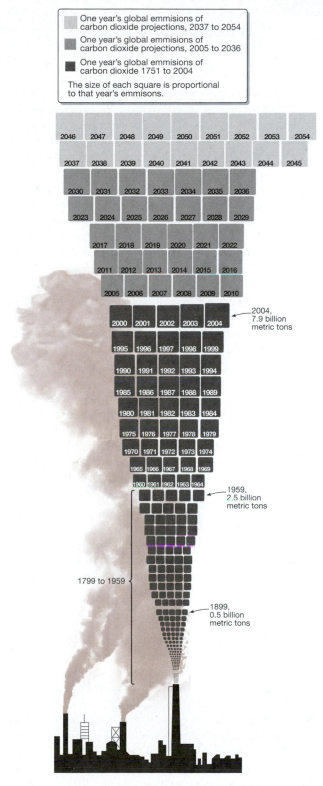

FIGURE 4.7 Global emissions of carbon dioxide Each square represents one year's global emissions of carbon dioxide, measured by the weight of carbon it contains. (*Source:* Adapted from *New York Times*, December 16, 2007. http://www.nytimes.com/interactive/2007/12/16/weekinreview/20071216_EMISSIONS_GRAPHIC.html, accessed May 16, 2011.)

understands humans to be superior to nature. In this view, nature is something to be tamed or dominated. But other views of nature have emerged that depart dramatically from the dominant view. These include the environmental philosophies that became popular

One of the most crucial elements in the relationship between people and their environment is water. We can't live without it and all of our economic practices depend on it. For instance, it takes 150 gallons of water to produce one newspaper; 1,000 gallons for one quart of orange juice, 2,500 gallons for 1 pound of beef, and 40,000 gallons of water for one new car. The water embedded in the production of the food or other things we consume is referred to as **virtual water** (**Figure 4.A**). Without abundant water, people in the core could not live the high-quality lives they currently enjoy; without clean water, those in the periphery die. It's therefore important to question what our lives would be like if the price of water was as volatile as, for example, oil. The price of oil has risen and fallen repeatedly due to civil unrest in the Middle East most recently, but also to the power of 12 countries—known as the Organization of Petroleum Exporting Countries (OPEC)—to set oil production schedules and prices. Privatized water would be subject to similar conditions and price effects.

While this may seem a far-fetched question for people in the core who enjoy artificially low water costs, in the periphery populations have been feeling the effects of escalating water prices for nearly a decade, and in some places far longer. The *2006 UN Human Development Report* states that 1.1 billion people in peripheral countries do not have adequate access to water, with 2.6 billion lacking basic sanitation. Almost two-thirds of the people lacking access to clean water survive on less than $2 a day. Access to piped household water is unequal, with 85 percent of the wealthiest 20 percent of the population having adequate access, compared to a meager 20 percent of the poorest 20 percent of the population with access. Perhaps most disturbing is the statistic that 1.4 million children will die each year from lack of access to safe drinking water and sanitation.

Experts around the world have begun to talk about a global "water crisis," which is occurring not because there is a shortage, but because water as a public good is rapidly turning into a privatized commodity. That is, the privatization of water (rather than the provision of water through publicly owned utilities)—touted as a way of bringing equitable access and efficiency to all water users—has instead led to increased prices and accessibility problems for poor and marginalized peoples not only in the global periphery but in peripheral areas of many core countries as well. Nearly two decades ago, many governments—national as well as local—looked to the privatization of water as a way of unburdening themselves of a relatively expensive service in response to the public's demand for smaller government. Corporations—mostly multi-national ones—began buying up municipal water providers and offering water provision based on profit-and-loss considerations.

An illustration of the kinds of problems that the privatization of water can create is the case of Cochabamba in Bolivia. Recently the subject of an internationally acclaimed motion picture, *Even the Rain/También la lluvia*, starring Gabriel García Bernal, explores the real-life conflict that erupted around a European water company's attempt to privatize water in this lush valley community at the entrance to the Amazon rain forest. In

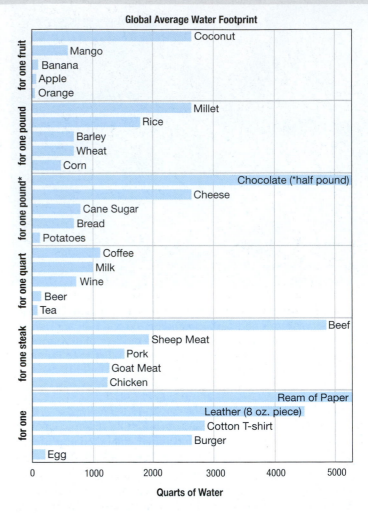

Global Average Water Footprint

FIGURE 4.A Water footprint This graph provides examples of the virtual water expended to produce some of our everyday foods. Notice how much processing adds to the total amount of virtual water involved. (*Source:* Data from Hoekstra, A.Y. and A.K. Chapagain (2008) "Globalization of water: Sharing the planet's freshwater resources." Oxford, UK: Blackwell Publishing; and www.waterfootprint.org.)

1998, the International Monetary Fund (IMF) approved a $138 million loan for Bolivia to help the country control inflation and bolster economic growth. In compliance with IMF-drafted structural reforms for the nation, Bolivia agreed to sell off a wide range of public utilities, including Cochabamba's local water agency, SEMAPA. Cochabamba in central Bolivia has a metropolitan population of about 1 million. It's a prosperous place with an economic base dedicated to commerce, industrial production, and services. It has a large indigenous population—Quechua people—and enjoys a reputation for being economically and socially progressive.

Water rates in most municipalities around the world are subsidized since water is absolutely essential to life and the "real" cost of water—acquisition, treatment, delivery, waste management—can be prohibitive. Moreover, subsidies exist not just for households but also for agriculture and industries. However, when water prices

FIGURE 4.B Water conflict in Cochabamba, Bolivia Pictured here are protesters running from tear gas fired by national police attempting to disperse thousands of demonstrators on the streets of Coachabamba, Bolivia, during the action against the hike in water rates.

increased in Cochabamba, due to the sale of its municipal water supply to Aguas del Tunari, a multinational consortium of private investors, including a subsidiary of the multinational Bechtel Corporation, no subsidies were provided to assist the population in affording the new, much higher water rates. Water rates doubled and tripled, forcing many residents to discontinue water service. Eventually, citizens assembled to protest against the difficult situation. The result was four days of conflict where members of The Coalition for the Defense of Water and Life (La Coordinadora) went on strike and erected roadblocks throughout the city (**Figure 4.B**). The protests in Cochabamba spread to Bolivia's administrative capital, La Paz, and other cities and outlying rural communities. Across the country, thousands clashed with riot police, erected roadblocks, and protested not only the water-rate hikes but also high unemployment. The violence culminated in a historic victory for the residents of Cochabamba and their supporters that guaranteed the withdrawal of Aguas del Tunari, granted control of Cochabamba's water to La Coordinadora, assured the release of detained protesters, and promised the repeal of water privatization legislation.

But stories similar to that of Cochabamba continue to unfold across municipalities globally. Large international water companies such as Paris-based Suez, the largest water company, Veolia (formerly Vivendi), also French and the world's second-largest water giant, and RWE, a German utility conglomerate, are competing for water supplies across the world. Since 2002 Lexington, Kentucky, for instance, has been fighting to regain control of its local water utility from RWE. The water privatization in New Orleans was predicted to be the largest private water contract in the nation when it was proposed in 2000. Vivendi (as it was then known) and Suez vied to profit from control of Crescent City's (California) water supply. The deal was predicted to be worth more than $1.5 billion over the next 20 years. Ultimately, the city was able to defeat the takeover.

Reports abound across the United States and the globe of private purchase of local water supplies and resulting rate hikes, negative economic impacts, inadequate customer service, and harm to natural resources. The bottom line is that water—clean, drinkable water—like lots of other natural resources, is becoming increasingly scarce and an attractive investment opportunity. Chile has become the most water-privatized country in the world. But unlike other places, it undertook privatization through careful staging that unfolded over many years. By the time private companies actually assumed control of water provision across Chile's urban areas, strict rules and regulations were in place to increase efficiency, improve service quality, and mobilize capital to extend wastewater treatment. Most importantly, in Chile the basic institutional structure of the water sector based on private service provision is underwritten by means-targeted subsidies and regulated by a public, autonomous entity.

One of the most significant challenges of the twenty-first century will be how to ensure the wide accessibility of adequate supplies of quality drinking water for the world's people. In response to the global water privatization movement, individuals as well as governments and nongovernmental organizations are increasingly arguing that access to safe and adequate supplies of water should be seen as a human right and that large corporations are not capable of guaranteeing that right. It is important to be aware that the conflict over who should provide safe water is occurring in cities and towns throughout the core as well as in the periphery. Besides issues of access and public safety, many of those opposing water privatization argue that conflicts over water are really about fundamental questions of democracy. In particular, if water is no longer a public good, who will make the decisions that affect our future access to it, and who will be denied access? This question is an especially significant one as climate change is expected to have dramatic impacts on water quality and quantity throughout the world.

in the nineteenth and early twentieth centuries and the more radical political views of nature that gained prominence in the late twentieth century. Among the latter are approaches based on ecotheology—including Christianity, Hinduism, Islam, and Judaism—which reject the long-standing Western tradition. We examine these and other well-known approaches to nature in this section.

Henry David Thoreau (1817–1862), an American naturalist and activist, perhaps best illustrates a view that began to emerge in the mid-nineteenth century in the United States—the incorporation of North American Indian conceptions of nature into approaches to the ecology. Thoreau lived and studied the natural world around the town of his birth, Concord, Massachusetts. He is most famous for his book *Walden*, which chronicles the two years he spent living and observing nature in solitude in a house he built at Walden Pond (**Figure 4.8**). Thoreau represented a significant alternative to the "humans-over-nature" approach that characterized his times. Many people regard him as the originator of a U.S. ecological philosophy.

Thoreau was impressed with the power of nature. He often described its unrestrained and sometimes explosive capacity, which he thought had the potential to overthrow human dominion if left unchecked. He also emphasized the interrelatedness of the natural world, where birds depended upon worms, fish depended upon flies, and so on along the food chain. Most notably, however, Thoreau regarded the natural world as an antidote to the negative effects of technology on the landscape and the American character. Concord was just 32 km (20 mi) west of Boston and an equal distance south of the booming mill towns of Lowell and Lawrence. Although he spent his life in a more or less rural setting, the Industrial Revolution was in full force all around Thoreau, and he was keenly aware of its impacts. In fact, Thoreau's approach to the natural world was very much a response to the impacts of the early forces of globalization. His research on the animals and plants that surrounded Concord was an attempt to reconstruct the landscape as it had existed before colonization and massive European immigration.

Thoreau's ideas embraced European notions of **romanticism**, a philosophy that emphasized the interdependence of humans and nature. In direct revolt against those who espoused a Judeo-Christian understanding of nature, the romantics believed that *all* creatures—human and otherwise—were infused with a divine presence that commanded respect and that humans were not exceptional in this scheme. Rather, human divinity issued from humble participation in the natural community.

A branch of American romanticism known as **transcendentalism** also influenced understandings of nature during the early 1800s. Transcendentalism was espoused most eloquently by a Unitarian minister turned poet and philosopher, Ralph Waldo Emerson, a neighbor and contemporary of Thoreau's. It encouraged people to attempt to rise above nature and the limitations of the body to the point where the spirit dominates the flesh and a mystical and spiritual life replaces a primitive and savage one.

Thoreau and Emerson are two of the most important influences on contemporary American ideas about the human–nature relationship. Another major influence on U.S. environmentalism was George Perkins Marsh, a native Vermonter, who in 1864 wrote a treatise entitled *Man and Nature, or Physical Geography as Modified by Human Action* (heavily revised and republished in 1874 as *The Earth as Modified by Human Action*). The first work to suggest that human beings are significant agents of environmental change, it is considered one of the most important advances in geography, ecology, and resource management in the nineteenth century. Marsh's ideas served as the foundation of the U.S. environmental movement in the twentieth century.

Early in the twentieth century, writers like Gifford Pinchot and politicians like Theodore Roosevelt drew on the ideas of Thoreau, Emerson, and Marsh to advocate the wise use of natural resources and the conservation of natural environments. Their view that nature should be conserved has survived to the present. **Conservation** holds that natural resources should be used thoughtfully and that humans should serve as stewards, not exploiters, of

FIGURE 4.8 Walden Pond today
Though a quiet spot during Thoreau's day, Walden Pond is now a popular summer retreat for local residents.

the natural world. Conservation implies responsibility to future generations as well as to the natural world itself in the utilization of resources. The writings of all these individuals eventually helped to inspire a wide range of environmental organizations, including the Environmental Defense Fund, World Watch Institute, the Nature Conservancy, and the Sierra Club.

Those who espouse a more radical approach to nature see the conservation approach as too passive to be truly effective in protecting the environment. Such individuals believe that conservation leaves intact the political and economic system that drives the exploitation of nature. They believe that nature is sacred and should be preserved, not used at all. This more extreme position, **preservation**, advocates that certain habitats, species, and resources should remain off-limits to human use, regardless of whether the use maintains or depletes the resource in question.

Greenpeace, yet another environmental organization, focuses on the battle against environmental polluters. Greenpeace is global in its reach, meaning that both its membership and its areas of emphasis are international. (We talk more about global environmental organizations later in this chapter.) Greenpeace undertakes direct opposition as well as formal international legal actions. In its membership—with world headquarters in Amsterdam and regional offices in most major industrial countries—as well as in its objectives—halting environmental pollution worldwide—Greenpeace articulates the belief that places are interdependent and what happens in one part of the globe affects us all.

Organizations like Greenpeace, and others like the Sea Shepherd Conservation Society—an international nonprofit organization dedicated to safeguarding the biodiversity of ocean ecosystems—are practical illustrations of approaches to understanding human interactions with nature that have developed since the publication of *Silent Spring* over 40 years ago (**Figure 4.9**). These, as well as other new approaches—including environmental ethics, ecofeminism, deep ecology, environmental justice, and ecotheology—take the view that nature is as much a physical universe as it is a product of social thought. Each provides a different way of understanding how society shapes our ideas about nature.

Environmental ethics is a philosophical perspective that prescribes moral principles as guidance for our treatment of nature. From this perspective society has a moral obligation to treat nature according to the rules of moral behavior that exist for human beings. An aspect of environmental ethics that has caused a great deal of controversy is the idea that animals, trees, rocks, and other elements of nature have rights in the same way that humans do. If the moral system of our society insists that humans have the right to a safe and happy life, then, it is argued, the same rights should be extended to nonhuman nature.

Ecofeminism holds that patriarchy—a system of social ideas that values men more highly than women—is at the center of our present environmental malaise. Because patriarchy has equated women with nature, it has promoted the subordination and exploitation of both. The many varieties of ecofeminism range from nature-based spirituality oriented toward a goddess to more political approaches that emphasize resistance and opposition to the dominant models that devalue what is not male. Some ecofeminists are also environmental ethicists. Not only a movement of the core, ecofeminism has also been widely embraced in the periphery, where women are primarily responsible for the health and welfare of their families in environments that are being rapidly degraded. The unifying objective in all of ecofeminism is to dismantle the patriarchal biases in Western culture and replace them with a perspective that values social, cultural, and biological diversity.

Deep ecology, which shares many points of view with ecofeminism, is an approach to nature revolving around two key components: self-realization and biospherical egalitarianism. Self-realization embraces the view that humans must learn to recognize that they are part of the nonhuman world. Biospherical egalitarianism insists that Earth, or the biosphere, is the central focus of all life and that all components of nature, human and nonhuman, deserve the same respect and treatment. Deep ecologists, like environmental

FIGURE 4.9 Sea Shepherd Society protest Using direct-action as a form of protest, the Sea Shepherd Conservation Society has been involved in scuttling and disabling whaling vessels and intervening in Canadian seal hunts as well as ramming other vessels, among other things, in an attempt to halt these practices. This photograph shows members of the organization on the near right interfering with the progress of a large Japanese whaling ship.

ethicists, believe that there is no absolute divide between humanity and everything else and that a complex and diverse set of relations constitutes the universe. The belief that all things are internally related should enable society to treat the nonhuman world with respect and not simply as a source of raw materials for human use.

Activists in the **environmental justice** movement consider the pollution of their neighborhoods by, for example, factories and hazardous-waste dumps to be the result of a structured and institutionalized inequality that is pervasive in both the capitalist core and the periphery. They see their issues as distinct from those of middle-class mainstream groups like the Sierra Club or even Greenpeace. For environmental justice activists, their struggles are uniquely rooted in their economic status. Thus, theirs are not quality-of-life issues, such as whether any forests will be left for recreation, but sheer economic and physical survival. As a result, the questions raised by environmental justice activists involve the distribution of economic and political resources. Such questions are not easily resolved in courts of law, but speak to more complex issues such as the nature of racism and sexism and of capitalism as a class-based economic system. Like ecofeminism, the environmental justice movement is not restricted to the core. Indeed, poor people throughout the world are concerned that the negative impacts of economic development consistently affect them more than the rich.

Ecotheology calls for a reevaluation of the Western relationship to nature. The term came into prominence in the late twentieth century—mainly in Christian circles, though it has since spread across religious boundaries—in association with the scientific field of ecology. Within religious circles, there is a fear that science may not be capable of inspiring the changes in behavior necessary to thwart continuing environmental destruction. Ecotheologists also argue that capitalist political and economic institutions actively contribute to environmental degradation. In their view, it has become necessary, therefore, to address the current environmental crisis through belief systems that will overcome the inadequacies of humanly created institutions. Ecotheology recognizes the value of other creatures and God's intent for the cosmos as the basis for developing ethical models that take into account politics, economics, and practical issues in the quest for intelligent environmental policies.

All of these approaches attest to a growing concern over the environmental effects of globalization. Acid rain, deforestation, the disappearance of species, nuclear accidents, and toxic waste have all been important stimuli for newly emerging philosophies about the relationships between society and nature within a globalizing world. While none of these philosophies is a panacea, each has an important critique to offer. More than anything, however, each serves to remind us that environmental crises are not simple, and simple solutions will not suffice.

APPLY YOUR KNOWLEDGE Provide three examples of how Thoreau's ideas about nature can be understood in terms of late-twentieth-century environmental philosophies such as environmental justice and ecotheology. (Hint: What is Thoreau saying about nature and what are the different environmental movements' views of nature? How are they similar?) Apply the same analysis to Marsh or Emerson. ∎

EUROPEAN EXPANSION AND GLOBALIZATION

The history of European expansion provides a powerful example of how a society possessing new environmental attitudes was able to radically transform nature. These new attitudes drew upon a newly emerging science and its contribution to technological innovation as well as the capitalist political and economic system.

Initially, European expansion was internal—largely contained within continental boundaries. The most obvious reason for expansion was population increase: from 36 million in 1000 to over 44 million in 1100, nearly 60 million in 1200, and about 80 million by 1300 (**Figure 4.10**). As population increased, more land was brought under cultivation. In addition, more forest land was cleared for agriculture, more animals were killed for food, and more minerals and other resources were exploited for a variety of needs. Forests originally covered upward of 90 percent of Western and Central Europe. At the end of the period of internal expansion, however—around 1300—the forested area amounted to only 20 percent.

The bubonic plague, also known as the Black Death, temporarily slowed population growth by wiping out over a third of Europe's people in the mid-fourteenth century. By then, agricultural settlement had taken up all readily available land and had even extended into less desirable areas. In England, Italy, France, and the Netherlands, for example, marshes and fens had been drained and the sea pushed back or the water table lowered to reclaim and create new land for agriculture and settlement.

In the fifteenth century, Europe underwent its second phase of expansion. This phase was external and not only changed the global political map but launched a period of environmental change that continues to this day. European external expansion—colonialism—was the response to several impulses, ranging from self-interest to altruism. Europeans were fast running out of land, and as we saw in Chapter 2, explorers were being dispatched by monarchs to conquer new territories, enlarge their empires, and collect tax revenues from new subjects. Many of these adventurous individuals were also searching for fame and fortune or avoiding religious persecution. Behind European external expansion was also the Christian impulse to bring new souls into the kingdom of God. Other forces behind European colonialism included the need to expand the emerging system of trade, which ultimately meant increased wealth and power for a new class of people—the merchants—as well as for the aristocracy.

Over the centuries, Europe came to control increasing areas of the globe. The two examples we will discuss in the next sections of the chapter illustrate how the introduction of European people, ideologies, technologies, plant species, pathogens, and animals changed the environments into which they were introduced and also the societies they encountered.

Disease and Depopulation in the Spanish Colonies

Historians generally agree that the European colonization of the New World was eventually responsible for the greatest loss of human life in history. Historians also agree that the primary reason for that loss was disease. New World populations, isolated for millennia from the Old World, possessed immune systems that had never encountered some of the most common European diseases. **Virgin soil epidemics**—in which the population at risk has no natural

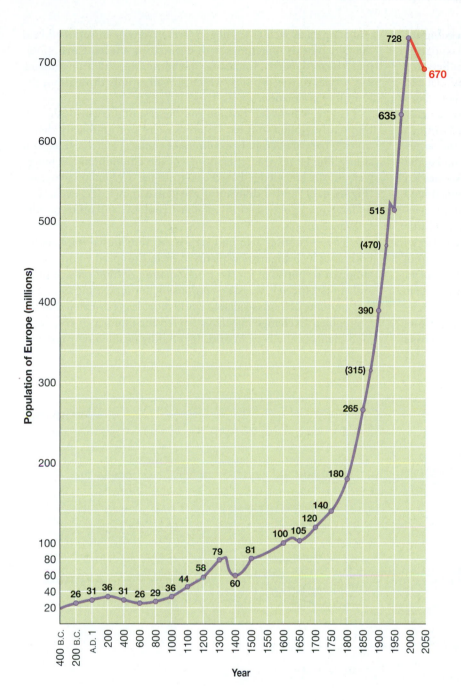

FIGURE 4.10 Population growth in Europe This graph shows growth and change in the European population from 400 B.C. to A.D. 2050. The growth in European population has been especially dramatic in the last 500 years as a result of capitalist globalization. The increase in human numbers at the beginning of the 1500s gave an important push to exploration and colonization beyond the confines of the continent. The dip in the graph from 1300 to 1500 is partially explained by the bubonic plague epidemic and food shortages. The dip around the mid-twentieth century shows the effect of the two world wars. European population is expected to continue to decline over the next four decades. (Updated from C. McEvedy and R. Jones, *Atlas of World Population History.* London: Allen Lane, 1978, Fig. 1.2, p. 18.)

case is Hispaniola (present-day Haiti and the Dominican Republic), to which Columbus's 1493 voyage probably brought influenza through the introduction of European pigs carrying swine fever. Subsequent voyages brought smallpox and other diseases that eventually led to the extinction of the island's Arawak population.

In a second example, in Central Mexico, Lovell writes of Hernán Cortés's contact with the Aztec capital of Tenochtitlán in the first decades of the sixteenth century. This interaction led to a devastating outbreak of smallpox among the local virgin soil population. In a third example, Lovell described the Jesuits' efforts to establish missions in northern Mexico during a slightly later period. Because these efforts gathered dispersed population groups into single locations, ideal conditions were created for the outbreak of disease. Contact with Spanish conquistadors in advance of the missionaries had already reduced native populations by perhaps 30 to 50 percent. When native groups were confined to smaller areas organized around a mission, mortality rates climbed to 90 percent. Eventually, disease was diffused beyond the initial area of contact as traders carried it across long-distance trade routes to the periphery of the Mayan empire in advance of the Spanish armies and missionaries.

immunity or previous exposure to the disease within the lifetime of the oldest member of the group—were common in the so-called Columbian Exchange. Though in this case mostly one-way, the **Columbian Exchange** was the interaction between the Old World (Europe) and the New World (the Americas) initiated by the voyages of Columbus. Diseases such as smallpox, measles, chicken pox, whooping cough, typhus, typhoid fever, bubonic plague, cholera, scarlet fever, malaria, yellow fever, diphtheria, influenza, and others were unknown in the pre-Columbian New World.

Geographer George Lovell has examined the role disease played in the depopulation of some of Spain's New World colonies from the point of initial contact until the early seventeenth century, using several cases to illustrate his argument.[1] The first

Lovell describes similar disease impacts in Mayan Guatemala and the Central Andes of South America that also led to devastating depopulation. Scholars refer to the phenomenon of near genocide of native populations as **demographic collapse**. The ecological effect of the population decline caused by the high rates of mortality was the transformation of many regions from productive agricultural land to abandoned land. Many of the Andean terraces, for example, were abandoned and dramatic soil erosion ensued. And large expanses of cleared land eventually returned to forest in areas such as the Yucatán.

Old World Plants and Animals in the New World

A second case study of the environmental effects of European colonization involves the introduction of Old World plants and animals into the New World, and vice versa. The introduction of

[1]W.G. Lovell, "Heavy Shadows and Black Night: Disease and Depopulation in Colonial Spanish America," *Annals, Association of American Geographers, 82,* 1992, pp. 426–443.

exotic plants and animals into new ecosystems is called **ecological imperialism**, a term now widely used by geographers, ecologists, and other scholars of the environment. The interaction between the Old and the New Worlds resulted in both the intentional and unintentional introduction of new crops and animals on both sides of the Atlantic. Europeans brought from their homelands many plants and animals that were exotics, that is, unknown to American ecosystems. For example, the Spanish introduced wheat and sugarcane, as well as horses, cattle, and pigs. These introductions altered the environment, particularly as the emphasis on select species led to a reduction in the variety of plants and animals that constituted local ecosystems. Inadvertent introductions of hardy exotic species included rats, weeds such as the dandelion and thistle, and birds such as starlings, which crowded out the less hardy indigenous species. As with the human population, the indigenous populations of plants, birds, and animals had few defenses against European plant and animal diseases and were sometimes seriously reduced or made extinct through contact.

Contact between the Old and the New Worlds was, however, an exchange—a two-way process—and New World crops and animals as well as pathogens were likewise introduced into the Old World, sometimes with devastating effects. Corn, potatoes, tobacco, cocoa, tomatoes, and cotton were all brought back to Europe; so was syphilis, which spread rapidly throughout the European population.

Contacts between Europe and the rest of the world, though frequently violent and exploitative, were not uniformly disastrous. There are certainly examples of contacts beneficial to both sides. The largely beneficial ones of the Columbian Exchange were mostly knowledge-based or nutritional. Columbus's voyages (**Figure 4.11**) added dramatically to global knowledge, expanding understanding of geography, botany, zoology, and other rapidly growing sciences.

The encounter also had significant nutritional impacts for both sides by bringing new plants to each. European colonization, although responsible for the extermination of hundreds of plant and animal species, was also responsible for increasing the types and amounts of foods available worldwide. It is estimated that the Columbian Exchange may have tripled the number of cultivable food plants in the New World. It certainly enabled new types of

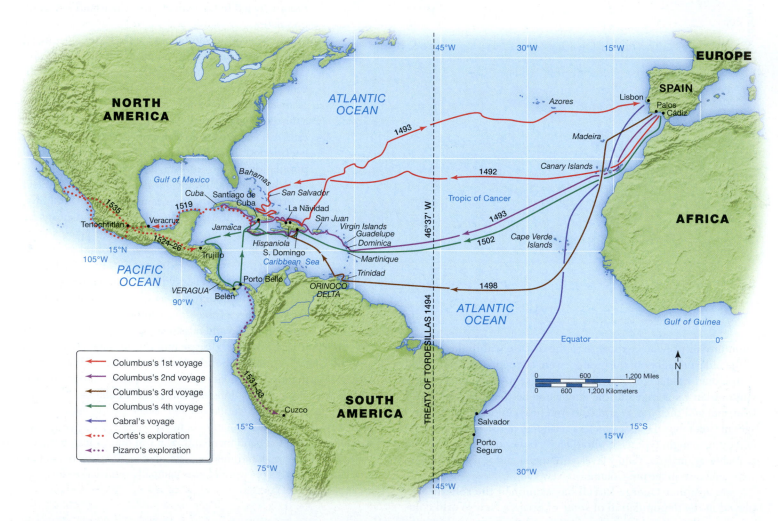

FIGURE 4.11 European voyages of exploration The lines on this map illustrate the voyages and missions of Columbus, Pizarro, Cabral, and Cortés. Departing from Portugal and Spain, Columbus encountered several of the islands of the Caribbean as well as the coastal area of present-day Honduras and Venezuela. (Adapted from *The Penguin Atlas of the Diasporas*, by G. Chaliand and J. P. Rageau, translated by A. M. Berrett. Translation copyright © 1995 by G. Chaliand and J. P. Rageau.)

food to grow in abundance where they had never grown before, and it introduced animals as an important source of dietary protein. The advantages of having a large variety of food plants are myriad. For instance, if one crop fails, another more than likely will succeed because not all plants are subject to failure from the same set of environmental conditions.

The introduction of animals provided the New World with both additional sources of protein and additional animal power. Before the Columbian Exchange, the only important sources of animal energy were the llama and the dog. The introduction of the horse, the ox, and the ass created a virtual power revolution in the New World. These animals also provided fibers and, after death or slaughter, hides and bones to make various tools, utensils, and coverings. Most significant in its environmental impact, however, was the ox. Land that had escaped cultivation because the indigenous digging sticks and tools were unable to penetrate the heavy soil and matted root surface became workable with an ox-drawn plow. The result was that the indigenous form of intensive agricultural production (small area, many laborers) was replaced by extensive production (large area, fewer laborers). This transformation was not entirely without negative impacts, such as soil destabilization and erosion.

When discussing the impact of exchange, it is important to acknowledge the impact of native New World peoples on their environment. The popular image of indigenous peoples living in harmony with nature, having only a minimal effect on their environment, is flawed. In reality, different groups had different impacts, and it is erroneous to conflate the thousands of groups into one romanticized caricature. In New England, for example, prior to European contact, some groups hunted for wild game and gathered wild foods. More sedentary types, living in permanent and semipermanent villages, cleared and planted small areas of land. Hunter-gatherers were mobile, moving with the seasons to obtain fish, migrating birds, deer, wild berries, and plants. Agriculturalists planted corn, squash, beans, and tobacco and used a wide range of other natural resources. The economy was a fairly simple one based on personal use or on barter (trading corn for fish, for example). The idea of a surplus was foreign: People cultivated or exploited only as much land and resources as they needed to survive. Land and resources were shared, without concepts such as private property or land ownership. Fire was used to clear land for planting as well as for hunting. Although vegetation change did occur, it was minimal and not irreversible.

Indians of South and Central America altered their environment as well, though perhaps in more dramatic ways. The Aztecs of Mexico and the Incas of Peru had developed complex urban civilizations—many of them more sophisticated and highly populated than any European cities—dependent on dense populations and employing intensive agricultural techniques on the urban outskirts (**Figure 4.12**). These groups were responsible for environmental modifications through cultivation techniques that included the irrigation of dry regions and the terracing of steep slopes. Irrigation over several centuries results in the salinization of soils. In the lowland tropics, intensive agricultural practices resulted in widespread deforestation as people cut and set fire to patches of forest, planted crops, and then moved on when soil fertility declined. A surplus was key to the operations of both societies, as tribute by ordinary people to the political and religious elite was required in the form of food, animals, labor, or precious metals. The construction of the sizable Inca and Aztec empires required the production of large amounts of building materials in the form of wood and mortar. Concentrated populations and the demands of urbanization meant that widespread environmental degradation existed prior to European contact.

APPLY YOUR KNOWLEDGE Conduct online or library research to find three plants and animals in your own region that, in your opinion, are examples of ecological imperialism. ∎

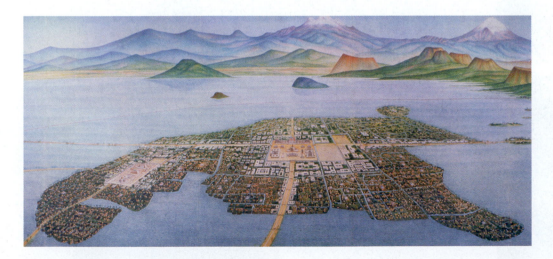

FIGURE 4.12 Tenochtitlán, circa 1500 This famous contemporary painting by Miguel Covarrubias of the capital city of the Aztecs, Tenochtitlán, illustrates the existence of dense social, cultural, and political activity in the core of the city with agricultural fields on the periphery, particularly to the north. Agricultural goods were also imported from the area surrounding the capital beyond the shores of the lake. With a population of two hundred thousand, it was roughly twice the size of Seville, then Spain's largest city. When Cortés came upon the capital he noted that ". . . in Spain there is nothing to compare with it."

HUMAN ACTION AND RECENT ENVIRONMENTAL CHANGE

No other transition in human history has had the impact on the natural world that industrialization has. When we couple industrialization with its frequent companion, urbanization, we have the two processes that, more than any others, have revolutionized human life and effected far-reaching ecological changes. The changes wrought by industrialization have moved beyond a local or a regional level to the entire globe. In order to mark the evidence and extent of the human activities emerging from industrialization that have had a significant global impact on the Earth's ecosystems, Nobel-prize-winning atmospheric chemist Paul Crutzen coined a term, **anthropocene**, to describe the modern era, a word meant to signal a new geological era in Earth's history. While some scientists assert, from atmospheric evidence, that the era began with the Industrial Revolution, others feel it would be more appropriate to set its onset simultaneously with the rise of agriculture. The concept is so new that these disagreements have not yet been settled. In this book we follow the lead of Crutzen, who coined the term, and we use it to explore some of the dramatic contemporary environmental impacts that industrial technology produced. In doing so, we highlight the two issues most central to environmental geography today: energy-use and land-use change.

The Impact of Energy Needs on the Environment

Certainly the most central and significant technological breakthrough of the Industrial Revolution was the discovery and utilization of fossil fuels: coal, oil, and natural gas. Although the very first factories in Europe and the United States relied on waterpower to drive the machinery, hydrocarbon fuels provided a more constant, dependable, and effective source of power. A steady increase in power production and demand since the beginning of the Industrial Revolution has been paralleled, not surprisingly, by an increase in resource extraction and conversion.

At present, the world's population relies most heavily for its energy needs on nonrenewable energy resources that include fossil fuels and nuclear ones, as well as renewable resources such as **biomass**, solar, hydroelectric, wind, and geothermal power. Fossil fuels are derived from organic materials and are burned directly to produce heat. Nuclear energy originates with isotopes that emit radiation. Most commercial nuclear energy is produced in reactors fueled by uranium. Biomass fuels are from biological material from living or recently living organisms and include wood, waste, gas, and alcohol fuels (**Figure 4.13**). Renewable sources of energy such as the sun, wind, water, and steam are captured in various ways and used to drive pumps, machines, and electricity generators.

Fossil Fuels

The largest proportion of the world's current consumption of energy resources, 34 percent, is from oil (see Box 4.2, "Window on the World: Peak Oil"); 25 percent is from coal; 21 percent from gas; 2 percent from hydropower (largely from dams); 7 percent from nuclear power; and 11 percent from combustible biomass and waste (which includes wood, charcoal, crop waste, and dung). The production and consumption of these available resources, however, are geographically uneven, as **Figure 4.14** shows. In 2007, 30 percent of the world's oil supplies were

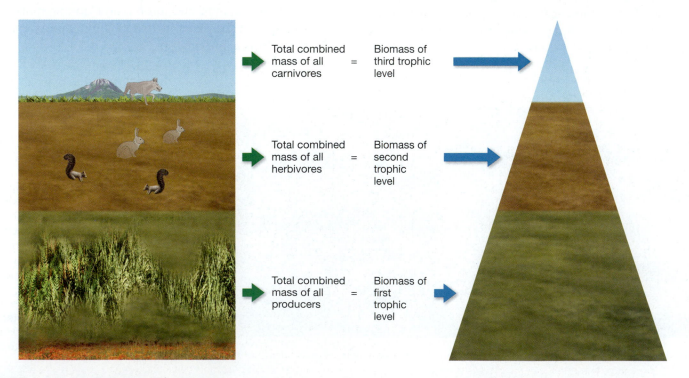

FIGURE 4.13 Biomass pyramid This graphic shows the extent of biomass per unit area within different levels in an ecological system. The bottom level of the representation is usually occupied by the producers, and carnivores are shown in the top levels. (Adapted from http://apesnature.homestead.com/files/fg02_013.jpg.)

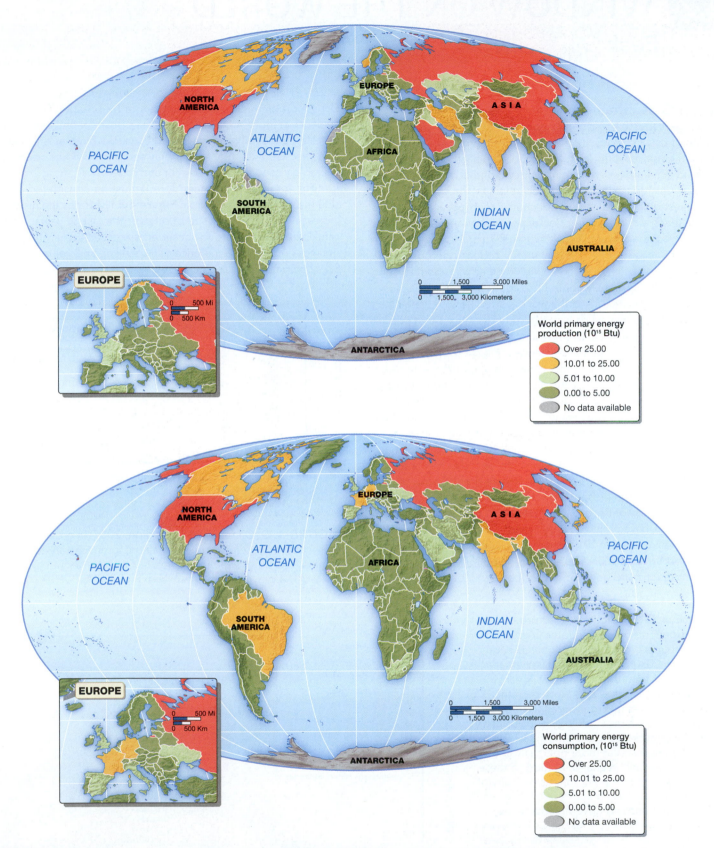

FIGURE 4.14 World production and consumption of energy, 2008 These paired maps provide a picture of the uneven distribution of the production and consumption of energy resources around the world. The United States is the largest producer and consumer of a range of energy resources. Notice that although the Middle East and North African countries as well as Nigeria are important producers of energy resources, their consumption (as well as that of the rest of the African continent, excluding South Africa) is very low. Japan produces a negligible amount of the total of world energy resources but consumes a relatively high share. (*Source:* Adapted from US Energy Information Administration, http://www.eia.gov/cfapps/ipdbproject/IEDIndex3.cfm#).

By Ian G. R. Shaw

Any finite resource has a midway point between the growth and decline of its availability. If we were to plot the rates of oil extraction on a graph, it would loosely resemble a bell-shaped curve and we could label the half-way point, or peak, of this imaginary graph as "Peak Oil." In this sense, Peak Oil refers to the point at which the highest global rate of oil production is reached. Many experts believe that if world consumption of petroleum-derived products is not slowed, the arrival of Peak Oil is likely to bring with it a worldwide energy crisis. Keep in mind, that in order to determine the point of Peak Oil it is necessary to know something about existing oil reserves. It is generally agreed that the summit of oil discovery was passed in the 1960s, and the world started using more oil than was contained in new fields in 1981. The calculations that predict the point of Peak Oil are premised not only on current use data but also on the known global oil reserves (**Figure 4.C**).

Oil is the lifeblood of most economies, and the increasing dependence on oil in core, semiperipheral, and peripheral countries—especially rapidly developing and highly populous ones like China and India—means that Peak Oil will dramatically affect the global economy, demanding proactive mitigation measures by businesses, governments, and individuals. Optimistic predictions place Peak Oil somewhere in the next few decades, whereas pessimistic predictions indicate that Peak Oil has already occurred—or is imminent. Two well-respected scientific networks, the Association for the Study of Peak Oil and Gas (ASPO) and Energy Watch Group (EWG), have projected fairly similar peaks. ASPO argues that Peak Oil was reached in 2010, whereas EWG believes Peak Oil production occurred in 2006. There are some others—mostly influential corporations and government entities—who believe that Peak

Oil is a problematic theory that cannot be substantiated by the evidence (**Figure 4.D**). Whatever one's position on Peak Oil, the other aspect of oil production and consumption that will unarguably affect both the global economy and our individual practices is the rapidly escalating demand for and price of oil.

The forecasts in demand for oil must be considered alongside the increasing cost of oil. **Figure 4.E** illustrates the history of increases in the barrel price of oil over the last 30 years. Significantly, the rising price of oil occurred over a relatively short period of time—in June 2003 the price of a barrel of crude oil was around $30, compared to a record-breaking $130 in June 2008—a rise

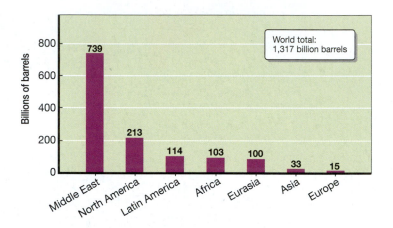

FIGURE 4.C Oil reserves by region, 2007 In order to reasonably project future oil production, it is necessary to have reliable data on known oil reserves. (*Source*: Adapted from "Worldwide Look at Reserves and Production," *Oil and Gas Journal*, 104(47), 2006, pp. 24–25.)

FIGURE 4.D Predictions of future oil production This figure illustrates Energy Watch Group's prediction for global oil supply through 2030. The prediction is dramatically different from the International Energy Agency's (IEA), which is also plotted on the graph, in red. This divergence reflects not only differences in methodology but also contradictory attitudes to future ambiguity—with the IEA opting for a far more optimistic and "business-as-usual" scenario. Muddying these predictions further is the notorious difficulty of obtaining accurate estimations of oil reserves from corporations and governments. (*Source*: Adapted from Energy Watch Group, *Crude Oil: The Supply Outlook*, 2007, p. 12. http://www.energywatchgroup.org/fileadmin/global/pdf/ EWG_Oilreport_10-2007.pdf, accessed May 16, 2011.)

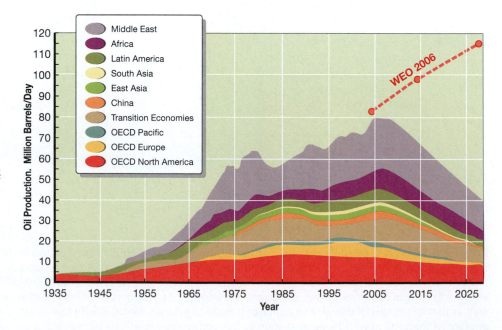

of over 333 percent. The price then began a precipitous decline in advance of the global financial crises of October 2008. More recently, the price of oil has increased again, though its movement is unsteady, with significant rises and declines occurring monthly due to both political and economic instability around the globe. This fluctuation in price illustrates how sensitively entwined oil production and worldwide economies are.

The underlying causes of rising oil prices are highly contested. First, are the economic explanations: Record demand for oil is outstripping available supplies, leading to surging prices. Currently, world consumption of oil is at around 83 million barrels per day, but this is predicted to jump to 118 million barrels in 2030, with two-thirds of this rise coming from increased use in the transportation sector. Other explanations for high prices vary. The Organization of the Petroleum Exporting Countries (OPEC), comprised of 13 countries and responsible for two-thirds of the international oil supply, has been negatively criticized by some for restricting the levels of supply. Many commentators also blame "speculators" for "betting" on rising prices of oil. We can add to our list of economic explanations the falling price of the U.S. dollar. Geopolitical explanations for rising prices are also important: There are fears about the political stability of oil-producing countries like Nigeria and Venezuela and mounting tensions between the United States and Iran. Wars and skirmishes can also drive the price of oil by compromising supply, as happened during the U.S.-led "War in Iraq." In 2002, production capacity in Iraq was over 6 million barrels per day. In 2011, it is about 2.4 million barrels per day. Given these economic and geopolitical driving forces, combined with increased population growth, urbanization, and rising living standards throughout the world, what are some of the impacts and responses we are likely to see?

The arrival of Peak Oil is likely to affect people and places differently—depending not so much on future policies and responses, but the responses made right *now*. This urgency is precipitated by the huge time lags involved with mitigation responses. The *Hirsch Report*, published in 2005 for the U.S. Department of Energy, investigated the likelihood of Peak Oil and the necessary mitigation responses. Crucially, if governments wait for Peak Oil production to arrive, the consequences will be severe, with liquid-fuel deficits for over two decades. Even a response 10 years in *advance* of Peak Oil would likely leave the world in deficit for over a decade. What is further cause for concern is the rapidity with which oil reserves are predicted to potentially dwindle once the Peak is reached. For example, supplies in United Kingdom oil fields might more or less "crash" after the Peak, rather than follow a smoother bell-shaped curve. All this points to the need to make a response *now* rather than later.

Peak Oil is not just a momentary "energy crisis." Right now, there are simply no alternatives for the modes of transportation we use every day, such as cars, trains, planes, and ships. Large-scale agriculture will become increasingly expensive due to the huge energy inputs required. This may lead to the mushrooming of smaller, local and organic farms and produce. Perhaps we will also see the reshaping of the urban landscape in the United States and other core countries and the so-called "end of suburbia." Low-density housing tracts built around the use of the automobile will become increasingly costly places to reside. Migration back to inner cities may occur or, at the very least, surging demand for public transportation. Given the mounting social unrest targeted at rising gas prices in Western Europe and much of the world, it is likely that the transition through Peak Oil will be difficult, contested, and fraught with social and political upheaval.

Ian Shaw is a postdoctoral researcher in the School of Geographical and Earth Sciences at the University of Glasgow.

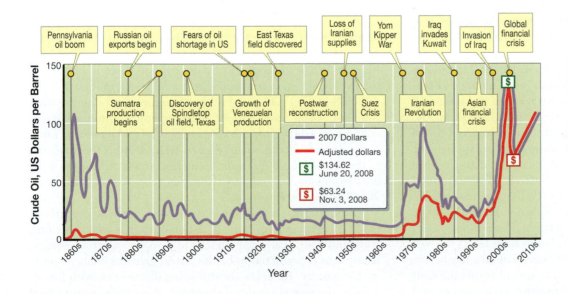

FIGURE 4.E Price of crude oil, 1986–2011 As this figure shows, the barrel price of oil was relatively stable until 2003. Production, consumption, and political factors all contributed to its recent instability. (*Source:* Adapted from M. Janofsky, "Forecasts of $250 Oil Finds Few Believers," *Arizona Daily Star*, June 22, 2008. http://www.azstarnet.com/allheadlines/244867.php, accessed June 22, 2008.)

from the Middle East, and most of the coal was from the Northern Hemisphere, with China accounting for 41 percent of world production, followed by the United States at 19 percent and the former Soviet Union at 8 percent. Nuclear reactors are a phenomenon of the core regions of the world. For example, France has 59 nuclear reactors operated by government-owned Electricité de France, supplying the country with over 430 billion kilowatt hours of electricity per year, which is 78 percent of the total generated there.

The consumption of energy also varies geographically. It has been estimated that current annual world energy consumption is equal to what it took about 1 million years to produce naturally. In one year, global energy consumption is equal to about 1.3 billion tons of coal. What is most remarkable is that this is 4 times what the global population consumed in 1950 and 20 times what it consumed in 1850. And according to the $I = PAT$ formula, the affluent core regions of the world far outstrip the peripheral regions in energy consumption. With nearly 4 times the population of the core regions, the peripheral regions account for less than one-third of global energy expenditures. Yet consumption of energy in the peripheral regions is rising quite rapidly as globalization spreads industries, energy-intensive consumer products such as automobiles, and energy-intensive agricultural practices into regions of the world where they were previously unknown. It is projected that within the next decade or so, the peripheral regions will become the dominant consumers of energy (**Figure 4.15**).

Most relevant to our discussion in this chapter, however, is that every stage of the energy conversion process—from discovery to extraction, processing, and utilization—has an impact on the physical landscape. In the coalfields of the world, from the U.S. Appalachian Mountains to western Siberia, mining results in a loss of vegetation and topsoil, in erosion and water pollution, and in acid and toxic drainage. It also contributes to cancer and lung disease in coal miners (**Figure 4.16**). In addition, coal burning is associated with relatively high emissions of environmentally harmful gases, such as carbon dioxide and sulfur dioxide (**Figure 4.17**).

The burning of home heating oil, along with the use of petroleum products for fuel in internal combustion engines, also launches harmful chemicals into Earth's atmosphere—causing air pollution and related health problems. The production and transport of oil have resulted in oil spills and substantial pollution of water and ecosystems. Media images of damage to seabirds and mammals after tankers have run aground and spilled oil have shown how immediate the environmental damage can be. Indeed, the oceans are acutely affected by the widespread use of oil for energy purposes. There is no better example of the damage of oil spills to the environment than the accident that occurred when British Petroleum's *Deepwater Horizon* oil rig exploded and caught fire on April 20, 2010, in the Gulf of Mexico, 400 km (250 miles) offshore of Houston. The explosion killed 11 people and injured a least two dozen more when a blowout preventer, intended to check the release of oil, failed to activate. As the well continued to leak an estimated 40,000 barrels of oil a day before the well was effectively capped, oil slicks spread across the Gulf from Florida to Texas, contaminating the environment and

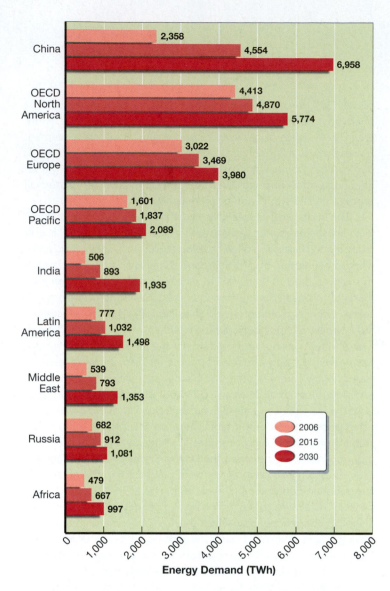

FIGURE 4.15 Increase in energy demand This graph looks at rising demand for energy across key world regions from 2006–2030. (*Source:* Adapted from "New Challenges in Power Generations," October 27, 2009. Eunergy – European Energy Trends. OECD/IEA (2008).)

seriously damaging the Gulf coastal fisheries and tourism economy, especially in Lousiana, Mississippi, Alabama, and northern Florida. When the well was finally capped three months after the explosion, it is estimated that a total of 4.9 million barrels, or 205.8 million gallons, of oil had been released into the environment. As the world largest accidental oil spill to date, it will continue to contaminate or damage marine and terrestrial life—both human and nonhuman (**Figure 4.18**). Despite the huge cost of the Gulf oil spill cleanup (estimated at $40 billion) to British Petroleum, the company registered a 17 percent profit in its 2011first-quarter earnings.

Natural gas is one of the least noxious of the hydrocarbon-based energy resources because its combustion is relatively clean. Now supplying nearly one-quarter of global commercial energy, natural gas is predicted to be the fastest-growing energy source in this century. Reserves are still being discovered, with Russia holding the largest amount—about one-third of the world's total

FIGURE 4.16 Coal mining This coal-mining operation in western Germany is typical of the surface-mining technologies used to exploit shallow deposits in most parts of the world. After deposits are located, the vegetation and overburden (rocks and dirt overlaying the coal seam) are removed by bulldozers and discarded as waste material. With the coal seam exposed, heavy equipment is used to mine the deposit. Some countries, including the United States, have laws that require restoration of newly mined landscapes. Successful restoration can make it difficult to tell that an area was once a mining site. Sites exploited before these laws were introduced remain unrestored, however. Unfortunately, many such sites are in arid or semiarid areas where soil and climate prevent full restoration. In addition to substantial land disturbance, the mining and processing of coal resources often cause soil erosion and water and air pollution.

FIGURE 4.17 Air pollution, Shanghai, China China has taken a lax approach to air pollution, even as it affects some of its largest cities, like Shanghai pictured here. Coal powered plants, like this one, pour SO_2 into the air. Volatile Organic Compounds (VOCs) have been recognized as a kind of important air pollutant contributing to urban photochemical smog and haze also shown in the photo. VOCs are highly hazardous to human health. One significant source of VOCs in Shanghai is the gas vapor from over 20 oil storages, more than 800 gas stations and above 200 oil tanker vehicles.

(**Figure 4.19**). While regarded as a preferred alternative to oil and coal, natural gas is not produced or consumed without environmental impacts. The risk of explosions at natural gas conversion facilities is significant; groundwater contamination and leakages and losses of gas from distribution systems contribute to the deterioration of Earth's atmosphere.

Nuclear Energy

At the midpoint of the twentieth century, nuclear energy was widely promoted as a clearly preferable alternative to fossil fuels. It was seen by many as the answer to the expanding energy needs of core countries, especially as the supply of uranium worldwide was thought to be more than adequate for centuries of use. Nuclear energy was also regarded as cleaner and more efficient than fossil fuels. Although nuclear war was a pervasive threat, and there were certainly critics of nuclear energy even in the early years of its development, the civilian "atomic age" was widely seen as a triumphant technological solution to the energy needs of an expanding global economic system. It was when serious accidents at nuclear power plants began to occur—such as those at

FIGURE 4.18 Oil spill caused by the explosion of the *Deepwater Horizon* Rig Pictured is an aerial view of crude oil on May 24, 2010 in Elmer Island Louisiana. Skimmer ships, floating containment booms, anchored barriers, and sand-filled barricades along shorelines as well as chemical dispersants were used to attempt to prevent hundreds of miles of beaches and wetlands from the spreading oil. In addition, underwater plumes of dissolved oil not visible at the surface as well as a 210 kilometer (80 square mile) "kill zone" surrounding the blown well were also reported by scientists.

FIGURE 4.19 Global natural gas reserves, 2010 Almost three-quarters of the world's natural gas reserves are located in the Middle East and Eurasia. Together, Russia, Iran, and Qatar account for about 55 percent of the world's natural gas reserves. Four major natural gas producers in the Middle East—Qatar, Iran, Saudi Arabia, and the United Arab Emirates—account for 84 percent of the natural gas produced in the Middle East. (*Source:* Adapted from BP World Energy Report.)

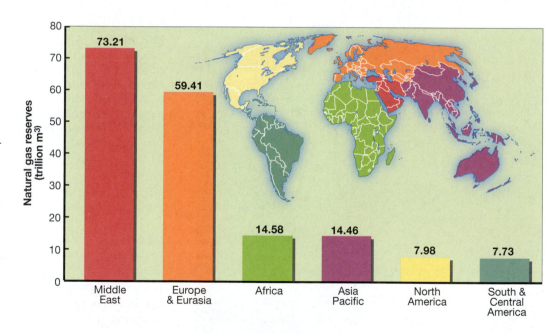

Windscale in Britain and Three Mile Island in the United States, Chernobyl in Russia, and now Fukushima in Japan—that the voices of concerned scientists and citizens were raised in opposition. Incontestable evidence points to the need to address the problems associated with nuclear energy production, such as how to ensure nuclear reactor safety and how to safely dispose of nuclear waste (which remains radioactive for tens of thousands of years). After the devastating accidents and meltdowns, many countries decided to go nuclear-free, such as Sweden, Denmark, Finland, and New Zealand. In light of the recent meltdown at the Fukushima plant in Japan, the safety of nuclear energy is again being publicly debated. France, for example, which derives 75 percent of its electricity from nuclear energy power, saw protests

against its continued use emerge in the spring of 2011. Other opposition to nuclear power occurred across Europe in Germany, Bulgaria, Italy, and the United Kingdom as well as in the United States, Asia, and Latin America.

And yet, while some core countries have moved away from nuclear energy because of the possibility of environmental disaster in the absence of fail-safe reactors, many more—and especially populous—developing countries are moving in the opposite direction, even in the wake of the Fukushima disaster (**Figure 4.20**). Given that world energy consumption is predicted to increase by 57 percent between now and 2030, it is not surprising that rapidly developing countries like India, South Korea, and China have growing nuclear energy programs. And

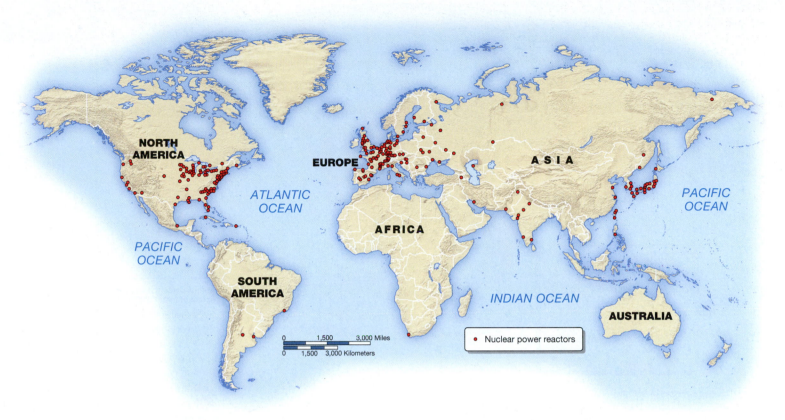

FIGURE 4.20 World distribution of nuclear reactors, 2011 Most of the dependence on nuclear power is concentrated in the core countries. South America and Africa together contain only four nuclear reactors. With the recent accident in Japan, some countries, such as France, that were very enthusiastic supporters of nuclear energy, are now experiencing popular opposition to them. Although just five years ago nuclear power was experiencing new popularity, a reconsideration of its ultimate value, when disasters cannot be 100% prevented, is taking place.

(*Source:* International Nuclear Safety Center—http://www.insc.anl.gov/pwrmaps/.)

developing Eastern European countries like Czechia and Romania, as well as Russia and Belarus, are continuing to invest in nuclear power.

Biomass Fuels

While nuclear power problems are still largely confined to the core, the periphery is not without its energy-related environmental problems. Because a large proportion of populations in the periphery rely on biomass for their energy needs, as the populations have grown, so has the demand for fuelwood. One of the most immediate environmental impacts of wood burning is air pollution, but the most alarming environmental problem is the rapid depletion of forest resources. With the other conventional sources of energy (coal, oil, and gas) being too costly or unavailable to most peripheral households, wood or other forms of biomass—any form of material that can be used as fuel, such as animal wastes, livestock operation residues, and aquatic plants—is the only alternative. The demand for fuelwood has been so great in many peripheral regions that forest reserves are being rapidly used up **(Figure 4.21)**.

Fuelwood depletion is extreme in the highland areas of Nepal, as well as in the Bolivian and Peruvian Andes Mountains. The clearing of forests for fuelwood in these regions has led to serious steep-slope soil erosion. In sub-Saharan Africa,

where 90 percent of the region's energy needs are supplied by wood, overcutting of the forests has resulted in denuded areas, especially around rapidly growing cities. And although wood gathering is usually associated with rural life, it is not uncommon for city dwellers to use wood and other forms of biomass to satisfy their household energy needs as well. In Niamey, the capital of Niger, the zone of overcutting is expanding as the city itself expands. It is estimated that city dwellers in Niamey travel from 50 to 100 kilometers (31 to 62 miles) to gather wood. The same goes for inhabitants of Ouagadougou in Burkina Faso, where the average haul for wood is also over 50 kilometers. And fuelwood use in Asian and South American highland regions as well as in sub-Saharan Africa is expected to continue to increase substantially (34 percent by some estimates) over the next decade, at least.

Hydropower

Hydroelectric power, also known as hydropower, was also once seen as a preferred alternative to the more obviously environmentally polluting and nonrenewable fossil-fuel sources. The wave of dam building that occurred throughout the world over the course of the twentieth century improved the overall availability, quality, cost, and dependability of energy as well as harnessed water resources for food production, energy generation, flood control,

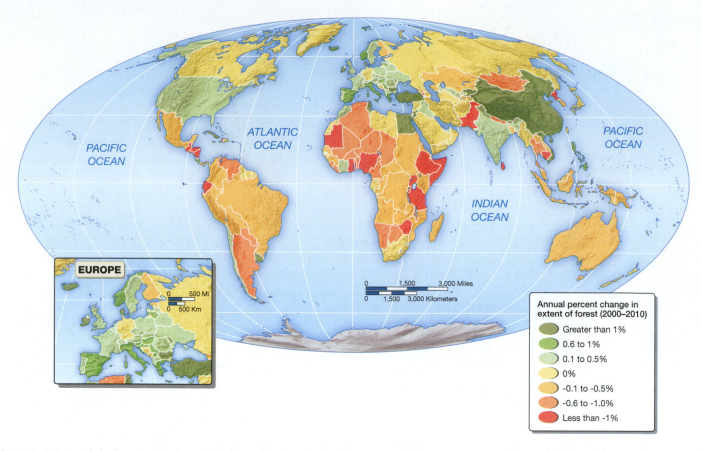

FIGURE 4.21 Global consumption of fuelwoods, 2010 Firewood, charcoal, and dung are considered traditional fuels, and although their availability is decreasing, dependence upon them is increasing. Dependence on traditional sources of fuel is especially high in the periphery, where, in Africa, for example, they are the most important energy source for cooking and heating. Wood and charcoal, although renewable sources, are replenished very slowly. Acute scarcity will be a certainty for most African households in the twenty-first century.

(*Source:* Data from UN Food and Agricultural Organization, *State of the World's Forests 2011*, Annex, Tab. 4, Rome, 2011.)

and domestic use (**Figure 4.22**). Unfortunately, however, dams built to provide hydroelectric power (as well as water for irrigation, navigation, and drinking) for the burgeoning cities of the core and to encourage economic development in the periphery and semiperiphery have also had profound negative environmental impacts. Among the most significant of these impacts are changes in downstream flow, evaporation, sediment transport and deposition, mineral quality and soil moisture, channeling and bank scouring, and aquatic biota and flora, as well as conditions threatening to human health.

Furthermore, the construction of dams dramatically alters terrain, often with serious consequences. For example, clearance of forests for dam construction often leads to large-scale flooding. Felled trees are usually left to decay in the impounded waters, which become increasingly acidic. The impounded waters can also incubate mosquitoes, which carry diseases such as malaria. The effects on human populations are also significant as dam projects often require the relocation of long-settled populations. These populations are often indigenous and their displacement disrupts dense social networks, generations-long connections to the land, as well as established livelihoods.

Despite these problems and in response to the rising cost of fossil fuels, the use of hydroelectricity is expected to continue to expand over the next two decades by 1.9 percent per year, particularly where its use is supported by government policies and incentives, such as in Asia and South America. The construction of the Three Gorges Dam, as well as several other large-scale hydroelectric projects in China, illustrates this trend (**Figure 4.23**). In contrast, most of the increase in core energy production is expected to be in the form of renewable resources, such as wind, solar, geothermal, municipal solid waste, and biomass.

Energy-Related Pollutants

One reason hydroelectric power continues to be appealing in the periphery is that it produces fewer atmospheric pollutants than fossil fuels. Indeed, coal and gas power stations as well as factories, automobiles, and other forms of transportation are largely responsible for the increasingly acidic quality of Earth's atmosphere. Although it is true that people as well as other organisms naturally produce many gases, including oxygen and carbon dioxide, increasing levels of industrialization and motor

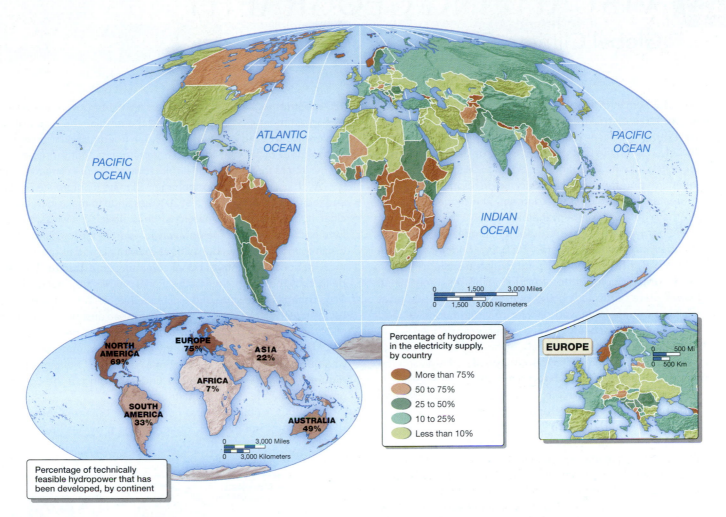

FIGURE 4.22 Percentage of hydropower in the electricity supply by country, 2010 Although the great dam-building era for core countries is now largely completed, many peripheral countries, in a bid to participate more actively in the world economy, are building dams. Only a few countries are almost exclusively dependent on the hydropower produced from dams. These include Norway, Nepal, Zambia, Ghana, Paraguay, and Costa Rica. Although the power produced by dams is environmentally benign, the construction of large dams can be extremely destructive to the environment and can dislocate large numbers of people. Still, given the increasing need for electricity by rapidly developing peripheral countries, hydropower is becoming a more attractive energy option for many of them because of the uncertain supply of oil in the future. The larger map shows the amount of hydropower that is currently available; the smaller one shows the potential for hydropower development, especially for peripheral regions. (*Source:* U.S. Energy Information Administration, U.S. Department of Energy.)

vehicle use have destabilized the natural balance of such gases, leading to serious atmospheric pollution (see Box 4.3, "Visualizing Geography: Global Climate Change"). Sulfur dioxide, nitrogen oxides, and hydrocarbons, among other gases, which are released into the atmosphere from motor vehicle exhaust, industrial processes, and power generation (based on fossil fuels), are increasing the level of acids in the atmosphere. If these gases reach sufficient concentrations and are not effectively dispersed in the atmosphere, acid rain can result.

Acid rain or acid deposition, as it is known scientifically, is the wet deposition of acids upon Earth through the natural cleansing properties of the atmosphere. The term *acid rain* also includes acid mists, acid fogs, and smog. Acid rain occurs as the water droplets in clouds absorb certain gases that later fall back

to Earth as acid precipitation. **Figure 4.24** illustrates the widespread nature of acid emissions. The problem first emerged as a health problem in the industrial countries of the Northern Hemisphere, especially the United Kingdom, Scandinavia, the United States, and Canada, in the 1950s and eventually came to be seen as an environmental concern as well. Acid deposition poisons soils and water bodies that become too acidic to support life. In urban areas, acid rain corrodes marble and limestone structures, affecting important buildings such as the Parthenon in Athens and St. Paul's Cathedral in London as well as others in Europe. By the late 1990s, however, acid deposition became far less of a problem in Europe and North America as decades of environmental regulations and cooperative agreements diminished the release of environmental pollutants into the atmosphere. The problem

The issue of a human-induced (anthropogenic) change in Earth's climate, which is distributed *unevenly* across the planet, has become a growing concern for scientists and policymakers since the 1980s. By 1992, at the Earth Summit in Rio de Janeiro, the international community had begun to seek a way to strike a balance between increasing the pace of economic development without further threatening the global environment. The biggest potential threat to the global environment then and now is the impact that increased energy use will have on global climate. Extensive climate changes may alter and threaten the living conditions of much of humankind. These changes may induce large-scale migration and lead to greater competition for the earth's resources. Such changes will place particularly heavy burdens on the world's most vulnerable countries. Increased danger of violent conflicts and wars, within and between states, is one possible consequence.

At the Rio Earth Summit, 167 nations ratified the UN Framework Convention on Climate Change with the aim of reducing the amount of **greenhouse gases (GHGs)** generated by energy use. A GHG is any gas that absorbs infrared radiation in the atmosphere, including, but not limited to, water vapor, carbon dioxide (CO_2), methane (CH_4), and nitrous oxide (N_2O). An equally critical aim of the convention is to ensure that the burden of protecting the environment is shared equitably across all nations.

In December 1997, the nations that attended the 1992 summit began to confront the problem of balancing global economic development and environmental protection more substantively by forging the Kyoto Protocol. The protocol marked the first attempt to limit the amount of greenhouse gas emissions generated by core countries. The aim of the protocol is to cut the combined emissions of greenhouse gases from core countries by roughly 5 percent from their 1990 levels by 2012. (Recall from earlier in the chapter that core countries account for a disproportionate amount of emissions.) In late 2007, the climate community met again, this time in Bali, Indonesia, for the UN Climate Change Conference. More than 10,000 participants attended, including representatives of over

180 countries together with observers from intergovernmental and nongovernmental organizations and the media. The result of the conference was the adoption of the Bali Roadmap, which consists of a number of forward-looking commitments that represent the various tracks that are essential to reaching a secure climate future.

As the subsequent Copenhagen Climate Conference of 2009 and the Cancún Agreements of 2010 continue to make increasingly clear, human-induced climate change is affecting a wide range of Earth systems and that the only way to address these changes is to begin to modify human behavior as individuals and as members of local, national, and international communities. **Climate change** is defined by the Intergovernmental Panel on Climate Change (IPCC) as "a change in the state of the climate that can be identified (e.g., using statistical tests) by changes in the mean and/or the variability of its properties, and that persists for an extended period, typically decades or longer. It refers to any change in climate over time, whether due to natural variability or as a result of human activity." **Figure 4.F** provides a schematic understanding of the complex process of climate change and its effects.

The IPCC's *Fourth Assessment Report: Climate Change Synthesis Report: Summary for Policymakers* provides helpful data

FIGURE 4.F Schematic framework of climate change drivers, impacts, and responses This diagram represents human-induced drivers, impacts of and responses to climate change, and their linkages. (*Source:* UNEP/GRID-Arendal, http://www.grida.no/graphic.aspx?f=series/vgclimate2/large/1.jpg))

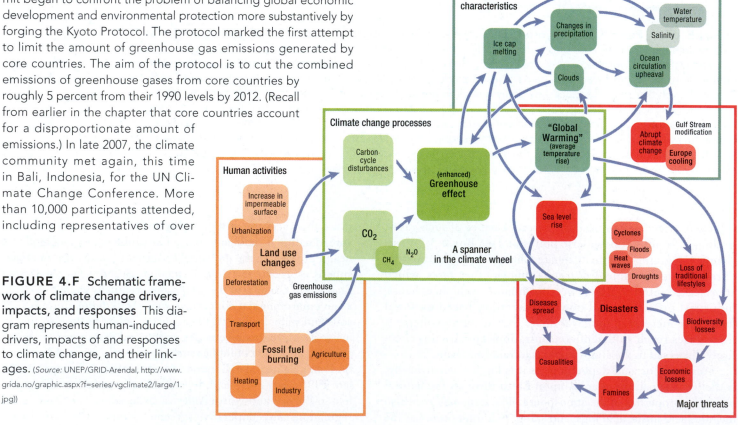

about the state of the world's climate in a format that is easily accessible to a lay audience. We draw directly from that report as well as synthesize some of its finding in the following content.

Observations of Climate Change

Eleven of the last twelve years (1995–2006) rank among the 12 warmest years in the instrumental record of global surface temperature (since 1850).

The temperature increase is widespread over the globe and is greater at higher northern latitudes. In the past 100 years, average Arctic temperatures have increased at almost twice the global average rate.

Increases in sea level are consistent with warming. Global average sea level rose at an average rate of 1.8 [1.3 to 2.3] mm per year over 1961 to 2003 and at an average rate of about 3.1 [2.4 to 3.8] mm per year from 1993 to 2003.

Satellite data since 1978 show that annual average Arctic sea ice extent has shrunk by 2.7 [2.1 to 3.3] percent per decade, with larger decreases in summer of 7.4 [5.0 to 9.8] percent per decade.

Mountain glaciers and snow cover have declined on average in both hemispheres.

Trends in precipitation amounts from 1900 to 2005 have been observed in many large regions. Over this period, precipitation increased significantly in eastern parts of North and South America, northern Europe, and northern and central Asia, whereas precipitation declined in the Sahel, the Mediterranean, southern Africa, and parts of southern Asia. Globally, the area affected by drought has likely increased since the 1970s.

Average Northern Hemisphere temperatures during the second half of the twentieth century were very likely higher than during any other 50-year period in the last 500 years and likely the highest in at least the past 1,300 years.

Some Projected Impacts

Continued GHG emissions at or above current rates will cause further warming and induce many changes in the global climate system during the twenty-first century that will very likely be larger than those observed during the twentieth century.

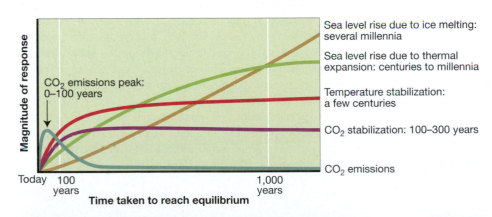

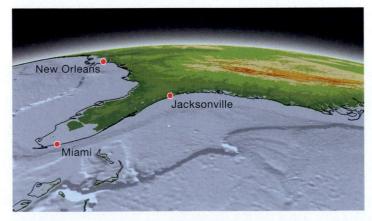

FIGURE 4.G Projected sea level rise This graph shows the changes in sea level rise over varying time scales. The maps show the impact of 5 meters of sea level rise on Florida (l) and Southeast Asia (r) (*Source: From Vital Geo Graphics,*
published by UNEP/GRID-Arendal based on UNEP's *Global Environment Outlook: environment for development (GEO4)*. Copyright 2009 © UNEP, UNEP/GRID-Arendal.)

Snow cover area is projected to contract. Widespread increases in thaw depth are projected over most permafrost regions. Sea ice is projected to shrink in both the Arctic and Antarctic.

Based on a range of models, it is likely that future tropical cyclones (typhoons and hurricanes) will become more intense, with larger peak wind speeds and more heavy precipitation associated with ongoing increases of tropical sea-surface temperatures.

Anthropogenic warming and sea level rise would continue for centuries due to the time scales associated with climate processes and feedbacks, even if GHG concentrations were stabilized (see **Figure 4.G**).

The resilience of many ecosystems is likely to be exceeded this century by an unprecedented combination of climate change, associated disturbances (e.g., flooding, drought, wildfire, insects, ocean acidification), and other global change drivers (e.g., land-use change, pollution, fragmentation of natural systems, overexploitation of resources).

Crop productivity is projected to increase slightly at mid- to high latitudes for local mean temperature increases of up to 1° to 3°C, depending on the crop, and then decrease beyond that in some regions. At lower latitudes, especially in seasonally dry and tropical regions, crop productivity is projected to decrease for even small local temperature increases (1° to 2°C), which would increase the risk of hunger.

Coasts are projected to be exposed to increasing risks, including coastal erosion, due to climate change and sea level rise.

The health status of millions of people is projected to be affected through, for example, increases in malnutrition; increased deaths, diseases, and injury due to extreme weather events; increased burden of diarrheal diseases; increased frequency of cardio-respiratory diseases due to higher concentrations of ground-level ozone in urban areas related to climate change; and the altered spatial distribution of some infectious diseases.

Climate change is expected to exacerbate current stresses on water resources from population growth and economic and land-use change, including urbanization.

Mitigation Efforts

There are many ways that individuals can begin to address the effects of climate change as described above, from using compact fluorescent light bulbs, to taking shorter showers, to buying local produce and using reusable water bottles instead of disposable ones. More concentrated efforts are the responsibility of corporations, governments, and large organizations.

One increasing popular strategy for reducing GHGs is emissions trading. Already an established practice, emissions trading allows governments to regulate the amount of emissions produced in aggregate by setting the overall cap but allowing corporations the flexibility of determining how and where the emissions reductions will be achieved. Corporations that want to limit their emissions are allocated allowances, with each allowance representing a ton of the relevant emission, such as carbon dioxide. Corporations can emit in excess of their allocation of allowances by purchasing allowances from the market. Similarly, a company that emits less than its allocation of allowances can sell its surplus allowances.

Emissions trading is believed to be a more desirable mitigation strategy than strict regulation because it gives companies the flexibility to meet emission reduction targets according to their own strategy, for example, by reducing emissions on-site or by buying allowances from other companies that have excess allowances. Theoretically, the environmental outcome is not affected because the amount of allowances allocated is fixed.

FIGURE 4.23 Three Gorges Dam, China In May 2006, the world's largest dam was completed in Yichang, central China's Hubei Province. The axis of the dam is 2,309 meters (1.4 miles), the longest in the world. Begun in 1993, the project proceeded with the backing of the communist leadership despite objections to its $22 billion cost and projected environmental and social impact. More than 1.3 million people have been relocated to make way for the dam and its reservoir; environmentalists and engineers warned that the reservoir risks becoming polluted with waste from cities and towns upriver. The shore of the reservoir has already collapsed in 91 places, a total of 36 km (22 mi), killing scores of people. Although developing countries in Asia and South America are increasingly looking to dams for the provision of hydroelectricity, there is still a great deal of environmental opposition to them globally. The inset map shows the 477 km (297 mi) extent of the reservoir that is impounded behind the dam.

has not gone away entirely, however, and is emerging with great negative effect in South Asia, especially China, where few governmental limits on industrial pollution are in place. In fact, because of rapid industrial growth, China has some of the world's most polluted cities and rivers. The steady emissions of sulfur dioxide from unregulated factories and vehicles has resulted in one-third of China's territory being affected by acid rain, posing a major threat to soil water and food safety.

Alternative Energy Sources

Before giving up all hope that the use of energy can ever be anything but detrimental to the environment, it is important to realize that alternatives exist to fossil fuels, hydroelectric power, and nuclear energy. Energy derived from the sun, the wind, Earth's interior (geothermal sources), and the tides has been found to be clean, profitable, and dependable. Japan, the United States, and Germany all have solar energy production facilities that are cheap and nonpolluting. Although contributing only small amounts to the overall energy supply, the production of energy from geothermal and wind sources has also been successful in a few locations around the globe. Italy, Germany, Iceland, the United States, Mexico, Denmark, and the Philippines all derive some of their energy production from geothermal or wind sources.

And since, as was previously mentioned, most of the world's energy consumption is for transportation, it's important to mention the trend toward hybrid and electric vehicles, which is still restricted largely to the core due to the high cost of this technology. City and county governments in the United States along with large corporations like Federal Express (FedEx) are already converting their transportation fleets to hybrids. Hybrid and electric automobiles are also very popular in Europe and the United Kingdom. By the spring of 2011, sales of new hybrid and electric cars were growing faster than sales of conventional vehicles in the United States, which has the largest vehicle market in the world. The growing hybrid and electric automobile market is increasingly being bolstered by the fleet market, which every year sees new commitments to all-electric delivery vehicles.

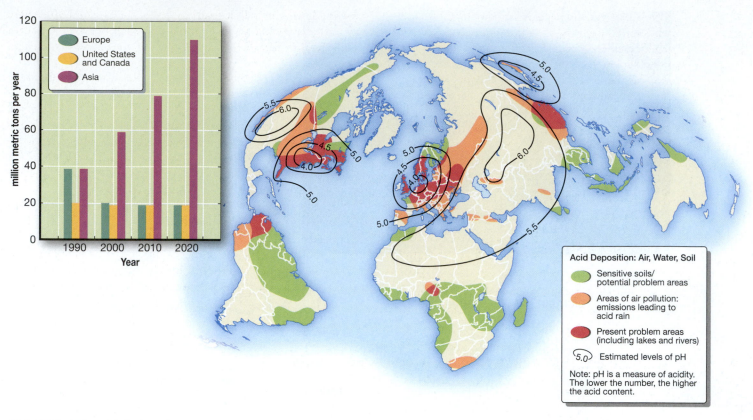

FIGURE 4.24 Global acid emissions Acid emissions affect various elements of the natural and the built environment. In some parts of the world, the damage to soils is especially severe. In others, acid emissions cause serious air pollution. Large amounts of acid-producing chemicals may be generated in one place but exported to another by prevailing winds. Most industrialized countries have cut sulfur dioxide emissions to help mitigate the damage of acid rain to ecosystems. But the acid rain problem is not yet solved, as it is emerging as a major problem in the developing world, especially in parts of the Asia and the Pacific region where energy use has surged. It is likely that the acid emissions experienced in the core countries in the twentieth century will be repeated in the periphery in the twenty-first century. (*Source:* Adapted from J.L. Allen, *Student Atlas of Environmental Issues,* Guilford, CT: Duskin/McGraw-Hill: 1997, p. 45; World Resources Institute, "Acid Rain: Downpour in Asia," *World Resources 1998–1999,* 1998; The International Energy Agency.)

Investment in the development of geothermal, solar, wind, and tidal energy is increasing as petroleum prices increase. In the United States, the Obama administration has funneled new monies into research and development activities as well as into rebates to taxpayers for employing alternative energy sources, especially solar and wind. Investment in geothermal is also increasing; the Australian, Canadian, Chilean, European, and U.S. Geothermal Energy associations have formed an international coalition to encourage further development of geothermal resources around the world (**Figure 4.25**).

One of the most significant changes in energy consumption is greater efficiency of use. In the United States, for example, average household energy use dropped about 31 percent between 1978 and 2005. More energy-efficient appliances—refrigerators, washing machines, heating and cooling systems—have helped as has better insulation and construction to reduce the escape of heat and cooling through the walls, roofs, and windows of homes. It should be pointed out, however, that the increased use of electronic devices—cell phones, televisions, computers—are partly offsetting these greater efficiencies. Energy consumption to fuel the increasing use of these devices is expected to double over the next 10 years.

APPLY YOUR KNOWLEDGE Conduct an Internet search to determine three ways that various countries other than the United States are addressing the problem of acid rain. ■

The Impact of Land-Use Change on the Environment

In addition to industrial pollution and steadily increasing demands for energy, the environment is also being dramatically affected by land use. The clearing of land for fuel, farming, grazing, resource extraction, highway building, energy generation, and war all have significant impacts. Geographers classify land into five categories: forest, cultivated land, grassland, wetland, and areas of settlement. Geographers also speak of land-use change as occurring in either of two ways: conversion or modification. *Conversion* is the wholesale transformation of land from one use to another (for example, the conversion of forest to settlement). *Modification* is an alteration of existing land use (for example, a grassland overlaid with railroad line or a forest thinned and not clear-cut).

One of the most dramatic impacts is loss or alteration of forest cover, which humans have cleared for millennia to make way for cultivation and settlement. Forests are cleared not only to obtain land to accommodate population increases but also to extract the vast timber resources they contain. The approximate chronology and estimated extent of the clearing of the world's forests are listed in **Table 4.1**, which shows that the forested area of the world has been reduced by about 8 million square kilometers (about 3 million square miles) since preagricultural times. Rapid clearance of the world's forests has occurred through logging, settlement, and agricultural clearing or through fuelwood cutting around urban areas. **Figure 4.26** shows the global extent of deforestation in recent years.

Forests

The permanent clearing and destruction of forests, **deforestation**, is currently occurring most alarmingly in the world's rain forests. The UN Food and Agricultural Organization has estimated that we are destroying rain forests globally at the rate of 0.40 hectare (1 acre) per second. Today, rain forests cover less than 7 percent of the land surface, half of what they covered only a few thousand years ago.

Destruction of the rain forests, however, is not just about the loss of trees, a renewable resource that is being eliminated more quickly than it can be regenerated. It is also about the loss of the biological diversity of an ecosystem. This also translates into the potential loss of biological compounds that may have great medical value. In addition, the destruction of rain forests can lead to the destabilization of the oxygen and carbon dioxide cycles of the forests, which may have long-term effects on global climate.

Much of the destruction of the South American rain forests is the result of peripheral countries' attempts at economic development. In the Bolivian Amazon rain forest the introduction of coca production has become an important source of revenue for farmers in the region and has led to the removal of small tracts of forest. The pressure of economic

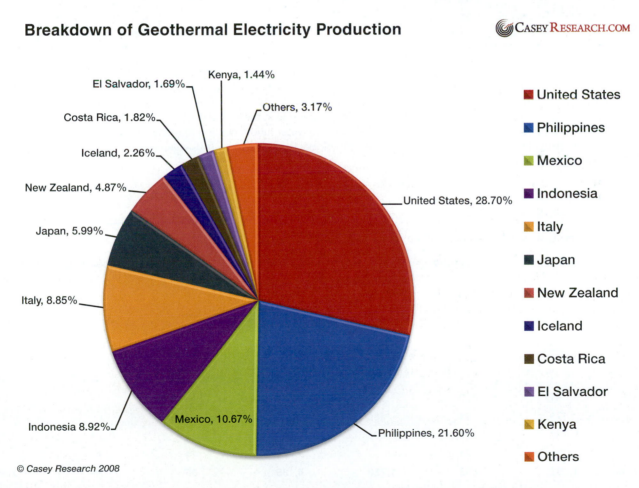

FIGURE 4.25 Geothermal energy Earth's geothermal energy originates from the formation of the planet, from radioactive decay of minerals, from volcanic activity, and from solar energy absorbed at the surface. The geothermal gradient (the difference in temperature between the core of the planet and its surface) drives a continuous conduction of thermal energy in the form of heat from the core to the surface. Where tectonic activity is high, it is normal to find good sources of geothermal energy. This is the reason the Philippines possesses major geothermal resources. (From "5 Huge Green-Tech Projects in the Developing World," Alexis Madrigal, *Wired Science*, March 3, 2009. Chart by Marin Katusa, Chief Investment Strategist, Casey Research Group. Reprinted with Permission.)

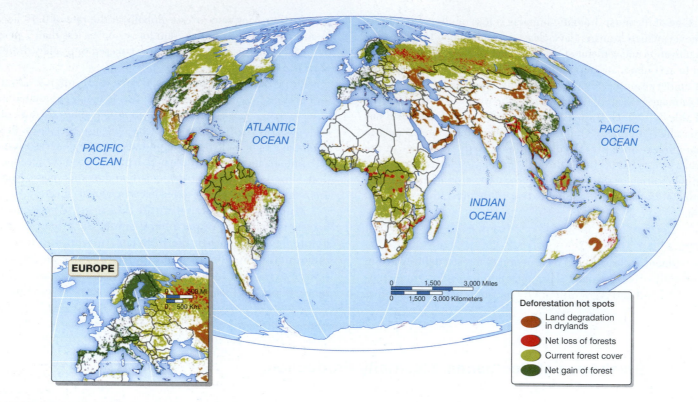

FIGURE 4.26 Global deforestation The world's forests are disappearing or being reduced or degraded everywhere, but especially in tropical countries. Since agriculture emerged about 10,000 years ago, human activities have diminished the world's forest resources by about 25 percent. Whereas forests once occupied about one-third of Earth's surface, they now take up only one-quarter. Playing an important role in the global ecosystem, forests filter air and noise pollution, provide a habitat for wildlife, and slow down water runoff, helping to recharge streams and groundwater. They also influence climate at local, regional, and global levels. (*Source:* World Resources Institute, http://images. wri.org/sdm-gene-02-deforestation.jpg or http://www.wri.org/image/view/10981/_original.)

TABLE 4.1   Estimated Area Cleared (31,000 km²)

Total Region or Country	Pre-1650	1650–1749	1750–1849	1850–1978	1990–2000	2000–2005	Estimate
North America	6	80	380	641	0	1	1,108
Central America	15	30	40	200	+4*	3	288
Latin America and Carribean	12	100	170	637	42	45	1,006
Oceania	4	5	6	362	4	4	385
Former USSR	61	150	260	575	—	—	1,046
Europe	176	54	186	18	+9	+7	481
Asia	732	190	606	1,220	8	10	2,766
Africa	126	24	72	469	44	40	775
Total Estimate	**1,132**	**633**	**1,720**	**4,185**	**85**	**96**	**7,855**

*"Plus" sign indicates a growth, not loss, of total forest area (31,000 km²).

(*Source:* Adapted from B. L. Turner II, W. C. Clark, R. W. Kates, J. F. Richards, J. T. Mathews, and W. B. Meyer, *The Earth as Transformed by Human Action: Global and Regional Changes in the Biosphere over the Past 300 Years.* Cambridge: Cambridge University Press, 1990, p. 180; UN Food and Agricultural Organization, *State of the World's Forests,* 2007, pp. 109–115.)

development persists in the region such that net rain forest and forest land more generally are continuing to decline in Central and South America at a faster rate than in the 1990s.

Great geographical variability exists with respect to human impacts on the world's forests. For most of the core regions, the forests that were once cleared have been replaced by regeneration. For most of the periphery, in contrast, clearance has accelerated to such an extent that one estimate shows a 50 percent reduction in the amount of forest cover since the early 1900s.

Cultivated Lands

Cultivation is another important component of global land use, which we will deal with extensively in Chapter 8. In this section, we briefly cover one or two points about the environmental impacts of cultivation that are particularly pertinent to our current discussion. During the past 300 years the land devoted to cultivation has expanded globally by 450 percent. In 1700, the global stock of land in cultivation took up an area about the size of Argentina. Today, it occupies an area roughly the size of the entire continent of South America. While the most rapid expansion of cropland since the mid-twentieth century has occurred in the peripheral regions, the amount of cropland has either held steady or been reduced in core regions. The expansion of cropland in peripheral regions is partly a response to growing populations and rising levels of consumption worldwide. It is also due to the globalization of agriculture (see Chapter 8), with some core-region production having been moved to peripheral regions.

The phenomenon of corporations and rich governments investing in the agricultural land of peripheral countries is often called a "land grab" in the popular press. While it is impossible to know exactly how much peripheral land is being sold to investors, because land deals are often secret, there are reliable estimates. The Land Coalition, a nongovernmental organization, estimates that by 2011 almost 80 million hectares have been subject to some sort of negotiation with a foreign investor, more than half in Africa (**Figure 4.27**). To appreciate just how much land this is, 80 million hectares is more than all the farmland of the United Kingdom, France, Germany, and Italy combined. Big investors are countries with concerns about feeding their own people that purchase farmland abroad as a guarantee of future food supplies. China is the largest of these investors, buying or leasing twice as much land as anyone else, usually in Africa. The most worrying aspect of these land grabs is the high levels of corruption among the buyers and low levels of benefit for the sellers.

Grasslands

Grasslands—distinct from agricultural land—are also used productively the world over, either as rangeland or pasture for livestock grazing. Most grasslands are found in arid and semiarid regions that are

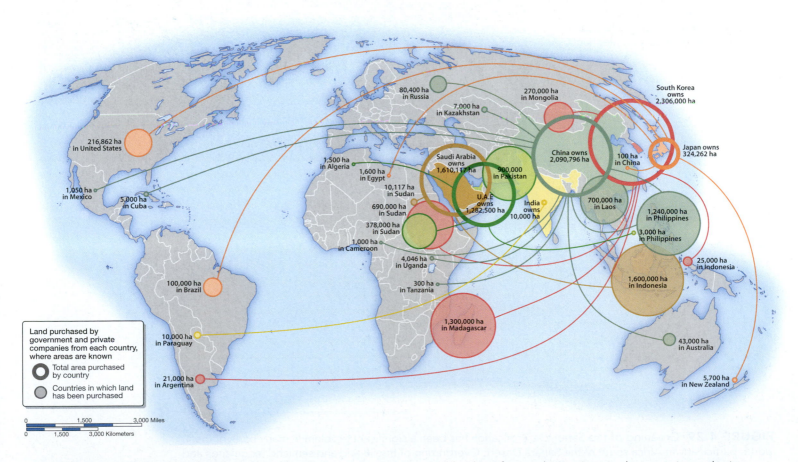

FIGURE 4.27 Global land grab This graphic illustrates the investment in land in Africa and Latin America by countries and private companies in Asia and the Middle East. The shades shown on the filled circles correspond to the country that has purchased the land. For example, the 324,262 hectares of land that Japan has purchased is shown in orange with orange lines connecting it to Brazil, the United States, Egypt, China and New Zealand (*Source*: Copyright Guardian News & Media Ltd. 2008. Reprinted with permission).

unsuitable for farming due to lack of water or poor soils (**Figure 4.28**). Some grasslands, however, occur in more rainy regions where tropical rain forests have been removed. Others are at the mid-latitudes, such as the tall- and short-grass prairies of the central United States and Canada. Approximately 68 million square kilometers (26 million square miles) of global land surface is currently taken up by grasslands.

Human impacts on grasslands are largely of two sorts. The first is the clearing of grasslands for other uses, most frequently settlement. As the global demands for beef production have increased, the use of the world's grasslands has intensified. Widespread overgrazing of grasslands has led to acute degradation. In its most severe form, overgrazing has led to desertification.

Desertification is the spread of desert conditions resulting from deforestation, overgrazing, and poor agricultural practices, as well as reduced rainfall associated with climatic change. Until recently, one of the most severe examples of desertification has been occurring in the Sahel region of Africa. The degradation of the grasslands bordering the Sahara Desert has not been a simple case of careless overgrazing by thoughtless herders, however. Severe drought, land decline, recurrent famine, and the breakdown of traditional systems for coping with disaster have all combined to create increased pressure on fragile resources, resulting in a loss of grass cover and extreme soil degradation since the 1970s. However, increasing evidence is suggesting that between 1982 and 2000 the desertification in this region has been declining (**Figure 4.29**). Scientists believe they are seeing signals that the Sahara Desert and surrounding regions are actually **greening**—adding biomass including grasses as well as trees—due to increasing rainfall. The increased rainfall may be the result of global climate change, which is leading to warmer temperatures. Hotter air can hold more moisture, which in turn creates more rain. Aerial photographs as well as ground studies have confirmed the greening phenomenon. The most optimistic projection of the various climate

FIGURE 4.28 **African grasslands**
Also known as savannas, grasslands include scattered shrubs and isolated small trees and are normally found in areas with high-to-average temperatures and low-to-moderate precipitation. They occur in an extensive belt on both sides of the equator. African tropical savannas, such as the one pictured here in Kenya, contain extensive herds of hoofed animals, including gazelles, giraffes, zebras, wildebeests, antelopes, and elephants.

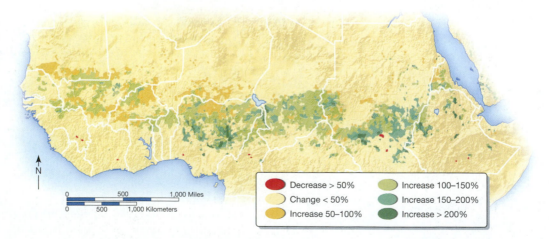

FIGURE 4.29 **Greening of the Sahel** Desertification has been a significant problem in many parts of the world, particularly in Africa south of the Sahara Desert. Overgrazing of fragile arid and semiarid rangelands and deforestation without reforestation have been the chief causes. Recently, however, there is growing evidence of increasing biomass production in this region as a result of increasing rainfall. This figure is based on trend analyses of time series over the Sahel region from 1982 to 1999. (*Source:* NOAA AVHRR NDVI-data from 1982 to 1999.)

models that attempt to better understand the impacts of the greening is that the rains could continue to revitalize drought-ravaged regions, providing new grazing lands for farming communities.

Wetlands

The wetlands category covers swamps, marshes, bogs, peatlands, and the shore areas of lakes, rivers, oceans, and other water bodies. Wetlands can be associated either with saltwater or freshwater. Most of Earth's wetlands are associated with the latter. The human impacts on wetland environments are numerous. The most widespread has been the draining or filling of wetlands and their conversion to other land uses, such as settlement or cultivation. One reliable estimate places the total area of the world's wetlands at about 8.5 million square kilometers (3.3 million square miles), with about 1.5 million square kilometers (0.6 million square miles) lost to drainage or filling. For example, Australia has lost all of its original 20,000 square kilometers (7,740 square miles) of wetlands to conversion. For the last 400 years or so, people have regarded wetlands as nuisances and as sources of disease. In core countries, technological innovation made modification and conversion of wetlands possible and profitable.

In San Francisco, California, for example, the conversion of wetlands in the mid-nineteenth century allowed speculators and real estate developers to extend significantly the central downtown area into the once marshy edges of San Francisco Bay. The Gold Rush in the Sierra Nevada sent millions of tons of sediment down the rivers into the bay, filling in its marshland and reducing its nearshore depth. It is estimated that in 1850 the San Francisco Bay system, which includes San Pablo Bay as well as Suisun Bay, covered approximately 315 square kilometers (about 120 square miles). One hundred years later only about 125 square kilometers (about 50 square miles) remained. By the 1960s the conversion and modification of the wetlands (as well as the effects of pollution pouring directly into the bay) had so dramatically transformed water quality and the habitats of fish, fowl, and marine life that the viability of the ecosystem was seriously threatened. Since then, restoration activities have been undertaken and parts of the bay have returned to something approximating their former state, but large parts are heavily urbanized and cannot be restored.

APPLY YOUR KNOWLEDGE Conduct an Internet search to find a core country and a peripheral country that share a common environmental concern relating to threats to forests, grasslands, or wetlands. Compare and contrast the different ways core and peripheral countries are addressing the same issue. ■

THE GLOBALIZATION OF THE ENVIRONMENT

The combustion of fossil fuels, the destruction of forest resources, the damming of watercourses, and the massive change in land-use patterns brought about by the pressures of globalization—industrialization being the most extreme phase—contribute to environmental problems of enormous proportions. It is now customary to speak of the accumulation of environmental problems we, as a human race, experience as global in dimension. Geographers and others use the term **global change** to describe the combination of political, economic, social, historical, and environmental problems with which human beings across Earth must currently contend. Very little, if anything, has escaped the embrace of globalization, least of all the environment.

In fact, no other period in human history has transformed the natural world as profoundly as the anthropocene. While we reap the benefits of a modern way of life, it is important to recognize that these benefits have not been without cost. Fortunately, the costs have not been accepted uncritically. Over the last two to three decades, responses to global environmental problems have been on the increase as local groups have mobilized internationally.

Global Environmental Politics

The increasing importance of flows and connections—economic, political, social, and cultural—means that contemporary globalization has resulted in an increasingly shrinking world. In addition to allowing people and goods to travel farther faster and to receive and send information more quickly, a smaller world means that political action has also become global. It can now move beyond the confines of the state into the global political arena, where rapid communications enable complex supporting networks to be developed and deployed, facilitating interaction and decision making. Good examples can be seen in the protests that occurred in Seattle, Washington, in 1999 at the World Trade Organization (WTO) meeting; in Genoa, Italy, in July 2001 at the Group of 8 (G8) summit; in Bangkok in 2003 to protest the Asian Pacific Economic Cooperation summit meeting; in June 2009 in Copenhagen, Denmark, around the UN climate negotiations conference; and again in December 2010 in Cancún, Mexico, during the UN climate conference. Telecommunications as well as Facebook, Twitter, and other social media sites enable protest leaders to organize and deploy demonstrators from interested groups all over the world. Such political protests reflect attempts to match the political reach of institutions like the WTO, the International Monetary Fund (IMF), the World Bank, and the United Nations.

One indication of the expanding influence and geographical extent of popular political groups is the growth of environmental organizations whose purview and membership are global. These organizations have emerged in response to the global impact of such contemporary environmental problems as fisheries depletion, global warming, genetically modified seeds, and the decline in global biodiversity. Since the 1990s, these groups—ranging from lobbying organizations and nongovernmental organizations (NGOs) to direct-action organizations and political parties like the Green Party in Europe and drawing on distinctive traditions and varying levels of resources—have become an important international force.

Increasingly, agreements and conventions protecting biodiversity are being created, and not a moment too soon. The decline in the diversity of simple foodstuffs, like lettuce, potatoes, tomatoes, and squash, occurred most dramatically over the course of the twentieth century. For instance, in 1903 there were 13 known varieties of asparagus; by 1983 there was just one, a decline of 97.8 percent. There were 287 known varieties of carrots in 1903; this number today is just 21, a fall of 92.7 percent. A decline in the diversity of foodstuffs means that different resistances to pests inherent in these different varieties have also declined, as have their different nutritional values and tastes. It should be pointed out, however, that a growing local food movement in the core countries is working hard to recover some of these lost varieties.

Moreover, new sources of medicine may be lost not only because of deforestation in tropical forests but also because of the decline in indigenous languages, cultures, and traditions. Recognizing that many indigenous people have extensive knowledge of local plants and animals and their medicinal uses, the Convention on Biological Diversity that emerged from the Rio Summit in 1992 is attempting to protect global biodiversity by preserving and protecting these cultures and traditions. Traditional knowledge and practices are being lost as globalization homogenizes languages and draws more and more people into a capitalist market system. The UN Environment Program devotes a great deal of its energies to biological and cultural diversity. Even the WTO has begun to recognize the value of indigenous knowledge and the promise of biodiversity through its advocacy of intellectual property rights of both corporations and indigenous peoples. For the latter, this means protection against **bioprospecting**, the scientific or commercial practice of searching for a useful application, process, or product in nature, often in extreme environments such as deserts, rain forests, and cold places like the Artic and Antarctic. (**Figure 4.30**).

Clearly, global environmental awareness is on the rise from both ends of the political spectrum, including the conservative (such as the WTO) and the progressive (such as programs devoted to preserving genetic diversity in seed strains). This increasing awareness is directly responsible for the staging of global environmental conferences like Rio in 1992, Kyoto in 1997, and Johannesburg in 2002 (and Rio+20, which is scheduled to occur in June 2012), which have not only affected international laws but continue to shape the debates about and responses to environmental problems. Most recently, these debates have centered on the concept of sustainability.

APPLY YOUR KNOWLEDGE Conduct an Internet or library search to find two different environmental policies that demonstrate cooperation on an environmental problem across state boundaries. How successful have they been? Provide statistical data to support your conclusion.

Environmental Sustainability

The interdependence of economic, environmental, and social problems, often located within widely different political contexts, means that some parts of the world are ecological time bombs.

Most environmental threats are greatest in the world's periphery, where daily environmental pollution and degradation amount to a catastrophe that will continue to unfold in slow motion in the coming years.

In the peripheral regions there is simply less money to cope with environmental threats. The poverty endemic to peripheral regions also adds to environmental stress. In order to survive, the rural poor are constantly impelled to degrade and destroy their immediate environment, cutting down forests for fuelwood and exhausting soils with overuse. In order to meet their debt repayments, governments generate export earnings by encouraging the harvesting of natural resources. In the cities of the periphery, poverty encompasses so many people in such concentrations as to generate its own vicious cycle of pollution, environmental degradation, and disease (see Chapter 11). Even climate change, an inherently global problem, seems to pose its greatest threats to poorer, peripheral regions.

A more benign relationship between nature and society has been proposed under the principle of *sustainable development*, a term that incorporates the ethic of intergenerational equity, with its obligation to preserve resources and landscapes for future generations. Sustainable development involves employing ecological, economic, and social measures to prevent environmental degradation while promoting economic growth and social equality. Sustainable development insists that economic growth and change

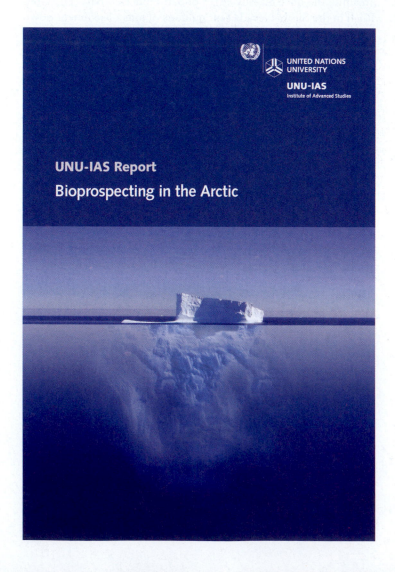

FIGURE 4.30 Bioprospecting in the Artic The combination of extreme temperatures and special light conditions has led to the evolution of organisms with unique properties and potentially valuable bioactive compounds in the Artic as well as the Antarctic. The United Nations is currently exploring bioprospecting in the cold regions of the globe in an attempt to establish governance standards that protect those resources from reckless exploitation.

should occur only when the impacts on the environment are benign or manageable and the impacts (both costs and benefits) on society are fairly distributed across classes and regions. This means finding less-polluting technologies that use resources more efficiently and managing renewable resources (those that replenish themselves, such as water, fish, and forests) to ensure replacement and continued yield. In practice, sustainable development policies of major international institutions, such as the World Bank, have promoted reforestation, energy efficiency and conservation, and birth control and poverty programs to reduce the environmental impact of rural populations. At the same time, however, the expansion and globalization of the world economy has resulted in increased use of resources and inequality, contradicting many of the goals of sustainable development.

APPLY YOUR KNOWLEDGE Research an international institution, such as the World Bank, that implements programs that strive to promote sustainable development. Evaluate the effectiveness of one of its programs in a peripheral region by identifying the reasons for both its successes and failures. ■

Future Geographies

The continued expansion of the global economy and the globalization of industry will undoubtedly boost the overall demand for raw materials and energy and continues to shape the relationship between people and nature. While the extraction of raw materials will be important in the future, the most concerning issue, by far, will be energy resources. World energy consumption has been increasing steadily. As the periphery is industrialized and its population increases further, the demand for energy will expand even more rapidly. Basic industrial development tends to be highly energy-intensive. The International Energy Agency, assuming (fairly optimistically) that energy in peripheral countries will be generated in the future as efficiently as it is in developed ones now, estimates that developing-country energy consumption will more than double by 2025, lifting total world energy demand by almost 50 percent. Unless peripheral countries are able to limit the degradation associated with energy use, the globe will continue to feel the negative public health and environmental effects of air, water, and terrestrial pollution.

Despite the threat to people and the environment, industrialization geared to meeting the growing worldwide market for consumer goods, such as automobiles, air conditioners, refrigerators, televisions, and household appliances, will continue (**Figure 4.31**). Without higher rates of investment in exploration and extraction than at present, production will be slow to meet the escalating demand. Many experts believe that current levels of production in fact represent "Peak Oil" (See Box 4.2, "Window on the World: Peak Oil"), and that the world is at the halfway point of depleting its finite reserves of crude oil.

The past 25 years have seen a growing public awareness of how continued globalization will affect the world in which we live. Increasingly, citizens, nongovernmental organizations, and environmental policymakers are expressing concern over the negative outcomes of rapid and enduring global economic growth. However, because growth is so critically tied to improving the lives of poor people around the world, governments are reluctant to limit it. The response from the global community, hammered out during international meetings, through academic publications, and in response to social protest, is to link globalization to governmental cooperation across states with the assumption that global challenges will require international political and economic cooperation.

FIGURE 4.31 Containers Recyclable paper and metal, half-processed forest products and other raw materials as well as agricultural products account for most of the U.S. goods that are shipped to China and other ports in Northeast Asia. Those that come from China to the United States are loaded with manufactured commodities such as clothing, shoes, toys, furniture, appliances and other house hold goods.

CONCLUSION

The relationship between people and nature is very much mediated by institutions and practices, from technology to religious beliefs. In this chapter, we have seen how the nature–society relationship has changed over time and how the globalization of the capitalist world economy has had a more widespread impact on attitudes and practices than any cultural or economic system that preceded it.

The expansion of European trade, followed by colonization and eventually industrialization, broadcast worldwide the belief that humans should take their place at the apex of the natural world. The Western attitude toward nature as manifested by the capitalist economic system is the most pervasive shaper of nature–society interactions today.

In addition to exploring the history of ideas about nature and contemporary environmental philosophies and organizations in the United States,

this chapter has also shown that people and nature are interdependent and that events in one part of the global environmental system affect conditions in the system elsewhere. Finally, we have shown that events in the past have shaped the contemporary state of society and nature.

In short, as economies have globalized so has the environment. We can now speak of a global environment in which not only the people but also the physical environments where they live and work are linked in complex and essential ways.

Along with the recognition of a globalized environment have come new ways of thinking about global economic development. Sustainable development, one of these new ways of thinking, has come to dominate the agenda of international institutions, as well as environmental organizations, as the new century unfolds.

Learning Outcomes Revisited

- **Recognize how people and nature form a complex relationship such that nature is both a physical realm and a social construct.**

  Nature is not only an object, it is a reflection of society in that the philosophies, belief systems, and ideologies people produce shape the way we think about and employ nature. Society is the sum of the inventions, institutions, and relationships created and reproduced by human beings across particular places and times. The relationship between nature and society is two-way: Society shapes people's understandings and uses of nature at the same time that nature shapes society.

- **Compare and contrast the many views of nature operating both historically and in society today, from the traditional Western approach to radical left and contemporary ecotheological ones.**

  In the contemporary world, views of nature are dominated by the Western (also known as Judeo-Christian) tradition that understands humans to be superior to nature. In this view, nature is something to be tamed or dominated. But other views of nature have emerged that depart dramatically from the dominant view. These include the environmental philosophies that became popular in the nineteenth and early twentieth centuries and the more radical political views of nature that gained prominence in the late twentieth century. Among the latter are approaches based on ecotheology, which reject the long-standing consumption-based Western tradition.

- **Assess how European colonization as well as contemporary globalization transformed nature in the New World on an unprecedented scale.**

  In the fifteenth century, Europe initiated territorial expansion that changed the global political map and launched dramatic environmental change. Europeans were fast running out of land, and explorers were dispatched to conquer new territories, enlarge their empires, and collect tax revenues from new subjects. European people, ideologies, technologies, plant species, pathogens, and animals changed the environments into which they were introduced and the societies they encountered.

- **Appraise how the globalization of the capitalist political economy has affected the environment so that environmental problems, often predicated on industrialization and its attendant energy needs, are increasingly global in scope.**

  No other transition in human history has had the impact on the natural world that industrialization and urbanization have. The combustion of fossil fuels, the destruction of forest resources, the damming of watercourses, and the massive change in land-use patterns brought about by the pressures of globalization contribute to environmental problems of enormous proportions. Geographers and others use the term *global change* to describe the combination of political, economic, social, historical, and environmental problems with which human beings across Earth must currently contend.

- **Evaluate the ways that sustainability has become a predominant approach to global economic development and environmental transformation.**

  Sustainable development involves employing ecological, economic, and social measures to prevent environmental degradation while promoting economic growth and social equality. Sustainable development insists that economic growth and change should occur only when the impacts on the environment are benign or manageable and the impacts (both costs and benefits) on society are fairly distributed across classes and regions. This means finding less-polluting technologies that use resources more efficiently and managing renewable resources (those that replenish themselves, such as water, fish, and forests) to ensure replacement and continued yield.

KEY TERMS

acid rain *(p. 129)*
anthropocene *(p. 120)*
biomass *(p. 120)*
bioprospecting *(p. 140)*
Columbian Exchange *(p. 117)*
conservation *(p. 114)*
cultural ecology *(p. 109)*
deep ecology *(p. 115)*

deforestation *(p. 135)*
demographic collapse *(p. 117)*
desertification *(p. 138)*
ecofeminism *(p. 115)*
ecological imperialism *(p. 118)*
ecosystem *(p. 106)*
ecotheology *(p. 116)*
environmental ethics *(p. 115)*

environmental justice *(p. 116)*
global change *(p. 139)*
greenhouse gases (GHGs) *(p. 130)*
greening *(p. 138)*
nature *(p. 107)*
political ecology *(p. 111)*
preservation *(p. 115)*
romanticism *(p. 114)*

society *(p. 107)*
technology *(p. 108)*
transcendentalism *(p. 114)*
virgin soil epidemics *(p. 116)*
virtual water *(p. 112)*

REVIEW AND DISCUSSION

1. Consider the relationship between industrialization and physical geography by examining mountain top removal in Appalachia. Visit the Beehive Collective graphic campaign entitled "The True Cost of Coal" at http://www.beehivecollective.org/english/coal.htm. Describe the environmental philosophies that the Beehive Collective most closely adheres to. (*Hint:* Is it the collective part of the environmental justice or ecofeminist movements? Or a combination of a variety of different theories?) Explain your choice. Be specific in the reasons you give by citing examples from the organization's Web site. List four examples of how the graphics on the site depict the complex relationship between nature and society.

2. Colleges and universities have become much more involved in sustainability over the last decade. In order to know where your institution fits into the larger college and university sustainability movement, visit the College Sustainability Report Card Web site at http://www.greenreportcard.org/about/faq and determine how your institution is rated. Discuss how it compares to other institutions and how your institution might improve or how sustainability practices engaged in at your institution could be transferred to other colleges or universities.

3. The concept of the anthropocene is being debated in scientific circles around the world. Geologists are divided on this issue. To mirror this situation, set up a debate in which half of the group is tasked with providing reasons for accepting the anthropocene as a new geological era and the other half of the group is arguing against its adoption

UNPLUGGED

1. Many communities have begun to produce an index of environmental stress, in essence a map of the toxic sites in a city or region. Plot a rudimentary map like this of your own community using the local phone book as a data source. Consult the Yellow Pages to identify the addresses of environmentally harmful and potentially harmful businesses, such as dry-cleaning businesses, gas stations, automotive repair and car-care businesses, aerospace and electronic manufacturing companies, agricultural supply stores, and other such commercial enterprises where noxious chemicals may be produced, sold, or applied. Alternatively consider your campus and where its most toxic sites are. Your school directory is the ideal place to start.

2. Read a natural history of the place where your college or university is located. What sorts of plants and animals dominated the landscape there during the Paleolithic period? Do any plants or animals continue to survive in altered or unaltered forms from that period? What new plants have been introduced and how extensive are they? What sorts of tending and maintenance do these new species require and how have they changed the landscape? Finally, with the information you have gathered, create a map of the paths that these specific plants have taken to arrive in your region.

3. Colleges and universities are large generators of waste, from plain-paper waste to biomedical and other sorts of wastes, that can have significant environmental impacts. Identify how your campus handles this waste stream and how you, as a member of the academic community, contribute to it. Where does the waste go when it leaves the school? Is it locally deposited? Does it go out of state?

Mastering GEOGRAPHY™

Log in to www.masteringgeography.com for MapMaster™ interactive maps, geography videos, RSS feeds, flash cards, Web links, an eText version of *Human Geography: Places and Regions in Global Context*, and self-study quizzes to enhance your study of people and nature.

MapMaster™ presents 13 Place Name and 13 Layered Thematic interactive maps to help students practice and master their geographic literacy, spatial reasoning, and critical thinking skills.

5 CULTURAL GEOGRAPHIES

Antarctica is simultaneously both deeply connected to the rest of the world as well as practically remote. Earth's southernmost continent, it is also its coldest place. The only plant and animal life that can survive in Antarctica is cold-adapted, including whales, penguins, and seals as well as tundra vegetation.

Antarctica's remoteness at the "bottom of the world" is linked to its climate where temperatures average −51°C (−60°F) during the 6-month winter. By September each year (at winter's end) half the surrounding ocean is frozen, creating a vast land and ice pack area combination extending approximately 26 million square kilometers (10 million square miles) with a thickness of more than 2 meters (6.6 feet). The cold, ice, lack of air pollution, and towering mountains render Antarctica one of the most visually striking landscapes in the world. In the clear, bright light, the unbounded snow and ice fields seem endless. In summer, the Antarctic pack ice breaks up, and the coastal glaciers calve off huge icebergs that drift and change shape by the hour.

There is growing concern that global warming is threatening Antarctica. Temperatures are warming and changing the ecology, and ice cover is disappearing. As a result, travel companies have begun to capitalize on the urge of tens of thousands of tourists to see Antarctica before it is forever changed.

Tourists photographing Emperor Penguin chicks at Snow Hill Island Antarctica in summer.

Tourism enables the traveler to experience new places by direct bodily contact with the sounds, sights, smells, and feel of a place. Tourists travel to Antarctica to appreciate firsthand this extreme landscape; but these visitors must transform themselves—"become polar"—in order to encounter the landscape without harming themselves. Special thermal clothing must be worn to guard the body against the extreme cold. This includes special eyewear. Snow blindness—a serious, painful, and possibly disabling condition—can occur when unprotected human eyes are exposed to excessive light generated where snow and ice reflect 85 percent of ultraviolet radiation. Alcohol consumption is often discouraged here, where unforeseen events may require swift, intelligent action, as the ability of a person to deal effectively with a mishap is reduced by intoxication.

Tourism in Antarctica illustrates a new way of thinking about culture and nature in human geography today. In this view, people and land (or culture and nature) are not to be understood as separate entities but entwined in such a way that in the encounter, each brings the other into being. From this perspective, we can see the Antarctic landscape not as some totality of air, land, and water that can be observed, but as a world that creates the tourist as a human subject seeking to know. In this way, the Antarctic makes tourists out of thrill seekers just as the thrill seeker as tourist creates an imaginary of the Antarctic that shapes his or her experience of it. ∎

CULTURE AS A GEOGRAPHICAL PROCESS

Anthropologists, geographers, and other scholars who study culture, such as historians, sociologists, and political scientists, agree that culture is a complex concept. Over time, our understanding of culture has been changed and enriched. A simple definition of culture is that it is a particular way of life, such as a set of skilled activities, values, and meanings surrounding a particular type of practice. Scholars also describe culture in terms of classical standards and aesthetic excellence in, for example, opera, ballet, or literature.

The term *culture* also describes the range of activities that characterize a particular group, such as working-class culture, corporate culture, or teenage culture. Although all these understandings of culture are accurate, for our purposes they are incomplete. Broadly speaking, **culture** is a shared set of meanings that is lived through the material and symbolic practices of everyday life. Our understanding in this book is that culture is not something that is necessarily tied to a place and thus a fact to be discovered. Rather, we regard the connections among people, places, and cultures to be social creations that can be altered and are therefore always changing, sometimes in subtle and other times in more dramatic ways. The "shared set of meanings" can include values, beliefs, practices, and ideas about religion, language, family, gender, sexuality, and other important identities **(Figure 5.1)**. These values, beliefs, ideas, and practices are routinely subject to reevaluation and redefinition and can be, and very frequently are, transformed from both within and outside a particular group.

In short, culture is a dynamic concept that revolves around and intersects with complex social, political, economic, and even historical factors. For much of the twentieth century, geographers, like anthropologists, have focused most of their attention on material culture, as opposed to its less tangible symbolic or spiritual manifestations. For example, they have explored the spatial extent of particular religious practices such as the global distribution of Buddhism, and the physical expression of religiosity, such as the practice of installing crosses along roadways where fatal traffic accidents have occurred **(Figure 5.2)**. This understanding of culture is part of a longer, evolving tradition within geography and other disciplines. We will look more closely at the development of the cultural tradition in geography in the following section, in which we discuss the debates surrounding culture within the discipline.

Like agriculture, politics, and urbanization, globalization has had complex effects on culture. Terms such as *world music* and *international television* are a reflection of the sense that the world seems a smaller place now as people everywhere are sharing aspects of the same culture through the widespread influence of television, the Internet, and other media. Yet, as pointed out in Chapter 2, although powerful homogenizing forces are certainly at work, the world has not become so uniform that place no longer matters. With respect to culture, just the opposite is true. Place matters more than ever in the negotiation of global forces, as local forces confront globalization and translate it into unique place-specific forms. For example, Box 5.1, "Geography Matters: The Culture of Hip-Hop," illustrates that while hip-hop music has spread globally, it is produced differently in different places.

The place-based interactions occurring between culture and global political and economic forces are at the heart of cultural geography today. **Cultural geography** focuses on the way space, place, and landscape shape culture at the same time that culture shapes space, place, and landscape. As such, cultural geography demarcates two important and interrelated parts. Culture is the ongoing process of producing a shared set of meanings and practices, while geography is the dynamic context within which groups operate to

FIGURE 5.1 Maori men with traditional tattoos These indigenous peoples of New Zealand are dancing a *haka*, a traditional war dance meant to display aggressiveness and fearlessness. The dancers use facial contortions to deride their enemy. The tattoos, known as *ta moko*, can signify identification, rank, genealogy, tribal history, eligibility to marry, beauty, and ferocity.

FIGURE 5.2 **Road accident memorial** Originating in Spain and exported to Mexico, these road shrines are known as descansos, memorials ... that mark the place where the soul has left the body in a fatal accident. Descansos are ubiquitous in Mexico and Spain and have begun, over the last several decades, to appear across the United States in non-Hispanically influenced areas as well as in the Philippines.

shape those meanings and practices and in the process to form an identity and act. Geography in this definition can be a space that is as small as the body and as large as the globe.

A contemporary example of this two-way relationship between geography and culture can be seen in the widespread popularity of online social networks designed to connect friends, family, co-workers, and people with similar interests. The most popular online social networking Web site Facebook—with around 750 million active users worldwide—allows members to share photos, blogs, music and videos, and a host of other personal information, as well as organize events, join groups, or seek relationship partners.

Facebook and other virtual networks have revolutionized the way people communicate with each other, as friendship groups and circles obscure the boundaries between online and offline domains. Cultural scholars who study them believe that online social networks are not turning us into "virtual creatures" existing only in the ether of the Internet but rather into hybrid creatures with a foot in both the virtual and the real world. This enables people to be in more than one place at any one time and to connect to others whom they may never meet in person, but with whom they may form significant relationships or simply passing ones. And in those interactions people are changed, perhaps only in small ways, but those ways can and often do add up to something quite substantial. Even virtual spaces are thus venues where cultural identities and practices can emerge and flourish.

Before we proceed any further in our discussion of culture, it is important to discuss a significant difference between our view and that of more traditional cultural geographers. Many introductory human geography texts divide culture into two major categories: folk and popular culture. **Folk culture** is seen by specialists as the traditional practices of small groups, especially rural people with a simple lifestyle (compared with modern, urban people), such as the Amish in Pennsylvania or the Roma (also known as Gypsies or

Travelers) in Europe, who are seen as homogeneous in their belief systems and practices. **Popular culture,** by contrast, is viewed by some cultural geographers as the practices and meaning systems produced by large groups of people whose norms and tastes are often heterogeneous and change frequently, often in response to commercial products. Hip-hop would be seen by these theorists as an example of popular culture,

In this text, we do not divide culture into categories. We see culture as an overarching process that is shaped by and shapes politics, the economy, and society and cannot be neatly demarcated by reference to the number of characteristics or degree of homogeneity of its practitioners. We see culture as something that can be enduring as well as newly created, but always influenced by a whole range of interactions as groups maintain, change, or even create traditions from the material of their everyday lives. For us, there is no purpose served in categorically differentiating between hip-hop and Hinduism, as both are significant expressions of culture and both are of interest to geographers.

APPLY YOUR KNOWLEDGE In your own words define cultural geography. Once you have done this, identify three aspects of your own culture and the ways that place and space have shaped it. ∎

BUILDING CULTURAL COMPLEXES

Geographers focus on the interactions between people and culture and among space, place, and landscape. One of the most influential geographers was Carl Sauer, who taught at the University of California, Berkeley, in the early to mid-twentieth century. Sauer was largely responsible for creating the "Berkeley School" of cultural geography. He was particularly interested in trying to understand the material expressions of culture by focusing on their

Hip-hop is a popular manifestation of contemporary cultural practices among U.S. big-city and inner-city youth. Characterized to some extent by graffiti art, and earlier on by break dancing, hip-hop is understood globally through rap music and a distinctive idiomatic vocabulary. Like most nations, hip-hop has its forebears. These include boxer Muhammad Ali, Jamaican Rastafarian and reggae musician Bob Marley, Black Panther Huey Newton, and funksters James Brown and George Clinton. The hip-hop nation has its recognized origins in the Bronx, in New York City. But it also has much older roots in the West African storytelling culture known as *griot*. Hip-hop in the United States enlarged upon those origins, and now hip-hop is both appreciated and produced on six continents.

Hip-hop is a cultural practice that has truly globalized, not because its practitioners have migrated far and wide but because its culture has migrated via telecommunications and the music and film industries. Hip-hop has become a cultural phenomenon that exists beyond geography in the music, the clothes, and the language of its practitioners.

Predominantly black and male, hip-hop also crosses color lines and includes women and gays, though the latter two groups have also been the targets of entrenched sexism and homophobia by a wide range of rappers. In addition to the founding fathers mentioned previously, its pioneers include white graffiti artists and Latinos who influenced break dancing as well as hip-hop DJ (disk jockey) and MC (rapper) styles.

Music is the heart and soul of hip-hop nation and the geography of U.S. hip-hop—where the most important music has come from—can be crudely divided into East Coast, West Coast, South Coast, and a region in and around Detroit where white rap-metal groups became popular. The East Coast includes the five boroughs of New York, Long Island, Westchester County, New Jersey, and Philadelphia. The West Coast includes Los Angeles, Compton, Long Beach, Vallejo, and Oakland. The South Coast region is made up of Atlanta, New Orleans, Miami, and Memphis (Figure 5.A).

Hip-hop is about how space and place shape the identities of rappers in particular but also African Americans more generally. It demonstrates how race, space, and place come together to produce the contradiction of "home" not only as a locus of roots and the foundation of personal history but also as a site of devaluation vis-à-vis the dominant white society.

Since its emergence in New York in the 1970s, rap has emerged in major cities across the United States, with each place producing a style that blends some of the foundations of original rap with local variations. For example, in addition to the distinctive style of rap metal that emerged in Detroit through Eminem and Kid Rock, hybrid forms of rap were also hatched in other Midwestern cities, including Chicago, Cleveland, Cincinnati, Kansas City, Madison, Milwaukee, Minneapolis, Omaha, and St. Louis, each with its own interesting history and style. And the Midwest is not the only region with unique rap artists and

FIGURE 5.A The sources and diffusion of U.S. rap This map portrays the centers of rap music in the United States today, showing how rap, which began in African American inner-city neighborhoods in New York City in the late 1970s, moved westward and then southward. In the 1990s, a hybrid form of rap metal emerged in the U.S. Midwest urban center of Detroit. Detroit and other Midwestern cities typically contain large numbers of both African Americans and working-class whites who lost their jobs in the restructuring of the automobile industry in the 1970s and early 1980s. What the rap-metal genre confirms is that although hip-hop culture has its roots in the African American experience, it derives much of its power from issues of poverty and class as well.

enthusiastic audiences. Distinctive rap styles have also emerged in the South, the Southwest, and the Pacific Northwest.

And just as hip-hop has broken out of its regional boundaries, it has transcended national boundaries as well. Hip-hop graffiti art can be found in urban areas as distant as Austria and South Africa. Rap music is as popular in the Philippines as it is in Paris. And as hip-hop cultural theorist Murray Forman argues, "Virtually all of the early descriptions of hip-hop practices identify territory and the public sphere as significant factors, whether in visible artistic expression and appropriation of public space via graffiti or b-boying [break dancing], the sonic impact of a pounding bass line, or the discursive articulation of urban geography in rap lyrics [and films]."[1]

The most controversial variant of U.S. hip-hop is "gangsta rap," pioneered in the late 1980s by artists like Ice T and groups such as N.W.A. Today, gangsta rap is one of the most popular and lucrative of all hip-hop genres in the United States. The element that has drawn the most condemnation from both the left and the right is lyrics, which can be interpreted as glorifying promiscuity, violence, misogyny (hatred of women), gang culture, rape, drug dealing, and other acts of criminal behavior both major and minor. Gangsta artists respond, however, by arguing that theirs is not glorification but merely a retelling of the inner-city life of young African Americans. Gangsta rap artists include Snoop Dog, Lil Wayne, Dr. Dre, and Jay Z, all of whom have achieved global success. Nicki Minaj, a more eclectic rapper, frequently incorporates gangsta lyrics as well (**Figure 5.B**).

Cultural critic bell hooks turns the condemnation of gangsta rap on its head by pointing out that gangsta rap does not exist in a vacuum but is an extension of white, male-dominated, capitalist society.[2] She argues that it is far easier to attack gangsta rap than the culture that produces and reproduces it (through white, middle-class consumption of the music). Simultaneously, hooks points out that while we are repelled by the misogyny promoted by gangsta rappers, we fail to see it everywhere else—including

FIGURE 5.B Rapper Nicki Minaj She was born in Trinidad and Tobago but grew up in Queens, New York, and is fast becoming one of the most popularly acclaimed female hip-hop artists in the United States. Her debut studio album *Pink Friday* (2010) became a commercial success, peaking atop U.S. *Billboard* 200 and was certified platinum a month after its release. That same year she became the first artist ever to have seven songs on the *Billboard* Hot 100.

---

[1] M. Forman, "Ain't No Love in the Heart of the City: Hip-Hop, Space, and Place," in M. Forman and M. A. Neal (eds.), *That's the Joint! The* Hip-Hop *Studies Reader.* New York: Routledge, 2004, p. 155.

[2] b. hooks, "Sexism and Misogyny: Who Takes the Rap?" *Z Magazine*, February 1994, pp. 26–29, http://race.eserver.org/misogyny.html, accessed June 29, 2008.

within mainstream U.S. culture. For instance, who's to say which is more misogynist, *Beverly Hills Housewives* or Lil Wayne?

Hip-hop, as a youth-oriented cultural product that has become widely commercialized in core countries by multinational corporations, is still in many parts of the world, a fully homegrown phenomenon. Interestingly, rather than homogenizing local cultures, the global spread of hip-hop from the United States outward has tended to create hybrid and synthetic forms of music, adjusting to local cultures and struggles. What is most consistent about the rap music that is being produced in the periphery is that it has become a focal point for the underprivileged around the world to unite and challenge the oppression of the status quo.

Social struggle seems to be the single strongest thread that weaves together the disparate sounds, cultures, and artists of global hip-hop. The British, Pakistani-born Muslim rapper Aki Nawaz is well-known for his controversial lyrics that attack the hypocrisy and immorality of the West. In Sudan, Emmanuel Jal, a former child soldier in Sudan's People's Liberation Army, uses rap as a way of educating the world about the struggles that oppressed people face in Africa: political injustice, government terror, war, as well as the hope of freedom from these tyrannies.

Indeed, rap may be the most pervasive music among young people the world over. There are Islamic rappers in the United Kingdom and France protesting the racist treatment of Muslims there. The children of Turkish immigrants in Germany rap about racism and the experiences of second-class citizenship there. In Sydney, Australia, the rap group Def Wish Cast composes its rap around an attempt to forge a white, Australian-accented, nationalistic hip-hop culture. There are also revolutionary rhetoric "combat" Italian rappers like Onda Rossa Posse and Assalti Frontali and rappers in mainland China, like Cui Jian, who question the 1997 handover of Hong Kong, and artists in Aotearoa in New Zealand who combine rap, soul, and reggae with traditional Maori music.

Worldwide sales of rap music have been dropping since 2000. It is not clear whether this signals a death knell for rap or whether audiences have simply gotten tired of the U.S.-dominated scene or whether this music is often downloaded from the Web illegally and the sales therefore only reflect part of its popularity. Whatever the reason or reasons, it is doubtful that rap as a musical form and hip-hop as a cultural complex will disappear entirely any time soon.

FIGURE 5.3 Masai village, Kenya The cultural landscape, as defined by Carl Sauer, reflects the way that cultural and environmental processes come together to create a unique product as they do in the small village pictured here, where herding is the major occupation. The village is enclosed by thorny brambles and branches harvested from the surrounding area. Within the enclosure, the dwellings are arranged in a unique circular pattern, with the animal pens in the middle of the settlement for easy observation by the residents.

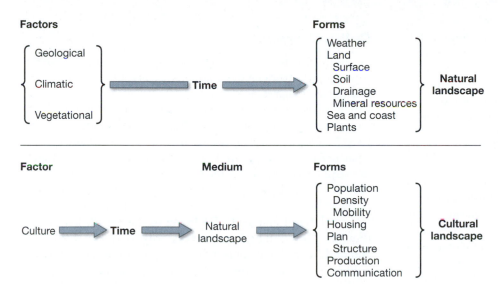

FIGURE 5.4 Sauer's cultural landscape This figure summarizes the ways the natural and cultural landscapes are transformed. Physical and climatic factors shape the natural landscape. Cultural practices also have an important impact upon it. The results of cultural factors are cultural forms, such as population distributions and patterns of housing. Over time, people—through culture—reshape the natural landscape to meet their needs. (Adapted from C. Sauer, "The Morphology of Landscape." in J. Leighly (ed.), *Land and Life: Selections from the Writings of Carl Ortwin Sauer.* Berkeley and Los Angeles: University of California Press, 1963, pp. 315–350.)

manifestations in the landscape **(Figure 5.3)**. This interest came to be embodied in the concept of the **cultural landscape,** a characteristic and tangible outcome of the complex interactions between a human group—with its own practices, preferences, values, and aspirations—and its natural environment. Sauer differentiated the cultural landscape from the natural landscape. He emphasized that the former was a "humanized" version of the latter, such that the activities of humans resulted in an identifiable and understandable alteration of the natural environment. **Figure 5.4** lists the differences between a natural and a cultural landscape.

For roughly five decades, interest in culture within the field of U.S. geography largely followed Sauer's important work. His approach to the cultural landscape was ecological, and his many published works reflect his attempts to understand the myriad ways that humans transform the surface of Earth. In his own words:

> The cultural landscape is fashioned from a natural landscape by a cultural group. Culture is the agent, the natural area is the medium, the cultural landscape is the result. Under the

influence of a given culture, itself changing through time, the landscape undergoes development, passing through phases, and probably reaching ultimately the end of its cycle of development. With the introduction of a different—that is an alien—culture, a rejuvenation of the cultural landscape sets in, or a new landscape is superimposed on remnants of an older one.[1]

In Europe, geographers interested in human interactions with the landscape took slightly different approaches. For example, in Great Britain the approach to understanding the human imprint on the landscape was termed *historical geography*, while in France it was conceptualized as *genre de vie*. **Historical geography,** very simply defined, is the geography of the past. Its most famous practitioner was H. C. Darby, who attempted to understand "cross sections" or sequences of evolution, especially of rural landscapes.

[1]C. Sauer, "The Morphology of Landscape," in J. Leighly (ed.), Land and Life: Selections from the Writings of Carl Ortwin Sauer. Berkeley, CA: University of California Press, 1964, pp. 315–350.

FIGURE 5.5 Market gardens in Corsica This image shows a rural setting in Corsica, an island nation in the Mediterranean, where commercial agriculture is being undertaken. Farming is a way of life—a *genre de vie*—that we can read from the landscape where extensive cultivated fields and isolated farmhouses constitute key elements.

Genre de vie, a key concept in Vidal de la Blache's approach to cultural geography in France, referred to a functionally organized way of life characteristic of a particular culture group. *Genre de vie* centered on the livelihood practices of groups that were seen to shape physical, social, and psychological bonds **(Figure 5.5)**. Although emphasizing some landscape components over others or giving a larger or smaller role to the physical environment, all of these approaches placed the cultural landscape at the heart of their study of human-environment interactions.

H. C. Darby most successfully implemented his historical approach to cultural geography and landscape by developing a geography based on the *Domesday Book,* a key historical document. William the Conqueror ordered the *Domesday* compiled in 1085 so he could have a list of his spoils of war. The book provides a rich catalog of the ownership of nearly every tract of land in England. It includes the names of landholders in each county; the manors they held and their values; the names of their subtenants; the names of many boroughs and details of their customs; the number of freemen, sokemen, (freemen landholders) unfree peasants, and slaves on each manor; and the resources of each manor. Geographers like Darby found such data invaluable for reconstructing the political, economic, and social forces that shaped past landscapes.

Vidal de la Blache emphasized the need to study small, homogeneous areas in order to uncover the close relationships that exist between people and their immediate surroundings. He constructed complex descriptions of preindustrial France that demonstrated how the various *genres de vie* emerged from the possibilities and constraints posed by local physical environments. Subsequently, he wrote about the changes in French regions brought on by industrialization, observing that regional homogeneity was no longer the unifying element. Instead, the increased mobility of people and goods produced new, more complex geographies where previously isolated *genres de vie* were integrated into a competitive industrial economic framework. Anticipating the widespread impacts of globalization, de la Blache also recognized how people in various places struggled to cope with the big changes transforming their lives.

Cultural Traits

Geographers also examine specific aspects of culture, ranging from single attributes to complex systems. One simple aspect of culture of interest to geographers is the idea of special traits, which include such things as distinctive styles of dress, dietary habits, and styles of architecture **(Figure 5.6)**. A **cultural trait** is a single aspect of the complex of routine practices that constitute a particular cultural group. For example, Canon Law for Catholics requires fasting during the holy season of Lent. This practice may be said to be a cultural trait of Catholic people. Geographers are interested in learning how cultural traits come together to form larger frameworks for living in the world. Ultimately, though, cultural traits are not necessarily unique to one group, and understanding them is only one aspect of the complexity of culture. For instance, there are a number of cultural groups, including Hindus, Muslims, and Jews, who avoid pork in their diet **(Figure 5.7)**.

Many cultures also recognize the passage from childhood into adulthood with a celebration or ceremony. Called **rites of passage,** these are acts, customs, practices, or procedures that recognize key transitions in human life—birth, menstruation, and other markers of adulthood such as sexual awakening and marriage. Such rites of passage are not uncommon among many of the world's cultures. Some non-Western cultures, for example, send adolescent boys away from the village to experience an ordeal—ritual scarring or circumcision, for example—or to meditate in extended isolation on the new roles they must assume as adults **(Figure 5.8)**.

Cultural Complexes and Regions

In Roman Catholicism, the passage of boys and girls into adulthood, traditionally around the age of 12, is celebrated by confirmation. In this religious ceremony, the confirmed chooses a new name to mark this important spiritual transition. Jews mark the passage of adolescent boys and girls into adulthood with

FIGURE 5.6 Tuareg men in Niger The Tuareg are also known as the Blue Men of the Sahara because of their distinctive indigo blue robes and veils. Tuareg women do not wear veils but men do when they reach maturity.

FIGURE 5.7 Halal butcher For Muslims, there is a prescribed method of slaughtering all animals except for fish and most sea-life. (Prohibited flesh includes pork, carnivorous animals, and birds of prey.) Flesh that is butchered in the prescribed manner is considered "halal" or "permissible" to eat.

FIGURE 5.8 A coming-of-age ceremony, South Korea The coming-of-age ceremony for girls in South Korea is held every May 15 in Seoul. There, young women who are turning 20 participate in dances and a citywide celebration that is meant to remind them of the responsibilities they face as adults. In South Korea, people who turn 60 are also ritually celebrated. This ritual is not so much a coming-of-age celebration as it is a coming-to-an-age celebration. Called *hwan-gap,* this celebration is considered one of the larger celebrations in a person's life. It is significant because it marks the day on which an individual has completed a full zodiacal cycle. But, more importantly, it is also celebrated, because to live to be 60 is seen as a great accomplishment. In the past, most people in South Korea died before their 60th birthday.

separate religious ceremonies: a *bar mitzvah* for boys and a *bat mitzvah* for girls. Although marking the passage into adulthood is a trait of both religious groups, they do not exhibit the trait in exactly the same way. This and other traits always occur in combination. The combination of traits characteristic of a particular group is known as a **cultural complex.** The avoidance of pork, the celebration of *bar* and *bat mitzvahs,* and other dietary, religious, and social practices constitute the cultural complex of Judaism, although it is important to note that even within the cultural complex of Judaism, variation exists among regions and sects.

Another concept key to traditional approaches in cultural geography is the cultural region. Although a cultural region may be quite extensive or very narrowly described and even discontinuous in its extension, it is the area within which a particular cultural system prevails. A **cultural region** is an area where certain cultural practices, beliefs, or values are more or less practiced by the majority of the inhabitants. For example, the state of Utah is considered to be a Mormon cultural region because the population of the state is dominated by people who practice the Mormon religion and presumably adhere to its beliefs and values. **Figure 5.9** illustrates the overall religious geography of the United States.

APPLY YOUR KNOWLEDGE Identify two traits that are characteristic of the cultural group to which you belong. Are the traits related to the country or region in which you live? Describe the relationship or explain why there is none. ■

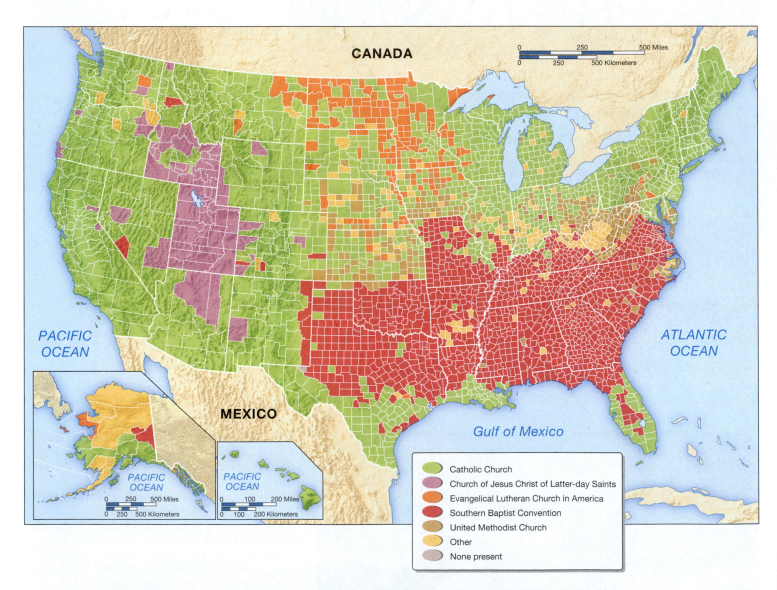

FIGURE 5.9 **Cultural systems: U.S. religious population distribution by county, 2000** This map shows the majority religion by county for the United States and illustrates the concept of cultural regions based on religion. However, at such a scale, it would be erroneous to assume too much homogeneity within these regions. In each aggregation of counties or region, there are likely to be substantial variations in belief systems and practices at the local level. (*Source: Adapted from D. E. Jones, S. Doty, C. Grammich, J. E. Horsch, R. Houseal, M. Lynn, J. P. Marcum, K. M. Sanchagrin, and R. H. Taylor, Religious Congregations and Membership in the United States 2000: An Enumeration by Region, State and County Based on Data Reported for 149 Religious Bodies. Nashville, TN: Glenmary Research Center, 2000, p. 562.*)

CULTURAL SYSTEMS

Broader than the cultural complex concept is the cultural system, a collection of interacting components that, taken together, shape a group's collective identity. A **cultural system** includes traits, territorial affiliation, and shared history, as well as other, more complex elements, such as language and religion. In a cultural system it is possible for internal variations to exist in particular elements at the same time that broader similarities lend coherence. For example, Christianity unites all Protestant religions (as well as Catholic ones), yet the practices of particular denominations—Lutherans, Episcopalians, and Quakers—vary. And while Mexicans, Bolivians, Cubans, and Chileans exhibit variations in pronunciation, pitch, stress, and other aspects of vocal expression, they all speak Spanish. This means they share a key element of a cultural system (which, for these nationalities, also includes Roman Catholicism and a Spanish colonial heritage).

Geography and Religion

Two key components of a cultural system for most of the world's people are religion and language. **Religion** is a belief system and a set of practices that recognize the existence of a power higher than humankind. Although religious affiliation is on the decline in some parts of the world's core regions, it still acts as a powerful shaper of daily life, from eating habits and dress codes to coming-of-age rituals and death ceremonies, holiday celebrations, and family practices, in both the core and the periphery. And, like language, religious beliefs and practices change as new interpretations are advanced or new spiritual influences are adopted.

The most important influence on religious change has been conversion from one set of beliefs to another. From the Arab invasions following Muhammad's death in 632, to the Christian Crusades of the Middle Ages, to the onset of globalization in the fifteenth century, religious missionizing—propagandizing and persuasion—as well as forceful and sometimes violent conversion have been key elements in changing geographies of religion. Especially in the 500 years since the onset of the Columbian Exchange (see Chapter 4), conversion of all sorts has escalated throughout the globe. Since 1492, traditional religions have become dramatically dislocated from their sites of origin not only through missionizing and conversion but also by way of diaspora and emigration. Whereas missionizing and conversion are deliberate efforts to change the religious views of a person or peoples, diaspora and emigration involve the involuntary and voluntary movement of peoples who bring their religious beliefs and practices to their new locales.

Diaspora is the spatial dispersion of a previously homogeneous group. The processes of global political and economic change that led to the massive movement of the world's populations over the last five centuries have also meant the dislodging and spread of the world's many religions from their traditional sites of practice. Religious practices have become so spatially mixed that it is a challenge to present a map of the contemporary global distribution of religion that reveals more than it obscures. This is because the globe is too gross a level of resolution to portray the wide variation that exists among and within religious practices. **Figure 5.10** identifies the contemporary distribution of what religious scholars consider to be the world's major religions because they contain the largest number of practitioners. As with other global representations, the map is useful in that it helps to present a generalized picture.

Figure 5.11 identifies the source areas of four of the world's major religions and their diffusion from those sites over time. The map illustrates how the world's major religions originated and diffused from two fairly small areas of the globe. The first, where Hinduism and Buddhism (as well as Sikhism) originated, is an area of lowlands in the subcontinent of India drained by the Indus and Ganges rivers (Punjab on the map). The second area, where Christianity and Islam (as well as Judaism) originated, is in the deserts of the Middle East.

Hinduism was the first religion to emerge, among the peoples of the Indo-Gangetic Plain, about 4,000 years ago. Buddhism and Sikhism evolved from Hinduism as reform religions, with Buddhism appearing around 500 B.C.E. and Sikhism developing in the fifteenth century. It is not surprising that Hinduism helped to produce new religions because India has long been an important cultural crossroads. As a result, ideas and practices originating in India spread rapidly. At the same time, other ideas and practices were brought to India from far-flung places and were absorbed and translated to reflect Indian needs and values.

For example, Buddhism emerged as a branch of Hinduism in an area not far from the Punjab. At first a very small group of practitioners surrounding Prince Siddhartha Gautama, the founder of the religion, were confined to northern India. Slowly and steadily Buddhism dispersed to other parts of India and was carried by missionaries and traders to China (100 B.C.E.–C.E. 200), Korea and Japan (C.E. 300–500), Southeast Asia (C.E. 400–600), Tibet (C.E. 700), and Mongolia (C.E. 1500) (see **Figure 5.12**). And as it spread, Buddhism developed many regional forms; Tibetan Buddhism, for example, is distinct from Japanese Buddhism.

Religious beliefs are organized and codified, often based on the teachings and writings of one or more founders. And it is important to recognize that each of the world's religions contain all sorts of variation. For example, Christians may be Catholics or Protestants, and even within these large groups there exists a great deal of variation. The same is true for all religions.

Hinduism, the oldest religion, is codified in the Veda and the Upanishads, its sacred scriptures, but *Hinduism* is a term that broadly describes a wide array of sects. Hindus strive, through the practice of yoga, adherence to Vedic scriptures, and devotion to a personal guru, for release from repeated reincarnation. Hinduism's most central deities are the divine trinity, representing the cyclical nature of the universe: Brahma the creator, Vishnu the preserver, and Shiva the destroyer.

An offshoot of Hinduism, Buddhism, was founded by Gautama—known as the Buddha (Enlightened One)—in southern Nepal. The central tenets of Buddhism are that meditation and the practice of good religious and moral behavior can lead to nirvana, the state of enlightenment. Before reaching nirvana, however, one must experience repeated lifetimes that are good or bad depending on one's actions (karma). There are four noble truths in Buddhism: (1) existence is a realm of suffering; (2) desire, along with the belief in the importance of one's self, causes suffering; (3) achievement of nirvana ends suffering; and (4) nirvana is attained only by meditation and by following the path of righteousness.

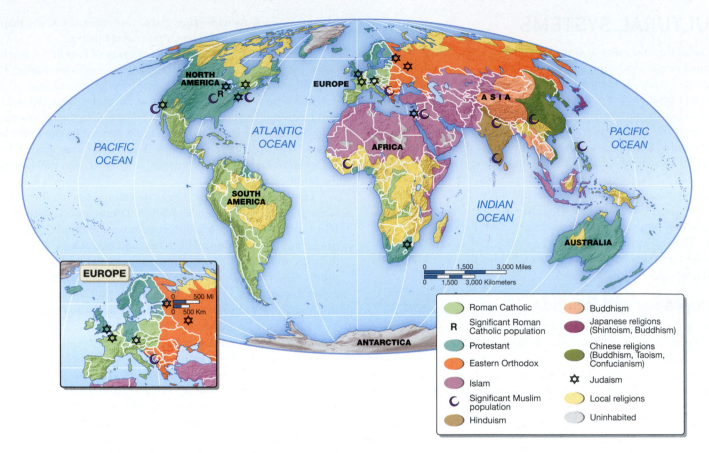

FIGURE 5.10 World distribution of major religions Most of the world's peoples are members of these religions. Not evident on this map are the local variations in practices, as well as the many smaller religions that are practiced worldwide.

FIGURE 5.11 Origin areas and diffusion of four major religions The world's major religions originated in a fairly small region of the world. Judaism and Christianity began in present-day Israel and Jordan. Islam emerged from western Arabia. Buddhism originated in India, and Hinduism in the Indus region of Pakistan. The source areas of the world's major religions are also the cultural hearth areas of agriculture, urbanization, and other key aspects of human development.

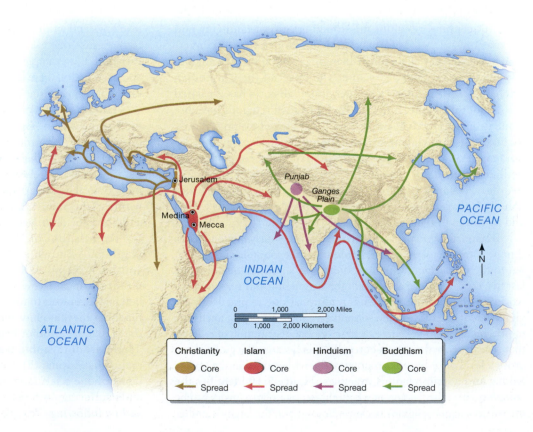

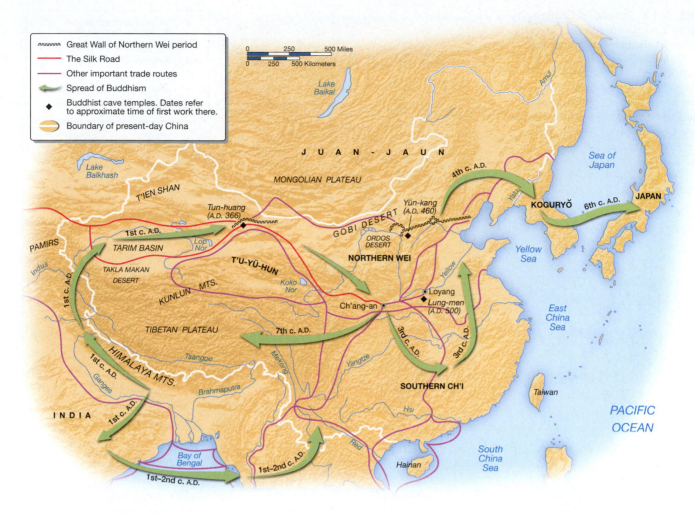

FIGURE 5.12 Spread of Buddhism Buddhism diffused from its source area in India to China and then from China on to Korea and Japan. Commercial routes, like the Silk Road, were important vectors for the spread of the religion from India to China. Missionaries were responsible for the spread of Buddhism from China to Korea and Japan. (Adapted from C. Schirokauer, *A Brief History of Chinese and Japanese Civilizations*, 2nd ed. Florence, KY: Wadsworth, 2005.)

Like Hinduism and Buddhism, Sikhism advocates the pursuit of salvation through disciplined, personal meditation in the name and message of God, who is regarded not as a person but as the Universe itself.

Christianity, Islam, and Judaism all developed among the Semitic-speaking people of the deserts of the Middle East. And like the Indo-Gangetic religions, these three religions are related. Although Judaism is the oldest of the three, it is the least widespread. Judaism originated about 4,000 years ago, Christianity about 2,000 years ago, and Islam about 1,300 years ago. Judaism developed out of the cultures and beliefs of Bronze Age peoples and was the first monotheistic (belief in one God) religion. Although Judaism is the oldest monotheistic religion, and one that spread widely and rapidly, it is numerically small because it does not seek converts.

Christians and Jews, along with Muslims and Bahá'í and Druze followers, share the same source as Abrahamic religions, which means that they all claim a strong connection to the prophet Abraham and the source area of the Middle East. Although sharing the same origin, Judaism is also based on the teachings of the prophets Isaac and Jacob as well as Abraham. Jews espouse belief in a God who leads his people by speaking through these prophets. They believe that God's word is revealed in the Hebrew Bible (or Old Testament), especially in that part known as the Torah, which contains a total of 613 biblical commandments, including the Ten Commandments, explicated in another part of the Hebrew Bible known as the Talmud. Jews believe that the human condition can be improved, that the letter and the spirit of the Torah must be followed, and that a Messiah will eventually bring the world to a state of paradise.

Christianity developed in Jerusalem among the disciples of Jesus of Nazareth, who proclaimed that he was the Messiah expected by the Jews. As it moved east and south from its hearth area, Christianity's diffusion was helped by missionizing and by imperial sponsorship. The diffusion of Christianity in Europe is illustrated in **Figure 5.13**. Although we discuss Islam and Islamism later in this chapter, it is important to point out here that Islam, like Christianity, was for centuries routinely spread by forced conversion, to save souls but also often for the purposes of political control.

FIGURE 5.13 Spread of Christianity in Europe Christianity spread through Europe largely because of missionary efforts. Monks and monasteries were especially important as hubs of diffusion in the larger network. Monasteries and other sorts of religious communities are indicated by dots on the map. The shaded areas are places where Christian converts dominated by the year 300. (Adapted from C. Park, *Sacred World*. New York: Routledge, 1994.)

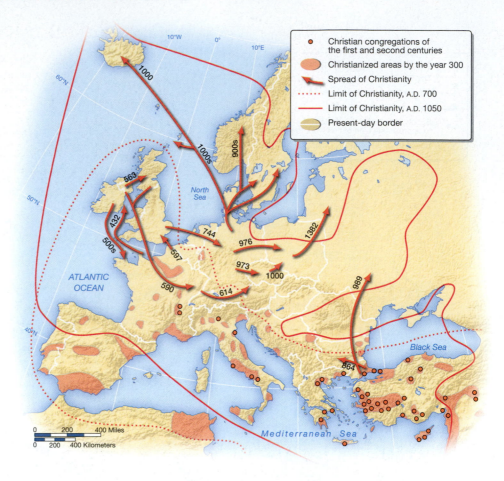

Christianity is a religion based on the life and teachings of Jesus as written in the New Testament of the Holy Bible and is the world's largest religion, with 1 to 2 billion adherents. Christians believe that Jesus is the son of God and the Messiah prophesied in the Old Testament, the holy scripture that is also significant to Judaism. As a central figure in all Christian religions, Jesus is revered as a teacher, the model of a pious life, the manifestation of God, and most importantly the savior of humanity who suffered, died, and was resurrected in order to bring about human salvation from sin. Christians maintain that upon his death, Jesus ascended into heaven and that he will return to judge the living and the dead and grant life everlasting to those who have followed his teachings.

Islam is an Arabic term that means "submission," specifically submission to God's will. A **Muslim** is a member of the community of believers whose duty is obedience and submission to the will of God. As a revealed religion, Islam recognizes the prophets of the Old and New Testaments of the Holy Bible, but to Muslims Muhammad is the last prophet and God's messenger on Earth. The Qur'an, the principal holy book of the Muslims, is considered the word of God as revealed to Muhammad by the Angel Gabriel beginning in about C.E. 610. There are two fundamental sources of Islamic doctrine and practice: the Qur'an and the Sunna. Muslims regard the Qur'an as directly spoken by God to Muhammad. The Sunna is not a written document but a set of practical guidelines to behavior. It is effectively the body of traditions that are derived from the words and actions of the prophet Muhammad. We talk at greater length about Islam in a later section of this chapter.

While these are the world's major religions, there are many others that figure prominently in the cultural lives of people around the world. Among them are Confucianism, Taoism, Shintoism, Mormonism, Zoroastrianism, and Jansenism, as well as Voodoo, Rastafarianism, and animism. The point is that faith, a trusting belief in a transcendent reality or a supreme being, is a profoundly powerful force. It can guide people's actions and attitudes and shape the worlds in which they live.

Religion and Globalization

Faith is a significant element of globalization. As disparate belief systems come into contact, in some cases causing tension and even violent conflict, new religions are introduced among populations who are adherents of a different one. An excellent illustration of the global forces behind the changing geography of religious faith is the Columbian contact with the New World. Before Columbus and later Europeans reached the continents of North and South America, the people living there practiced, for the most part, various forms of animism and related rituals. They viewed themselves holistically, as one part of the wider world of animate and inanimate nature. Shamanism, in which spiritually gifted individuals are believed to possess the power to control preternatural forces, is one important aspect of the belief system that existed among Native American populations at the time of European contact **(Figure 5.14)**.

Perhaps what is most interesting about the present state of the geography of religion is how during the colonial period religious missionizing and conversion flowed from the core to the periphery. In

FIGURE 5.14 San Xavier del Bac mission near Tucson, Arizona The influence of missionaries on native populations was profound, as indigenous rituals became mixed with Christian ones. Even today in missions such as San Xavier, native peoples practice Christianity along with components of their native belief system.

FIGURE 5.15 Tibetan monks protest in Nepal China's refusal to acknowledge an independent Tibet has sent the fourteenth Dalai Lama on numerous international tours to broadcast the plight of Tibetans, who are experiencing extreme persecution. Pictured here are Tibetan monks in exile in Nepal protesting outside the Chinese embassy to mark the anniversary of the failed 1959 uprising against China and to demand that China grant sovereignty to Tibet.

the postcolonial period, however, an opposite trend has occurred. For example, the fastest-growing religion in the United States today is Islam, and it is in the core countries where Buddhism is making the greatest numbers of converts. The recently deceased Pope John Paul was one of the most widely traveled leaders in Roman Catholic history. And the same can be said for the Dalai Lama, who, as the spiritual leader of Tibetan Buddhists and the head of the Tibetan government-in-exile, has been a tireless world traveler. Pope John Paul's efforts were mostly directed at maintaining Roman Catholic followers and attempting to dissuade their conversion to other religions, such as evangelicalism in the United States and Latin America. The Dalai Lama promotes conversion to Buddhism; he carries its message to new places, especially in the core, and advocates sovereignty for Tibet, a region once independent but currently controlled by China **(Figure 5.15)**.

One other impact of globalization upon religious change occurs through the electronic media. The rise of television

evangelism, or *televangelism*—especially in the United States—has led to the conversion of large numbers of people to Christian fundamentalism, which is a term popularly used to describe strict adherence to Christian doctrines based on a literal interpretation of the Holy Bible. Christian televangelism is also widespread in other countries, including Brazil, Argentina, and Chile, as well as India, Kenya, and China. (And Muslim televangelism is just beginning to emerge.) The Christian reference to *fundamentalism* derives from a late nineteenth- and early twentieth-century transdenominational Protestant movement that opposed the accommodation of Christian doctrine to modern scientific theory and philosophy, especially Darwinian theories about the origin of life on Earth. It began with the publication of a series of pamphlets between 1910 and 1915 entitled "The Fundamentals: A Testimony to the Truth." Today, Christian fundamentalism is strong and growing stronger with a trend

FIGURE 5.16 Megachurch, South Barrington, Illinois Willow Creek Community Church is typical of the megachurch phenomenon that has come to characterize fundamental Christian sects in many places across the United States. A Sunday service at Willow Creek often draws 20,000 or more. According to a recent survey, there are 1,200 Protestant churches that claim more than 2,000 weekly worship attendees in the United States. There are megachurches in 45 out of 50 states; Texas has 174 (14 percent), California has 169 (13.7 percent), Florida 83 (6.7 percent), and Georgia 64 (5.2 percent). Houston and Dallas alone host 56 megachurches, or 4.5 percent of the total. The following states do not have megachurches: Maine, New Hampshire, Rhode Island, South Dakota, Vermont, Wyoming.

toward megachurches as popular sites of worship in the United States and elsewhere around the world **(Figure 5.16)**.

Geography and Language

Languages are another aspect of cultural systems that interest geographers. Language is an important focus for study because it is a central aspect of cultural identity. Without language, cultural accomplishments could not be transmitted from one generation to the next. And language itself reflects the ways that different groups regard and interpret the world around them. The distribution and diffusion of languages tell much about the changing history of human geography and the impact of globalization on culture. Before looking more closely at the geography of language, it is necessary to become familiar with some basic vocabulary.

Language is a way of communicating ideas or feelings by means of a conventionalized system of signs, gestures, marks, or articulate vocal sounds. Communication is symbolic, based on commonly understood meanings of signs or sounds. Within standard languages (also known as official languages because they are maintained by offices of government such as schools and the courts), regional variations, known as **dialects,** exist. Dialects feature differences in pronunciation, grammar, and vocabulary that are place-based in nature.

For the purposes of classification, languages are grouped into families, branches, and groups. A **language family** is a collection of individual languages believed to be related in their prehistorical origin. About 50 percent of the world's people speak a language that is in the Indo-European family. A **language branch** is a collection of languages that possesses a definite common origin but has split into individual languages. A **language group** is a collection of several individual languages that is part of a language branch, shares a common origin in the recent past, and has relatively similar grammar and vocabulary. Spanish, French, Portuguese, Italian, Romanian, and Catalan are a language *group,* classified under the Romance *branch* as part of the Indo-European language *family.*

Traditional approaches in cultural geography have identified the source areas of the world's languages and the paths of diffusion of those languages from their places of origin. Carl Sauer identified the origins of certain cultural practices with the label "cultural hearth." **Cultural hearths** are the geographic origins or sources of innovations, ideas, or ideologies. Language hearths are a subset of cultural hearths; they are the source areas of languages. **Figure 5.17** shows the locations of the world's major languages and language families.

In the recent population census in India, people reported the use of over 1,600 languages, which the government collapsed into 200 related ones. These translate into a broad regional grouping of four major language families. The Indo-European family of languages, introduced by the Aryan herdsmen who migrated from Central Asia between 1500 and 500 B.C.E., is prevalent in the northern plains region, Sri Lanka, and the Maldives. This language family includes Hindi, Bengali, Punjabi, Bihari, and Urdu. Munda languages are spoken among the tribal hill peoples who still inhabit the more remote hill regions of peninsular India. Dravidian languages (which include Tamil, Telegu, Kanarese, and Malayalam) are spoken in southern India and the northern part of Sri Lanka. Finally, Tibeto-Burmese languages are scattered across the Himalayan region **(Figure 5.18)**.

In India, the boundaries of many of the country's constituent states were established after partition on the basis of language. Overall, no single language is spoken or understood by more than 40 percent of the people. There have been efforts since the country became independent to establish Hindi, the most prevalent language, as the national language, but this has been resisted by many of the states within India, whose political identity is now closely aligned with a different language. In terms of popular media and literature, there is a thriving Hindi and regional language press. Film and television are dominated by Hindi and Tamil, with some Telegu programming.

English, spoken by fewer than 6 percent of the people, serves as the link language between India's states and regions. As in other

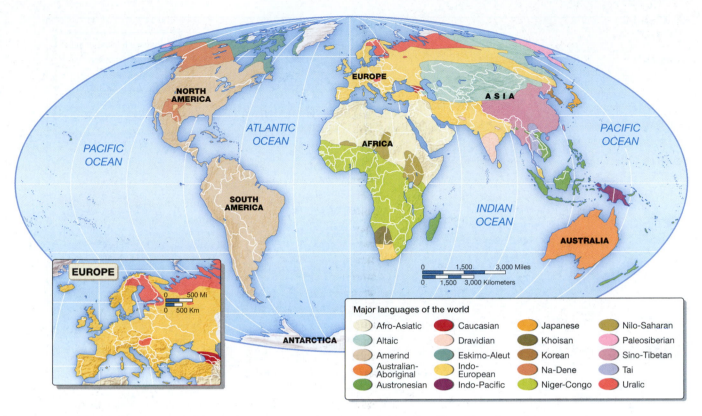

FIGURE 5.17 World distribution of major languages and major language families Classifying languages by family and mapping their occurrence across the globe provide insights about human geography. For example, we may discover interesting cultural linkages between seemingly disparate cultures widely separated in space and time. We may also begin to understand something about the nature of population movements across broad expanses of time and space. (Adapted from E. F. Bergman, *Human Geography: Cultures, Connections, and Landscapes.* Upper Saddle River, NJ: Prentice Hall, 1994; Western Hemisphere after J. H. Greenberg, *Language in the Americas.* Stanford, CA: Stanford University Press, 1987; Eastern Hemisphere after D. Crystal, *The Cambridge Encyclopedia of Language.* Cambridge: Cambridge University Press, 1997.)

former British colonies in South Asia, English is the language of higher education, the professions, and national business and government. Without English, there is little opportunity for economic or social mobility. Most children who do attend school are taught only their local language and so are inevitably restricted in their prospects. Guards, sweepers, cooks, or drivers who speak only Hindi or Urdu will likely do the same work all their lives. In contrast, those who can speak English—by definition the upper middle classes—are able to practice their profession or do business in any region of their country and in most parts of the world.

With globalization, the geography of language has become even more dynamic. The plethora of languages and dialects in a region like South Asia makes communication and commerce among different language speakers difficult. It may, furthermore, create problems in governing a population. For this reason developing states create one national language to facilitate communication and enable the efficient conduct of state business. In general, where official languages are put into place, indigenous languages are threatened **(Figure 5.19)**. Yet the actual unfolding of globalizing forces—such as official languages—works differently in different places and in different times. The overall trend appears to be toward the loss of indigenous language (and other forms of culture), as **Figure 5.20** shows. Language can also serve as an important means of challenging the political, economic, cultural, and social forces of globalization as they occur in places like Belgium (among the Flemish people), Spain

(with the Basque separatist movement), Canada (around the Québecois movement), and other countries.

APPLY YOUR KNOWLEDGE Conduct an Internet search to identify the different dialects spoken in your native language. (Be aware that there are distinct dialects in American English.) On a map identify where these dialects are spoken and consider these regions in terms of culture and power. Do certain dialects have different cultural power? Does any one dialect dominate the way culture is communicated, say in newspapers or on television? ■

Culture and Society

Through its influence on social organization, culture has an important impact. Social categories like kinship, gang, or generation, or some combination of categories, can figure more or less prominently, depending upon geography. Moreover, the salience of these social categories may change over time as the group interacts with people and forces outside of its boundaries.

For example, countries in the Middle East and North Africa have a complex social organization—as complex as that of any other region of the globe—shaped by cultural ties and meaning systems that highlight gender, tribe, nationality, kinship, and family. Global media technologies, such as satellite television and the Internet, are

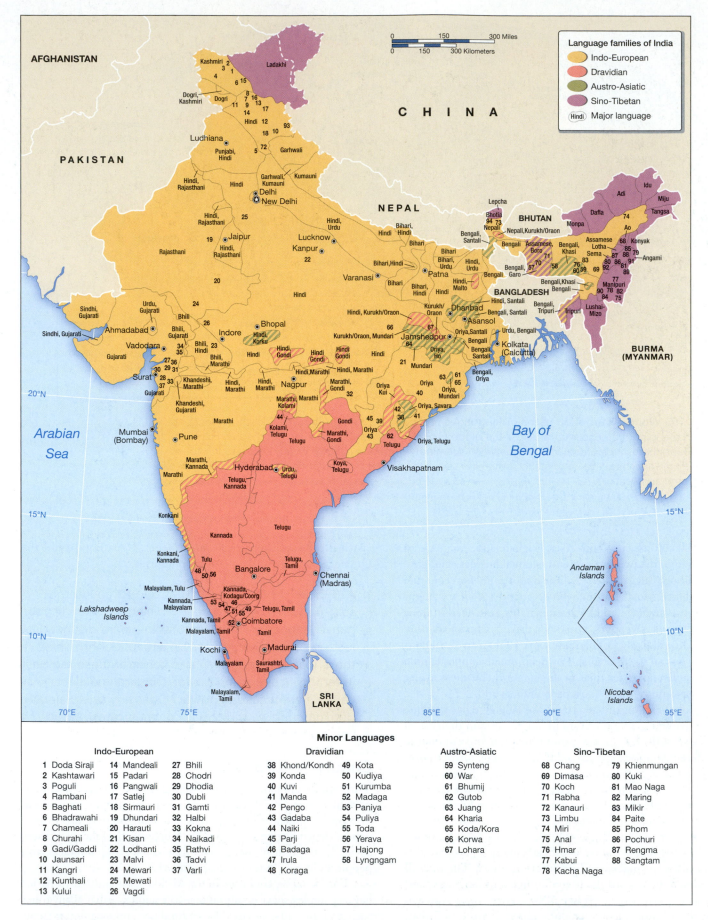

FIGURE 5.18 Language map of India India's linguistic landscape is complex, with hundreds of distinct languages in use.

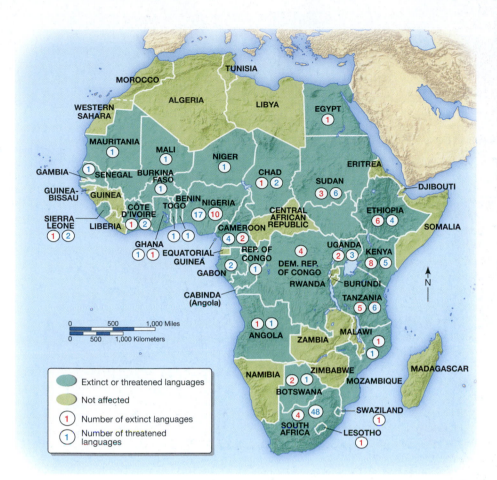

FIGURE 5.19 African countries with extinct and threatened languages It is not absolutely certain how many languages are currently being spoken worldwide, but the estimates range between 4,200 and 5,600. While some languages are being created through the fusion of an indigenous language with a colonial language, such as English or Portuguese, indigenous languages are mostly dying out. Although only Africa is shown in this map, indigenous languages are dying throughout the Americas and Asia as well.

increasingly penetrating the region, however, and the potential for new social forms to emerge and old ones to be reconfigured is increasing. Generally, the predominant forms of social organization in the region have persisted for hundreds of years. It would be incorrect, however, to assume that both subtle and dramatic changes within these forms have not already occurred.

Kinship is a form of social organization that is particularly central to the culture system of the Middle East and North Africa. **Kinship** is normally thought of as a relationship based on blood, marriage, or adoption. This definition needs to be expanded, however, to include a *shared notion of relationship* among members of a group. The point is that not all kinship relations are understood by social groups to be exclusively based on biological or marriage ties. While biological ties, usually determined through the father, are important in the Middle East, they are not the only important ties that link individuals and families.

Kinship is such a valued relationship for expressing solidarity and connection that it is often used to assert a feeling of group closeness and as a basis for identity even where no "natural" or "blood" ties are present. This notion of kinship might also be seen displayed in America among fraternity "brothers," sorority "sisters," gang members, and even police officers and firefighters who feel a strong familial bond with co-workers. In the Middle East and North Africa, kinship is even an important factor in shaping the spatial relationships of the home, as well as outside the home, determining who can interact with whom and under what circumstances. This is especially the case in the interaction of gender and kinship, where women's and men's access to public and private space is sharply differentiated.

The idea of the tribe is also central to understanding the sociopolitical organization of the Middle East and North Africa, as well as other regions of the world. Although tribally organized populations appear throughout the region, the tribe is not a widespread form of social organization. Generally speaking, a **tribe** is a form of social identity created by groups who share a set of ideas about collective loyalty and political action. The term *tribe* is a highly contested concept and one that should be treated carefully. For instance, it is often seen as a negative label applied by colonizers to suggest a primitive social organization throughout Africa. Where it is adopted in the Middle East and North Africa, however, tribe is seen as a valuable element in sustaining modern national identity. Tribes are grounded in one or more expressions of social, political, and cultural identities created by individuals who share those identities. The result is the formation of collective loyalties that result in primary allegiance to the tribe.

APPLY YOUR KNOWLEDGE Create a kinship map of your family. Once you have done this, identify a kinship system from the Middle East or North Africa. Compare and contrast that system with your own. What aspects of that kinship organization are similar to yours? What aspects are different and why? ■

Culture and the Nation

The protection of regional languages as a way of resisting globalization is just one part of a larger movement that interests geographers and other scholars. The movement, known

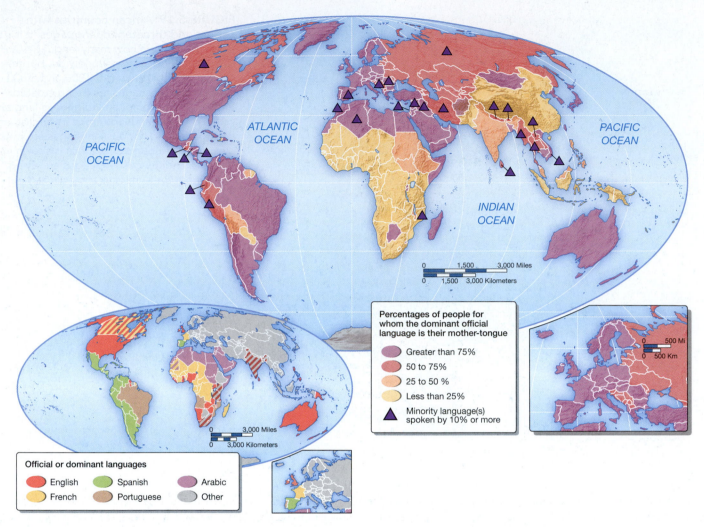

FIGURE 5.20 Tongue-tied Another way to examine the geography of language is to consider the proportion of people around the world whose mother tongue is not their country's official language. In places like China, where the dominant language is often not the mother tongue, simple communication may pose considerable challenges. Mandarin is the official language, but there are 6 major regional dialects of Mandarin as well as another 41 distinct languages spoken by ethnic minorities in China. In other cases, the disparity between the official language and the mother tongue is the result of colonization. For instance, in South Africa before apartheid was dismantled, there were 2 official languages, English and Afrikaans—both languages of the colonizers—but over 50 indigenous languages. Since the election of Nelson Mandela in the early 1990s, the number of official languages has grown to 11. The increase gives more residents access to public information and is intended to expand participation in civil society as well as preserve the cultural diversity of the country. (*Source:* Adapted from M. Kidron and R. Segal, *The New State of World Atlas*, 4th ed. New York: Simon and Schuster, 1991.)

as **cultural nationalism,** is an effort to protect regional and national cultures from the homogenizing impact of globalization, especially from the penetrating influence of U.S. culture. **Figures 5.21** and **5.22** picture two widespread aspects of U.S. culture—films and television. Many other U.S. products also travel widely outside its borders. Although many of these products are welcomed abroad, many others are not. France, for example, has been fighting for years against the "Americanization" of its language, food, and films. Nations can respond to the homogenizing forces of globalization and the spread of U.S. culture in any number of ways. Some groups attempt to seal themselves off from undesirable influences. Other groups attempt to legislate the flow of foreign ideas and values. Most recently, Muslim countries have been especially resistant to the

effects of Western culture propelled by globalization on their beliefs and practices.

The Islamic world includes very different societies and regions, from Southeast Asia to Africa. Muslims comprise over 85 percent of the populations of Afghanistan, Algeria, Bangladesh, Egypt, Indonesia, Iran, Iraq, Jordan, Pakistan, Saudi Arabia, Senegal, Tunisia, Turkey, and most of the independent republics of Central Asia and the Caucasus (including Azerbaijan, Turkmenistan, Uzbekistan, and Tajikistan). In Albania, Chad, Ethiopia, and Nigeria, Muslims make up 50 to 85 percent of the population. In India, Burma (Myanmar), Cambodia, China, Greece, Slovenia, Thailand, and the Philippines, significant Muslim minorities exist. After Christianity, Islam possesses the next largest number of adherents worldwide—about 1 billion.

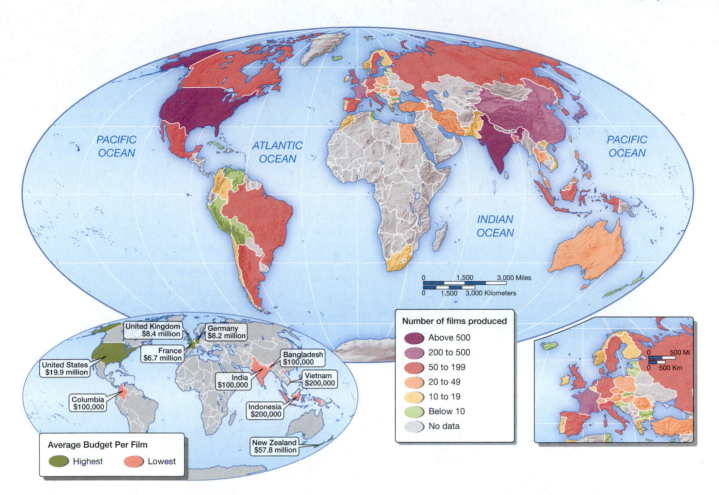

FIGURE 5.21 World film production, 2001 This map shows the number of films produced by different countries around the world. As expected, core countries are much more likely to be film producers. They also import fewer films as a result. For instance, the United States produces four times more films than it imports. In the semiperiphery, India, a country with an enormously popular film industry, produces eight times more films than it imports. Despite popular beliefs, the United States is not the largest film producer in the world. India far exceeds U.S. production, and China and Hong Kong as well as the Philippines are also film production giants. (Adapted from UNESCO, *Survey on National Cinematography, 2001*, available at http://www.unesco.org/culture/industries/cinema/html_eng/prod.shtml. *World Culture Report*, 2000, table 4, http://www.unesco.org/culture/worldreport/html_eng/stat2/table4.pdf. All rights reserved.)

The map in **Figure 5.23** shows the relative distribution of Muslims throughout Europe, Africa, and Asia as well as the heartland of Islamic religious practice.

One of the most widespread cultural forces in the world today is Islamism, more popularly, although incorrectly, known as Islamic fundamentalism. Whereas *fundamentalism* is a general term that describes the desire to return to strict adherence to the fundamentals of a religious system, **Islamism** is an anticolonial, antiimperial political movement. In Muslim countries, Islamists resist core, especially Western, forces of globalization—namely modernization and secularization. (See Box 5.2, "Window on The World: Fashionable Veiling.") Not all Muslims are Islamists, although Islamism is the most militant movement within Islam today.

The basic intent of Islamism is to create a model of society that protects the purity and centrality of Islamic precepts through the return to a universal Islamic state—a state that would be religiously and politically unified. Islamists object to modernization because they believe the corrupting influences of the core place the rights of the individual over the common good. They view the popularity of Western ideas as a move away from religion to a more secular (nonreligious) society. Islamists desire to maintain religious precepts at the center of state actions, such as introducing principles from the sacred law of Islam into state constitutions.

Islamism—a radical and sometimes militant movement—should not be regarded as synonymous with the practices of Islam, any more generally than Christian fundamentalism is with Christianity. Islam is not a monolithic religion, and even though all adherents accept the basic pillars of the faith, specific practices vary according to the different histories of countries, nations, and tribes. Some expressions of Islam allow for the existence and integration of Western styles of dress, food, music, and other aspects of culture, while others call for the complete elimination of Western influences.

APPLY YOUR KNOWLEDGE Locate online the video entitled, "Lifting the Veil: Muslim Women Explain Their Choice." Before watching, write down your current understanding and knowledge of the hijab. After viewing the segment, describe how your impressions changed or stayed the same in response to the content of the program. ■

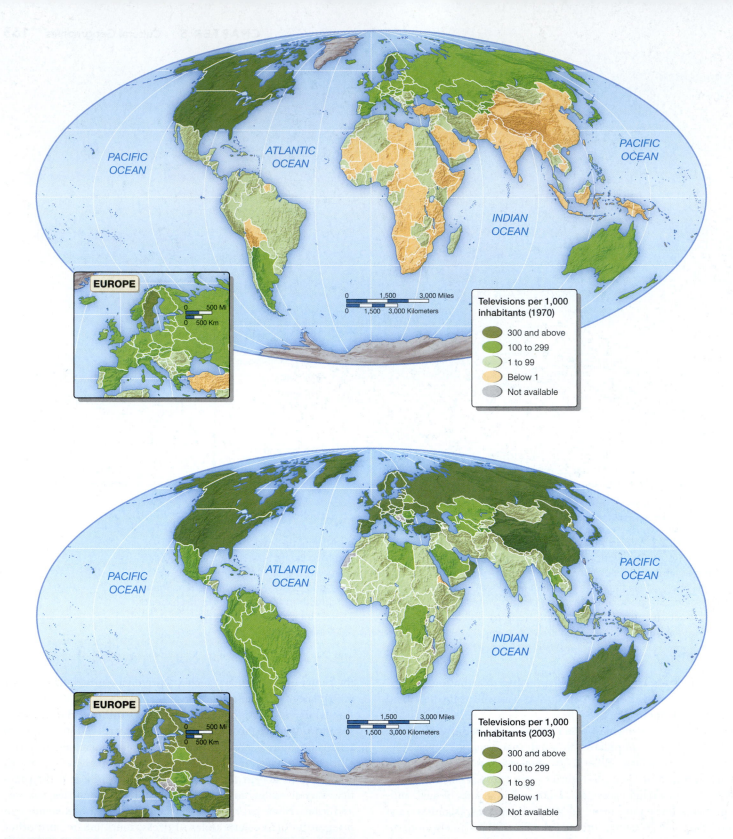

FIGURE 5.22 World distribution of TV sets, 1970 and 2003 The ability to receive television broadcasts no matter who is exporting or producing them depends, of course, on access to a television set. In some areas of the periphery—for example, in some of the wealthier South American countries—there are over 300 television sets per 1,000 people. In much of sub-Saharan Africa, however, there are fewer than 10 sets for every 1,000 people. In the core, in countries such as the United States and Japan, there is on average 1 color television set per household, as well as a high level of ownership of videocassette recorders. Cable subscription opens up a market to even more commercial television products. With respect to television programming exports, the core vastly overwhelms the periphery in the export of television programs, with the leaders being the United States and the United Kingdom. (Adapted from J. M. Rubenstein, *The Cultural Landscape: An Introduction to Human Geography*, 6th ed., © 1999. Update from World Bank, *World Development Indicators 2002*, Washington, DC: World Bank.)

FIGURE 5.23 Muslim world Like the Spanish colonial effort, the rise and growth of Muslim colonization were accompanied by the diffusion of the colonizers' religion. The distribution of Islam in Africa, Southeast Asia, and South Asia that we see today testifies to the broad reach of Muslim cultural, colonial, and trade activities. But the heart of the Muslim culture remains the Middle East, the original cultural hearth. It is also in this area that Islamism is most militant. (Adapted from D. Hiro, *Holy Wars*. London: Routledge, 1989.)

Legend: **Percent Muslim population**
- 51 to 100%
- 26 to 50%
- 2 to 25%
- 1% and fewer
- Islamic cultural hearth

CULTURE AND IDENTITY

In addition to exploring cultural forms, such as religion and language, and movements, such as cultural nationalism, geographers have increasingly begun to ask questions about other forms of identity. This interest largely has to do with certain long-established and some more recently self-conscious cultural groups that are beginning to use their identities to assert political, economic, social, and cultural claims.

Sexual Geographies

One identity that has captured the attention of cultural geographers is sexuality. **Sexuality** is a set of practices and identities that a given culture considers related to each other and to those things it considers sexual acts and desires. One of the earliest and most effective examples of the geographic study of sexuality is the examination of the spatial expression of prostitution. Research on prostitution in California found that sex work—as it is now more commonly known—is spatially differentiated based upon the target clientele as well as systems of surveillance. More typical contemporary work on sexuality explores the spatial constraints on homosexuality and the

ways in which lesbian, gay, bisexual, and transgendered (LGBT) people respond to and reshape them. Two particular areas of research have emerged: gay and lesbian consumerism, and the body as a site of performance of sexuality and gender identities.

One approach to understanding homosexuality, particularly as it is represented in contrast to heterosexuality, is research and writing on the body and the ways in which individuals literally perform their sexual identities through clothing, attitude, body language, and grooming, such as hairstyles and bodily adornments. Research on the body has also been used to decode the social and cultural practices that express other kinds of social identity, such as gender and class. But the work has been most fully developed in its application to sexuality.

The central theoretical position of this research is that sexual identities are learned and performed through standard, taken-for-granted practices. Using gender as a starting point and recognizing it as a key feature of all human identity that marks and defines humans as "sexed" beings, researchers in this area contend that gender is not something people essentially are (because of a given set of physical characteristics), but something people do (something we enact by the way we present our bodily selves to the world).

Fashionable Veiling

Why do some Muslim women wear headscarves or veils when they are out in public? In large part because the Muslim holy book, the Qur'an, instructs them, as well as men, to be modest in their public comportment. Some Muslim women wear headscarves to demonstrate their piety and publicly identify themselves as people who adhere to Muslim religious beliefs. For them, covering one's head is an expression of religious conviction. But being religious doesn't mean a Muslim woman must also be unfashionable. And just because many Muslim women wear headscarves (or other forms of more modest dress such as full-length skirts and long sleeves) does not mean that they all dress the same.

In fact, fashionable Muslim dress has grown increasingly popular since the 1980s. This move toward more fashion choices for Muslim women is connected to globalization and the growth in consumer culture in increasingly liberal democracies in Muslim nations that encourage more personal freedom. A complex, widespread Islamic consumer culture includes well-educated, upwardly mobile Islamic young women who are active in moderate Islamist politics and possess personal disposable income.

Geographers Banu Gökariksel and Anna Secor have studied the veiling fashion industry in Turkey, a particularly interesting place to explore because of the over 200 firms operating there. With retail outlets in the Middle East, Europe, and North America, these firms clearly have a global reach. One such firm, Tekbir (meaning "God is great"), has outlets in Germany, the Netherlands, France, England, Belgium, Austria, Switzerland, Bosnia Herzegovina, Macedonia, Azerbaijan, Dubai, Lebanon, Jordan, Syria, Palestine, Libya, Egypt, the Sudan, Algeria, South Africa, the United States, Canada, and Australia. Moreover, in addition to Istanbul, other Islamic fashion centers include Beirut, Cairo, Dakar, Dubai, Kuala Lumpur, and Jakarta.

Turkish veiling-fashion designers visit the annual fashion shows in Paris to find inspiration for their new styles in emerging Western trends of fabric, color, cut, and style. Veiling-fashion is little different from Western fashion in that it is part of a larger system of production, consumption, and changing cultural tastes. The Turkish veiling-fashion industry takes these elements and combines them with a commitment to modest dress and head covering to produce women's everyday wear, though some haute couture (high fashion) is also retailed.

While fashionable veiling in Turkey looks to the modern Western fashion world for inspiration, some Muslim women are looking in the opposite direction. An interesting contrast is found in Indonesia, where fashionable young women who previously did not cover their hair are now wearing headscarves and those who have been wearing a headscarf for most of their lives have started to adopt additional fashionable styles of covered dress. Indeed, fashionable veiling is a trend that exists throughout most of the Islamic world, from the Islamic states of Saudi Arabia or Oman to Muslim immigrant communities in the United States, Europe, Canada, Australia, the United Kingdom, Japan, and the Philippines.

In the Netherlands, Islamic fashion researcher Annelies Moors explored the trend of young women, children of immigrant guest workers, moving toward wearing headscarves. Though it is important to appreciate that not all Muslim women wear headscarves, these young woman, who are usually better educated than their mothers, have taken an interest in Islamic dress in response to an increasing commitment to their religion. In exploring this phenomenon, Moors interviewed three Muslim women in order to understand their reasons for these choices.

While all three women are immediately recognizable as Muslim, wearing a headscarf that covers the hair completely, the oufits that they currently wear are strikingly different. Feride [who is in her mid-thirties and is the daughter of working class Turkish immigrant parents] is most comfortable in elegant suits, combinations of floor-length skirts and well-cut jackets of high-quality materials, often brought from Turkey. She pays much attention to wearing matching headscarves, but does not like the style that many younger Turkish girls have adopted, who use underscarves and other materials to produce a high and voluminous head shape. Her style is more personal, playing with the various ways that are common in eastern Turkey. Lisa [a 22-year-old college student whose father is from Pakistan and mother is a Hindu from Suriname who converted to Islam], in contrast, is not interested in elegance, sophistication, or, for that matter, high-heeled shoes. She describes her style as casual, sporty, urban, and cool. She usually wears jeans with a tunic or a blouse over them, [a headscarf], and "always a hoodie, combined with a cool bag and Nikes." It is only on special occasions amongst Pakistanis that she can be spotted wearing shalwar qamiz [a set consisting of a long shirt or tunic and a pair of loose trousers] often brought from Pakistan. Malika [a 24-year-old college student whose parents were Moroccan immigrants] wears an outfit that shows the least variety. She wears a long, loose, all-covering dress, made by a seamstress who specializes in such oufits, and combines this with a three-quarter-length, all-enveloping veil (khimar); sometimes she also wears gloves. Underneath, however, she wears very fashionable styles, including brand-name jeans.

Although Feride considers herself very much a religious person, she also explained that she had to grow into wearing these styles of covered dress. "You have your own personal taste, and also fashion plays a role. If long splits are fashionable, you buy skirts with long splits . . . Covering is a form of

FIGURE 5.C Islamic veiling fashion These images make it clear that veiling fashion among Muslim women is highly variable Figure 5.C(a) shows a style created by Indonesian designer. The silken fabric of the body covering is feminine and attention getting, reflecting both sartorial trends and spiritual dictates. Figure 5.C(b) shows teenage girls with brightly colored veils worn with one or two underscarves wrapped around the neck. Figure 5.C(c) shows young women entering Regent's Park Mosque in London, England. Though they are dressed in full length body covering, their veils are only loosely worn exposing their hair, which would not be an acceptable style among the more conservative and older Syrian women shown in Figure 5.C(d). Notice, however, that the young woman in the middle of the image, is modestly but also more fashionably veiled than the older others.

worship, and you are not really supposed to draw attention to yourself or to make yourself beautiful. That is a thin line . . ." Feride is outspoken about why she wears covered dress. "For me, covering is a form of worshipping the Creator, a form of devotion. Some would say that you need to cover to avoid arousing sexual feelings in men, but I do not do it for that reason. If they have a problem with their feelings, that is their problem, not mine.

Malika began wearing her veil almost a year ago. Although to an outsider this shift in clothing may look like a huge change, she said: "I do not think it is that much of a change. I still love fashion, but now it is halal [permissible according to Islamic law]. I simply wear a long dress over my jeans. Sometimes people who know me from before say what a pity, all those curls, what happened, you were always so free. Then I say that was my appearance, but what you see is not the same as what is inside. I had always been practicing. I used to pray off and on, but I started to do this more consistently during the last three years, especially last year during Ramadan. Now I fast extra days, do not listen to music, and do not shake hands with men. For me, this has become normal, but people around me find it rather extreme. I first wanted to change my behavior and only when I had done that, to start to veil. My veil has made it [the transition] complete.

Lisa . . . likes brand names: "G-Star, for instance. The quality is much better." She often goes to the smaller stores where they have different brands. "But for simple things, I go to Vero Moda or H&M." Thinking about what she finds Islamically acceptable, she says: "I would not say everything except the face and the hands, but I do cover my chest and the section between my waist and my knees. Also, it should not be too tight or too short. If I wear skirts, it is a long skirt, or I wear trousers underneath. With skinny jeans, I make sure that they do not fit too tightly around my legs. I wear such styles with my headscarf, because of [my] love for God. It is not that it makes you less attractive, because your eyes or your face can also do that." Lisa stresses that in the first place it is "a personal relation with God." At the same time, it is an attempt not to attract attention "so that men do not look [at you] twice. But nowadays they also think that the headscarf is beautiful and they make comments about that. But I do feel better. It is clear that I have committed myself and that is visible to all."[1]

The stories these very different women tell about themselves and their personal choices indicate that traditional cultural traits, such as dressing the female body.

Consistent with religious strictures, are never static. Fashion, as a mass-produced cultural product, is always open to interpretation. As the different veiling styles shown in the composite figure (**Figure 5.C**) demonstrate, Muslim women are as capable of producing their own sense of style as manufacturers are of capitalizing on them. Indeed, the Web sites of Islamic fashion designers make it clear that while a great deal of inspiration for their work comes from international fashion trends, they also look to the street and what everyday women are wearing for their new ideas.

[1]From Annelies Moors, 2009, "Islamic Fashion in Europe: Religious Conviction, Aesthetic Style and Creative Consumption," *Encounters*, 1:175–200. Reprinted by permission of *Encounters* and the author.

FIGURE 5.24 Gay pride parade, São Paulo, Brazil Almost 2 million LGBT people and their supporters—many in lavish Carnival costumes and waving rainbow-colored flags—paraded in Brazil's biggest city in May 2005 to celebrate gay pride and call for the legalization of civil unions between homosexuals. In comparison, San Francisco's gay pride parade typically attracts tens of thousands of people, and the 2004 World gay pride day celebrations in Berlin attracted between 200,000 and 500,000 participants.

Research on sexuality and the body in geography recognizes that space is central to understanding. Where gay or straight identities can be performed plays a central role in who occupies those spaces, how they are occupied, and even what sexual identity people might be performing if they occupy one space or another. But issues of sexuality and space go beyond the performance of sexual identity to the emergence of new political cultures that are being constructed to protect the rights of lesbian, gay, and transgendered people in particular national and international spaces **(Figure 5.24)**.

Ethnicity and Territory

Ethnicity is another area in which geographers explore cultural identity. **Ethnicity** is a socially created system of rules about who belongs to a particular group based upon actual or perceived commonalities, such as language or religion. A geographic focus on ethnicity is an attempt to understand how it shapes and is shaped by space and how ethnic groups use space with respect to mainstream culture as well as other ethnic groups. For cultural geographers, territory is also a basis for ethnic group cohesion (see Chapter 9 for more on territory). For example, cultural groups—ethnically identified or otherwise—may be spatially segregated from the wider society in ghettos, ethnic enclaves, homelands, and tribal areas.

In China, for example, 91 percent of the population is Han. That there are 55 different ethnic groups in addition to the Han. These are mostly residual groups of indigenous people such as the Miao, the Dong, Li, Naxi, and Qiang, who are economically disadvantaged and found in relatively remote border regions, removed from central authority in Beijing. Tensions exist between the dominant Han and several of the larger minority groups such as the Tibetans and the Uyghurs. The latter are a Turkic Muslim minority group living in western China. They demand greater control over their territory in order to practice self-governance and to follow their own cultural traditions and not those of mainstream (Han-dominated) society. The Uyghur ethnic identity is fragmented, however. Some Uyghur support a Pan-Islamic vision, exemplified in the East Turkestan Islamic Movement; others support a Pan-Turkic vision, as in the East Turkestan Liberation Organization; and a third group—the East Turkestan independence movement—promotes a "Uyghurstan" state. As a result, Uyghurs do not speak with one voice and members of each of these groups have committed violence against other Uyghurs who should be whom, not who they view as too assimilated to Chinese or Russian society or not religious enough **(Figure 5.25)**.

Ethnicity is a complex cultural category constituted through shared history and often through language, religion, and an attachment to a particular place or homeland. And though there are many examples of ethnic groups who share a cohesive sense of community and belonging, not all do. As with any identity, variation exists in practice.

APPLY YOUR KNOWLEDGE One indicator of ethnicity and territory is the geographic distribution of places of worship across an urban area. Churches, temples, mosques, meeting houses, and chapels are usually located close to the communities they serve. Using your local phone book, map out the listings of your town or city's places of worship, and speculate on what this distribution might tell you about the local ethnic geography. ■

Race and Place

Geographers also use prevailing ideas and practices with respect to race to understand places.[2] **Race** is a problematic classification of human beings based on skin color and other physical

[2]Adapted from K. Anderson, "The Idea of Chinatown: The Power of Place and Institutional Practice in the Making of a Racial Category," Annals of the Association of American Geographers 77(4), 1987, pp. 580–598.

FIGURE 5.25 Uyghur protest Uyghur residents protest in the city of Urumqi, following bloody riots in the capital of China's Xinjiang region. 156 people died and over 1,400 people were arrested.

FIGURE 5.26 Chinatown, San Francisco Chinatowns have been portrayed as places voluntarily chosen by immigrant Chinese. But they can also be seen as places of exclusion and racism created by the dominant society that prefers to segregate people different from the wider society.

characteristics. **Racialization** is the practice of creating unequal castes based on the norm of whiteness. Biologically speaking, no such thing as race exists within the human species. Yet consider the categories of race and place that correspond to "Chinese" and "Chinatown." Powerful Western ideas about Chinese as a racial category have enabled the emergence and perpetuation of Chinatowns in many North American cities **(Figure 5.26)**. In this and other cases, the visible characteristics of hair, skin, and bone structure made race into a category of difference that was (and still is) widely accepted and often spatially expressed.

The mainstream approach views neighborhood as a spatial setting for systems of affiliation more or less chosen by people with similar skin color. Cultural geographers have overturned this approach in order to see neighborhoods as spaces that affirm the dominant society's sense of identity. For example, from the perspective of white society, nineteenth-century Chinatowns in the United States and Canada were the physical expression of what set the Chinese apart from whites. The distinguishing characteristics revolved around the way the Chinese looked, what they ate, their non-Christian religion, opium consumption, gambling habits, and other "strange" practices. Place—Chinatown—maintained and manifested differences between white and Chinese society. Place continues to be a mechanism for creating and preserving local systems of racial classification and for containing geographical difference within defined geographical confines. The homelands of South Africa and the dismantling of apartheid there also illustrate the interaction of race and place on a much larger scale.

Unexamined in the kinds of perceptions and practices that create racialized places as different as Chinatown and the homelands of South Africa is the taken-for-grantedness of whiteness. Whiteness is seen to be the norm or the standard against which all other visible differences are compared. But whiteness, as geographers and other scholars have shown, is itself a category of difference that depends on visible distinctions, not biological ones, and is always constructed in relation to other categories. Recently, researchers in the humanities and social sciences have begun to "denaturalize" whiteness by investigating the way whiteness has been constructed as a social category in different periods and places as well as the ways that whiteness operates in particular sites such as classrooms, on the street, and in boardrooms. It is only when we begin to challenge the naturalness of whiteness and the spaces that enable it that we can truly begin to undo racist practices.

Gender and Other Identities

Gender has received a great deal of attention from cultural geographers within the last three decades. Gender is a category reflecting the social differences between men and women. As with other forms of identity, gender implies a socially created difference in power between groups. In the case of gender, the power difference gives an advantage to males over females and is not biologically determined but socially and culturally created. Gender interacts with other forms of identity and can intensify power differences among and between groups. The implications of these differences are played out differently in different parts of the world.

For example, although gender differences play an important part in shaping social life for men and women in the Middle East, as elsewhere around the globe, there is no single Islamic, Christian, or Jewish notion of gender that operates exclusively in the region. Many in the West have formed stereotypes about the restricted lives of Middle Eastern women because of the operation of rigid Islamic traditions. It is important to understand, however, that these do not capture the great variety in gender relations that exists in the Middle East and North Africa across lines of class, generation, level of education, and geography (urban versus rural origins), among other factors.

In South Asia, gender is greatly complicated by class such that among the poor, women bear the greatest social and economic burden and the most suffering. Generally speaking, South Asian society—India, Pakistan, Afghanistan, Nepal, Bhutan, and Bangladesh—is intensely patriarchal, though the form that patriarchy takes varies by region and class. The common denominator among the poor throughout South Asia is that women not only have the constant responsibilities of motherhood and domestic chores but also have to work long hours in informal-sector occupations **(Figure 5.27)**. In many poor communities, 90 percent of all production occurs outside of formal employment, more than half of which is the result of women's efforts. In addition, women's property rights are curtailed, their public behavior is restricted, and their opportunities for education and participation in the waged labor force are severely limited.

The picture for women in South Asia, the Middle East, and elsewhere is not entirely negative, however, and one of the most significant developments has been the increasing education of girls. Because women's education is so closely linked to improvements in economic development, more and more countries are investing in their education. As more and more women become educated, they have fewer children and greater levels of economic independence and political empowerment.

In discussing questions of geography and identity, one last point should be made. In the 1990s, cultural theorists began to point out that identities were not fixed and a more nuanced and

FIGURE 5.27 Indian women in the informal sector Many women in South Asia are self-employed as small vendors in the daily markets. Others do home-based work such as weaving and dyeing cloth or embroidering or sewing garments. Nearly all workers in the informal sector, whether male or female, lack any sort of social protection such as health or unemployment insurance.

subtle understanding of the formation of human beings as cultural subjects—man, woman, homosexual, heterosexual, etc.—was required. In order to appreciate this flexibility in identity the term *hybridity* was introduced. **Hybridity** is meant to convey a mixing of different types. In cultural geography, hybridity is most often associated with movements across a binary of, for instance, the racial categories of black and white such that identities are more multiple and ambivalent. U.S. President Barack Obama, as a child of a white mother and a black father, characterizes this notion of hybridity; he is a mixture of cultural locations and his identity can be seen neither one nor the other but both (though he does self-identify as black).

> **APPLY YOUR KNOWLEDGE** Make a list of gender constructions that are characteristic of your culture. Now research a culture that is different from your own and identify its specific gender constructions. Are they different from those of your own culture? Explain your observations. ■

EMERGENT CULTURAL GEOGRAPHIES

Since the arrival of the new century, cultural geography has experienced, again, a transformation in the way its practitioners think about the relationship between people and their worlds. These new ways of conceptualizing culture and space are still developing but they hold the promise of opening up a wide range of previously ignored aspects of daily life to advance our attempts at explanation.

Actor-Network Theory

Perhaps the most important influence on cultural geography in the twenty-first century has been **actor-network theory** (ANT), which is actually less of a theory and more of an orientation. Actor-network theory views the world as composed of "heterogeneous things," including humans and nonhumans and objects. What makes ANT so interesting is that the approach attributes to nonhumans and objects as much force in the composition of social life as humans have. Rather than elevating humans as the superior species that determines all social practice and action, ANT recognizes that humans coexist with nonhumans (who may be other living species or inert objects) in a network that includes all sorts of social and material bits and pieces. Things are just as important to social life as are humans in an actor network.

An example of an actor network is a family. According to ANT, a family is both a network *and* an actor that hangs together and for certain purposes acts as a single entity. A family is connected through all sorts of objects: cell phones, houses, automobiles, family dinners, the family pet, etc. Nonhumans help constitute the fabric of "family" and enable it to persist, though its contours are always changing. It is not only the humans who make the family what it is, it is also the nonhumans that help the assemblage to come together as a recognizable and coherent entity. Agency—the ability to make things happen—resides not only in the human members of the assemblage but also in the nonhuman (the house, the dog, the barbeque grill, the flat-screen television) (**Figure 5.28**).

Non-Representational Theory

One of the criticisms of ANT has been that it treats humans as undifferentiated and lacking in affect and emotion as they go about the business of social life. In attempting to go further into what it means to be human and particularly into what is present in human experience, cultural geographers have become increasingly interested in non-representational theory. In brief, **non-representational theory** (NRT) understands human life as a process that is always unfolding, always becoming something different, even if only slightly so. It recognizes that much of this becoming occurs outside of

FIGURE 5.28 Family watching television in China Grandparents, parents, and their one child gather around their flat screen television to enjoy time together.

conscious thought. Because much of human existence is precognitive (decisions are made before the conscious self is aware of them), NRT is interested in those moments of indeterminacy when events emerge that produce new orderings that may persist or give way to older, more settled ones. These are moments we have all experienced, when something happens that no one expected and yet in which everyone participated in enabling the new ordering. This focus on indeterminacy means that NRT's task is a difficult one in that it attempts to attend to things that words (as representations) cannot express.

Geographer Ben Anderson has used non-representational theory to explore memory and music. His work is concerned with how the process of listening to music produces both remembering and forgetting within the context of ordinary living. He describes a young woman (whom he interviewed) listening to music as background to her getting ready for work. As she sits flipping through a magazine and eating breakfast with a Frank Sinatra tune playing in the background, she suddenly hears her mother (who died several years before) singing along to the music. It's a moment from her past; her mother would frequently sing along to all music, and this song was one of her favorites. And in that moment, the young woman is transported from her apartment back in space and time to her family home. The woman notes that when this happened she got "shivery feelings really suddenly."[3] Anderson identifies that "shivery feeling" as **affect**—emotions that are embodied reactions to the social and physical environment. Affect is also about the power of these emotions to result in or enable action.

NRT is keenly interested in events in which things suddenly shift and something involuntary occurs. And it is particularly interested in events when whole groups experience something not anticipated in advance. These are moments that hold significant political potential. For example, in Tunisia in the spring of 2011 the self-immolation of a fruit seller ignited a city's anger at their political dispossession, launching a political event that spread across the Middle East. In Madison, Wisconsin, thousands of people swarmed the state capitol to protest Governor Scott Walker's intention to eliminate collective bargaining rights for public workers. (See Box 5.3, "Visualizing Geography: Geographies of Protest and Care.")

Materialism

Both ANT and non-representational theory share an interest in matter and materiality. **Materialism** emphasizes that the material world—its objects and nonhuman entities—is at least partly separate from humans and possesses the power to affect humans. A materialist approach is about attempting to understand the ways that specific properties of material things affect the interactions between humans and nonhuman entities. What materialism adds to NRT and ANT is a commitment to understanding the material world as it unfolds in unpredictable ways. Both theories point to real physical and mental entities as significant to our attempts to explain the world. And they both recognize that humans are not separate from or in any way superior to the world of things. The aim of these new ways of thinking about cultural geography is to rethink how

material objects work in the world so that they come to be seen as "lively"—having real force and intensity in the world.

When seen in terms of cultural geography, the materialism that is at the center of these new ways of thinking directs our attention to how cultural beliefs and values gain permanence and power through material form. Buildings, symbols, commodities, or rituals—the twin towers of the World Trade Center, the Star of David, the diamond ring, the Thanksgiving meal—shape us and affect the way we are recognized by others. Consider the cell phone as an example. This mobile device connects people to family and friends, nearly instantaneously and certainly in ways that a landline never could. The materiality of that object—its thingness—is not inert; it has an active forcefulness that can produce significant outcomes such that we are often at a loss without them. In a way, cell phones have become objects that make us do things we might not or could not otherwise.

APPLY YOUR KNOWLEDGE Identify an object (other than a cell phone, which has already been discussed in the text) that has a powerful influence over the way you operate in the world. How does that object work on you? How does it affect your relationships to others? Are there particular times and places in which the object is effective or ineffective? Why might that be? ■

GLOBALIZATION AND CULTURAL CHANGE

Anyone who has ever traveled between major world cities will have noticed the many familiar aspects of contemporary life in settings that until recently were thought of as being quite different from one another. Airports, offices, and international hotels have become notoriously alike, and their similarities of architecture and interior design become reinforced by the near-universal dress codes of the people who frequent them. For example, the business suit, especially for males, has become the norm for office workers throughout much of the world. Jeans, T-shirts, and sneakers, meanwhile, have become the norm for young people, as well as those in lower-wage jobs.

Americanization and Globalization

It is these commonalities—as well as others, such as the same automobiles, television shows, popular music, food, and global brands like Sony, Coca-Cola, and Microsoft—that provide a sense of familiarity to core travelers abroad. From the point of view of cultural nationalism, the "lowest common denominator" of this familiarity is often seen as the culture of fast food and popular entertainment that emanates from the United States. Popular commentators have observed that cultures around the world are being Americanized (see Chapter 2). This process represents the beginnings of a single global culture based on material consumption, with the English language as its medium.

There is certainly some evidence to support this point of view, not least in the sheer numbers of people around the world who view

[3]B. Anderson, "Recorded Music and Practices of Remembering," Social and Cultural Geography, 5(1), 2004, pp. 3–18.

The 2011 Wisconsin Protests
by Mario Bruzzone, Abigail H. Neely, Sigrid Peterson, and Keith Woodward

Throughout February and March 2011, hundreds of thousands of people journeyed to the State Capitol building in Madison, WI, to voice opposition to State Senate Bill 11—the "Budget Repair Bill." Hastily announced and scheduled to be voted on with little time for debate, the bill sought to strip most collective bargaining rights from public sector employees, to curtail public health care benefits for poor and elderly Wisconsinites, and to radically loosen restrictions on no-bid sales of publicly owned power plants. Within weeks, the daily demonstrations at the Capitol grew into one of the largest sustained protests in the U.S. since the Vietnam War. By March 13th, a month into the protests, attendance swelled to an estimated 100,000—an extensive mobilization for a state with a population of 5.3 million people.

For cultural geographers, the significance of these protests extends beyond their historic size. Positioned in the center of a large public square and frequently traversed by residents and visitors, the Capitol Building came to exemplify the fraught relation between popular protest and public space. During the first week of protests, state workers, students, teachers, farmers, activists, organizers, and concerned citizens who had made their way to the Capitol began to camp over night on its marble floors. This peaceful "occupation"—and the "rotunda community" it created—offered a glimpse at how political cultures can grow spontaneously out of novel organizational forms and unique uses of places and spaces (**Figure 5.D(a)**). Normally the location of legislative meetings and grade-school field trips, the spaces inside the Capitol—its walls, hallways, and rotunda area—were re-imagined and transformed by citizens into living, sleeping, and working spaces suited to the practical needs of the community, complete with food preparation and distribution areas, meeting centers, an information station, a first-aid table, and a family space for children to play.

Many different groups and individuals made up the Capitol community. Some came from established organizations, including members of the Teaching Assistants Association, the graduate student union at the University of Wisconsin-Madison, which had helped organize the first protest and worked around the clock in a borrowed Capitol office space. Others came from undergraduate groups at UW-Madison; from political movements such as Move to Amend, Code Pink, and Greenpeace; and from groups that grew out of the protests, such as A People's Movement and the Autonomous Solidarity Organization. These coalitions provided a living portrait of self-organization and self-governance by people, many who did not know each other previously. They collectively developed rules, etiquette, decision-making structures, shared labor, and maintained and cared for the common space.

The Capitol thus became a new cultural space. Once bare, polished marble, the walls were now covered with signs posted by citizens. These signs not only voiced protests but also offered strategies for peaceful protest, announced collective decisions concerning practices for keeping the space clean, displayed the general schedule for Capitol living (announcing "quiet" sleeping times for example), provided instructions on where to get reliable information, and suggested opportunities to volunteer. Other signs were more colorful; messages of solidarity, jokes, and letters of support from elsewhere provided a sense of welcome to first-time visitors and inspired those who were in the occupation for the long haul.

Some of the most striking elements of the Wisconsin protests were the visual and sensory challenges to the traditional, political characteristics of the Capitol space. State capitols almost invariably represent power through the prominent visual display of statues, figures, paintings, murals, and other objects that blend political ideology with local and national cultural ideals. Capitol spaces often cultivate an almost-reverential atmosphere, instituting rules or norms that proscribe noisy behaviors like yelling, singing, and loud talking. By contrast, in Madison, the visual sensation of the signs, the massive number of protesters filling the multiple floors of the Capitol building, and their loud chants of "Whose House? Our House!" echoing through the hallways, presented a radically different sense of what democratic practice could look and sound like.

Other transformations in the sensory experience of the Capitol space during the protests included "solidarity dance parties," drum circles, sing-alongs, string quartet performances, and a "people's microphone," which allowed anyone to sign up and directly address the crowds assembled in the building. Thus, Capitol space was transformed from a scene of representational democratic governance to a site of direct democratic practice.

Citizens also re-imagined and transformed the Capitol through their practical use of its spaces. Throughout February, daytime crowds packed open spaces on the building's first three floors. In the evenings, the rotunda area on the mezzanine became the site of meetings, teach-ins, and, later in the night, sleeping spaces. Such alternative uses made space for a variety of encounters. The often disorienting Capitol architecture was transformed by new activities: a drum circle and speaking space sprouted up on the first floor; a family space, a medical area, and a peaceful protest training area on the second; and the TAA workspace on the third. Even sleeping spaces were organized practically; a large group of college-aged Capitol residents, for example, came to be known as the "cuddle puddle" because they slept closely together to stave off the cold of the marble floors. (**Figure 5.D(b)**)

The intimacy of the Capitol spaces became a key component of these struggles facilitating a sense of empowerment. Additionally, the dissemination of protest events through social and popular media entangled the emerging Capitol culture with more "global" struggles. The Wisconsin protests unfolded during the same period that the "Arab Spring" was at its height in

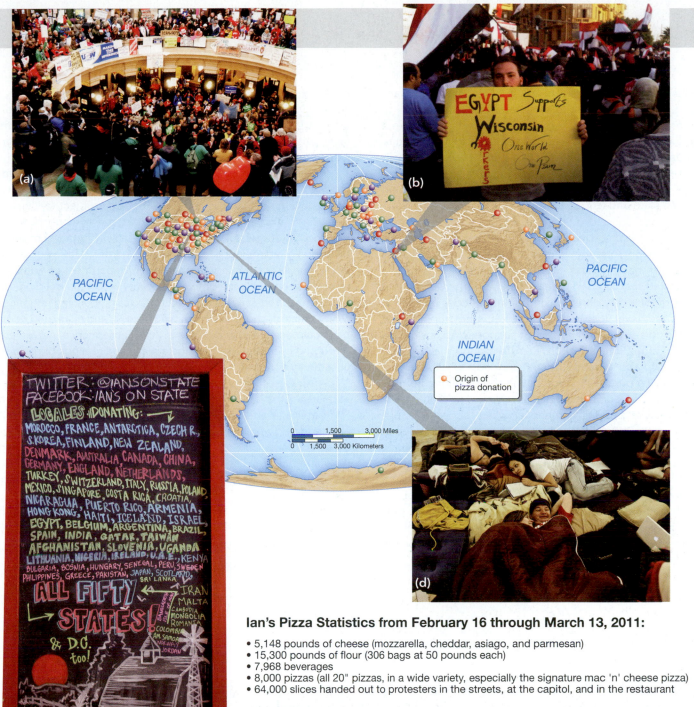

Ian's Pizza Statistics from February 16 through March 13, 2011:

- 5,148 pounds of cheese (mozzarella, cheddar, asiago, and parmesan)
- 15,300 pounds of flour (306 bags at 50 pounds each)
- 7,968 beverages
- 8,000 pizzas (all 20" pizzas, in a wide variety, especially the signature mac 'n' cheese pizza)
- 64,000 slices handed out to protesters in the streets, at the capitol, and in the restaurant

FIGURE 5.D (a) The Capitol Demonstrators fill the Capitol Building in Madison, WI on the fourth day of protests, Thursday, February 17, 2011 **(b) Solidarity message from Egypt** One of hundreds of thousands of demonstrators who gathered in Tahrir Square to celebrate the success of the Egyptian revolt on Friday, February 18, 2011, Muhammad Saladin Nusair carries a sign expressing solidarity with the Wisconsin workers' protest. This image quickly "went viral," becoming a ubiquitous image at the Capitol and leading some journalists to draw connections between the workers' struggles in Wisconsin and Egypt. **(c) Geography of support for the Wisconsin protests** Throughout the Wisconsin protests, Ian's Pizza—located just off of Capitol Square in Madison—continuously updated a restaurant chalkboard listing the countries from which thousands of pizza donations for the Capitol-bound protesters originated. **(d) The Cuddle Puddle** By February 23, 2011, the occupation of Wisconsin's Capitol had transformed much of the first two floors to accommodate the lives of the protesters. Likewise, they had to adapt to the strange living conditions of the space. Here, many gather together in a nightly "cuddle puddle," sleeping close together to fight off the cold of the building's marble floors.

the Middle East and across North Africa. While diverse in many ways, all of these struggles shared in the effort to bring back to the forefront questions of workers' rights, strategies of international worker solidarity, and concerns about the changing nature of class-based exploitation. Indeed, the network of global solidarity that enabled worldwide connection to and communication with far-off activists emboldened activists in Wisconsin, helping to anchor the state and its Capitol building in transnational workers' struggles.

From Egypt, the union leader Kamal Abbas posted an expression of solidarity on YouTube. Soon after, amidst the outpouring of support from national and international union leadership, a photo appeared that rapidly spread throughout the Capitol and became a sign of a very different kind of international solidarity, one at the street level. It showed a protester in Egypt posing for a photo holding a sign that read "Egypt supports Wisconsin: One World, One Pain" (**Figure 5.D(c)**). People around the U.S. and around the world also showed support and solidarity with the Wisconsin protests by purchasing food online and by phone from local grocers, bakeries, and restaurants. Ian's Pizza, less than a block from the Capitol, processed international orders for over 8000 pizzas donated to protestors after "tweeting" about such an order on their Twitter account (**Figure 5.D(d)**).

Finally, the transformation of Capitol space can also be understood through the novel kinds of bodily movements that the occupation enabled and restricted. How bodies move—or are moved—through a place often informs how that place is perceived, embraced, or contested. For protestors, free movement through the Capitol allowed them to come into contact with other activists, coordinate activities, and find projects and other ways to volunteer. Likewise increasing familiarity with the Capitol's intricacies and quirks through movement gave protestors increasing feelings of ownership and pride in the space they had helped create.

At the same time, the movement of bodies through the building was also connected to practices of governance and security. While government spaces in the U.S. became subject to increasing securitization after September 11, 2001, the Wisconsin Capitol building had successfully maintained a relatively low level of security that allowed the public to come and go freely. However, as the 2011 protests progressed, the Wisconsin Department of Administration responded by heightening security within the building, introducing airport-style metal detectors, closing building entrances (which created long lines for building access), and restricting items that could be carried in by citizens. As a consequence, movement through the Capitol slowed substantially and feelings of risk and uncertainty increased with the police presence.

After the protests, the community was evicted and security escalated. Displaced protesters constructed a tent city named "Walkerville"—after the Wisconsin State Governor, Scott Walker, and modeled upon the Hoovervilles of the 1930s and Wisconsin's "Solidarity City" of the 1970s—on the square surrounding the Capitol. The space of the Capitol has continued to undergo transformation while remaining a space for transformative political practice.

Mario Bruzzone, Abigail H. Neely, Sigrid Peterson, and Keith Woodward are all either students or faculty members at the University of Wisconsin, Madison.

The Bachelor, drink Coca-Cola, and eat in McDonald's franchises or similar fast-food chains. Largely through consumer goods, U.S. culture is increasingly embraced by local entrepreneurs around the world. It seems clear that U.S. products are consumed as much for their symbolism of a particular way of life as for their intrinsic value. McDonald's burgers, along with Coca-Cola, Hollywood movies, rock music, and NFL and NBA insignia, have become associated with a lifestyle package that features luxury, youth, fitness, beauty, and freedom.

It is important to recognize, however, that U.S. products often undergo changes when they travel across the globe. For instance, Revlon, a U.S.-based international corporation and a leading mass-market producer of cosmetics, skin-care products, and fragrances, is very much aware of the need to vary its products for their consumption in non-U.S. markets. Modi Revlon, a joint venture between Revlon and the Indian company Modi Mundipharma, provides a product line of color cosmetics that complement Indian women's skin tones and appeal to upper-income markets. The success of Modi Revlon, which has captured 80 percent of the Indian cosmetics market, is remarkable.

The economic success of the U.S. entertainment industry has also helped reinforce the idea of an emerging global culture based on Americanization. Today the entertainment industry is the leading source of foreign income in the United States, with a trade surplus of over $25 billion. The originals of over half of all the books translated in the world (more than 25,000 titles) are written in English, the majority of which are produced by U.S. publishers. In terms of international flows of everything from mail and phone calls to press-agency reports, television programs, radio shows, and movies, a disproportionately large share originates in the United States.

Neither the widespread consumption of U.S. and U.S.-style products nor the increasing familiarity of people around the world with global media and international brand names, however, adds up to the emergence of a single global culture. Rather, what is happening is that processes of globalization are exposing the world's inhabitants to a common set of products, symbols, myths, memories, events, cult figures, landscapes, and traditions. People living in Tokyo or Tucson, Turin or Timbuktu, may be perfectly familiar with these commonalities without necessarily using or responding to them in uniform ways. It is also important to recognize that cultural flows take place in all directions, not just outward from the United States. Think, for example, of European fashions in U.S. stores; of Chinese, Indian, Italian, Mexican, and Thai restaurants in U.S. towns and cities; and of U.S. and European stores selling exotic craft goods from the periphery.

A Global Culture?

The answer to the question of whether there is a global culture must therefore be "no," or at least there is no indisputable sign of it yet. While people around the world share an increasing familiarity with a common set of products, symbols, and events (many of which originate in the U.S. culture of fast food and popular entertainment), these commonalties are configured in different ways in different places, rather than constituting a single global culture. The local interacts with the global, often producing hybrid cultures. Sometimes traditional local cultures become the subject of global consumption; sometimes it is the other way around.

An illustration of the absence of a homogenous global culture is **world music**, the musical genre defined largely by the surge of non–English-language recordings released in the United Kingdom and the United States during the 1980s. The term is one employed primarily by the media and record stores, and it includes such diverse sources as Tuvan throat singers and Malian griots. (Throat singing is a performance style originating in south-central Russia where the vocal sounds are projected from the throat, allowing two to four tones to be simultaneously produced by the singer. Griots were ancient praise singers of West Africa and also important advisors to royalty.)

There are at least two major views on the effect of globalization on indigenous musical productions. The first emphasizes how the Western music industry has enabled indigenous music to be more widely disseminated and therefore more widely known and appreciated. This position sees local roots mixing with Western popular musical styles, with a hybrid sound resulting. The second view worries that the influence of Western musical styles and the Western music industry have transformed indigenous musical productions to the point where their authenticity has been lost and global musical heterogeneity diminished. Despite their fundamental disagreement, holders of both positions recognize that world music has enabled cultural diversity to flourish and hope that indigenous performers will be able to resist the power of the Western music industry to homogenize their work and that at the same time hybrid forms will emerge that are satisfying to a wide audience **(Figure 5.29)**.

Another example of the variation that exists despite the globalization of commercial cultural products can be found in the film industry. The commercial film industry has become global; the sites of production vary, and the products themselves are significantly different aesthetically and practically. Most Americans are aware of Bollywood films, made in India and distributed widely because of the large global diaspora of Indian peoples. No one would mistake a Bollywood film for one made in Hollywood, largely because Indian audiences have different expectations for a film experience. Less well-known among Americans, however, but arguably more globally popular than either Hollywood or Bollywood, is Nollywood, the film industry centered in Lagos, Nigeria. Nollywood produces over 1,000 films a year that are viewed avidly not only by Nigerians but across the globe by the large diasporic African population **(Figure 5.30)**. Unlike Hollywood films, with their high production values, Nollywood video films are made on inexpensive Chinese cameras with short production schedules and a minimum of editing. Within about a three-week period, the video film is shot and edited and is available for retail consumption. Americans watching the films might complain about the quality of the lighting and sound, among other things, but Africans appreciate the video films as reflections of their own lives and experiences. As with music, the world film industry translates into anything but homogeneity in this instance.

APPLY YOUR KNOWLEDGE Visit your local video rental store either in person or via the Web to see if films are available from Indian or Nigerian production companies. Read the descriptions of the films for ten to twelve productions from each country. How are the issues they deal with the same or different from mainstream U.S. films? What do these differences and similarities reflect about national cultures? ■

FIGURE 5.29 Bossa nova dancers Both a musical style and a dance, *bossa nova* (Portugese for "new style" or "new beat") originated in Brazil in the late 1950s and was a fad until the mid-1960s, when it faded out. The music style is a blend of cool jazz and hot Latin rhythms accompanied by intimate vocals and instrumental improvisation. It is well suited for listening but failed to become widespread dance music—many felt it was too slow to dance to—despite heavy promotion. Yet in the twenty-first century, *bossa nova* has been experiencing resurgence as a popular dance form as shown here. It has even been merged with four wall country line dancing performed as a group, rather than in couples.

FIGURE 5.30 On the set of Nollywood Nollywood video films are popular the world over and rival Hollywood and Bollywood in numbers produced. The video films are unlike those of either of the other two leading film industries, using different production techniques and invoking different kinds of stories, ultimately resulting in a different aesthetic. Despite the name, Nollywood is *not* a cheap imitation of Hollywood, but a unique cultural form.

Future Geographies

In 2005, the United Nations adopted the Convention on the Protection and Promotion of the Diversity of Cultural Expressions. The convention "is a legally binding international agreement that ensures artists, cultural professionals, practitioners, and citizens worldwide can create, produce, disseminate, and enjoy a broad range of cultural goods, services, and activities, including their own."[4] This UN convention is just one of the ways that governing bodies are actively working to encourage and protect culture and creativity in a rapidly changing world. Given this growing commitment to recognizing and proactively appreciating the relevance of cultural diversity to the global community, it is fair to predict that culture—in both its material as well as less tangible manifestations—will continue to be the focus of formal attention and support well into the future (**Figure 5.31**).

[4]UNESCO, accessed May 22, 2011, from http://www.unesco.org/new/en/culture/themes/cultural-diversity/2005-convention/the-convention/.

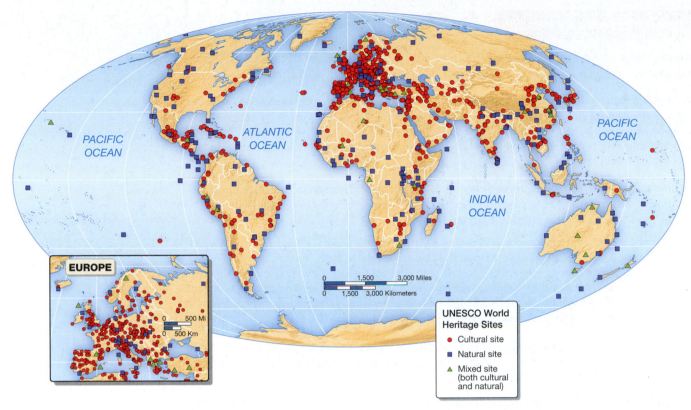

FIGURE 5.31 UNESCO world heritage sites The World Heritage List includes 936 properties forming part of the cultural and natural heritage which the World Heritage Committee considers as having outstanding universal value. (Source: UNESCO World Heritage List, http://whc.unesco.org/pg.cfm?cid=31&l=en&&&&mode=table.)

Importantly, connections across global space are making it increasingly possible for distant groups to share their ideas, cultural practices, and products in a way that helps both to circulate and sustain them. Geographers who work with culture as their research focus are beneficiaries of these trends as they seek to understand how culture is produced and maintained and how it morphs as it is practiced and shared across distances that once would have prohibited contact.

CONCLUSION

Culture is a complex and exceedingly important concept within the discipline of geography. A number of approaches exist to understanding culture. It may be understood through a range of elements and features, from single traits to complex systems. Cultural geography recognizes the complexity of culture and emphasizes the roles of space, place, and landscape and the ecological relationships between cultures and their environment. It distinguishes itself from other disciplinary approaches, providing unique insights that reveal how culture shapes the worlds we live in at the same time that the worlds we inhabit shape culture.

Cultural geography is a diverse subfield that incorporates three general approaches. The first, traditional cultural geography, is a reflection of the work of Carl Sauer, who understood landscape as the definitive unit of geographic study. The second approach is centered on identity, ideology, power, and meaning. Gender and sexuality, race and ethnicity, as well as media are frequent empirical objects of this approach. Most recently, cultural geography has been influenced by what has come to be known as the "non-representational" approach. Cultural geographers who are non-representationalists reject what they see as static views of space inherent in more conventional approaches and turn instead to an empirical focus on embodied practices and dynamic processes and how best to comprehend them and their effects without sacrificing their dynamism.

Cultural geographers continue to embrace all of these approaches in their aims to understand how culture is the product of human's relationships with the world around them, a world that is alive with symbols, artifacts, practices, and discourses.

- Interpret how place and space shape culture and, conversely, how culture shapes place and space.

A simple understanding of culture is that it is a particular way of life, such as a set of skilled activities, values, and meanings surrounding a particular type of economic practice. Geographers understand culture to be shaped by the places in which people live and make meaning from their lives. This means that social relations, politics, and economy all play a role in the production of cultural practices by different groups in different places.

- Compare and contrast the different ways that contemporary approaches in cultural geography interpret the role played by politics and the economy in establishing and perpetuating cultures and cultural landscapes.

Culture is not something that is necessarily tied to a place and thus a fact waiting to be discovered. Rather, the connections among people, places, and cultures are social creations that can be altered by new impulses from the economy or politics, for example, and are therefore always changing, sometimes in subtle and other times in more dramatic ways. As a result, a particular ethnic landscape may change dramatically after only a decade as the economy improves or declines and members of the group have access to additional or fewer resources that then shape their homes, vehicles, businesses, etc.

- Probe the ways that differences—especially gender, class, sexuality, race, and ethnicity— are both products of and influences upon geography, producing important variations within, as well as between, cultures.

Like most social scientists, geographers understand that cultural groups are not homogeneous. All women are not alike anymore than all working-class people are. Where people live can have an important impact on their sexual identity, for instance, when they are living in a place that is homophobic.

- Appreciate the conceptual changes that are taking place in cultural geography that include actor-network theory and non-representational theory.

Over the last decade, cultural geography has experienced a dramatic change in the way its practitioners think about the relationship between people and their worlds. These new ways of conceptualizing culture and space are still developing, but they focus on the importance of objects and material practices and how they shape the ways we experience and conduct our daily lives.

- Show that globalization does not necessarily mean the world is becoming more homogeneous, and recognize that in some ways, globalization has made the local even more important than before.

While globalization is undoubtedly reshaping the world and bringing different cultural groups closer together than they have ever been previously, there is no conclusive evidence that globalization leads to cultural homogenization. Instead, globalization seems to be a differential process, which means that it is deployed differently in different places and experienced and responded to differently by the people who live in those places.

KEY TERMS

actor-network theory (p. 174)
affect (p. 175)
cultural complex (p. 154)
cultural geography (p. 146)
cultural hearths (p. 160)
cultural landscape (p. 151)
cultural nationalism (p. 164)
cultural region (p. 154)

cultural system (p. 155)
cultural trait (p. 152)
culture (p. 146)
dialects (p. 160)
diaspora (p. 155)
ethnicity (p. 171)
folk culture (p. 147)
genre de vie (p. 152)
historical geography (p. 151)

hybridity (p. 174)
Islam (p. 158)
Islamism (p. 165)
kinship (p. 163)
language (p. 160)
language branch (p. 160)
language family (p. 160)
language group (p. 160)
materialism (p. 175)
Muslim (p. 158)

non-representational theory (p. 174)
popular culture (p. 147)
race (p. 171)
racialization (p. 172)
religion (p. 155)
rites of passage (p. 152)
sexuality (p. 167)
tribe (p. 163)
world music (p. 179)

REVIEW AND DISCUSSION

1. Make a list of movies that members of your group have viewed recently. Pick two of the films on which to conduct research. If possible, watch the trailers for each film. While watching the trailers, identify three different aspects of culture that are depicted. In terms of cultural geography, how do the aspects of culture you identified interact with space, place, and landscape?

2. Variations in gender identities are often the result of the different spaces in which they are enacted. In your group, discuss what this means by creating a list of how gender might be performed in different spaces by members of your group. For instance, in what spaces are you likely to project a highly feminized or masculinized identity?

In what spaces might you project an identity that is not traditionally feminine or masculine? How does the space—for example, a venue for job interview, a date, a sports competition, a place of worship—shape your gender performance?

3. Find a description of a coming-of-age ceremony in any part of the world other than the United States in your library or online. Summarize that description and then compare it to one you have either experienced directly or observed in the United States. What are the differences and the similarities between your experience and the one you read about? What might be some of the reasons for these?

UNPLUGGED

1. Using *Billboard Magazine* (a news magazine of the record industry), construct a historical geography of the top 20 singles over the last half century in order to determine how different regions of the country (or the globe) have risen and fallen in terms of musical significance. Determine an appropriate interval for sampling—3 to 5 years is generally accepted. You may use the hometowns of the recording artists or the headquarters of the recording studios as your geographic variables. Once you have organized your data, answer the following questions: How has the geography you have documented changed? What might be the reasons for these changes? What do these changes mean for the regions of the country (or globe) that have increased or decreased in terms of musical prominence?

2. Ethnic identity is often expressed spatially through the existence of neighborhoods or business areas dominated by members of a particular group. One way to explore the spatial expression of ethnicity in a place is to look at newspapers over time. Look at ethnic change in a particular neighborhood over time by using your library's holdings of local or regional newspapers. Examine change over at least a four-decade period. To do this, identify an area of the city in which you live

or some other city for which your library has an extensive newspaper collection. Trace the history of an area you know is now occupied by a specific ethnic group. How long has the group occupied that area? What aspects of the group's occupation of that area have changed over time (school, church, or sports activities, or the age of the households)? If different groups have occupied the area, what might be the reasons for the changes?

3. College and university campuses generate their own cultural practices and ideas that shape behaviors and attitudes in ways that may not be so obvious at first glance. Observe a particular practice that occurs routinely at your college or university. (For example, fraternity and sorority initiations are important rituals of college life, as are sports events and class discussions.) Who are the participants in this practice? What are their levels of importance? Are there gender, age, or status differences in those who carry out this practice? What are the time and space aspects of the practice? Who controls its production? What are the intended outcomes? How does the practice contribute to or detract from the maintenance of order in the larger culture?

6 INTERPRETING PLACES AND LANDSCAPES

Maya Yang Lin was an undergraduate architecture student at Yale University when she won the public design competition for the Vietnam Veterans War Memorial in 1981. The memorial wall is composed of polished black granite with the names of the over 58,000 individuals killed or missing carved into its face. It is V-shaped; one wall points to the Lincoln Memorial and the other to the Washington Monument, each of which is reflected in the wall's face. Sunk into the ground, it symbolizes a wound in the earth—a wound created by the loss of so many service people. Visitors can see their reflections superimposed on the engraved names, thus becoming connected to the deaths. Rather than hailing the fallen soldiers as heroes, the wall reminds those who visit it of the emotional and psychological impacts of their deaths.

The memorial design was highly controversial from the start. Many saw its horizontal expression—in contrast to traditional vertical war memorials—as a political statement, an admission of shame about an unvictorious war. Surprisingly, although controversy swirled from its design phase to its unveiling, the memorial has become one of the most visited and appreciated sites in the United States. The actual *experience* of the Vietnam Veterans War

Members of Rolling Thunder–an organization dedicated to recovering missing-in-action soldiers from the Viet Nam War–visits the Memorial Wall on its annual Memorial Day, "Ride to the Wall".

Memorial caused all the controversy about its design to disappear. The reason why is, quite simply, the over 58,000 names of war casualties etched into the stone.*

Lin insisted that the names be entered on the wall chronologically so that the story of the war could be seen in the swelling casualties that occurred each day. It is possible to see in certain groups of names a particular ambush or a significant battle. This chronology allows veterans who visit the wall to have a spatial reference for their own experience as they locate their fallen comrades along the 500-foot walkway. Because of this, the memorial has given all veterans of U.S. wars, but especially the Vietnam War, a place to gather and speak about the grief of loss and the bittersweet feelings of survival.

The Vietnam Veterans Memorial is a compelling illustration of the power of landscape to affect us. Landscape can be coded through elements like design, use of space, light, and topography, with a profound impact on the viewer. In this chapter, we take up the concepts and processes that are captured in the example of the wall and explore places and landscapes as the manifestation of multifaceted relationships among people and space. ■

*Marita Sturken, "The Wall, the Screen, and the Image: The Vietnam Veterans Memorial," *Representations, 35* (Special Issue: Monumental Histories, Summer, 1991), pp. 118–142.

BEHAVIOR, KNOWLEDGE, AND HUMAN ENVIRONMENTS

In addition to attempting to understand how the environment shapes and is shaped by people, geographers also seek to identify how it is perceived and understood by people. Asserting that there is an interdependence between people and places, geographers explore how individuals and groups acquire knowledge of their environments and how this knowledge shapes their attitudes and behaviors. Some geographers focus their research on natural hazards as a way of addressing environmental knowledge, while others try to understand how people ascribe meaning to landscape and places. In this chapter, we consider the key geographical concepts of place, landscape, and space and explore the ways in which people understand them, create them, and operate within them. Our goal is to understand how individuals and groups experience their environments, create and struggle within places, and find meaning in the landscapes they create.

In their attempts to understand environmental perception and knowledge, geographers share a great deal with other social scientists, but especially with psychologists. Human cognition and behavior are at the center of psychology. What makes environmental knowledge and behavior uniquely geographical is their relation to both the environmental context and the humans who struggle to understand and operate within it. Much of what we as humans know about the environment we live in is learned through direct and indirect experience. Our environmental knowledge is also acquired through a filter of personal and group characteristics, such as race, gender, stage of the life cycle, religious beliefs, and where we live (**Figure 6.1**).

For instance, children have interesting and distinct relationships to the physical and cultural environment. How do children acquire knowledge about their environments? How do boys and girls differ in the ways they learn about and negotiate their environments? What kind of environmental knowledge do children acquire, and how do they use it? What role do cultural influences play in the process? What happens when larger social, economic, and environmental changes take place?

Geographer Cindi Katz conducted research in rural Sudan to find answers to these questions. Working with a group of 10-year-old children in a small village, she sought to discover how they acquired environmental knowledge. What she also learned was how the transformation of agriculture in the region changed not only the children's relationships to their families and community but also their perceptions of nature.[1] In this Sudanese village, as in similar communities elsewhere in the periphery, children were important contributors to subsistence activities, especially planting, weeding, and harvesting.

FIGURE 6.1 Conflicting environmental perceptions At the University of California, Berkeley, eight individuals sat for weeks in the trees of an oak grove scheduled for removal by the university. The university needed the space the trees occupied for a new student athletic training center. Although City of Berekeley law prohibits removing any live oak with a trunk larger than six inches, the University proceeded to destroy the grove amidst widespread protest. The protest ran from December 2006 to September 2008.

[1]C. Katz, *Growing Up Global: Economic Restructuring and Children's Everyday Lives*. Minneapolis: University of Minnesota Press, 2004.

The villagers were strict Muslims and thus had stringent rules about what female members of the community were allowed to do and where they were allowed to go. Many of the subsistence activities that required leaving the family compound were customarily delegated to male children. Within the traditional subsistence culture, boys predominated in all agricultural tasks except planting and harvesting and were responsible for herding livestock as well. Many boys (and occasionally girls) were also responsible for fetching water and helping to gather firewood. Both boys and girls collected seasonal foods from lands surrounding the village. Work and play were often mixed together, and play, as well as work, provided a creative means for acquiring and using environmental knowledge and for developing a finely textured sense of the home area (**Figure 6.2**).

What happens when the agricultural production system is thoroughly changed, as it was in this village when irrigated cash-crop cultivation was introduced? Through an international development scheme, with the financial assistance of outside donors, a Sudanese government project transformed the agriculture of the region from subsistence livestock raising and cultivation of sorghum and sesame to cultivation of irrigated cash crops such as cotton. The new cash-crop regime, which required management of irrigation works, application of fertilizers, herbicides, and pesticides, and more frequent weeding, required children as well as adults to work longer and harder. Parents were often forced to keep their children out of school because many of the tasks had to be done during the school term.

Destruction of local forests required children to range farther afield to procure fuelwood and to make gathering trips more frequently. Soon wealthier households began buying wood rather than increasing the demands on their children or other household members. For the children of more marginalized households, the selling of fuelwood, foods, and other items provided a new means for earning cash to support their families but also placed increasing demands on their energies and resourcefulness and changed their whole experience of their world.

For the children of this village, globalization (in the form of the transition to cash-crop agriculture) altered their relationships with their environments and with their futures. The kinds of skills the children had learned for subsistence production were

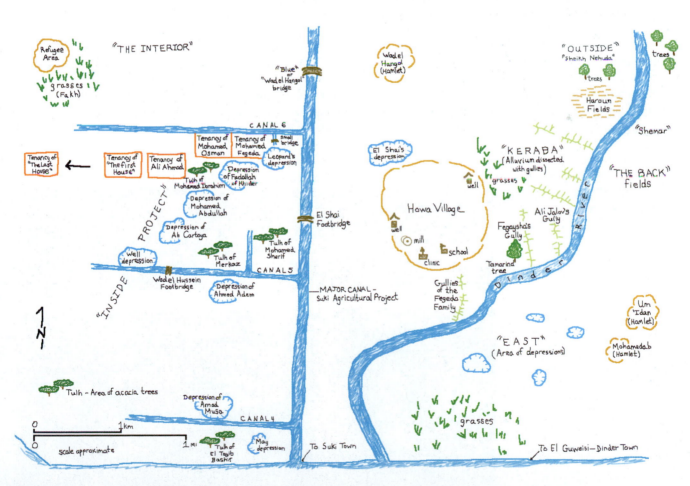

FIGURE 6.2 Shepherd's map This map, drawn by a 10-year-old Sudanese boy, shows the area over which the sheep are herded. It illustrates the detailed environmental knowledge Sudanese children possess of the landscape that surrounds their village. The village is an Islamic one, and norms determine the kinds of tasks in which boys and girls can participate. Only boys are allowed to tend sheep, which requires a particular environmental knowledge about grazing areas and water availability. (*Source:* Image courtesy of C. Katz.)

no longer useful for cash-crop production. As they played less and worked longer hours in more specialized roles, their experience of their environment became narrowed. At the same time, as their roles within the family changed, they attended school less and learned less about their world through formal education. As a result, their perceptions of their environment transformed, along with their values and attitudes toward the landscape and the place they knew as home.

APPLY YOUR KNOWLEDGE Scrutinize how globalization has shaped the environment you operate in as a student. How has it shaped you—the buildings, the people, the climate, the social life, etc.—and how do you shape it by interacting with it? ■

PLACE-MAKING

Places are socially constructed—given different meanings by different groups for different purposes. Most people identify with places as part of their personal identity, drawing on particular images and particular histories of places in order to lend distinctiveness to both their individuality and their sense of community. But identifying with place may also imply the exclusion of other people and the stereotyping of other places. People often reinforce their sense of place and of who they are by contrasting themselves with places and people they feel are very different from them. Seen in this context, place-making stands at the center of issues of culture and power relations, a key part of the systems of meaning through which we make sense of the world.

Territoriality

Some social scientists believe that wanting to have a place where you feel you belong is a natural human attribute, part of a strong territorial instinct. Humans, it is argued, have an innate sense of territoriality, just like many other species. The concept of **territoriality** refers to the persistent attachment of individuals or peoples to a specific location or territory. The concept is important to geographers because it can be related to fundamental place-making forces.

The specific study of people's sense of territoriality is part of the field of **v**, the scientific study of the formation and evolution of human customs and beliefs. The term also refers to the study of the behavior of animals in their natural environments. According to ethologists, humans carry genetic traits resulting from our species' need for territory. Territory provides a source of physical safety and security, a source of stimulation (through border disputes), and a physical expression of identity. These needs add up to a strong territorial urge that can be seen in people's behavior: claims to space in reading rooms or on beaches, for example, and claims made by gangs to neighborhood turf (**Figure 6.3**). Ethologists argue that the territorial urge also can be observed when people become frustrated because of overcrowding. They become stressed and, in some circumstances, begin to exhibit aggressive or deviant behavior. Ethologists and environmental psychologists link crowding to everything from vandalism and assault to promiscuity, listlessness, and clinical depression.

While such claims are difficult to substantiate, as is the whole notion that humans have an inborn sense of territoriality, the idea of territoriality as a product of *culturally* established meanings is supported by a large body of scientific evidence. Some of this evidence comes from the field of **proxemics**, the study of the social

FIGURE 6.3 Graffiti as territorial markers Graffiti are used by neighborhood gangs to establish and proclaim their identity.

and cultural meanings that people give to personal space. These meanings make for unwritten territorial rules that can be seen in the microgeography of people's behavior. It has been established, for example, that people develop unwritten protocols about how to claim space. One common protocol is simply regular use (think of students' habits in classroom seating arrangements). Another is through the use of spatial markers such as a newspaper or a towel to fix a space in a reading room or on a beach. There are also bubbles, or areas, of personal space that we try not to invade (or allow to be invaded by others). Varying in size and shape according to location and circumstance, these bubbles tend to be smaller in public places and in busier and more crowded situations; to be larger among strangers and in situations involving members of different social classes; and to vary from one social class or cultural group to another.

On larger spatial scales, territoriality is mostly a product of political relations and cultural systems. This aspect of territoriality underpins a great deal of human geography. All social organizations and the individuals who belong to them are bound at some scale or another through formal or informal territorial limits. Many organizations—nations, corporations, unions, clubs—actually claim a specific area of geographic space to be under their influence or control. In this context, territoriality can be defined as any attempt to assert control over other people, resources, or relationships over a specific geographic area. Territoriality also is defined as any attempt to fulfill socially produced needs for identity, defense, and stimulation. Territoriality covers many phenomena, including the property rights of individuals and private corporations; the neighborhood covenants of homeowners' associations; the market areas of commercial businesses; the heartlands of ethnic or cultural groups; the jurisdictions of local, state, and national governments; and the reach of transnational corporations and supranational organizations.

Territoriality thus provides a means of meeting three social and cultural needs:

- The regulation of social interaction
- The regulation of access to people and resources
- The provision of a focus and symbol of group membership and identity

Territoriality also gives tangible form to power and control but does so in a way that directs attention away from the personal relationships between the controlled and the controllers. In other words, rules and laws become associated with particular spaces and territories rather than with particular individuals or groups. Finally, territoriality allows people to create and maintain a framework through which to experience the world and give it meaning. Bounded territories, for example, make it easier to differentiate "us" from "them."

APPLY YOUR KNOWLEDGE Describe the relationship between the field of ethology and the concept of territoriality. List and evaluate three examples that you experience in everyday life of proxemics as a territorializing force (beyond the classroom). How are they an expression of power and culture? ∎

People and Places, Insiders and Outsiders

Places are constantly under social construction as people respond to the opportunities and constraints of their particular locality. What this means is that, as people live and work in places, they gradually impose themselves on their environment, modifying and adjusting it to suit their needs and express their values. But at the same time they gradually accommodate both to their physical environment and to the values, attitudes, and comportment of people around them. People are constantly modifying and reshaping places, and places are constantly coping with change and influencing their inhabitants.

Places are both centers of meaning for people and the frameworks for their actions and behavior. It is important to remember that places are constructed by their inhabitants from their own subjective point of view and that they are simultaneously constructed and seen as an external "other" by outsiders. A neighborhood, for example, is both an area of special meanings to its residents as well as an area containing houses, streets, and people that others may view from an outsider's ("decentered") perspective.

As we saw in Chapter 1, a key concept is that of the *lifeworld*, the taken-for-granted pattern and context through which people conduct their day-to-day lives without having to make it an object of conscious attention. People's familiarity with one another's vocabulary, speech patterns, dress codes, gestures, and humor, and with shared experiences of their physical environment, often carries over into people's attitudes and feelings about themselves and their locality and to the symbolism they attach to that place. When this happens, the result is a collective and self-conscious "structure of feeling": a sociocultural frame of reference generated among people as a result of the experiences and memories that they associate with a particular place.

Experience and Meaning

The interactions between people and places raise some fundamental questions about the meanings that people attach to their experiences: How do people process information from external settings? What kind of information do they use? How do new experiences affect the way they understand their worlds? What meanings do particular environments have for individuals? How do these meanings influence behavior? How do people develop and modify their sense of a place, and what does it mean to them? The answers to these questions are by no means complete. It is clear, however, that people filter information from their environments through neurophysiological processes and also draw on personality and culture to produce cognitive images of their environment, representations of the world that can be called to mind through the imagination (**Figure 6.4**). Cognitive images are what people see in the mind's eye when they think of a particular place or setting.

Cognitive images both simplify and distort real-world environments. Research has suggested, for example, that many people tend to organize their cognitive images of particular parts of their world in terms of several simple elements (**Figure 6.5**):

Paths: The channels along which they and others move; for example, streets, walkways, transit lines, canals.

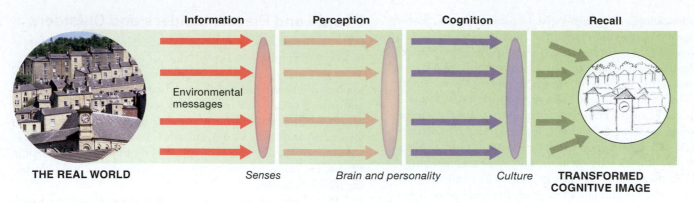

Information **Perception** **Cognition** **Recall**

Environmental messages

THE REAL WORLD *Senses* *Brain and personality* *Culture* **TRANSFORMED COGNITIVE IMAGE**

FIGURE 6.4 The formation of cognitive images People form cognitive images as a product of information about the real world, experienced directly and indirectly, and filtered through their senses, brain, personality, and the attitudes and values they have acquired from their cultural background. (*Source:* Adapted from R. G. Golledge and R. J. Stimpson, *Analytical Behavior Geography.* Beckenham, UK: Croom Helm, 1987, Fig. 3.2, p. 3.)

FIGURE 6.5 Cognitive image of Boston This map was compiled by Kevin Lynch from interviews with a sample of Boston residents. Lynch found that the residents of Boston tended to structure their cognitive images of the city with the same elements. He produced ingenious maps, such as this one, to demonstrate the collective "mental map" of the city, using symbols of different boldness or color to indicate the proportion of respondents who had mentioned each element. (*Source:* Adapted from K. Lynch, *The Image of the City.* Cambridge, MA: M.I.T. Press, 1960, p. 146.)

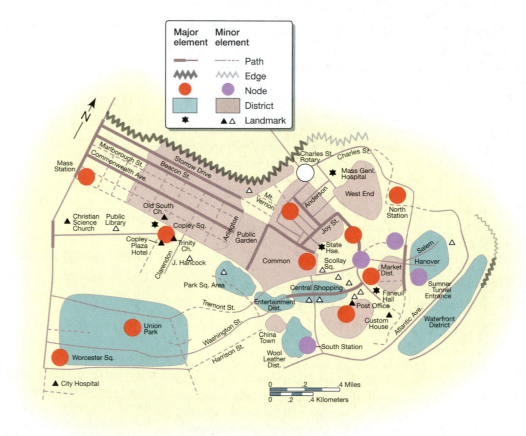

Edges: Barriers that separate one area from another; for example, shorelines, walls, railroad tracks.

Districts: Areas with an identifiable character (physical and/or cultural) that people mentally "enter" and "leave"; for example, a business district or an ethnic neighborhood.

Nodes: Strategic points and foci for travel; for example, street corners, traffic junctions, city squares.

Landmarks: Physical reference points; for example, distinctive landforms, buildings, or monuments.

Individual landscape features may function as more than one kind of cognitive element. A freeway, for instance, may be perceived

as both an edge and a path in a person's cognitive image of a city. Similarly, a railroad terminal may be seen as both a landmark and a node.

Distortions in our cognitive images are partly the result of incomplete information. Once we get beyond our immediate living area, we know few spaces in complete detail. Yet our worlds—especially those of us in developed countries who are directly tied to global networks of communication and knowledge—are increasingly large in geographic scope. As a result, these worlds must be conceived, or understood, with limited direct stimuli. We have to rely on fragmentary and often biased information from other people, from books, magazines, television, and the Internet. Distortions in cognitive images are also partly the result of our own biases. What we remember about places, what we like or dislike, what we think is significant are all functions of our personalities, our experiences, and the cultural influences to which we have been exposed.

APPLY YOUR KNOWLEDGE Use the five elements noted in this section of the chapter to map out your image of your college campus. What are the key paths, edges, districts, nodes, and landmarks that form your cognitive image? ∎

Images and Behavior

Cognitive images are compiled, in part, through behavioral patterns. Environments are "learned" through experience. Meanwhile, cognitive images, once generated, influence behavior. Via this two-way relationship, cognitive images are constantly changing. Each of us also generates—and draws on—different kinds of cognitive images in different circumstances.

Elements such as districts, nodes, and landmarks are important in the kinds of cognitive images that people use to orient themselves and to navigate within a place or region. The more of these elements an environment contains—and the more distinctive they are—the more legible that environment is to people and the easier it is to get oriented and navigate. In addition, the more firsthand information people have about their environment and the more they are able to draw on secondary sources of information, the more detailed and comprehensive their images are.

This phenomenon is strikingly illustrated in **Figure 6.6**, which shows the collective image of Los Angeles as seen by the residents of three different neighborhoods: Westwood, an affluent

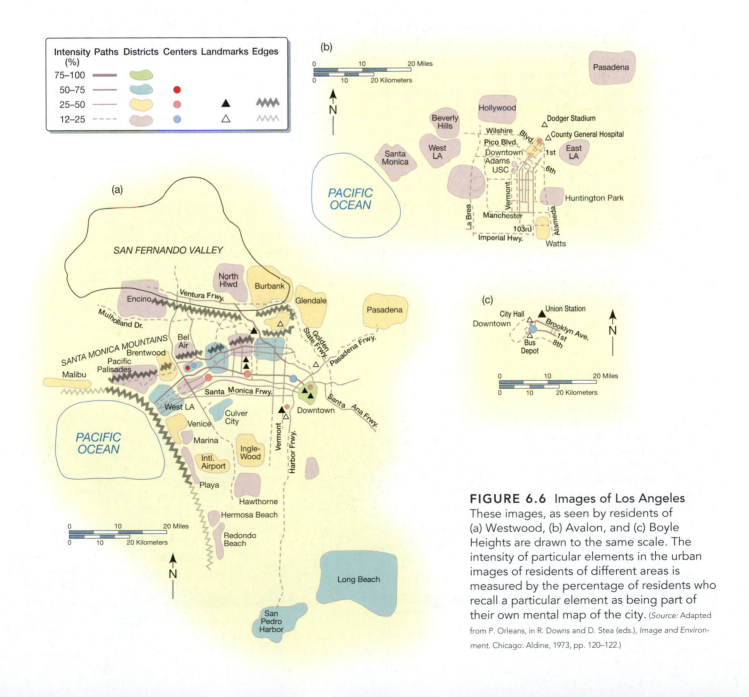

FIGURE 6.6 Images of Los Angeles These images, as seen by residents of (a) Westwood, (b) Avalon, and (c) Boyle Heights are drawn to the same scale. The intensity of particular elements in the urban images of residents of different areas is measured by the percentage of residents who recall a particular element as being part of their own mental map of the city. (*Source:* Adapted from P. Orleans, in R. Downs and D. Stea (eds.), *Image and Environment.* Chicago: Aldine, 1973, pp. 120–122.)

neighborhood; Avalon, a poor, inner-city neighborhood; and Boyle Heights, a poor, immigrant neighborhood. The residents of Westwood have a well-formed, detailed, and comprehensive image of the entire Los Angeles basin. At the other end of the socio-economic spectrum, residents of the black ghetto neighborhood of Avalon, near Watts, have a much more vague image of the city, structured only by the major east–west boulevards and freeways and dominated by the gridiron layout of streets between Watts and the city center. The Spanish-speaking residents of Boyle Heights—even less affluent, less mobile, and more isolated by language—have an extremely restricted image of the city. Their world consists of a small area around Brooklyn Avenue and First Street, bounded by the landmarks of city hall, the bus depot, and Union Station.

The importance of these images goes beyond people's ability simply to navigate around their environments. The narrower and more localized people's images are, for example, the less they will tend to venture beyond their home area. Their behavior becomes circumscribed by their cognitive imagery. People's images of places also shape particular aspects of their behavior. Research on shopping in cities, for example, has shown that customers do not necessarily go to the nearest store or to the one with the lowest prices; they are influenced by the configuration of traffic, parking, and pedestrian circulation within their imagery of the environment. The significance of this clearly has not been lost on the developers of shopping malls, who always provide extensive space for free parking and multiple entrances and exits.

In addition, shopping behavior, like many other aspects of behavior, is influenced by people's values and feelings. A district in a city, for example, may be regarded as attractive or repellent, exciting or relaxing, fearsome or reassuring, or, more likely, a combination of such feelings. As is the case with all cognitive imagery, such images are produced through a combination of direct experience and indirect information, all filtered through personal and cultural perspectives. Images such as these often exert a strong influence on behavior. Returning to the example of

consumer behavior, one of the strongest influences on shopping patterns relates to the imagery evoked by retail environments—something else that has not escaped the developers of malls, who spend large sums of money to establish the right atmosphere and image for their projects.

Shopping behavior is one narrow example of the influence of place imagery on behavior. Additional examples can be drawn from every aspect of human geography and at every spatial scale. The settlement of North America, for example, was strongly influenced by the changing images of the Plains and the West. In the early 1800s, the Plains and the West were perceived as arid and unattractive, an image reinforced by early atlases. When the railroad companies wanted to encourage settlement in these regions, they changed people's image of the Plains and the West with advertising campaigns that portrayed them as fertile and hospitable regions. The images associated with different regions and localities continue to shape settlement patterns. People draw on their cognitive imagery, for example, in making decisions about migrating from one area to another. **Figure 6.7** shows the composite image of the United States held by a group of Virginia Tech students, based on the perceived attractiveness of cities and states as places in which to live.

Another example of the influence of cognitive imagery on people's behavior is the way that people respond to environmental hazards, such as floods, droughts, earthquakes, storms, and landslides, and come to terms with the associated risks and uncertainties. Some people attempt to change the unpredictable into the knowable by imposing order where none really exists (resorting to folk wisdom about weather, for example), while others deny all predictability and take a fatalistic view. Some tend to overestimate both the degree and the intensity of natural hazards, while others tend to underestimate them.

Finally, one aspect of cognitive imagery is of special importance in modifying people's behavior: the sentimental and symbolic attributes ascribed to places. Through their daily lives and the cumulative effects of cultural influences and significant personal events, people develop bonds with places. They do this simultaneously at different geographic scales: from the home, through the

FIGURE 6.7 Preference map of the United States This isoline map illustrates collective preferences for cities in the coterminous United States as places in which to live and work, as expressed by architecture students at Virginia Tech in 1996. It is a generalization based on the scores the students gave to the 150 largest cities in the country.

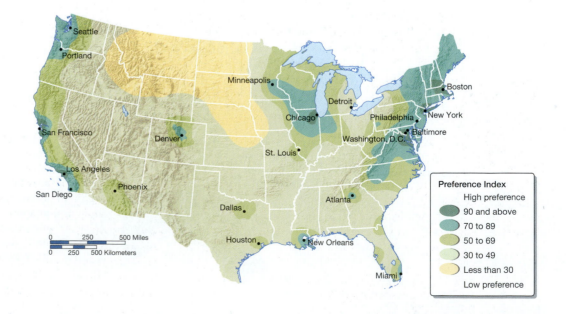

neighborhood and locality, to the national state. The tendency for people to do this has been called topophilia. **Topophilia** literally means "love of place." Geographers use the term to describe the complex emotions and meanings associated with particular places that, for one reason or another, have become significant to individuals. Most people have a home area, hometown, or home region for which they have a special attachment or sense of identity and belonging.

> **APPLY YOUR KNOWLEDGE** Place-making involves a wide range of cultural, social, and psychological processes. Choose a specific location in your town or university and elaborate on three ways that people make sense of this place through territoriality, proxemics, or cognitive images. Assess how age, race, gender, or class might influence this meaning making. ■

LANDSCAPES AS HUMAN SYSTEMS

Landscape is a term that means different things to different people. For some, the term brings to mind the design of formal gardens and parks, as in landscape architecture. For others, landscape signifies a bucolic countryside or even the organization of residences and public buildings. For still others, landscape calls to mind the artistic rendering of scenery, as in landscape painting.

Geographers study vernacular (ordinary) landscapes because they reflect the distinctive attributes of particular places or regions. They also study "landscapes of power," such as clusters of corporate tower blocks, and "landscapes of despair," such as homeless encampments and **derelict landscapes**. The latter are landscapes that have experienced abandonment, misuse, disinvestment, or vandalism. Additionally, geographers study symbolic landscapes because they reflect certain values or ideals—either those intended by the builders or financiers of particular places or those perceived by other groups. Some individual buildings and structures are so powerfully symbolic that they come to stand for entire cities: the Eiffel Tower in Paris, the Colosseum in Rome, and the Sugarloaf in Rio de Janeiro, for example (**Figure 6.8**). It is generic landscapes of different kinds that are most interesting to geographers, however.

Ordinary Landscapes

Some landscapes come to symbolize entire nations or cultures. Quite ordinary landscapes can be powerfully symbolic because they are understood as being a particular kind of place. The stereotypical New England townscape, for example (**Figure 6.9a**), is widely taken to represent not just a certain type of regional architecture but the best that Americans have known "of an intimate, family-centered, God-fearing, morally conscious, industrious, thrifty, democratic *community*."[2] Another ordinary townscape with powerful symbolic connotations is the typical

Main Street of Middle America (**Figure 6.9b**). It is "middle" in several respects: between the frontier to the west and the cosmopolitan seaports to the east, between agricultural regions and industrial metropolises, between affluence and poverty. It has come to symbolize places with a balanced community, populated by property-minded, law-abiding citizens devoted to free enterprise and a certain kind of social morality.

A more recent example is provided by the landscapes of contemporary American suburbia. They are landscapes of bigness and ostentation, characterized by packaged developments and simulated settings. These conservative utopias of themed and fortified subdivisions of private master-planned developments (**Figure 6.10**) reflect a presumed reciprocity between size and social superiority.

The key point is that ordinary landscapes, as geographers such as Don Mitchell have established, are instruments of social and cultural power that naturalize political-economic structures as if they were simply given and inevitable. As powerful complexes of signs, they perform vital functions of social regulation. Many of the landscapes of contemporary American suburbia, for example, have naturalized an ideology of competitive consumption and disengagement from civic affairs.

Conceptualization of Landscape

Since 1925, when Carl Sauer advocated the study of the *cultural landscape* as a uniquely geographical pursuit, new generations of geographers have been expanding the concept. The fact that different people comprehend the landscape differently is central to the **humanistic approach** in geography. This approach places the individual—especially individual values, meaning systems, intentions, and conscious acts—at the center of analysis. As the Sudanese example given earlier in the chapter suggests, children's perceptions of their worlds are different from those of their parents, and girls may see their world differently from boys, even in the same family.

Environmental perception and its close relative, behavioral geography, are interdisciplinary, drawing together geographers, landscape architects, psychologists, architects, and others. Professionals in these disciplines investigate what preferences people have in landscapes, how they construct cognitive images of their worlds, and how they find (or fail to find) their way around in various settings. The humanistic approach's focus on the perceptions of individuals is an important counterweight to the tendency to talk about a social group or society more generally. Nevertheless, some critics argue that humanistic research has limited utility because individual attitudes and views do not necessarily add up to the views held by a group or a society.

One alternative to the humanistic approach explores both the role of larger forces, such as culture, gender, and the state, and the ways in which these forces enhance or constrain individuals' lives. Much recent cultural geographical work, therefore, conceptualizes the relationship of people and the environment as interactive, not one-way, and emphasizes the role that landscapes play in shaping and reinforcing human practices. This most recent conceptualization of landscape is more dynamic and complex than the one Carl Sauer advanced, and it encourages geographers to look outside their own discipline—to anthropology, psychology, sociology, and even history—to fully understand its complexity.

[2]D. W. Meinig (ed.), *The Interpretations of Ordinary Landscapes: Geographic Essays.* New York: Oxford University Press, 1979, p. 27.

FIGURE 6.8 Famous landmarks Some cities are immediately recognizable because of certain buildings, landmarks, or cityscapes that have come to symbolize them. These examples—the Colosseum in Rome, the Sugarloaf in Rio de Janeiro, the Eiffel Tower in Paris, and the Houses of Parliament in London—are known worldwide.

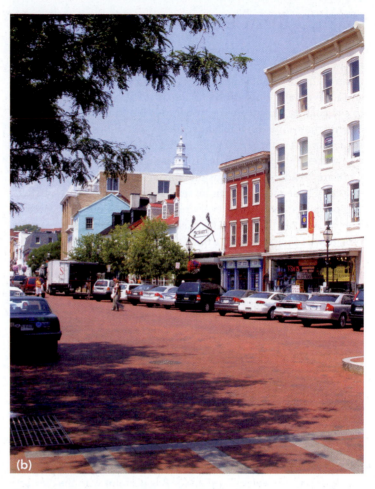

FIGURE 6.9 Ordinary landscapes Some ordinary cityscapes are powerfully symbolic of particular kinds of places. The New England village and the Main Street of Middle America are in this category, so much so that they have been taken as symbolizing the United States.

FIGURE 6.10 Vulgaria The dominant theme in upscale residential development in the United States is size and ostentation.

CODED SPACES

A dynamic and complex approach to understanding landscape is based on the conceptualization of **landscape as text**, by which we mean that, like a book, landscape can be read and written by groups and individuals. This approach departs from traditional attempts to systematize or categorize landscapes based on the different elements they contain. The landscape-as-text view holds that landscapes do not come ready-made with labels on them. Rather, there are "writers" who produce landscapes and give them meaning, and there are "readers" who consume the messages embedded in landscapes. Those messages can be read as signs about values, beliefs, and practices, though not every reader will take the same message from a particular landscape (just as people may differ in their interpretation of a passage from a book).

In short, landscapes both produce and communicate meaning, and one of our tasks as geographers is to interpret those meanings. In order to interpret or read our environment, we need to understand the language in which it is written. We must learn how to recognize the signs and symbols that go into the making of landscape. The practice of writing and reading signs is known as **semiotics**.

Semiotics asserts that innumerable signs are embedded or displayed in landscape, space, and place, sending messages about identity, values, beliefs, and practices. These signs may have different meanings for those who produce them and those who read, or interpret, them. Some signs are so subtle as to be recognizable only when pointed out by a knowledgeable observer; others may be more readily available and more ubiquitous. For example, semiotics enables us to recognize that by the way they dress, people, including college students, send messages about who they are and what they value. For some of us, certain types, such as the "skater" or the "emo," are readily identifiable by their clothes, hairstyle, or footwear, by the books they carry, or even by the food they eat.

Commercial Spaces

Semiotics, however, is not only about the signs that people convey by their mode of dress. Messages are also deployed through the landscape and embedded in places and spaces. Consider the very familiar landscape of the shopping mall. Although there is certainly a science to the size, scale, and marketing of a mall based on demographic research as well as environmental and architectural analysis, there is more to the mall than these concrete features. The placement and mix of stores and their interior design, the arrangement of products within stores, the amenities offered to shoppers, and the ambient music all combine to send signals to the consumer about style, taste, and self-image (**Figure 6.11**). Called by some "palaces of consumption," malls are complex semiotic sites, directing important signals not only about what to buy but also about who should shop there and who should not.

And yet, as much as Americans seem to enjoy shopping, a great many express disdain for shopping and the commercialism and materialism that accompany it. Shopping is a complicated activity about which people feel ambivalence. It is not surprising, therefore, that developers have promoted shopping as a kind of tourism. The mall is a "pseudoplace" meant to encourage one sort of activity—shopping—by projecting the illusion that something else besides shopping (and spending money) is actually going on.

The South Coast Plaza in Orange County, California, is a highly popular retail center in the United States, and with almost 3 million square feet of enclosed space covering 128 acres, it is one of the largest malls in the country. It is the most profitable too, boasting over $1.5 billion annually in sales. The mall contains 250 retail outlets, including luxury goods stores like Gucci, Versace, Chanel, Tiffany, Jimmy Choo, and Cartier as well as more popular, less pricey upscale stores like Bloomingdales and Nordstrom and finally middlebrow stores like Macy's and Sears. The latter stores anchor the mall at its outside corners, while the more expensive and luxurious shops and boutiques occupy interior locations. Thus the central stretches of the mall convey signs of affluence and luxury appealing to upper-class patrons while the periphery is more oriented to necessity and practicality for middle- and lower-middle-class patrons.

However complex the messages that malls send, one focus is consistent across class, race, gender, age, ethnicity, and other cultural boundaries: consumption, a predominant aspect

FIGURE 6.11 The Forums Mall, Caesar's Palace, Las Vegas Casinos in Las Vegas, such as this one, have invested heavily in creating opulent malls that are situated close to the gaming rooms.

of globalization. Indeed, malls are the early twenty-first century's spaces of consumption, where just about every aspect of our lives has become a commodity. Consumption—or shopping—defines who we are more than ever before, and what we consume sends signals about who we want to be. Advertising and the mass media tell us what to consume, equating ownership of products with happiness, a good sex life, and success in general. Within the space of the mall these signals are collected and resent. The architecture and design of the mall are an important part of the semiotic system shaping our choices and molding our preferences. As architectural historian Margaret Crawford writes:

> All the familiar tricks of mall design—limited entrances, escalators placed only at the end of corridors, fountains and benches carefully positioned to entice shoppers into stores—control the flow of consumers through the numbingly repetitive corridors of shops. The orderly processions of goods along endless aisles continuously stimulate the desire to buy. At the same time, other architectural tricks seem to contradict commercial consideration. Dramatic atriums create huge floating spaces for contemplation, multiple levels provide infinite vistas from a variety of vantage points, and reflective surfaces bring near and far together. In the absence of sounds from the outside, these artful visual effects are complemented by the "white noise" of MUZAK and fountains echoing across enormous open courts.[3]

Malls, condominium developments, neighborhoods, university campuses, and any number of other possible geographic sites possess codes of meaning. By thinking about their complex makeup and the individuals and groups they seek to influence, it is possible to interpret them and understand the implicit messages they contain.

APPLY YOUR KNOWLEDGE Apply what you just learned about codes and provide a description of the systems of signification that operate in your neighborhood. Appraise what information about itself your neighborhood conveys through coded means. ■

Sacred Spaces

Religious places can also be read and decoded. **Sacred spaces** are areas of the globe recognized by individuals or groups as worthy of special attention because they are the sites of special religious experiences and events. They do not occur naturally; rather, they are assigned sanctity through the values and belief systems of particular groups or individuals. Geographer Yi-Fu Tuan insists that what defines the sacredness of a space goes beyond the obvious shrines and temples. Sacred spaces are those that rise above the commonplace and interrupt ordinary routine.

In almost all cases, sacred spaces are segregated, dedicated, and hallowed sites that are maintained as such generation after generation. Believers—including mystics, spiritualists, religious followers, and pilgrims—recognize sacred spaces as being endowed with

divine meaning. The range of sacred spaces includes sites as different as an elegant and elaborate temple in Cambodia, Angor Wat, sacred to Hindus (**Figure 6.12**), and the Black Hills of South Dakota, the sacred mountains of the Lakota Sioux.

Often, members of a specific religion are expected to journey to especially important sacred spaces to renew their faith or to demonstrate devotion. A pilgrimage is a journey to a sacred space, and a pilgrim is a person who undertakes such a journey. In India many of the sacred pilgrimage sites for Hindus are concentrated along the seven sacred rivers: the Ganges, the Yamuna, the Saraswati, the Naramada, the Indus, the Cauvery, and the Godavari. The Ganges is India's holiest river, and many sacred sites are located along its banks (**Figure 6.13**). Hindus visit sacred pilgrimage sites for a variety of reasons, including seeking a cure for sickness, washing away sins, and fulfilling a promise to a deity. Perhaps the most well-known pilgrimage is the **hajj**, the obligatory once-in-a-lifetime journey of Muslims to Mecca. For one month every year, the city of Mecca in Saudi Arabia swells from its base population of 150,000 to over 1,000,000 as pilgrims from all over the world journey to fulfill their obligation to pray in the city and receive the grace of Allah. **Figure 6.14** shows the principal countries that send pilgrims to Mecca.

Pilgrimages to sacred sites are made all over the world, including Christian Europe. The most visited sacred site in Europe is Lourdes, at the base of the Pyrenees in southwest France, not far from the Spanish border (**Figure 6.15**). Another sacred site that attracts pilgrims throughout the world is the city of Jerusalem, and the Holy Land more generally, which is visited by Jews, Orthodox, Catholics, Protestants, Christian Zionists, and followers of many other religions (see Box 6.1, "Geography Matters: Jerusalem, the Holy City"). As with most sacred spaces, the codes that are embedded in the landscape of the Holy Land may be read quite differently by different religious and even secular visitors. Two students of pilgrimage observed:

> Each group brings to Jerusalem their own entrenched understandings of the sacred; nothing unites them save their sequential—and sometimes simultaneous—presence at the same holy sites. For the Greek Orthodox pilgrims, indeed, the precise definition of the site itself is largely irrelevant; it is the icons on display which are the principal focus of attention. For the Roman Catholics, the site is important in that it is illustrative of a particular biblical text relating to the life of Jesus, but it is important only in a historical sense, as confirming the truth of past events. Only for the Christian Zionists does the Holy Land itself carry any present and future significance, and here they find a curious kinship with indigenous Jews.[4]

APPLY YOUR KNOWLEDGE What is semiotics and how does it help us to understand important information about landscapes? Choose a religious or spiritual landscape and dissect the code that helps to explain how it operates. ■

[3]M. Crawford, "The World in a Shopping Mall," in M. Sorkin (ed.), *Variations on a Theme Park.* New York: Noonday Press, 1992, p. 14.

[4]J. Eade and M. Sallnow (eds.), *Contesting the Sacred.* London: Routledge, 1991, p. 14.

FIGURE 6.12 Angor Wat, Cambodia Built for the king Suryavarman II in the early twelfth century, Angor Wat was dedicated to the Hindu god Vishnu. In the late thirteenth century, this state temple became a sacred site for Buddhists, which continues to this day.

FIGURE 6.13 Sacred sites of Hindu India India's many rivers are holy places within the Hindu religion, so sacred sites are located along the country's many riverbanks. Shrines closer to the rivers are regarded as holier than those farther away.

(*Source:* Adapted from Ismail Ragi al Farugi and David E. Sopher, *Historical Atlas of the Religions of the World.* New York: Macmillan, 1974.)

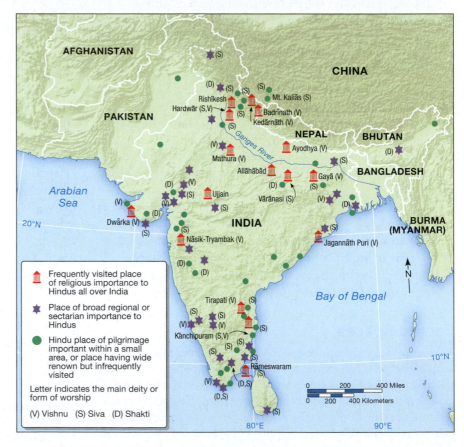

PLACE AND SPACE IN MODERN SOCIETY

While memorials, malls, and sacred places illustrate the way that places are coded within specific settings or contexts, there are also broader dimensions of social, economic, and political organizations that shape the ways people think about themselves and about the places they inhabit. For more than a century the philosophy of modernity has been a major influence on the interdependencies among culture, society, space, place, and landscape throughout much of the world.

Modernity is a forward-looking view of the world that emphasizes reason, scientific rationality, creativity, novelty, and progress. Its origins can be traced to the European Renaissance and the emergence of the world-system of competitive capitalism in the sixteenth century. At that time, scientific discovery and commerce began to displace traditional sociocultural views of the world that emphasized mysticism, romanticism, and fatalism. These origins

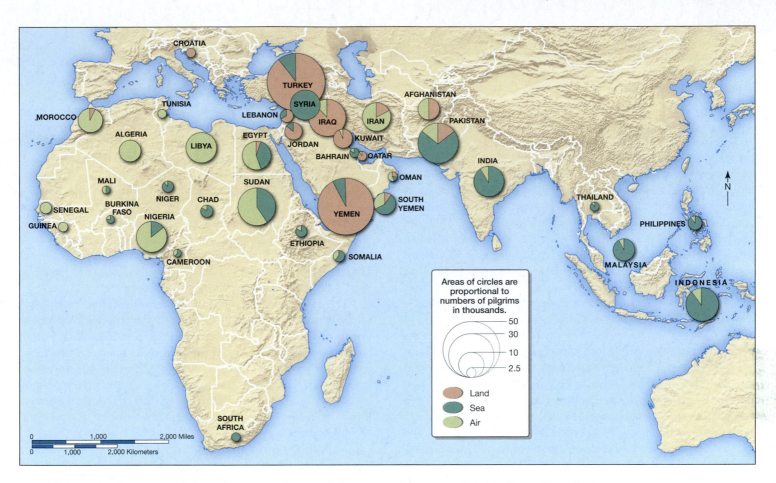

FIGURE 6.14 Source areas for pilgrims to Mecca Islam requires that every adult Muslim perform the pilgrimage to Mecca at least once in a lifetime. This obligation is deferred for four groups of people: those who cannot afford to make the pilgrimage; those who are constrained by physical disability, hazardous conditions, or political barriers; slaves and those of unsound mind; and women without a husband or male relative to accompany them. The pattern of actual pilgrimages to Mecca (which is located close to the Red Sea coast in Saudi Arabia) suggests a fairly strong distance-decay effect, with most traveling relatively short distances from Middle Eastern Arab countries. More distant source areas generally provide smaller numbers of pilgrims, though Indonesia and Malaysia are notable exceptions. (*Source:* After C. C. Park, *Sacred Worlds.* London: Routledge, 1994, p. 268.)

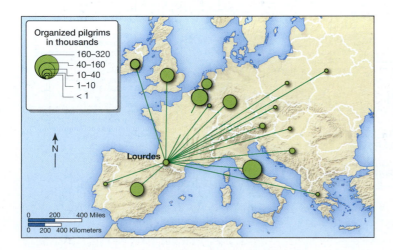

FIGURE 6.15 Source areas for pilgrims to Lourdes This map shows the points of origin of European group-organized pilgrims to Lourdes in 1978. These represent only about 30 percent of all pilgrims to Lourdes, most of whom travel to the shrine on their own. Improved transportation (mainly by train) and the availability of organized package trips have contributed to a marked increase in the number of pilgrims visiting. Many of the 5 million pilgrims who visit the town each year do so in the hope of a miraculous cure for medical ills at a grotto where the Virgin Mary is said to have appeared before 14-year-old Bernadette Soubirous in a series of 18 visions in 1858. (*Source:* After C. C. Park, *Sacred Worlds.* London: Routledge, 1994, p. 284.)

Although many cities in the world have been the object of struggle and conflict over the centuries, none has been as endlessly beset as Jerusalem. Visitors, writers, and residents believe Jerusalem to be the most beautiful city in the world. If there are other serious competitors to that coveted beauty title, Jerusalem certainly has few rivals for the title of the most sacred city in the world, possessing as it does an unmatched Christian, Jewish, and Islamic history. Jerusalem began as a small settlement on the slopes of Mount Moriah. In 997 B.C.E., it was captured by David, king of the Israelites, who made Jerusalem the capital of Israel. Solomon, David's son and successor, built the Great (First) Temple on Mount Moriah to commemorate the place where Abraham offered to sacrifice his son. Though the temple was destroyed centuries ago, the site is central to the Jewish faith.

The history and the map of the city reflect the histories of the various empires that dominated and were succeeded by new and yet more powerful empires (**Figure 6.A**). Nebuchadnezzar, king of Babylon, destroyed the Great Temple in 586 B.C.E. and banished the Jews. But the Babylonian control of Jerusalem eventually gave way to the Persians, under whose rule the Jews were allowed to return and rebuild their temple, known as the Second Temple. The Persian occupation of Jerusalem was swept aside by Alexander the Great (356–323 B.C.E.), the king of Macedonia and one of the world's greatest military leaders, who helped spread the Greek Empire from the southern shores of the Caspian Sea into central Asia. The Romans entered the scene around 63 B.C.E., and Herod the Great was eventually installed to command the Roman Kingdom of Judea from Jerusalem.

During the early Christian period in Jerusalem, the Jews revolted openly against the Roman occupiers. In 132 C.E., the Romans responded by destroying the Second Temple and banishing all Jews from Jerusalem and Palestine. As a result, the Jews scattered north into Babylon and later into Europe and North Africa. The popular myth is that this ancient Jewish diaspora remained in exile until 1948, when the state of Israel was created.

The major Christian influence on Jerusalem, the "Holy City," began when Constantine I (285–337 C.E.), the emperor of the Eastern Roman Empire, converted to Christianity in 313 C.E. This event led to the construction of churches and other buildings dedicated to celebrating the life of Jesus Christ. But Christian influence over the city ceased when Jerusalem eventually succumbed to Islam. In 638 C.E., Jerusalem was designated a holy city of Islam because it was believed that Muhammad's spirit had once made a visit to heaven while he was in the city.

Although for several centuries Jews, Christians, and Muslims were all allowed access to the city of Jerusalem, by the tenth

FIGURE 6.A Jerusalem, the Holy City This map of Jerusalem demarcates the main sections of the city. Over the many years of the Israeli-Palestinian peace process, numerous proposals have been advanced about how to divide the city to satisfy the wishes of both Palestinians and Israelis. The disposition of Jerusalem is one of the major issues in the ongoing peace process. (*Source:* Redrawn from *The Guardian,* 14 October 2000, p. 5.)

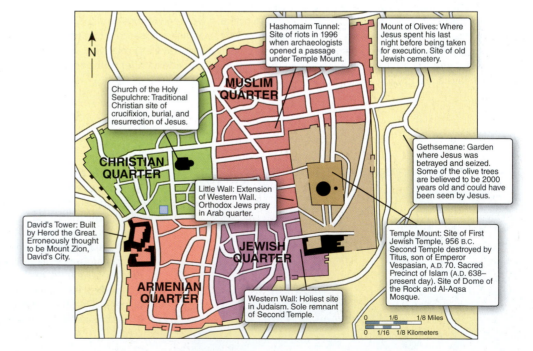

century, the persecution of non-Muslims became common. From the eleventh century until the thirteenth, European Christians undertook military expeditions—called the Crusades—to the Holy Land in an attempt to wrest control of Jerusalem from the Muslims. In 1099, Crusaders captured the city; Christians lost it again to the Muslim military leader Saladin in 1187. In 1517, Jerusalem was absorbed into the Ottoman Empire, and the city was ruled from Istanbul for more than 400 years. The Ottomans, however, had little interest in Jerusalem, and Jewish immigrants began returning to the city and Palestine starting in the mid-nineteenth century.

The contemporary history of the city derives from the political and geographical implications of a British statement of policy, which stipulated that Jerusalem should be an international city with no one state claiming it as entirely its own. Today, Jerusalem is a highly contested city as Palestinians, Christians, Muslims, and Jews fight for control of it. An example of this contest for control is the continuing dispute over the Dome of the Rock, which was constructed between C.E. 688 and 691. Muslims claim the Dome as their most sacred site (**Figure 6.B**). Yet the Dome sits on a site sacred to the Jews, the Temple Mount, the site where the Great Temple and Second Temple were built and later destroyed. Indeed, the Dome is believed to enclose the sacred rock upon which

Abraham prepared to sacrifice his son, according to Jewish tradition, and, according to Islamic tradition, is the same rock from which the prophet Muhammad launched his spirit on a heavenly visit. Also located at the Temple Mount is the Al-Aqsa Mosque, a central sacred site to Muslims.

While nationalist Israelis maintain that Jerusalem will be the "eternal and undivided capital" of Israel, Palestinians believe that Jerusalem is the future capital of the Palestinian state. In fact, the 1992 Oslo peace accords that lead to a declaration of principles between Israel and Palestine hint at the possibility of negotiating the future of Jerusalem. Further peace negotiations, including those at Wye River, Maryland, in 1998; at Sharm al-Shaykh in the Egyptian Sinai Desert in 1999; in Camp David, Maryland, in 2000; in Washington, D.C. in 2001; and in Geneva in 2003 are cause for hope that a resolution to the future of Jerusalem that includes some control by the Palestinians can be achieved. At present, Jerusalem is entirely controlled by Israel. Yet in accordance with the Oslo peace agreements, in May 2000 Israel conceded three villages to the control of the Palestinian Authority. All three villages are part of greater Jerusalem in the West Bank, and all but one, Abu Dis, are actually a few meters beyond the Jerusalem municipal border. From Abu Dis, the Old City of Jerusalem is clearly visible, particularly the Dome of the Rock.

FIGURE 6.B Dome of the Rock Located in Haram Al-Shaif, or Temple Mount, the Dome of the Rock sits in the part of Jerusalem that is technically neither Jewish nor Muslim, but both. The Muslims want this entire holy site. The Israelis also want it because the Wailing Wall, sacred to Jews, occupies part of the Temple Mount.

were consolidated into a philosophical movement during the eighteenth century, when the so-called Enlightenment established the widespread belief in universal human progress and the sovereignty of scientific reasoning.

At the beginning of the twentieth century, this philosophy developed into a more widespread intellectual movement. A series of sweeping technological and scientific developments not only triggered a new round of spatial reorganization but also transformed the underpinnings of social and cultural life. These developments included the telegraph, the telephone, the X-ray, the motion picture, the radio, the bicycle, the internal combustion engine, the airplane, the skyscraper, relativity theory, and psychoanalysis. Universal human progress suddenly seemed to be a realistic prospect.

Nevertheless, the pace of economic, social, cultural, and geographic change was unnerving and the outcomes uncertain. The intellectual response, developed among a cultural avantgarde of painters, architects, novelists, and photographers, was a resolve to promote Modernity through radical changes in culture. These ideas were first set out in the "Futurist Manifesto," published in 1909 in the Paris newspaper *Le Figaro* in the form of a letter from the Italian poet Filippo Marinetti. Gradually, the combination of new technologies and radical design contributed to the proliferation of landscapes of Modernity. Among the most striking of these were Modernist urban landscapes (**Figure 6.16**). Indeed, in a general sense almost all of the place-making and landscapes of the early and mid-twentieth century are the products of Modernity.

Throughout this period a confident and forward-looking Modernist philosophy remained virtually unquestioned. Places and regions everywhere were heavily shaped by people acting out their notions of rational behavior and progress. Rural regions, for example, bore the imprint of agricultural Modernization. The hedgerows of traditional European field patterns were torn up to make way for landscapes of large, featureless fields in which heavy machinery could operate more effectively (**Figure 6.17**). On a different scale, peripheral countries sought to remake traditional landscapes through economic Modernization. Economic development and social progress were to be achieved through a modern infrastructure of highways, airports, dams, harbors, and industrial parks.

Globalization and Place-Making

The spread of Modernity to peripheral regions can be seen as part of globalization. These globalization processes have not only brought about a generalization of forms of industrial production, market behavior, trade, and consumption but reinforced and extended the commonalities among places. Three factors are especially important in this context.

First, mass communications media have created global culture markets in print, film, music, television, and the Internet. Indeed, the Internet has created an entirely new *kind* of space—cyberspace—with its own "landscape" (or technoscape) and its own embryonic cultures (see Box 6.2, "Visualizing Geography: The Cultural Geography of Cyberspace and Social Networking"). The instantaneous character of contemporary communications has also made possible the creation of a shared, global consciousness from the staging of global events such as the Olympic Games

and the World Cup. Second, mass communications media have diffused certain values and attitudes toward a wide spectrum of sociocultural issues, including citizenship, human rights, child rearing, social welfare, and self-expression. Third, international legal conventions have increased the degree of standardization and level of harmonization not only of trade and labor practices but of criminal justice, civil rights, and environmental regulations.

These commonalities have been accompanied by the growing importance of material consumption within many cultures.

FIGURE 6.16 Landscape of Modernity This photograph shows Modernist urban landscape at La Défense, in Paris: an office district designed to be symbolic of contemporary France and emblematic of the aspirations of Paris to compete as a major world city in the global economy.

FIGURE 6.17 Modernized rural landscape Circular bales of hay after being harvested in the East Anglia region of the United Kingdom.

Increasingly, people around the world are eating the same foods, wearing the same clothes, and buying the same consumer products. Yet the more people's patterns of consumption converge, the more fertile the ground for countercultural movements. The more transnational corporations undercut the authority of national and local governments to regulate economic affairs, the greater the popular support for regionalism. The more universal the diffusion of material culture and lifestyles, the more local and ethnic identities are valued. The faster the information highway takes people into cyberspace, the more they feel the need for a subjective setting—a specific place or community—they can call their own. The faster the pace of life in search of profit and material consumption, the more people value leisure time. And the faster their neighborhoods and towns acquire the same generic supermarkets, gas stations, shopping malls, industrial estates, office parks, and suburban subdivisions, the more people feel the need for enclaves of familiarity, centeredness, and identity. The United Nations Center for Human Settlements (UNCHS) notes:

> In many localities, people are overwhelmed by changes in their traditional cultural, spiritual, and social values and norms and by the introduction of a cult of consumerism intrinsic to the process of globalization. In the rebound, many localities have rediscovered the "culture of place" by stressing their own identity, their own roots, their own culture and values and the importance of their own neighborhood, area, vicinity, or town.[5]

One example of the impulse for people to recover a sense of place is provided by the slow city (CittaSlow) movement. The CittaSlow movement is a grassroots response to globalization and is closely related to the longer-established and better-known slow food movement. The aims of the two movements are different but complementary: In broad terms, both organizations are in favor of local, traditional cultures, a relaxed pace of life, and conviviality. Both are a response to the increasing pace of everyday life associated with the acceleration of money around local, national, and global circuits of capital. Both are hostile to big business and globalization, though their driving motivation is not so much political as ecological and humanistic. Slow food is devoted to a less hurried pace of life and to the true tastes, aromas, and diversity of good food (**Figure 6.18**). The movement also serves as a rallying point against globalization, mass production, and the kind of generic fast food represented by U.S.-based franchises like McDonald's, Burger King, Pizza Hut, Taco Bell, and Kentucky Fried Chicken. By 2011, the slow food movement, based in Bra, near Turin in northern Italy, had established convivia (local branches) in more than 100 countries, with over 100,000 members worldwide. Its campaigns cover a range of specific causes, from protecting the integrity of chocolate to promoting the cultivation of traditional crop varieties and livestock breeds and opposing genetically engineered foods.

The slow city movement was formed in October 1999, when Paolo Saturnini, mayor of Greve-in-Chianti, a Tuscan hill town, organized a meeting with the mayors of three other municipalities (Orvieto, Bra, and Positano) to define the attributes that might characterize a *città lente*—slow city. At their founding meeting in Orvieto, the four mayors committed themselves to a series of principles that included working toward calmer and less polluted physical environments, conserving local aesthetic traditions, and fostering local crafts, produce, and cuisine. They also pledged to use technology to create healthier environments, to make citizens aware of the value of more leisurely rhythms to life, and to share their experience in seeking administrative solutions for better living. The goal is to foster the development of places that enjoy a robust vitality based on good food, healthy environments, sustainable economies, and the seasonality and traditional rhythms of community life (see Box 6.3, "Window on the World: Waldkirch, Germany").

FIGURE 6.18   Slow food A man takes some vegetables inside "slow food" restaurant, Gusto, in central Rome.

[5]United Nations Center for Human Settlements, *Global Report on Human Settlements 2001*. London: Earthscan, 2001, p. 4.

The rapid growth of the Internet and of social networking sites like Facebook and Twitter is of great cultural significance. It has created the basis for a massive shift in patterns of social interaction, a seedbed for new forms of human consciousness, and a new medium for cultural change. Culture is fundamentally based on communication, and in cyberspace we now have an entirely new form of communication: uncensored, multidirectional, written, visual, and aural.

At face value, the Internet represents the leading edge of the globalization of culture. The Internet portends a global culture based on English as the universal world language, with a heavy emphasis on core-area cultural values, such as novelty, spectacle, fashionableness, material consumption, and leisure. It is unlikely, however, that the Internet will simply be a new medium through which core-area values and culture are spread.

To begin with, the impact of the Internet is likely to be highly uneven because of the **digital divide**. Moreover, there is resistance in some places and regions to the cultural globalization associated with cyberspace. The French and French Canadian authorities, for example, already sensitive about the influence of English-language popular culture, have actively sought ways to give Francophone cybernauts access to the Internet without submitting to English, the dominant language of Web sites. The French government has also subsidized an all-French alternative to the Internet: Minitel, an online videotext terminal that plugs into French telecommunication networks. A free Minitel terminal is available to anyone who stops by a France Telecom office.

In much of Asia, the Internet's basic function as an information-exchange medium clashes with local cultures in which information is a closely guarded commodity. Whereas many U.S. World Wide Web sites feature lengthy government reports and scientific studies, as well as lively debates about government policy, comparable Asian sites typically offer little beyond public relations materials from government agencies and corporations. In puritanical Singapore, political leaders, worried that the Internet will undermine

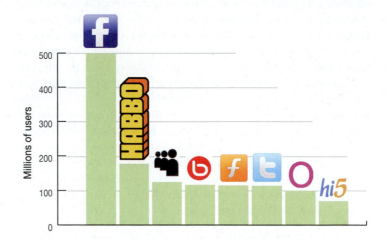

FIGURE 6.C Estimated number of worldwide users for each of the top 8 social networks. (*Source:* Numbers from the 2010 Social Networking Map by Ethan Bloch, August 5, 2010. http://www.flowtown.com/blog/the-2010-social-networking-map)

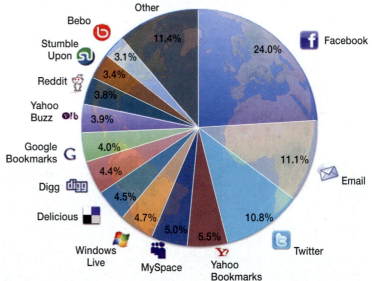

FIGURE 6.D How people share content on the web Data showing which sites are seeing the most shared content according to AddToAny, makers of a widget that allows sharing across any platform. Percentages represent portions of AddToAny users who share with a particular service. (From "Chart of the Day: How People Share Content on the Web" by Nicholas Carlson and Kamelia Angelova, July 21, 2009. Source: AddToAny. Reprinted by permission of Lockerz, Inc, owner and operator of www.lockerz.com and affiliated sites.)

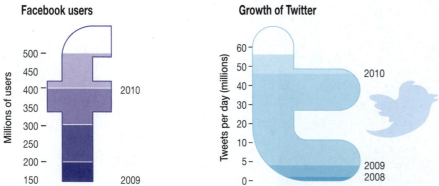

FIGURE 6.E Growth of Facebook and Twitter.

Top ten social cities in the United States
(larger size = more social media usage)

Seattle

Boston

San Francisco
San Jose

Salt Lake City

Denver

New York

Los Angeles

Atlanta

Austin

FIGURE 6.F Top 10 U.S. cities on the Netprospex Social Index (NPSI) The size of each city's name is porportional to its score on the NPSI. The NetProspex Social Index is determined by a business person's social connectedness, social activity, friendliness, and reach across top social networks, including Facebook, Twitter, and LinkedIn. Research conducted from the NetProspex database of business contacts. (Adapted from data in Netprospex Social Report Fall 2010. pdf available at http://netprospex.com/np/social)

What 20 minutes on Facebook looks like:

1,000,000 shared links
1,323,000 tagged photos
1,484,000 event invites sent
1,587,000 wall posts
1,851,000 status updates
1,972,000 friend requests accepted
2,716,000 photos uploaded
10,208,000 comments
4,632,000 messages

(from Democracy UK on Facebook)

morality, have taken to reading private e-mail as part of an all-out effort to beat back the menace of online pornography. Chinese authorities fear that the Internet will foment political rebellion, so officials have limited access to ensure that the Chinese portion of the Internet can easily be severed from the world in the event of political unrest. The reluctance of major Asian organizations to put important information on their Web sites—along with the need for Westerners to use special software to read any local-language documents that do exist—has resulted in a largely one-way flow of information, from America to Asia.

Nevertheless, the greatest potential of the Internet and social networking in terms of cultural change resides in the liberating and empowering potential of their vast resource of knowledge and information. By its very nature the Internet empowers individuals (rather than social groups or institutions), allowing millions to say whatever they want to each other, free (for the first time in history) from state control. As such, it is an important vehicle for the spread of participatory democracy to much of the world, as demonstrated by the political upheavals in Tunisia, Egypt, and much of the Middle East in 2011.

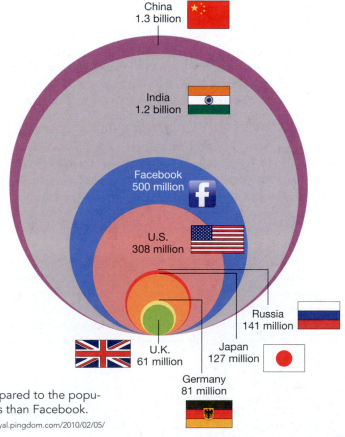

China
1.3 billion

India
1.2 billion

Facebook
500 million

U.S.
308 million

Russia
141 million

Japan
127 million

U.K.
61 million

Germany
81 million

FIGURE 6.G Facebook as a country Number of Facebook users compared to the populations of selected countries. Only China and India have larger populations than Facebook.

(Adapted from Facebook, social media juggernaut (infographic), posted February 5, 2010 by Pingdom. http://royal.pingdom.com/2010/02/05/facebook-social-media-juggernaut-infographic/)

Waldkirch, a small town of 12,500 inhabitants, sits astride the valley of the River Elz, in southwestern Germany, commanding the entrance to the narrow part of the valley as the river exits the Black Forest headed toward the Rhine. To the southwest of the town are the broad floodplains of the Elz and its tributaries, a region of intensive and prosperous agriculture—mainly orchards, viticulture, and market gardening. To the north, east, and south of the town is the Black Forest, dotted with prosperous and immaculately kept farmsteads set in meadows bordered by fruit trees.

Waldkirch was once famous for the manufacture of barrel organs, and the name of the town was carried around the world on its products. The industry began, along with clock-making, as a winter craft for Black Forest farmers in the late 1700s. It grew to become internationally famous between 1850 and the 1930s, with several firms exporting all kinds of finely crafted organs—from small drawing-room models to larger versions for churches, fairgrounds, music halls, theaters, and ocean liners—all over the world.

Today, Waldkirch is a prosperous and lively town with a strong sense of place and a collective commitment to the quality of life of its residents. Just 15 kilometers north of Freiburg, and connected to that city by light rail, Waldkirch has acquired a growing commuter population that has contributed to a high level of prosperity. Its vitality is reflected in its shops, services, clubs, and community organizations. Despite its relatively small population, there are scores of clubs and volunteer organizations, and the town is well endowed with stores—there are several optometrists, for example, plus a high-end furniture store (with customers from Freiburg and beyond), three shoe stores, a health food store, a good bookstore, a sportswear store, and several bakers, butchers, and pharmacists.

In addition to this general prosperity, Waldkirch has an unusually energetic and progressive group of community leaders that includes elected officials, bureaucrats, business owners, and company managers. In 1998, the city began a visioning process that resulted in a strong emphasis on quality of life and a parallel commitment to urban planning. Waldkirch joined the Cittaslow movement in 2002 as a natural outcome of this strategy. Policies that have been put in place to commit to the Cittaslow philosophy include a strong emphasis on using streets as social and play spaces. Priority is given to pedestrians wherever possible, and on residential streets there is a 7 kilometers per hour (4.3 mph) limit with radar enforcement; drivers caught traveling in excess of 14 kilometers per hour (8.6 mph) temporarily lose their license. Storefronts on the main street, Lange Strasse, are being gradually restored to traditional windows and façades (from 1960s and 1970s plate glass modernizations); and the local public utility company has installed a new energy system for public buildings, using wood chips as the fuel source.

Complementing the Cittaslow philosophy is the town's involvement in the Austrian-based *Lebens Qualität durch Nähe* ("quality of life with proximity") program. "Live Here; Buy Local" is the movement's slogan. The program has three elements: social networks and sense of place; lifestyle, food, and nutrition; and employment. The broad idea is to strengthen basic services in small settlements and to build local consciousness about the connections between quality of life and the availability of services and products that are locally produced and sold. An additional goal is to support surrounding farms and agriculture in order to prevent land reverting to forest. As a result, Waldkirch has developed special zoning requirements to ensure minimum percentages of arable land and pastureland.

FIGURE 6.H Waldkirch, Germany The heart of the town is he Marktplatz, a broad street that slopes down from the site of the original settlement, which dates from the early Middle Ages. Waldkirch began as a small ecclesiastical center and country market that developed on the southern shoulder of the valley, guarded by a small castle built above town.

FIGURE 6.I Waldkirch farmers' market A long-standing tradition, the twice-weekly farmers' market is an important element in sustaining the rhythm of the town and its residents' sense of place.

FIGURE 6.J Social bonds Waldkirch is an enthusiastic member of the Slow City movement, which endeavours to support and encourage conviviality. Opportunities for regular casual encounters among pedestrians are an important means of developing a sense of conviviality in the town.

The town's leadership has also placed a great deal of emphasis on social sustainability. A recent project involved the revitalization of a house that had been serving as a residence for the homeless in a neglected and run-down neighborhood, where the adjacent public space had been turned into an auto junkyard. The city government spent about US$1.6 million in renovating the house—now known as the "Red House" because of its bright red façade. Today, the house functions as a local meeting place and houses the office of a neighborhood social worker and a kitchen that serves meals to the neighborhood. A farmers' market with local vendors takes place once a week in front of the house. Stalls feature fresh fruit and vegetables, bread, and fair trade products. Since the "Red House" opened, neighborhood crime and vandalism have dropped and neighborhood residents of all ages and ethnic groups are building social networks. To connect these social efforts with economic opportunities for the residents, Waldkirch initiated a program that provides job opportunities for residents who have been long-term unemployed.

Places as Objects of Consumption

In much of the world, people's enjoyment of material goods now depends not just on their physical consumption or use. It is also linked more than ever to the role of material culture as a social marker. A person's home, automobile, clothing, reading, viewing, eating and drinking preferences, and choice of vacations are all indicators of that person's social distinctiveness and sense of style. This pressures individuals to continuously search for new and distinctive styles. The wider the range of foods, products, and ideas from around the world—and from past worlds—the greater the possibilities for establishing such styles.

Given that material consumption is so central to the repertoire of symbols, beliefs, and practices of postmodern cultures, the "culture industries"—advertising, publishing, communications media, and popular entertainment—have become important shapers of spaces, places, and landscapes. Because the symbolic meanings of material culture must be advertised (in the broadest sense of the word) in order to be shared, advertising (in its narrower sense) has become a key component of contemporary culture and place-making. In addition to stimulating consumer demand, advertising has always had a role in teaching people how to dress, how to furnish a home, and how to signify status through groupings of possessions.

In the 1970s and 1980s, however, the emphasis in advertising strategies shifted away from presenting products as newer, better, more efficient, and more economical to identifying them as the means to self-awareness, self-actualization, and group stylishness. Increasingly, products are advertised in terms of their association with a particular lifestyle rather than in terms of their intrinsic utility. Many of these advertisements deliberately draw on international or global themes, and some entire advertising campaigns (for Coca-Cola, Benetton, and American Express, for example) have been explicitly based on the theme of cultural globalization. Many others rely on stereotypes of particular places or kinds of places (especially exotic, spectacular, or "cool" places) in creating the appropriate context or setting for their product. Images of places therefore join with images of global food, architecture, pop culture, and consumer goods in the global media marketplace. Advertisements both instruct and influence consumers about products and also about spaces, places, and landscapes.

One result of these trends is that contemporary cultures rely much more than before not only on material consumption but also on *visual* and *experiential* consumption: the purchase of images and the experience of spectacular and distinctive places, physical settings, and landscapes. Visual consumption can take the form of magazines, television, movies, sites on the World Wide Web, tourism, window shopping, people-watching, or visits to galleries and museums. The images, signs, and experiences that are consumed may be originals, copies, or simulations.

The significance of the increased importance of visual consumption for place-making and the evolution of landscapes is that settings such as theme parks, shopping malls, festival marketplaces, renovated historic districts, museums, and galleries have all become prominent as centers of cultural practices and activities.

Examples of the re-creation and refurbishment of historic districts and settings are widespread and have become a mainstay of the "heritage industry." This industry, based on the commercial exploitation of the histories of peoples and places, is now worldwide, as evidenced by the involvement of the Economic, Social, and Cultural Organization of the United Nations (UNESCO) in identifying places for inclusion on World Heritage lists. In countries like the United Kingdom, with a high density of historic districts and settings, place marketing relies heavily on the heritage industry. In the United Kingdom, more than 90 million people pay to view about 650 historic properties each year, and millions more visit free-of-charge heritage sites such as cathedrals, field monuments, remote ruins, and no longer useful industrial waterways. In 2010, more than 200 million tourists visited designated heritage sites in the United Kingdom, spending about $53 billion on entry fees, retail sales, travel, and hotel accommodations.

One important consequence of the heritage industry is that urban cultural environments are vulnerable to a debasing and trivializing "Disneyfication" process. The United Nations Center for Human Settlements (UNCHS) *Global Report on Human Settlements 2001* notes:

> The particular historic character of a city often gets submerged in the direct and overt quest for an international image and international business. Local identity becomes an ornament, a public relations artifact designed to aid marketing. Authenticity is paid for, encapsulated, mummified, located, and displayed to attract tourists rather than to shelter continuities of tradition or the lives of its historic creators.[6]

As a result, contemporary landscapes contain increasing numbers of inauthentic settings—what geographer David Harvey has called the "degenerative utopias" of global capitalism. These are as much the product of contemporary material and visual culture as they are of any cultural heritage. Particularly influenced by images and symbols derived from movies, advertising, and popular culture, examples include the re-creation of the Wild West in the fake cowboy town of Old Tucson in Arizona; the Bavarian village created in Torrance, California; and the simulation of other cities and tourist hot spots in Las Vegas, Nevada (**Figure 6.19**). The town of Helen, Georgia, boasts a full-scale reconstruction of a Swiss mountain village, complete with costumed staff and stores selling Swiss merchandise. In Japan, a reproduction village called British Hills (**Figure 6.20**), complete with church, pub, and school, has been constructed north of Tokyo.

Another interesting aspect of the increasing trend toward the consumption of experiences is the emergence of restaurants as significant cultural sites. Restaurants often represent a synthesis of the global and the local, and they can be powerful cultural symbols in their own right (**Figure 6.21**). The dining experience in a particular restaurant can be an important symbolic good. By the same token, the social standing or celebrity status of customers contributes to the value of the dining experience offered by a restaurant. Restaurants themselves can be both theater and performance—particularly in big cities like

[6]Ibid., p. 38.

FIGURE 6.19 Las Vegas The boundaries between the heritage industry and the leisure and entertainment industries have become increasingly blurred and a great deal of investment has been channeled toward the creation of inauthentic "historic" settings whose characteristics owe more to movies and popular stereotypes than historic realities. Shown here is New York-New York, located on the Las Vegas Strip. The replica of the Statue of Liberty is 150 ft (46 m) tall.

FIGURE 6.20 Thames Town A neighborhood combining commercial, residential and cultural elements Shanghai, China.

FIGURE 6.21 Consumption in style Shown here is the Bed Restaurant in South Beach, Miami, Florida.

New York and Los Angeles, where underpaid or out-of-work actors, dancers, and other artists form an important part of the restaurant labor force. Restaurants bring together a global and a local labor force (immigrant owners, chefs, and waiters, as well as locals) and clientele (tourists and business travelers, as well as locals). Finally, restaurant design also contributes to a city's visual style as architects and interior designers, restaurant consul-tants, and restaurant-industry magazines adapt global trends to local styles.

APPLY YOUR KNOWLEDGE Compare and contrast modernity and postmodernity. Give specific examples of what each term means and the ways they are different. ■

Future Geographies

Globalization has already brought a significant degree of homogenization of culture through the language of consumer goods. This is the material culture of the West, enmeshed in Airbus jets, CNN, music video channels, cell phones, and the Internet and swamped by Coca-Cola, Budweiser, McDonald's, GAP clothing, Nikes, iPods, PlayStations, Toyotas, Disney franchising, and formula-driven Hollywood movies (**Figure 6.22**). Furthermore, sociologists have recognized that a distinctive culture of "global metropolitanism" is emerging among the transnational elite. This is simply homogenized culture at a higher plane of consumption (French wines instead of Budweiser, molecular gastronomy and international cuisine instead of McDonald's, Hugo Boss clothes instead of Levis, BMWs instead of Toyotas, and so on). The members of this new culture are people who make international conference calls, make decisions and transact investments that are transnational in scope, edit the news, design and market international products, and travel the world for business and pleasure, and it is likely to grow significantly in size and influence.

The idea of the emergence of a single global culture is too simplistic, but the pervasive emphasis on material, visual, and experiential consumption means that many aspects of contemporary culture will increasingly transcend local and national boundaries. More and more of the world's population are world travelers—either directly or via the TV in their living room—so that many are knowledgeable about aspects of others' cultures. This contributes to **cosmopolitanism**, an intellectual and aesthetic openness toward divergent experiences, images, and products from different cultures.

Cosmopolitanism is an important geographic phenomenon because it fosters a curiosity about all places, peoples, and cultures, together with at least a rudimentary ability to map, or situate, such places and cultures geographically, historically, and anthropologically. It also suggests an ability to reflect upon, and judge aesthetically between, different places and societies. Optimists would speculate that this bodes well for global peace and understanding: We can, perhaps, more easily identify with people who use the same products, listen to the same music, and appreciate the same sports stars that we do. At the same time, however, focusing attention on material consumption obscures the emergence of other trends. As people's lives are homogenized through their jobs and their material culture, many of them want to revive subjectivity, reconstruct

FIGURE 6.22 McDonald's Buddha The McDonald's franchise has always been aggressive in marketing their brand but many felt it went too far in Shanghai when in the mid 1990s a new branch of the fast-food restaurant was opened in the Zhongshan Guang Chang district featuring a gigantic fiberglass Ronald McDonald figure in the form of the Buddha seated in the lotus position. At the time, their attempt at coopting the holy man (and dedicated vegetarian) to market McDonald's hamburgers was widely criticized and the icon was removed within one month. In 2005, however, the Ronald McDonald figure greeting guests with the traditional Thai "wai" gesture, pictured here, was introduced to restaurants around Thailand with no strenuous objections.

we/us feelings, and reestablish a distinctive cultural identity. One feasible outcome of this is an increased probability of cultural and territorial conflict.

APPLY YOUR KNOWLEDGE Elaborate on three examples of how globalization of culture shapes places and landscapes. ■

CONCLUSION

Geographers study the interdependence between people and places and are especially interested in how individuals and groups acquire knowledge of their environments and how this knowledge shapes their attitudes and behaviors. People ascribe meanings to landscapes and places in many ways, and they also derive meanings from the places and landscapes they experience. Different groups of people experience landscape, place, and space differently. For instance, the experience that rural Sudanese children have of their landscapes and the ways in which they acquire knowledge of their surroundings differ from how middle-class children in an American suburb learn about and function in their landscapes. Furthermore, both landscapes elicit a distinctive sense of place that is different for those who live there and those who simply visit.

As indicated in previous chapters, the concepts of landscape and place are central to geographic inquiry. They are the result of intentional and unintentional human action, and every landscape is a complex reflection of the operations of the larger society. Geographers have developed categories of landscape to help distinguish the different types that exist. Ordinary landscapes, such as suburban neighborhoods, are ones that people experience in the course of their everyday lives. By contrast, symbolic landscapes represent the particular values and aspirations that their developers and financiers want to impart to a larger public. An example is Mount Rushmore in the Black Hills of South Dakota, designed and executed by sculptor Gutzon Borglum. Chiseling the heads of George Washington, Thomas Jefferson, Theodore Roosevelt, and Abraham Lincoln into the granite face of the mountain, Borglum intended to construct an enduring landscape of nationalism in the wilderness.

More recently, geographers have come to regard landscape as a text, something that can be written and read, rewritten, and reinterpreted. This concept suggests that a landscape can have more than one author, and different readers may derive different meanings from what is written there. The idea that landscape can be written and read is further supported by the understanding that the language in which the landscape is written is a code. To understand the significance of the code is to understand its semiotics, the language in which the code is written. The code may be meant to convey many things, including a language of power or of playfulness, a language that elevates one group above another, or a language that encourages imagination or religious devotion and spiritual awe.

The global transition from Modernity to postmodernity has altered cultural landscapes, places, and spaces differently as individuals and groups have struggled to negotiate the local impacts of this widespread shift in cultural sensibilities. The shared meanings that insiders derive from their place or landscape have been disrupted by the intrusion of new sights, sounds, and smells as values, ideas, and practices from one part of the globe have been exported to another. The Internet and the emergence of cyberspace have meant that new spaces of interaction have emerged that have neither a distinct historical memory attached to them nor a well-established sense of place. Because of this, cyberspace carries with it some unique possibilities for cultural exchange. It remains to be seen, however, whether access to this new space will be truly open—or whether the Internet will become another landscape of power and exclusion.

Learning Outcomes Revisited

■ Investigate how environment shapes people and how people shape environments.

People not only filter information from their environments through neurophysiological and psychological processes but draw on personality and culture to produce cognitive images of their environment— pictures or representations of the world that can be called to mind through the imagination. Thus human–environment relationship results in a variety of ways of understanding the world around us as well as different ways of being in the world as information about our environment is filtered by people.

■ Recognize that place-making stands at the center of issues of culture and power relations and is a key part of the systems of meaning through which humans make sense of the world.

Places are the result of a wide range of forces from economic to social. Economically, places emerge through all sorts of transactions that result from the complexities of the land market. But places are also more than just real estate. They can reflect tensions between social groups as well as harmonious interaction.

■ Identify how different cultural identities and status categories influence the ways people experience and understand landscapes, as well as how they are shaped by—and able to shape—landscapes.

Among the most important relations are the cultural identities of race, class, gender, ethnicity, and sexuality. Often these identities come together in a group, and their influence in combination becomes central to our understanding of how group identity shapes space and is shaped by it.

■ Understand how codes signify important information about landscapes, a process known as semiotics.

To interpret our environment, we must learn how to read the codes that are written into the landscape. Landscapes as different from each other as shopping malls and war memorials can be understood in terms of their semiotics, although it is important to appreciate that even when certain landscapes have intended meanings by those who have created them, those who perceive them may make their own sense of that landscape.

■ Describe how globalization has occurred in parallel with a transition from Modernity to Postmodernity and assess how those two periods differ.

Material consumption has become central to the repertoire of symbols, beliefs, and practices of postmodern cultures. As a result, the "culture industries"—advertising, publishing, communications media, and popular entertainment—have also become important shapers of spaces, places, and landscapes as have global products. In this way, landscapes can become familiar in foreign places like a McDonald's in Bangkok, Thailand, or a fancy shopping mall in Istanbul, Turkey, that contains a wide array of global brands.

KEY TERMS

cosmopolitanism *(p. 210)*

derelict landscapes *(p. 193)*

digital divide *(p. 204)*

ethology *(p. 189)*

hajj *(p. 197)*

humanistic approach *(p. 193)*

landscape as text *(p. 196)*

modernity *(p. 198)*

postmodernity *(p. 184)*

proxemics *(p. 188)*

sacred spaces *(p. 197)*

semiotics *(p. 196)*

territoriality *(p. 188)*

topophilia *(p. 193)*

REVIEW AND DISCUSSION

1. Consider the environment of your own town or campus. Conduct research exploring how the environment has been shaped or reshaped over the last 5 to 10 years. (*Hint:* You might want to research the building of new parking structures or sports complexes or consider how your university has expanded and bought more land or is not using all of its buildings.) Are there opposing perceptions regarding how your university has been shaped?

2. Conduct research to identify and explore the different elements (paths, edges, districts, nodes, and landmarks) within your campus or town. Begin by making a list of the different elements and who occupies these spaces. What are the territorial, social, and cultural markers of each space? List three of the proxemics you observe in these spaces. (*Hint:* You might start by considering the different coffee shops, bars, or restaurants at your university. Who hangs out at these different locations? List and assess the types of social interactions that are taking place. List the symbols or cultural behaviors that are associated with the location and the people who occupy it.) Once you have the information compiled, draw a territorial cognitive image map like the ones in Figures 6.5 and 6.6.

3. As a student on a campus, in what ways might your environmental perception of the campus landscape differ from that of the faculty on campus or from that of the janitorial staff who work on campus? Give three examples of how these different groups will have different perceptions of the landscape. Also provide a detailed sketch of how each group would read the landscape-as-text. For example, what is the meaning of a classroom, hallway, bathroom, or office for each group? List the different semiotics that is involved for each group.

UNPLUGGED

1. Write a short essay (500 words or two double-spaced pages) that describes, from your personal perspective, the sense of place that you associate with your hometown or county. Write about the places, buildings, sights, and sounds that are especially meaningful to you.

2. Draw up a list of the top 10 places in the United States in which you would like to live and work; then draw up a list of the bottom 10. How do these lists compare with the map of student preferences shown in Figure 6.7? Why might your preferences be different from theirs?

3. On a clean sheet of paper and without reference to maps or other materials, sketch a detailed map of the town or city in which you live. When you have finished, compare your sketch to Figure 6.5. Does your sketch contain nodes? Landmarks? Edges? Districts? Paths? How does "your" cognitive image map compare to your "real" town or city?

4. Using telephone books that can be accessed at your university library and a map of your local area, map the sacred landscape of your locality by identifying the locations of different churches, temples, mosques, and other places of worship. What patterns do you see? How do they fit with your knowledge of particular areas? What other sorts of sacred sites might exist beyond the institutionalized religious sites?

Mastering GEOGRAPHY™

Log in to www.masteringgeography.com for MapMaster™ interactive maps, geography videos, RSS feeds, flash cards, Web links, an eText version of *Human Geography: Places and Regions in Global Context*, and self-study quizzes to enhance your study of interpreting places and landscapes.

 MapMaster™ presents 13 Place Name and 13 Layered Thematic interactive maps to help students practice and master their geographic literacy, spatial reasoning, and critical thinking skills.

MapMaster™

7

GEOGRAPHIES OF ECONOMIC DEVELOPMENT

Learning Outcomes

- Scrutinize the nature and degree of unevenness in patterns of economic development at national and international scales.

- Analyze how geographical divisions of labor have evolved with the growth of the world-system and the accompanying variations in economic structure.

- Interpret how regional cores of economic development are created through the operation of several basic principles of spatial organization.

- Explain how spirals of economic development can be arrested in various ways, including the onset of disinvestment and deindustrialization.

- Demonstrate how globalization has resulted in patterns and processes of local and regional economic development that are open to external influences.

A trip in China on National Highway 321 east from Chengdu in Sichuan province to Shenzhen in Guangdong is a journey through economic development. Migrating workers who travel these highways often leave their families behind. But they also help their families escape poverty and propel China upward through the ranks of middle-income countries. As they travel eastward, they leave an agrarian realm with few jobs and miserable wages and enter the realm of "agglomeration economies," in which a big labor market attracts manufacturers who offer better wages.

Shenzhen attracts young workers—90 percent of its 8 million residents are of working age, between 16 and 65 years old. Cao Bin, aged 20, from Chengdu in Sichuan left for Shenzhen in 2008 and hasn't been back. Chengdu is considered one of the most livable cities in China, but Cao Bin thought it was boring. "That town is too lazy," he says. "I wanted to go somewhere where life is faster." Peng Chunxia, 21, migrated from Hunan at 17, following her elder sister. "Where we are from, most people leave for work . . . I was young, and I thought it'd be fun to come here." Li Chunying, 34, started working at a toy factory when she was 16. She now works as a line manager for an LED maker and looks forward to her day off, when she can spend time shopping with friends or eating in restaurants that serve food from their native Hunan. "When was I happiest? I don't know. There were so many times here that I've

Migrant workers queue to get on a specially arranged train from Beijing to Chongqing to go back home for the Spring Festival.

been happy," she says. Still, she longs for home, and would go back with her husband and two daughters if she could only be assured of finding a job.

With a ready supply of skilled and semiskilled young workers, Shenzhen is investing in better education and research facilities to ensure that the city supplies what industries need. The area specializes in electronic goods and it makes them in enormous quantities. In 2006, its exports exceeded India's, making the Shenzhen seaport the fourth busiest in the world. The port ships in intermediate inputs and ships out final products. It boasts expensive facilities, such as top-notch container ports and convention centers, and matches workers to the growing number of jobs as firms rapidly expand their operations. Proximity to Hong Kong provides access to finance, though Shenzhen is also home to a rapidly expanding financial sector of its own. And competition for customers among the multiple suppliers of inputs produces cost savings. In many ways, what is happening in Shenzhen is a direct reflection of the local, regional, and international processes of economic change that we analyze in this chapter.

Based on *World Development Report 2009*, The World Bank, Washington, D.C., 2009, p. 13; and A. Ramzy and J. Jiang, "Person of the Year 2009: The Chinese Worker," *Time Magazine*, 174, December 28, 2009, p. 8 ■

PATTERNS OF ECONOMIC DEVELOPMENT

We often discuss economic development in terms of levels and rates of change in prosperity, as reflected in bottom-line statistical measures of productivity, incomes, purchasing power, and consumption. Increased prosperity is only one aspect of economic development, however. For human geographers and other social scientists, the term *economic development* refers to processes of change involving the nature and composition of the economy of a particular region as well as to increases in the overall prosperity of a region. These processes can involve three types of changes:

- changes in the structure of the region's economy (for example, a shift from agriculture to manufacturing);

- changes in forms of economic organization within the region (for example, a shift from socialism to free-market capitalism);

- changes in the availability and use of technology within the region (see Box 7.1, "Visualizing Geography: Technological Change and Economic Development").

Economic development is also expected to bring with it some broader changes in the economic well-being of a region. The most important of these are changes in the capacity of the region to improve the basic conditions of life (through better housing, health care, and social welfare systems) and to improve the physical framework, or infrastructure, on which the economy rests.

The Unevenness of Economic Development

Geographically, the single most important feature of economic development is that it is *uneven*. At the global scale, this unevenness takes the form of core-periphery contrasts within the evolving world-system (see Chapter 2). These global core-periphery contrasts are the result of a competitive economic system that is heavily influenced by cultural and political factors. The core regions within the world-system—North America, Europe, and Japan—have the most diversified economies, the most advanced technologies, the highest levels of productivity, and the highest levels of prosperity. They are commonly referred to as *developed regions* (though processes of economic development are, of course, continuous, and no region can ever be regarded as fully developed).

Other countries and regions—the periphery and semiperiphery of the world-system—are often referred to as *developing* or *less developed*. Indeed, the nations of the periphery are often referred to as LDCs (less developed countries). Another popular term for the global periphery, developed as a political label but now synonymous with economic development in popular usage, is the *Third World*. This term had its origins in the early Cold War era of the 1950s and 1960s, when the newly independent countries of the periphery positioned themselves as a distinctive political bloc, aligned with neither the First World of developed, capitalist countries nor the Second World of the Soviet Union and its satellite countries.

At the global scale, levels of economic development are usually measured by economic indicators such as gross domestic product and gross national income. **Gross domestic product (GDP)** is an estimate of the total value of all materials, foodstuffs, goods, and services that are produced by a country in a particular year. To standardize for countries' varying sizes, total GDP is normally divided by total population,

which gives an indicator, *per capita* GDP, that is a good yardstick of relative levels of economic development. **Gross national income (GNI)** is a measure of the income that flows to a country from production wherever in the world that production occurs. For example, if a U.S.-owned company operating in another country sends some of its income (profits) back to the United States, this adds to the U.S. GNI.

In making international comparisons, GDP and GNI can be problematic because they are based on each nation's currency. As a result, it is now common to compare national currencies based on **purchasing power parity** (PPP). In effect, PPP measures how much of a common "market basket" of goods and services each currency can purchase locally, including goods and services that are not traded internationally. When we use PPP-based currency values to compare levels of economic prosperity, we usually see lower GNI figures in wealthy countries (because of the generally higher cost of living) and higher GNI figures in poorer nations (because of the generally lower cost of living). Nevertheless, even with this compression between rich and poor, economic prosperity is very unevenly distributed across nations.

As **Figure 7.1** shows, most of the highest levels of economic development are to be found in northern latitudes (very roughly, north of 30° N), which has given rise to another popular shorthand for the world's economic geography: the "North" (the core) and the "South" (the periphery). Viewed in more detail, the global pattern of per capita GNI (measured in the "international dollars" of PPP) in 2009 is a direct reflection of the core-semiperiphery-periphery structure of the world-system. In many of the core countries of North America, northwestern Europe, and Japan, annual per capita GNI (in PPP) exceeds $30,000. The only other countries that match these levels are Australia and Singapore, where annual per capita GNI in 2009 was $38,210 and $49,850, respectively. Semiperipheral countries such as Brazil, Russia, and Thailand have an annual per capita GNI ranging between $7,000 and $9,000. In the rest of the world—the periphery—annual per capita GNI (in PPP) is typically less than $5,000.

The gap between the highest per capita GNIs ($57,640 in Luxembourg and $56,050 in Norway) and the lowest ($290 in Liberia, $300 in the Democratic Republic of the Congo) is huge. The gap between the world's rich and poor is also getting wider. In 1970, the average GNI per capita of the ten most prosperous countries in the world was 50 times greater than the average GNI per capita of the ten poorest countries. By 2009, the relative gap had increased to a factor of 67. Overall, more than 80 percent of the world's population lives in countries where income differentials are widening.

APPLY YOUR KNOWLEDGE What kind of statistics besides gross national income provide an indication of international disparities in economic development? Find data on one such indicator and propose two possible reasons why the variations exist in the data you found. (*Hint:* Good sources are the World Bank, http://data.worldbank.org/, and the United Nations Development Programme, http://hdr.undp.org/en/statistics/.) ■

Resources and Development

Current patterns of economic development are the result of many different factors. One of the most important is the availability of key resources such as cultivable land, energy, and valuable minerals.

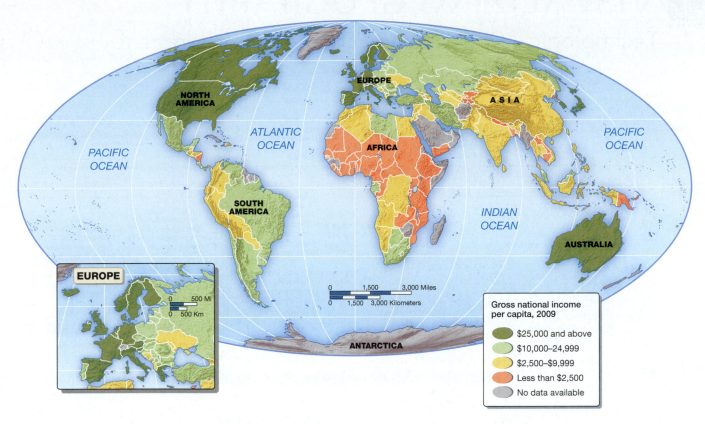

FIGURE 7.1 Gross national income (GNI) per capita GNI per capita is one of the best single measures of economic development. This map, based on 2009 data, shows the tremendous gulf in affluence between the core countries of the world economy—like the United States, Norway, and Switzerland, with annual per capita GNI (in PPP "international dollars") of more than $25,000—and peripheral countries like Angola, Haiti, and Mali, where annual per capita GNI was less than $2,500. In semiperipheral countries like South Korea, Brazil, and Mexico, per capita GNI ranged between $5,000 and $10,000 (*Source: The Fuller Projection™ Map design is a trademark of the Buckminster Fuller Institute, © 1938, 1967 & 1992. All rights reserved. www.bfi.org. Updated with data from World Bank, International Comparison Program database.*).

Unevenly distributed across the world are both key resources and—equally important—the *combinations* of energy and minerals crucial to economic development. A lack of natural resources can, of course, be remedied through international trade (Japan's success is a prime example of this). For most countries, however, the resource base remains an important determinant of development.

Energy

One particularly important resource in terms of the world's economic geography is energy. The major sources of commercial energy—oil, natural gas, and coal—are unevenly distributed across the globe. Most of the world's core economies are reasonably well off in terms of energy *production,* the major exceptions being Japan and parts of Europe. Most peripheral countries, on the other hand, are energy poor. The major exceptions are Algeria, Ecuador, Gabon, Indonesia, Libya, Nigeria, Venezuela, and the Gulf states—all major oil producers.

Because of this unevenness, energy has come to be an important component of world trade. As more of the world becomes industrialized and developed, demand for oil continues to increase. Yet, as we saw in Chapter 4, it is generally agreed that the peak of oil *discovery* was passed in the 1960s and that the world started using more oil than was contained in new fields in 1981 (**Figure 7.2**). Oil prices have risen sharply: more than 300 percent between 2005 and 2010. Oil is now the most important single commodity in world trade, making up more than 20 percent of the total by value in 2010.

FIGURE 7.2 Offshore oil drilling An oil exploration rig working in a shallow estuary off the coast of Sumatra in the Malacca Straits, Indonesia.

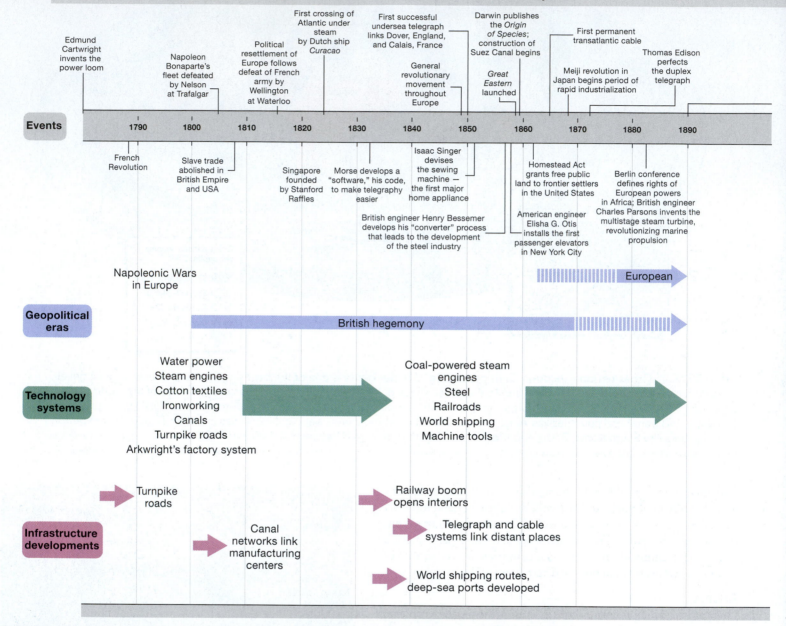

FIGURE 7.A Technological change and economic development The Industrial Revolution, which began in England at the end of the eighteenth century, was driven by a technology system based on water power and steam engines, cotton textiles, ironworking, river transport systems, and canals. It eventually resulted not only in the complete reorganization of the geography of the original European core of the world-system but also in an extension of the world-system core to the United States and Japan. Since then there have been several more technology systems, each opening new geographic frontiers and rewriting the geography of economic development while shifting the balance of advantages between regions. Overall, the opportunities for development created by each new technology system have been associated with distinctive economic epochs and long-term fluctuations in the overall rate of change of prices in the economy.

Beginning in the late eighteenth century, a series of technological innovations in power and energy, transportation, and manufacturing processes resulted in crucial changes in patterns of economic development. Each of these major clusters of technological innovations created new demands for natural resources as well as new labor forces and markets. The result was that each major cluster of technological innovations—called technology systems—tended to favor different regions and different kinds of places. Technology systems are clusters of interrelated energy, transportation, and production technologies that dominate economic activity for several decades at a time—until a new cluster of improved technologies evolves. What is especially remarkable about technology systems

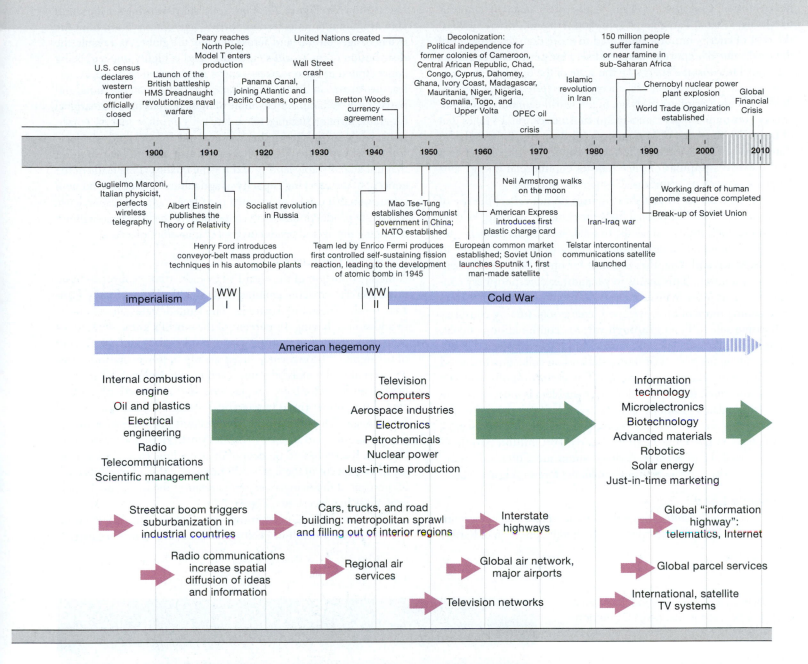

is that so far they have come along at about 50-year intervals. Since the beginning of the Industrial Revolution, we can identify four of them:

1790–1840: early mechanization based on water power and steam engines; development of cotton textiles and ironworking; development of river transport systems, canals, and turnpike roads.

1840–1890: exploitation of coal-powered steam engines; steel products; railroads; world shipping; and machine tools.

1890–1950: exploitation of the internal combustion engine; oil and plastics; electrical and heavy engineering; aircraft; radio and telecommunications.

1950–1990: exploitation of nuclear power, aerospace, electronics, and petrochemicals; development of limited-access highways and global air routes.

A fifth technology system, still incomplete, began to take shape in the 1980s with a series of innovations that are now being commercially exploited:

1990 onward: exploitation of solar energy, robotics, microelectronics, biotechnology, nanotechnology, advanced materials (fine chemicals and thermoplastics, for example), and information technology (digital telecommunications and geographic information systems, for example).

Each of these technology systems has rewritten the geography of economic development as it has shifted the balance of advantages between regions (**Figure 7.A**). From the mid-1800s, industrial development spread to new regions. The growth of those regions then became interdependent with the fortunes of other regions through a complex web of production and trade.

For many peripheral countries the cost of importing energy is a heavy burden. Consider, for example, the predicament of countries like India, Ghana, Paraguay, Egypt, and Armenia, where in 2010 the cost of energy imports amounted to more than one-quarter of the total value of exported merchandise. Few peripheral countries can afford to consume energy on the scale of the developed economies, so patterns of commercial energy *consumption* mirror the fundamental core-periphery cleavage of the world economy. In 2008, energy consumption per capita in North America was 14 times that of India, 18 times that of Mozambique, and nearly 50 times that of Bangladesh. The world's high-income countries, with 15 percent of the world's population, use half its commercial energy and 10 times as much per capita as low-income countries.

It should be noted that these figures do not reflect the use of firewood and other traditional fuels for cooking, lighting, heating and, sometimes, industrial needs. In total, such forms probably account for around 20 percent of total world energy consumption. In parts of Africa and Asia, they account for up to 80 percent of energy consumption. This points to yet another core-periphery contrast. Whereas massive investments in exploration and exploitation are enabling more of the developed, energy-consuming countries to become self-sufficient through various combinations of coal, oil, natural gas, hydroelectric power, and nuclear power, 1.5 billion people in peripheral countries depend on collecting fuelwood as their principal source of energy. The collection of wood fuel causes considerable deforestation. The problem is most serious in densely populated locations, arid and semiarid regions, and cooler mountainous areas, where the regeneration of shrubs, woodlands, and forests is particularly slow. Nearly 100 million people in 22 countries (16 of them in Africa) cannot meet their minimum energy needs even by overcutting remaining forests (**Figure 7.3**).

Cultivable Land

The distribution of cultivable land is another important factor in international economic development. Much more than half of Earth's land surface is unsuitable for any productive form of arable farming, as shown in **Figure 7.4**. Poor soils, short growing seasons, arid climates, mountainous terrain, forests, and conservation limit the extent of agricultural land across much of the globe. As a result, the distribution of the world's cultivable land is highly uneven, being concentrated in Europe, west-central Russia, eastern North America, the Australian littoral, Latin America, India, eastern China, and parts of sub-Saharan Africa. Some of these regions may be marginal for arable farming because of marshy soils or other adverse conditions, while irrigation or other factors sometimes extends the local frontier of productive agriculture. We also have to bear in mind that not all cultivable land is of the same quality. This leads to the concept of the **carrying capacity** of agricultural land: the maximum population that can be maintained in a place at rates of resource use and waste production that are sustainable in the long term without damaging the overall productivity of that or other places.

Industrial Resources

A high proportion of the world's key industrial resources—basic raw materials—are concentrated in Russia, the United States, Canada, South Africa, and Australia. The United States, for example, in addition to having 42 percent of the world's known resources of hydrocarbons (oil, natural gas, and oil shales) and 38 percent of the lignite ("brown coal," used mainly in power stations), has 38 percent of the molybdenum (used in metal alloys), 21 percent of the lead (used for batteries, gasoline, and construction), 19 percent of the copper (used for electrical wiring and components and for coinage), 18 percent of the bituminous coal (used for fuel in power stations and in the chemical industry), and 15 percent of the zinc. Russia has 68 percent of the vanadium (used in metal alloys), 50 percent of the lignite, 38 percent of the bituminous coal, 35 percent of the manganese, 25 percent of the iron, and 19 percent of the hydrocarbons (**Figure 7.5**).

The concentration of known resources in just a few countries is largely a result of geology, but it is also partly a function

FIGURE 7.3 **Deforestation** Rainforest logging operation in southeast Cameroon.

FIGURE 7.4 **Agricultural land cover** Some countries, like the United States, are fortunate in having a broad range of cultivable land, which allows for many options in agricultural development. Many countries, though, have a much narrower base of cultivable land and must rely on the exploitation of one major resource as a means to economic development. (Adapted from *World Resources 2000–2001: People and Ecosystems.* Washington, DC: World Resources Institute, 2000, p. 57. Originally from Wood et al., 2000. The map is based on Global Land Cover Characteristics Database Version 1.2 (Loveland et al. [2000]) and USGS/EDC (1999a). The figure is based on FAOSTAT (1999).

FIGURE 7.5 **Mineral exports** Rare earth being loaded at Lianyungang dock, Jiangsu province, China.

of countries' political and economic development. Political instability in much of postcolonial Africa, Asia, and Latin America has seriously hindered their exploration and exploitation of resources. In contrast, the relative affluence and great political stability of the United States have led to a much more intensive exploration of resources. We should also bear in mind that the significance of particular resources is often tied to particular technologies. As technologies change, so do resource requirements, and the geography

of economic development is "rewritten." One important example of this was the switch in the manufacture of mass-produced textiles from natural fibers like wool and cotton to synthetic fibers in the 1950s and 1960s. When this happened, many farmers in the U.S. South had to switch from cotton to other crops.

Regions and countries that are heavily dependent on one particular resource are vulnerable to the consequences of technological change. They are also vulnerable to fluctuations in the price set for their product on the world market. These vulnerabilities are particularly important for countries whose economies are dependent on nonfuel minerals, such as the Democratic Republic of the Congo (copper), Mauritania (iron ore), Namibia (diamonds), Niger (uranium), Sierra Leone (diamonds), Togo (phosphates), and Zambia (copper).

Resources and Sustainability

The ideal of **sustainable development** is one that achieves a balance among economic growth, the environmental impacts of that growth, and the fairness, or social equity, of the distribution of the costs and benefits of that growth. The importance of sustainability is cogently illustrated by the concept of an **ecological footprint**, which is a measure of the human pressures on the natural environment from the consumption of renewable resources and the production of pollution. It represents a quantitative assessment of the biologically productive area required to produce the resources (food, energy, and materials) and to absorb the wastes of an individual, city, region, or country. The ecological footprint of a country or region changes in proportion to population size, average consumption per person, and the resource intensity of the technology being used.

Humanity's footprint first grew larger than global biocapacity in the 1980s, and this overshoot has been increasing every year since. In 2006, demand exceeded supply by about 40 percent. This means that it took almost a year and five months for Earth to produce the ecological resources we used in that year. At 9.0 hectares (23.6 acres) per person, the United States currently has the sixth-largest per capita ecological footprint on the planet, just behind the United Arab Emirates, Qatar, Bahrain, Denmark, and Belgium. Other countries with extremely large ecological footprints include Australia, Canada, the Netherlands, Finland, and Sweden. Countries with the smallest ecological footprints, between 0.5 and 0.75 hectares (1.2 to 1.9 acres) per person, include Afghanistan, Bangladesh, Haiti, and Malawi.

Sustainable development means using renewable natural resources in a manner that does not eliminate or degrade them—by making greater use, for example, of solar and geothermal energy and recycled materials. It means managing economic systems so that all resources—physical and human—are used optimally. It means regulating economic systems so that the benefits of development are distributed more equitably (if only to prevent poverty from causing environmental degradation). It also means organizing societies so that improved education, health care, and social welfare can contribute to environmental awareness and sensitivity and an improved quality of life. A final and more radical aspect of sustainable development involves moving away from wholesale globalization toward increased "localization": a return to more locally based economies where production, consumption, and decision making are oriented to local needs and conditions (**Figure 7.6**).

Defined this way, sustainable development sounds eminently sensible yet impossibly utopian. A succession of international summit meetings on the topic has revealed deep conflicts of interest between core countries and peripheral countries. One of the most serious obstacles to prospects for sustainable development is continued heavy reliance on fossil fuels as the fundamental source of energy for economic development. This not only perpetuates international inequalities but also leads to transnational problems such

FIGURE 7.6 Promoting local economies Increasing awareness of the benefits of locally produced foods has encouraged many supermarkets, like this one in Lugano, Switzerland, to feature local products.

as acid rain, global warming, climatic changes, deforestation, health hazards, and, many would argue, war. The sustainable alternative—renewable energy generated from the sun, tides, waves, winds, rivers, and geothermal features—has been pursued half-heartedly because of the commercial interests of the powerful corporations and governments that control fossil-fuel resources.

A second important challenge to the possibility of sustainable development is the rate of demographic growth in peripheral countries. Sustainable development is feasible only if population size and growth are in harmony with the changing productive capacity of the ecosystem. It is estimated that 1.2 billion of the world's 6.9 billion people are undernourished and underweight.

But the greatest single obstacle to sustainable development is the inadequacy of institutional frameworks. Sustainable development requires economic, financial, and fiscal decisions to be fully integrated with environmental and ecological decisions. National and local governments everywhere have evolved institutional structures that tend to separate decisions about what is economically rational and what is environmentally desirable. International organizations, while better placed to integrate policy across these sectors and better able to address economic and environmental "spillovers" from one country to another, have (with the notable exception of the European Union) not acquired sufficient power to promote integrated, harmonized policies. Without radical and widespread changes in value systems and unprecedented changes in political will, "sustainable development" is likely to remain an embarrassing contradiction in terms.

> **APPLY YOUR KNOWLEDGE**   Use the Internet to find three different examples of renewable energy projects in the United States. List three ways they can change the nation's ecological footprint. Also consider a reason that their growth may be impeded. (*Hint:* You might want to consider what their funding source is and whether it is adequate.) ■

THE ECONOMIC STRUCTURE OF COUNTRIES AND REGIONS

The relative share of primary, secondary, tertiary, and quaternary economic activities determines the *economic structure* of a country or region. **Primary activities** are those concerned directly with natural resources of any kind; they include agriculture, mining, fishing, and forestry. **Secondary activities** are those that process, transform, fabricate, or assemble the raw materials derived from primary activities or that reassemble, refinish, or package manufactured goods. Secondary activities include steelmaking, food processing, furniture production, textile manufacturing, automobile assembly, and garment manufacturing. **Tertiary activities** are those involving the sale and exchange of goods and services; they include warehousing, retail stores, personal services such as hairdressing, and commercial services such as accounting, advertising, and entertainment. **Quaternary activities** are those dealing with the handling and processing of knowledge and information. Examples include data processing, information retrieval, education, and research and development (R&D).

Geographical Divisions of Labor

Variations in economic structure—according to primary, secondary, tertiary, or quaternary activities—reflect *geographical divisions of labor*. Geographical divisions of labor are national, regional, and locally based economic specializations that have evolved with the growth of the world-system of trade and politics and with the locational needs of successive technology systems. They represent one of the most important dimensions of economic development. For instance, countries whose economies are dominated by primary-sector activities tend to have a relatively low per capita GDP. The exceptions are oil-rich countries such as Saudi Arabia, Qatar, and Venezuela. Where the **international division of labor** (the specialization, by countries, in particular products for export) has produced national economies with a large secondary sector, per capita GDP is much higher (as, for example, in Argentina and South Korea). The highest levels of per capita GDP, however, are associated with economies that are *postindustrial*: economies where the tertiary and quaternary

sectors have grown to dominate the workforce, with smaller but highly productive secondary sectors.

As **Figure 7.7** shows, the economic structure of much of the world is dominated by the primary sector. In much of Africa and Asia, between 50 and 75 percent of the labor force is engaged in primary-sector activities such as agriculture, mining, fishing, and forestry. In contrast, the primary sector of the world's core regions is typically small, occupying only 5 to 10 percent of the labor force.

The secondary sector is much larger in the core countries and in semiperipheral countries, where the world's specialized manufacturing regions are located. In 2010, core countries accounted for almost three-quarters of world manufacturing value added (MVA). MVA is the net output of secondary industries; it is determined by adding up the value of all outputs and subtracting the value of all intermediate inputs. This share has been slowly decreasing, however. The core countries had an average annual growth rate for MVA of around 2 percent during 1990–2010, while the growth rate in the rest of the world was closer to 7 percent.

This growth has been concentrated in semiperipheral, **newly industrializing countries** (NICs). Newly industrializing countries are countries, formerly peripheral within the world-system, that have acquired a significant industrial sector, usually through **foreign direct investment**. Of the 20 biggest manufacturing countries in 2010, 7 were NICs: China, South Korea, Mexico, Brazil, India, Indonesia, and Thailand (listed here in order of importance). Indeed, China is now the world's second-largest exporter of manufactured goods; India and Brazil ranked ninth and tenth, respectively, in 2010. The vast majority of peripheral countries have a very small manufacturing output. For example, the share of world MVA for Africa has changed little over the last two decades, remaining at about 1 percent.

In terms of individual countries, the United States remains the most important source of manufactured goods, accounting for just over 22 per cent of global MVA in 2010. Just five countries—the United States, Japan, Germany, China, and the United Kingdom—together produced over 60 percent of the world total MVA. Another important aspect of secondary activities concerns *productivity*. In general, the highly capitalized manufacturing industries of the developed countries have been able to maintain high levels of worker productivity, with the result that the contribution of manufacturing to their GDP has remained relatively high even as the size of their manufacturing labor forces has decreased.

Within the framework of this continuing dominance of the advanced industrial economies there are several important trends. Although the United States has retained its leadership as the world's major producer of manufactured goods, its dominance has been significantly reduced. In the 1960s, its share of world manufacturing output was 40 percent, compared to its current share of around 22 percent. Meanwhile, Japan increased its share from less than 6 percent in the 1960s to about 18 percent by 2010. The emergence of a dozen or so NICs as settings for manufacturing is particularly striking. Several of these are in Latin America (Brazil, Mexico, and Argentina), but when it comes to the rate of manufacturing growth, the Asian NICs are most impressive.

China has experienced a dramatic increase in manufacturing production, achieving annual average growth rates during the

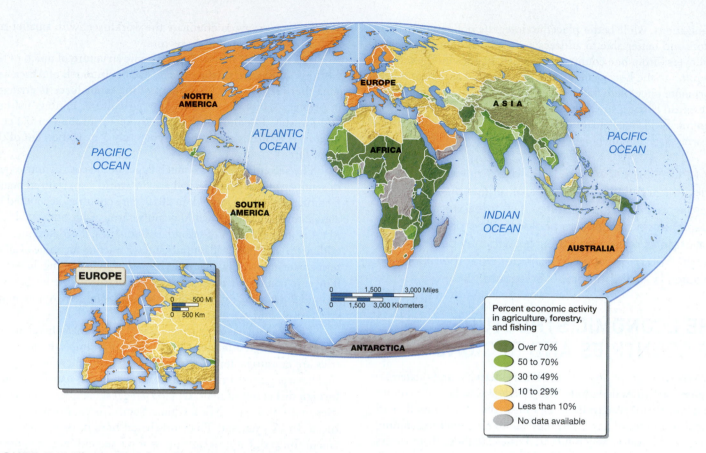

FIGURE 7.7 **The geography of primary economic activities** Primary economic activities are those that are concerned directly with natural resources of any kind. They include agriculture, mining, forestry, and fishing. The vast majority of the world's population, concentrated in China, India, Southeast Asia, and Africa, is engaged in primary economic activities. This map shows the percentage of the labor force in each country that was engaged in primary employment in 2002. In some countries, including China, primary activities account for more than 70 percent of the workforce. In contrast, primary activities always account for less than 10 percent of the labor force in the world's core countries, and often for less than 5 percent.

1970s, 1980s, 1990s, and 2000s of about 8 percent, 11 percent, 14 percent, and 9 percent, respectively (see Box 7.2, "Window on the World: China's Economic Development"). Of the four Asian "Tigers"—South Korea, Hong Kong, Taiwan, and Singapore— South Korea enjoyed the most spectacular increase in manufacturing production (**Figure 7.8**), achieving annual average growth rates of almost 18 percent in the 1960s, 17 percent in the 1970s, 12 percent in the 1980s, and over 7 percent in the 1990s and around 6 percent in the 2000s. More recently, other Pacific Rim NICs, such as Malaysia and Thailand, have experienced rapid growth in manufacturing production.

These shifts are part of a globalization of economic activity that has emerged as the overarching component of the world's economic geography. As we shall see, it has been corporate strategy, particularly the strategies of large transnational corporations (TNCs), that has created this globalization of economic activity. Transnational corporations are companies that participate not only in international trade but also in production, manufacturing, and/or sales operations in several countries.

The tertiary and quaternary sectors are significant only in the most affluent countries of the core. In the United States, for example, the primary sector in 2010 accounted for less than 2 percent of the labor force, the secondary sector for about 22 percent, the

tertiary sector for just over 50 percent, and the quaternary sector for 25 percent. In every core country, the tertiary sector has grown significantly in the past several decades as consumption and marketing became the hallmarks of postindustrial economies. More recently, globalization has meant that knowledge-based activities have become a critical aspect of economic development, resulting in the rapid growth of quaternary industries.

For the world's core economies, knowledge has become more important than physical and human resources in determining levels of economic well-being. More than half of the GDP of major core countries is based on the production and distribution of knowledge. In the United States, more workers are engaged in producing and distributing knowledge than in making physical goods.

For the world's peripheral economies, lack of knowledge— along with a limited capacity to absorb and communicate knowledge—is an increasingly important barrier to economic development. Poor countries have fewer resources to devote to research, development, and the acquisition of information technology. They also have fewer institutions for providing high-quality education, fewer bodies that can enforce standards and performance, and only weakly developed organizations for gathering and disseminating the information needed for business transactions. As a result, economic productivity tends to fall relative to the performance of places and

FIGURE 7.8 Manufacturing in South Korea Employees work on an assembly line in a Samsung semiconductor plant in Suwon, south of Seoul.

regions in core economies, where new knowledge is constantly generated and rapidly and effectively disseminated.

APPLY YOUR KNOWLEDGE Consider a product that you own, such as a T-shirt, a pair of shoes, or your smart phone or mp3 player, and map out the product's development through the primary, secondary, and tertiary activities. ■

International Trade

The geographical division of labor on a world scale means that the geography of international trade is very complex. One significant reflection of the increased economic integration of the world-system is that global trade has grown much more rapidly over the past few decades than global production. Between 1985 and 2008, the average annual growth rate of the value of world exports was twice that of the growth of world production and several times greater than that of world population growth.

The fundamental structure of international trade is based on a few **trading blocs**—groups of countries with formalized systems of trading agreements. Most of the world's trade takes place within four trading blocs:

- Western Europe, together with some former European colonies in Africa, South Asia, the Caribbean, and Australasia;
- North America, together with some Latin American states;
- the countries of the former Soviet world-empire; and
- Japan, together with other East Asian states and the oil-exporting states of Saudi Arabia and Bahrain.

Nevertheless, a significant number of countries exhibit a high degree of **autarky** from the world economy. That is, they do not contribute significantly to the flows of imports and exports that constitute the geography of trade. Typically, these are smaller, peripheral countries, such as Bolivia, Burkina Faso, Ghana, Malawi, Samoa, and Tanzania.

Patterns of world trade have been shifting rapidly, however, in response to several factors. In general, the trend has been toward an intensification of the long-standing domination of trade within and between core regions at the expense of trade between core countries and peripheral countries—with the major exception of trade in oil. Innovations in transport, communications, and manufacturing technology have diminished the importance of the distance.

Shifts in global politics have also affected the geography of trade. One important shift was the breakup of the former Soviet Union. Other significant geopolitical changes include the trend toward the political as well as economic integration of Europe and the increasing participation of China in the world economy. But perhaps the most important shift in global politics in relation to world trade has been the shift toward open markets and free trade through neoliberal policies propagated by core countries (**Figure 7.9**). **Neoliberal policies** are economic policies predicated on a minimalist role for the state that assume the desirability of free markets not only for economic organization but also for political and social life.

The globalization of economic activity has created new flows of materials, components, information, and finished products. As a global system of manufacturing has emerged, significant quantities of manufactured goods are now imported *and* exported across much of the world through complex commodity chains (as we saw in the case of blue jeans in Box 2.2. "Visualizing Geography: Commodity Chains"); no longer do developed economies export manufactures and peripheral countries import them. African countries are an important exception, with many of them barely participating in world trade in manufactures.

The most striking aspect of contemporary patterns of trade is the persistence of the dependence of peripheral countries on trade with core countries that are geographically or geopolitically close. For example, the United States is the central focus for the exports and the origin of the bulk of the imports of most Central American countries, while France is the focus for commodity flows to and from French ex-colonies such as Algeria, Cambodia, Benin, and the Ivory Coast. These flows, however, represent only part

China's Economic Development

Under the leadership of Deng Xiaoping (1978–1997), China embarked on a thorough reorientation of its economy, dismantling Communist-style central planning in favor of private entrepreneurship and market mechanisms and integrating itself into the world economy. Saying that he did not care whether the cat was black or white as long as it caught mice, Deng Xiaoping established a program of "Four Modernizations" (industry, agriculture, science, and defense) and an "open-door policy" that allowed China to be plugged in to the interdependent circuits of the global economy. As a result, China has completely reorganized and revitalized its economy. Agriculture has been decollectivized, with Communist collective farms modified to allow a degree of private profit-taking. State-owned industries have been closed or privatized, and centralized state planning has been dismantled in order to foster private entrepreneurship.

In the 1980s and early 1990s, when the world economy was sluggish, China's manufacturing sector grew by almost 15 percent each year. For example, almost all of the shoes once made in South Korea or Taiwan are now made in China. More than 60 percent of the toys in the world, accounting for more than $10 billion in trade, are also made in China. Since 1992, China has extended its open-door policy, permitted foreign investment aimed at Chinese domestic markets, and normalized trading relationships with the United States and the European Union. In 2001, China was admitted to the World Trade Organization, allowing China to trade more freely than ever before with the rest of the world.

China's increased participation in world trade has created an entirely new situation within the world economy. Chinese manufacturers, operating with low wages, have imposed a deflationary trend on world prices for manufacturers. The Chinese economy's size makes it a major producer, and its labor costs stay flat year after year because there is an endless supply of people who will work for 60 cents an hour. Meanwhile, the rapid expansion of consumer demand in China has begun to drive up commodity prices in the world market. Overall, China's economy is already the third largest in the world after those of the United States and Japan.

Nowhere have China's "open-door" policies had more impact than in South China, where the Chinese government has deliberately built upon the prosperity of Hong Kong, the former British colony that was returned to China in 1997. The coastline of South China provides many protected bays suitable for harbors and a series of ports, including Quanzhou, Shantou, Xiamen, and, on either side of the mouth of the Zhu Jiang (Pearl River), Macão and Hong Kong. These ports made possible South China's emergence as a core manufacturing region by providing an interface with the world economy. The established trade and manufacturing of Macão—a Portuguese colony that was returned to China in 1999—and Hong Kong provided another precondition for success.

When Deng Xiaoping established his "open-door" policy, a third factor kicked in: capital investment from Hong Kong, Taiwan, and the Chinese diaspora. By 1993, more than 15,000 manufacturers from Hong Kong alone had set up businesses in neighboring Guangdong Province, and a similar number established subcontracting relationships, contracting out processing work to Chinese companies. Today, the cities and special economic zones of South China's "Gold Coast" provide a thriving export-processing platform that has driven double-digit annual economic growth for much of the past two decades. The population of Shenzhen (**Figure 7.B**)

FIGURE 7.B Shenzhen The city of Shenzhen, just across the border from the Special Administrative Region of Hong Kong.

FIGURE 7.C New affluence A saleswoman waits for customers at an outlet of the French fashion brand Hermes in Shanghai. China's economic boom has led to a rapid increase in the size of the country's middle class, up nearly 25 percent since 2008, from 65 to 80 million people.

FIGURE 7.D Real estate boom In the rapidly growing regions of coastal China, house price inflation has risen as high as 9.5 percent a month. Shown here is a small fishing village near Sanya Harbor in Hainan province, where new luxury buildings are displacing the older homes and their inhabitants.

has grown from just 19,000 in 1975 to 8.5 million in 2007, with an additional 2 million in the surrounding municipalities. Such growth has generated a substantial middle class with significant spending power (Figure 7.C); it has also created significant inflation, especially in real estate values (Figure 7.D).

Much of China's manufacturing growth has been based on a strategy of import substitution (see page 230). In spite of China's membership in the World Trade Organization (which has strict rules about intellectual property), a significant amount of China's industry is based on counterfeiting and reverse engineering (making products that are copied and then sold under different or altered brand names) and piracy (making look-alike products passed off as the real thing). Copies of everything from DVDs, movies, designer clothes and footwear, drugs, motorcycles, and automobiles to high-speed magnetic levitation (maglev) cross-country trains save Chinese industry enormous sums in research and development and licensing fees, while saving the country even greater sums in imports.

Foreign investors, meanwhile, have been keen to develop a share of China's rapidly expanding and increasingly affluent market. The automobile market is particularly attractive to Western manufacturers. Volkswagen was the first to establish a presence in China, in 1985. By 2003, Volkswagen had claimed around 40 percent of China's annual production of almost 4 million cars and light trucks. General Motors, in partnership with Shanghai Automotive Industry Corporation, has a 10 percent market share. Other foreign

manufacturers operating in China include Honda, Toyota, Nissan, and, most recently, BMW and Mercedes.

Overall, most foreign investment in China comes from elsewhere within East Asia. Japan, Taiwan, and South Korea, having developed manufacturing industries that undercut those of the United States, now face deindustrialization themselves through the inexorable process of "creative destruction" (see page 237). More than 10,000 Taiwanese firms have established operations in China, investing an estimated $150 billion. Pusan, the center of the South Korean footwear industry that in 1990 exported $4.3 billion worth of shoes, is full of deserted factories. South Korean footwear exports are down to less than $700 million, while China's footwear exports have increased from $2.1 billion in 1990 to $29 billion in 2010. Several Japanese electronics giants, including Toshiba Corp., Sony Corp., Matsushita Electric Industrial Co., and Canon, Inc., have expanded operations in China even as they have shed tens of thousands of workers at home. Olympus manufactures its digital cameras in Shenzhen and Guangzhou. Pioneer has moved its manufacture of DVD recorders to Shanghai and Dongguan.

Toshiba's factory in Dalian illustrates the logic. Toshiba is one of about 40 Japanese companies that built large-scale production facilities in a special export-processing zone established by Dalian in the early 1990s with generous financial support from the Japanese government and major Japanese firms. By shifting production of digital televisions from its plant in Saitama, Japan, in 2001, Toshiba cut labor costs per worker by 90 percent.

FIGURE 7.9 World Economic Forum
China's Commerce minister Chen Deming looks on prior to an informal meeting of Ministers from countries in the World Trade Organization during a meeting in Davos, Switzerland, in 2011.

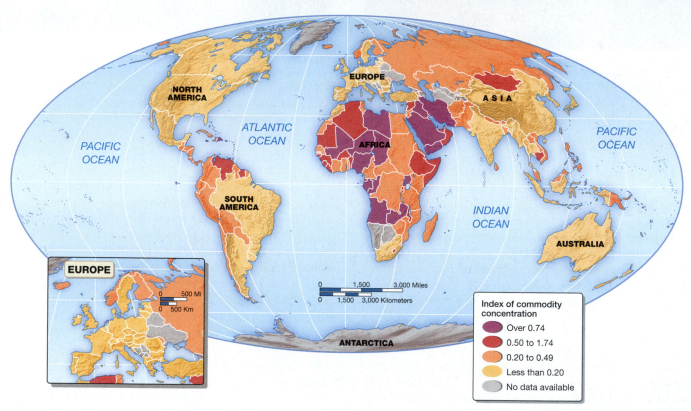

FIGURE 7.10 Index of commodity concentration of exports, 2002.

of the action for the core economies, whose trading patterns are dominated by flows to and from other core countries.

One implication of this situation is that the smaller, peripheral partners in these trading relationships are highly dependent on developed economies. An aspect of dependency, in this context, is the degree to which a country's export base lacks diversity. **Dependency** involves a high level of reliance by a country on foreign enterprises, investment, or technology. External dependence for a country means that it is highly reliant on levels of demand and the overall economic climate of other countries. Dependency for a peripheral country can result in a narrow economic base in which

the balancing of national accounts and the generation of foreign exchange depend on the export of one or two agricultural or mineral resources.

Figure 7.10 shows one reflection of dependency: the index of commodity concentration of exports. Countries with low scores on this index have diversified export bases. They include Argentina, Brazil, China, India, and North and South Korea, as well as most of the core countries. At the other extreme are peripheral countries where the manufacturing sector is poorly developed and the balancing of national accounts and the generation of foreign exchange depend on the export of one or two agricultural or mineral resources, as is the case in Angola, Chad, the Dominican Republic, Iran, Iraq, Libya, and Nigeria, for example.

Patterns of International Debt

In many peripheral countries, debt service—the annual interest on international debts—is a significant handicap to economic development. For every $1 that developing countries receive in aid, they end up having to return $5 in debt service payments to core countries. In many countries, 20 percent or more of all export earnings are swallowed up by debt service (**Figure 7.11**).

At the root of the international debt problem is the structured inequality of the world economy. The role inherited by most peripheral countries within the international division of labor has been one of producing primary goods and commodities for which both the elasticity of demand and price elasticity are low. The **elasticity of demand** is the degree to which levels of demand for a product or service change in response to changes in price. Where a relatively small change in price induces a significant change in demand, elasticity is high; where levels of demand remain fairly stable in spite of price changes, demand is said to be inelastic. Demand for the products of peripheral countries in their principal markets (the more developed countries) has a low elasticity: It tends to increase by relatively small amounts in response to significant increases in the incomes of their customers. Similarly, significant reductions in the price of their products tend to result in only a relatively small increase in demand.

Consider, for example, the cocoa-producing regions of West Africa (**Figure 7.12**). No matter how they improve productivity in order to keep prices low, and no matter how much more affluent their customers in core countries become, there is a limit to the demand for cocoa products. In contrast, the elasticity of demand and price elasticity of high-tech manufactured goods and high-order services (the specialties of core economies within the international division of labor) are both high. As a result, the **terms of trade** are stacked against the producers of primary goods.

The terms of trade are determined by the ratio of the prices at which exports and imports are exchanged. When the price of exports rises relative to the price of imports, the terms of trade reflect

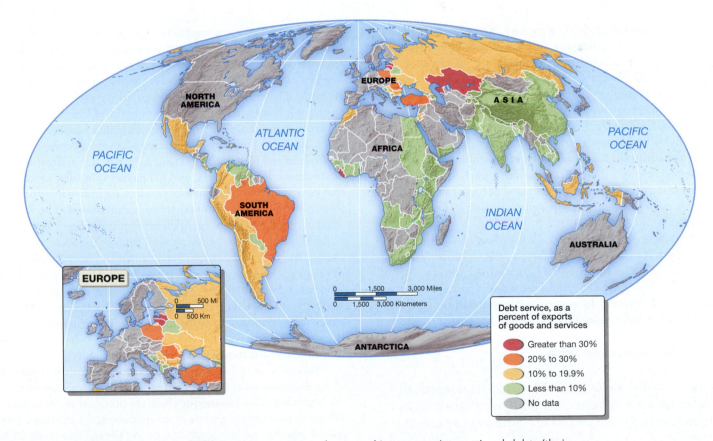

FIGURE 7.11 The debt crisis in 2008 In some countries, the annual interest on international debts (their "debt service") accounts for more than 20 percent of the annual value of their exports of goods and services. Many countries first got into debt trouble in the mid-1970s, when Western banks, faced with recession at home, offered low-interest loans to the governments of peripheral countries rather than being stuck with idle capital. When the world economy heated up again, interest rates rose and many countries found themselves facing a debt crisis. The World Bank and the International Monetary Fund (IMF), in tandem with Western governments, worked to prevent a global financial crisis by organizing and guaranteeing programs that eased poor countries' debt burdens. Western banks were encouraged to swap debt for equity stakes in nationalized industries, while debtor governments were persuaded to impose austere economic policies. These policies have helped ease the debt crisis, but often at the expense of severe hardship for ordinary people. In dark humor, among radical development theorists IMF came to stand for "imposing misery and famine." (*Source:* World Bank, Global Development Finance.)

FIGURE 7.12 Cocoa production Workers spread cocoa to dry in a Ghana hamlet.

an improvement for the exporting country. No matter how efficient primary producers may become, or how affluent their customers, the balance of trade is tilted against them. Quite simply, they must run in order to stand still.

An obvious counterstrategy for peripheral countries is to attempt to establish a new role in the international division of labor, moving away from a specialization in primary commodities toward a more diversified manufacturing base. This strategy is known as **import substitution**. It is a difficult strategy to pursue, however, because building up a diversified manufacturing base requires vast amounts of start-up capital. With the terms of trade running against them, it is extremely difficult for peripheral countries to accumulate this capital; so they have to borrow.

The debt problem has led to calls for affluent lending countries to provide debt relief to some of the poorest countries. In 2005, the world's richest countries—the G8 group—agreed to write off $40 billion in debts owed by 18 of the world's poorest countries, most of them in Africa. Addressing the full magnitude of LDC debt, however, will require continued substantial and sustained efforts if the poorest countries are to break free of the crushing financial obligations of their accumulated debts.

APPLY YOUR KNOWLEDGE Conduct research to determine the current debt of three different countries. List two main factors contributing to each of the specific countries' debt. ■

Fair Trade

The Fair Trade movement highlights the interdependencies involved in international trade. The movement is a result of increasing awareness within developed countries of the weak bargaining position of many small producers at the beginning of the commodity chains that underpin the global economy. Fair Trade has become part of the "mobilization against globalization," an attempt to raise consumers' consciousness about the relationships embodied in their purchases.

The Fair Trade movement is a global network of producers, traders, marketers, advocates, and consumers focused on building equitable trading relationships between consumers and the world's most economically disadvantaged artisans and farmers. As such, it is fundamentally a strategy for poverty alleviation and sustainable development. The key principles of Fair Trade include (1) creating opportunities for economically disadvantaged producers, (2) capacity building, (3) ensuring that women's work is properly valued and rewarded, (4) ensuring a safe and healthy working environment for producers, and (5) payment of a fair price—one that covers not only the costs of production but enables production that is socially just and environmentally sound. The result is often very modest in relation to the retail price of Fair Trade goods in developed countries—only about 9 cents of the $3.10 cost of an average fairly traded 100-gram bar of chocolate goes to producers in developing countries, the rest being accounted for by core-country food processors, designers, packagers, photographers, marketing staff, advertisers, shopkeepers, and tax authorities. Nevertheless, Fair Trade demonstrably helps producers in developing countries (**Figure 7.13**) and typically involves democratic decision making over how the extra money earned is to be distributed.

Several types of Fair Trade organizations perform various roles along the commodity chains linking producers to consumers. At the beginning of the commodity chain are producer organizations—village or community groups or cooperatives, for example, often joined together under export marketing organizations. In 2009, there were more than 750 Fair Trade–certified producer groups (including many umbrella bodies) in 58 countries. These organizations typically sell their products to a second kind of Fair Trade organization: registered importers and wholesalers in more developed countries. In 2009, these existed in about 60 countries. They in turn sell to Fair Trade retailers: "world shops" and catalog- or Internet-based retailers (**Figure 7.14**). In some countries, Fair Trade products are now mainstream, available in major supermarkets and independent shops and beginning to gain market share. In Switzerland, for example, Fair Trade bananas account for

20 percent of the retail market. In the United States, it is coffee that is the most important certified Fair Trade product, accounting for up to $600 million in sales in 2009.

There are also Fair Trade organizations that are focused on labor practices, such as the Ethical Trading Initiative (ETI), which evolved from Fair Trade campaigns run by British aid organizations. ETI involves a multiagency grouping of companies, nongovernmental organizations (NGOs), and trades unions that have together devised a basic code of labor practices covering the right to collective bargaining, safe and hygienic working conditions, living wages, and a standard working week of no more than 48 hours.

Interpretations of International Patterns of Development

The overall relationship between economic structure and levels of prosperity makes it tempting to interpret economic development in distinctive stages. Each developed region or country, in other words, might be thought of as progressing from the early stages of development, with a heavy reliance on primary activities (and relatively low levels of prosperity), through a phase of industrialization and on to a "mature" stage of postindustrial development (with a diversified economic structure and relatively high levels of prosperity). This, in fact, has been a commonly held view of economic development, first conceptualized by a prominent economist, W. W. Rostow (**Figure 7.15**).

Rostow's model, however, is too simplistic to be of much help in understanding human geography. The reality is that places and regions are now interdependent. The fortunes of any given place are increasingly tied up with those of many others. Furthermore, Rostow's model perpetuates the myth of "developmentalism," the idea that every country and region will eventually make economic progress toward "high mass consumption" provided that they compete to the best of their ability within the world economy. The main weakness of developmentalism is that it is simply not reasonable to compare the prospects of late starters to the experience of those places, regions, and countries that were among the early starters. For these early starters the horizons were clear: free of effective

FIGURE 7.13 Fair Trade coffee Susan Nangobi, aged 14, works with her father, a coffee farmer in the Kamuli region of Uganda. Their freshly picked coffee beans are ready to be dried before being taken to the Kulika Sustainable Organic Agricultural Training Program center and then sold to Ibero Coffee company under Fair Trade guidelines.

FIGURE 7.14 Fair Trade retailing A Fair Trade shop in Canterbury, England.

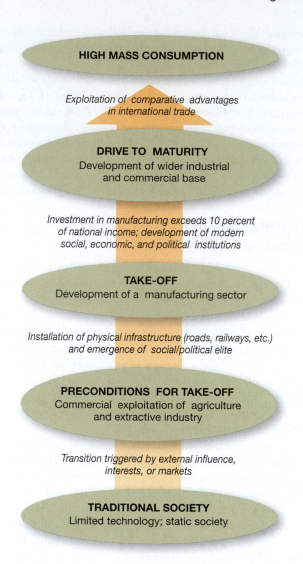

HIGH MASS CONSUMPTION

Exploitation of comparative advantages in international trade

DRIVE TO MATURITY
Development of wider industrial and commercial base

Investment in manufacturing exceeds 10 percent of national income; development of modern social, economic, and political institutions

TAKE-OFF
Development of a manufacturing sector

Installation of physical infrastructure (roads, railways, etc.) and emergence of social/political elite

PRECONDITIONS FOR TAKE-OFF
Commercial exploitation of agriculture and extractive industry

Transition triggered by external influence, interests, or markets

TRADITIONAL SOCIETY
Limited technology; static society

FIGURE 7.15 Stages of economic development This diagram illustrates a model based on the idea of successive stages of economic development. Each stage is seen as leading to the next, though some regions or countries may take longer than others to make the transition from one stage to the next. According to this view, now regarded as overly simplistic, places and regions follow parallel courses within a world that is steadily modernizing. Late starters eventually make progress, but at speeds determined by their resource endowments, their productivity, and the wisdom of their people's policies and decisions.

competition, free of obstacles, and free of precedents. For the late starters the situation is entirely different. Today's less developed regions must compete in a crowded field while facing numerous barriers that are a direct consequence of the success of some of the early starters.

Indeed, many writers and theorists of international development claim that the prosperity of the core countries in the world economy has been based on *under*development and squalor in peripheral countries. Peripheral countries, it is argued, could not "follow" the previous historical experience of developed countries in stages-of-development fashion because their underdevelopment (that is, exploitation) was a structural requirement for development

elsewhere. The development of Europe and North America, in other words, required the systematic underdevelopment of peripheral countries. By means of unequal trade, exploitation of labor, and profit extraction, the underdeveloped countries became increasingly rather than decreasingly impoverished.

The writing of economic historian and sociologist André Gunder Frank exemplifies the explanations of international economic change that arose from this critique. Frank rejected the idea that underdevelopment is an original condition, equivalent to "traditionalism" or "backwardness." To the contrary, he argues, it is a condition created by integration into the worldwide system of capitalism. The world economy, Frank argues, has been unequally structured since Europeans first ventured out into the world in the sixteenth century. Although the form of the dominance of core over periphery has changed from colonialism and imperialism to neocolonialism, an overall transfer of wealth from periphery to core continues to fuel growth in some places at the expense of others.

Frank's approach is an example of "dependency theory." This has been a very influential approach in explaining global patterns of development and underdevelopment. Dependency theory states, essentially, that development and underdevelopment are reverse sides of the same process: *Development somewhere requires underdevelopment somewhere else.* Immanuel Wallerstein's world-system theory (see Chapter 2) takes this kind of dependency into account. According to this perspective, the entire world economy is to be seen as an evolving market system with an economic hierarchy of states—a core, a semiperiphery, and a periphery. The composition of this hierarchy is variable: Individual countries can move from periphery to semiperiphery, core to periphery, and so on.

REGIONAL ECONOMIC DEVELOPMENT

Unevenness in economic development often has a regional dimension. Initial conditions are a crucial determinant of regional economic performance. Scarce resources, a history of neglect, lack of investment, and concentrations of low-skilled people all combine to explain the lagging performance of certain areas. In some regions, initial extreme disadvantages constrain the opportunities of individuals born there. A child born in the Mexican state of Chiapas, for example, has much bleaker prospects than a child born in Mexico City. The child from Chiapas is twice as likely to die before age 5, less than half as likely to complete primary school, and 10 times as likely to live in a house without access to running water. On reaching working age, he or she will earn 20 to 35 percent less than a comparable worker living in Mexico City and 40 to 45 percent less than one living in northern Mexico.

Other examples of regional inequality can be found throughout the world. Gansu, China, with an income per capita 40 percent below the national average, is one of the poorest and most remote regions in the nation. With poor soils highly susceptible to erosion, low and erratic rainfall, and few off-farm employment opportunities, a high proportion of its inhabitants live in poverty. Chaco Province, in Argentina, has a GDP per capita that is only 38 percent of the national average. Low educational attainment and lack of infrastructure, especially roads, explain much of this deviation.

Even when growth in some parts of a country's economy is strong, relative and absolute regional poverty can persist for long periods. In the seventeenth century, for example, the weakening of the sugar economy pushed northeastern Brazil into a decline from which it has never fully recovered. In the United States, the fortunes of coal-mining West Virginia waned with the collapse of the coal industry and the increased importance of oil and gas in energy production. It remains one of the poorest regions in the United States to this day. In Thailand, rapid development has failed to reach the northern hill people. Fewer than 30 percent of their villages have schools, and only 15 percent of the hill people can read and write. Their average annual income is less than a quarter of Thailand's overall GNI per capita.

Globalization has been associated with increasing regional inequality within many countries since the 1980s. In China, for example, disparities have widened dramatically between the interior and the export-oriented regions of the coast. The transition economies of the countries of the former Soviet Union and its Eastern European satellites have registered some of the largest increases in regional inequality, and some core countries—especially Sweden, the United Kingdom, and the United States—have also registered significant increases in regional inequality since the 1980s. At this regional scale, as at the global scale, levels of economic development often exhibit a fundamental core-periphery structure. Indeed, within-country core-periphery contrasts are evident throughout the world: in core countries such as France and the United States, in semiperipheral countries such as South Korea, and in peripheral countries such as Nigeria and Indonesia.

Patterns of regional economic development are historical in origin and cumulative in nature. Recognizing this, geographers are interested in **geographical path dependence**, the relationship between present-day activities in a place and the past experiences of that place. When spatial relationships and regional patterns emerge through the logic of fundamental principles of spatial organization, they do so in ways guided and influenced by preexisting patterns and relationships.

These observations lead to an important principle of regional economic development, the principle of initial advantage. **Initial advantage** highlights the importance of an early start in economic development. It represents a special case of external economies. Other things being equal, new phases of economic development take hold first in settings that offer **external economies**: existing labor markets, existing consumer markets, existing frameworks of fixed social capital, and so on. External economies are cost savings that result from advantages beyond a firm's organization and methods of production. These initial advantages are consolidated by **localization economies**—cost savings that accrue to particular industries as a result of clustering together at a specific location—and so form the basis for continuing economic growth.

Examples of localization economies include sharing a pool of labor with special skills or experience, supporting specialized technical schools, joining to create a marketing organization or a research institute, and drawing on specialized subcontractors, maintenance firms, suppliers, distribution agents, and lawyers. If such advantages lead to a reputation for high-quality production, localization is intensified because more producers want to cash in on the reputation. Among many examples are the electronics and software industries of Silicon Valley; recording companies in Los Angeles; Sheffield steel and steel products, especially cutlery, from England; the U.S. auto industry in Detroit; Swiss watches made in the towns of Biel, Geneva, and Neuchâtel; Belgian lace from Bruges, Brussels, and Mechelen; English worsted in Bradford and Huddersfield; Stoke-on-Trent English china; Irish linen from Athlone; and French perfume made in Grasse.

For places and regions with a substantial initial advantage, therefore, the trajectory of geographical path dependence tends to be one of persistent growth. This pattern reinforces, in turn, the core-periphery patterns of economic development found in every part of the world and at every spatial scale. That said, geographers recognize there is no single pathway to development. The consequences of initial advantage for both core and peripheral regions can be—and often are—modified. Old core-periphery relationships can be blurred, and new ones can be initiated.

APPLY YOUR KNOWLEDGE Identify an example of localized economics functioning in your own community. What industries are clustering together and how can this result in cost savings? ■

The Development of Regional Economic Cores

Regional cores of economic development are created cumulatively, following some initial advantage, through the operation of several of the basic principles of economic geography. Commercial and industrial location decisions all take place within complex webs of *functional interdependence*. These webs include the relationships and linkages between different kinds of industries, different kinds of stores, and different kinds of offices. Particularly important here are the **agglomeration effects** associated with various kinds of economic linkages and interdependencies. These interdependencies include the cost advantages that accrue to individual firms because of their location among functionally related activities. The trigger for these agglomeration effects can be any kind of economic development—the establishment of a trading port or the growth of a local industry or any large-scale enterprise. The external economies and economic linkages generated by such developments are the initial advantages that stimulate a self-propelling process of local economic development.

A number of interrelated effects come into play when new economic activity begins in an area. *Backward linkages* develop as new firms arrive to provide the growing industry with components, supplies, specialized services, or facilities. *Forward linkages* develop as new firms arrive to take the finished products of the growing industry and use them in their own processing, assembly, finishing, packaging, or distribution operations. Together with the initial growth, the growth in these linked industries helps to create a threshold of activity large enough to attract **ancillary industries** and activities (maintenance and repair, recycling, security, and business services, for example).

The existence of these interrelated activities establishes a pool of specialized labor with the kinds of skills and experience that make the area attractive to still more firms. Meanwhile, the linkages among all these firms help to promote interaction between professional and technical personnel and allow the area to support R&D (research and development) facilities, research institutes, and so on, thus increasing the likelihood of local inventions and innovations that might further stimulate local economic development.

Another aspect of local economic growth results from the increase in population represented by the families of employees. Their presence creates a demand for housing, utilities, physical infrastructure, retailing, personal services, and so on—all of which generate additional jobs. This expansion, in turn, helps to create populations large enough to attract an even wider variety and more sophisticated kinds of services and amenities. Last—but by no means least—the overall growth in local employment creates a larger local tax base. The local government can then provide improved public utilities, roads, schools, health services, recreational amenities, and so on—all of which serve to intensify agglomeration economies and so enhance the competitiveness of the area in attracting further rounds of investment.

Swedish economist Gunnar Myrdal, the 1974 Nobel Prize winner, was the first to recognize that any significant initial local advantage tends to be reinforced through geographic principles of agglomeration and localization. He called the process **cumulative causation (Figure 7.16)**. Cumulative causation refers to the spiraling buildup of advantages that occurs in specific geographic settings as a result of the development of external economies, agglomeration effects, and localization economies. Myrdal also pointed out that this spiral of local growth tends to attract people—enterprising young people, usually—and investment funds from other areas. According to the basic principles of spatial interaction, these flows tend to be strongest from nearby regions and areas with the lowest wages, fewest job opportunities, or least attractive investment opportunities.

In some cases this loss of entrepreneurial talent, labor, and investment capital is sufficient to trigger a cumulative negative spiral of economic disadvantage. With less capital, less innovative energy, and depleted pools of labor, industrial growth in peripheral

regions tends to be significantly slower and less innovative than in regions with an initial advantage. This in turn tends to limit the size of the local tax base, making it difficult for local governments to furnish a competitive **infrastructure** of roads, schools, and recreational amenities. Myrdal called these negative impacts on a region (or regions) of the economic growth of some other region **backwash effects**. Negative impacts take the form, for example, of out-migration, outflows of investment capital, and the shrinkage of local tax bases. Backwash effects are important because they help to explain why regional economic development is so uneven and why core-periphery contrasts in economic development are so common.

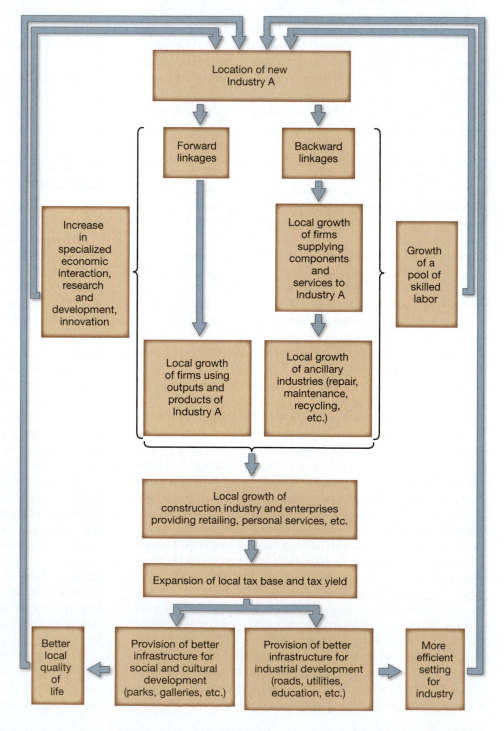

FIGURE 7.16 Processes of regional economic growth Once a significant amount of new industry becomes established in an area, it creates a self-propelling process of economic growth. As this diagram shows, the geographic principles of agglomeration and localization reinforce the initial advantages of industrial growth. The overall process is known as *cumulative causation*.

The Modification of Regional Core-Periphery Patterns

Although very important, cumulative causation and backwash effects are not the only processes affecting the geography of economic development. If they were, the world's economic geography would be even more starkly polarized than it is now. There would be little chance for the emergence of new growth regions, like Guangdong in Southeast China, and there would be little likelihood of stagnation or decline in once-booming regions, like northern England.

Myrdal himself recognized that peripheral regions do sometimes emerge as new growth regions and partially explained them in what he called *spread* (or trickle-down) effects. **Spread effects** are the positive impacts on a region (or regions) from the economic growth of some other region, usually a core region. Growth creates levels of demand for food, consumer products, and other manufactures that are so high that local producers cannot satisfy them. This demand gives investors in peripheral regions (or countries) the opportunity to establish a local capacity to meet the demand. Entrepreneurs who participate are also able to exploit the advantages of cheaper land and labor in peripheral regions. If strong enough, spread effects can enable peripheral regions to develop their own spiral of cumulative causation, thus changing the interregional geography of economic patterns and flows. The economic growth of South Korea, for example, is partly attributable to the spread effects of Japanese economic prosperity.

Another way in which peripheral regions can develop their own spiral of cumulative causation is through the process of *import substitution*. In this process, goods and services previously imported from core regions come to be replaced by locally made goods and locally provided services. Some things are hard to copy because of the limitations of climate or natural resources. However, many products and services *can* be copied by local entrepreneurs, thus capturing local capital, increasing local employment opportunities, intensifying the use of local resources, and generating profits for further local investment. The classic example is in Japan, where import substitution, especially for textiles and heavy engineering, played an important part in the transition from a peripheral economy to a major industrial power in the late nineteenth century. Import substitution also figured prominently in the Japanese "economic miracle" after World War II, featuring the automobile industry and consumer electronics. Today, countries like Brazil, Peru, and Ghana are seeking to follow the same sort of strategy, subsidizing domestic industries and protecting them from outside competitors through tariffs and taxes.

Core-periphery patterns and relationships can change as a result of internal changes in core regions that can slow or modify the spiral of cumulative causation. The main factor that can have this effect is the development of **agglomeration diseconomies**. Agglomeration diseconomies are the negative economic effects of urbanization and the local concentration of industry, including the higher prices that must be paid by firms competing for land and labor; the costs of delays resulting from traffic congestion and crowded port and railroad facilities; the increasing costs of solid waste disposal; and the burden of higher taxes that eventually have to be levied by local governments in order to support services and amenities previously considered unnecessary—traffic police, city planning, and transit systems, for example.

Diseconomies imposed through taxes can often be passed on by firms to consumers in other regions and other countries in the form of higher prices. Charging higher prices, however, decreases the competitiveness of a firm in relation to firms operating elsewhere. Agglomeration diseconomies that cannot be "exported"—noise, air pollution, increased commuting costs, and increased housing costs, for example—require local governments to tax even more of the region's wealth in attempts to compensate for a deteriorating quality of life. In California, this issue became so pressing in the 1990s that a report prepared jointly by the state, the Bank of America, the Greenbelt Alliance, and the Low Income Housing Fund concluded that "California businesses cannot compete globally when they are burdened with the costs of sprawl. An attractive business climate cannot be sustained if the quality of life continues to decline and the cost of financing real estate development escalates."[1] California's response has been to enact some of the nation's toughest traffic and environmental regulations.

Deindustrialization and Creative Destruction

The most fundamental cause of change in the relationship between initial advantage and cumulative causation is longer-term shifts in technology systems and the competition between states within the world-system. The innovations associated with successive technology systems generate new industries that are not yet tied down by enormous investments in factories or allied to existing industrial agglomerations. Combined with innovations in transport and communications, this creates *windows of locational opportunity* that can result in new industrial districts, with new generations of small towns or cities growing into dominant metropolitan areas through new rounds of cumulative causation.

Equally important as a factor in how core-periphery patterns change are the shifts in the profitability of old, established industries in core regions compared to the profitability of new industries in fast-growing new industrial districts. As soon as the differential is large enough, some disinvestment takes place within core regions. This disinvestment can take place in several ways. Manufacturers can reduce their wage bill by cutting back on production; they can reduce their fixed costs by closing down and selling off some of their factory space and equipment; or they can reduce their spending on research and development for new products. This disinvestment, in turn, leads to **deindustrialization** in formerly prosperous industrial core regions.

Deindustrialization involves a relative decline (and in extreme cases an absolute decline) in industrial employment in core regions as firms scale back their activities in response to lower levels of profitability (**Figure 7.17**). This is what happened to the Manufacturing Belt (sometimes called the "Rustbelt") of the United States in the 1960s

[1]*Beyond Sprawl*, Sacramento, CA: State of California-Resources Agency, 1995.

FIGURE 7.17 Regional economic decline
When the locational advantages of manufacturing regions are undermined for one reason or another, profitability declines and manufacturing employment falls. This can lead to a downward spiral of economic decline, as experienced by many of the traditional manufacturing regions of Europe and North America during the 1960s, 1970s, and 1980s. (Reprinted with permission of Prentice Hall, from P. L. Knox, *Urbanization*, © 1994, p. 55.)

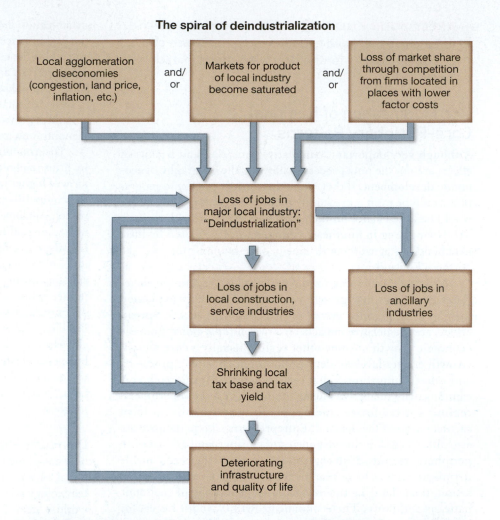

The spiral of deindustrialization

Local agglomeration diseconomies (congestion, land price, inflation, etc.)

and/ or

Markets for product of local industry become saturated

and/ or

Loss of market share through competition from firms located in places with lower factor costs

Loss of jobs in major local industry: "Deindustrialization"

Loss of jobs in local construction, service industries

Loss of jobs in ancillary industries

Shrinking local tax base and tax yield

Deteriorating infrastructure and quality of life

FIGURE 7.18 Deindustrialization This abandoned Packard automobile plant in Detroit, Michigan, is testament to the downward economic spiral in what was once one of the world's most important heavy manufacturing regions.

and 1970s (**Figure 7.18**). It also occurred in many of the traditional industrial regions of Europe during the 1960s, 1970s, and 1980s. In France, Belgium, the Netherlands, Norway, Sweden, and the United Kingdom, manufacturing employment decreased by between one-third and one-half from 1960 to 1990. The most pronounced example of this deindustrialization has been in the United Kingdom, where a sharp decline in manufacturing employment has been accompanied by an equally sharp rise in service employment.

Meanwhile, the capital made available from disinvestment in these core regions becomes available for investment by entrepreneurs in new ventures based on innovative products and production technologies. Old industries—and sometimes entire old industrial regions—have to be "dismantled" (or at least neglected) in order to help fund the creation of new centers of profitability and employment. This process is often referred to as creative destruction, something that is inherent to the dynamics of capitalism. **Creative destruction** involves the withdrawal of investments from activities (and regions) that yield low rates of profit in order to reinvest in new activities (and new regions). The concept of creative destruction provides a powerful image, helping us to understand the entrepreneur's need to withdraw investments from activities (and regions) yielding low rates of profit in order to reinvest in new activities (and, often, in new regions). In the United States, for example, the deindustrialization of the Manufacturing Belt provided the capital and the locational flexibility for firms to invest in the Sunbelt of the United States and in semiperipheral countries like Mexico and South Korea.

The process does not stop there, however. If the deindustrialization of the old core regions is severe enough, the relative cost of their land, labor, and infrastructure may decline to the point where they once again become attractive to investors. As a result, a seesaw movement of investment capital occurs, which over the long term tends to move from developed to less developed regions—then back again, once the formerly developed region has experienced a sufficient relative decline. "Has-been" regions can become redeveloped and revitalized, given a new lease on life by the infusion of new capital for new industries. This is what happened, for example, to the Pittsburgh region in the 1980s, resulting in the creation of a postindustrial economy out of a depressed industrial setting. USX, the U.S. Steel group of companies, reduced its workforce in the Pittsburgh region from more than 20,000 to less than 5,000 between 1975 and 1995. These losses have been more than made up, however, by new jobs generated in high-tech electronics, specialized engineering, and finance and business services.

Government Intervention

In addition to the processes of deindustrialization and creative destruction, core-periphery patterns can also be modified by government intervention. National governments realize that regional planning and policy can be an important component of broad economic strategies to stabilize and reorganize their economies, as well as to maximize their overall competitiveness. Without regional planning and policy, the resources of peripheral regions can remain underutilized, while core regions can become vulnerable to agglomeration diseconomies. For political reasons, too, national governments are often willing to help particular regions adjust to changing economic circumstances. At the same time, most local governments take responsibility for stimulating economic development within their jurisdiction, if only in order to increase the local tax base.

The nature and extent of government intervention has varied over time and by country. In some countries, special government agencies have been established to promote regional economic development and reduce core-periphery contrasts. Among the best-known examples are the Japanese MITI (Ministry of International Trade and Industry), the Italian Cassa del Mezzogiorno (Southern Development Agency, replaced in 1987 by several smaller agencies), and the U.S. Economic Development Administration. Some governments have sought to help industries in declining regions by undertaking government investment in infrastructure and providing subsidies for private investment; others have sought to devise tax breaks that reduce the cost of labor in peripheral regions. Still others have sought to deal with agglomeration diseconomies in core regions through increased taxes and restrictions on land use (**Figure 7.19**).

While each approach has its followers, one of the most widespread governmental approaches to core-periphery patterns involves the exploitation of the principle of cumulative causation through the creation of growth poles. **Growth poles** are places of economic activity deliberately organized around one or more high-growth industries. Economists have noted, however, that not all industries are equal in the extent to which they stimulate economic growth and cumulative causation. The ones that generate the most pronounced effects are known as "propulsive industries," and they have received a great deal of attention from geographers and economists who are interested in helping to shape strategic policies that might promote regional economic development. In the 1920s, shipbuilding was a propulsive industry. In the 1950s and 1960s, automobile manufacturing was a propulsive industry, and today biotechnology and digital technologies are propulsive industries. The basic idea is for governments to promote regional economic growth by fostering propulsive industries in favorable locations. These locations are intended to become growth poles—places that, given an artificial start, develop a self-sustaining spiral of economic prosperity.

Many countries have used the growth-poles approach as a basis for regional development policies. For example, French governments have designated certain locations as *technopoles*—sites for the establishment of high-tech industries (such as computers and biotechnology)—under the assumption that these leading-edge activities will stimulate further development. In southern Italy various heavy industries were located in a number of remote areas after World War II in order to stimulate ancillary development. In the United States a whole series of growth centers were designated within Appalachia in an attempt to establish a manufacturing base

FIGURE 7.19 Congestion charges The London congestion charge was introduced in 2003 to reduce traffic congestion in the center of the city. The standard charge is £10 ($16) for each day, for each non-exempt vehicle that travels within the zone between 7:00 am and 6:00 pm (Monday–Friday only); a penalty of between £60 ($96) and £180 ($288) is levied for non-payment.

in the region. The Appalachian Regional Commission, established by Congress in 1965 to promote the economic development of the region, initiated a massive program of road building in order to open up Appalachia for modern economic development. The commission also designated over 200 growth centers that were intended to attract industrialists by furnishing a new physical infrastructure and offering economic incentives in the form of tax breaks. The new industries, it was hoped, would trigger the geographic process of cumulative causation.

The results of such policies have been mixed, however. The French technopoles have been fairly successful because the French government invested large sums of money in establishing propulsive industries in favorable locations. But the Italian and U.S. growth-pole efforts, like many others, have been disappointing. In practice, governments often fail to invest in the right industries, and they nearly always fail to invest heavily enough to kick-start the process of cumulative causation. In Appalachia, for example, the growth centers were too small and too numerous. As a consequence, none of them has been able to generate self-sustaining processes of local economic growth.

APPLY YOUR KNOWLEDGE Identify two regions that have experienced deindustrialization in addition to the regions referred to in the text. List the industries each lost. Did the regions you found experience creative destruction? If so, what industries were dismantled and what ones were created? Has government intervention played a role in core-periphery patterns in the regions you selected? ■

GLOBALIZATION AND ECONOMIC DEVELOPMENT

In the past 30 years, regional core-periphery patterns have been increasingly influenced by globalization and the economic interdependence of major world regions. This globalization has been caused by four important and interrelated factors: a new international division of labor, a new technology system, a homogenization of international consumer markets, and an internationalization of finance.

A New International Division of Labor

A major wave of corporate globalization took place in the 1970s, led by manufacturing giants like General Motors and General Electric that wanted to reduce labor costs, outflank national labor unions, and increase overseas market penetration. This new international division of labor has resulted in three main changes. The first of these is that the United States has declined as an industrial producer. A second and related change is that manufacturing production has been decentralized from the world's core regions to semiperipheral and peripheral countries. For example, in 2009, U.S.-based companies employed about 31 million workers overseas, 80 percent of whom were in manufacturing jobs. An important reason for this trend has been the prospect of keeping production costs low by exploiting the huge differential in wage rates around the world.

The disparity between wages in the countries where these industries are headquartered and where their goods are manufactured is hard to overstate:

Nike pays USD $16 million (13 million Euro) a year to the Brazilian national football team and Adidas pays USD $1.8 million (1.5 million Euro) per year to French player Zinedine Zidane. Meanwhile, the Asian workers who make the football boots and other sports gear worn by players are paid as little as 47 cents Euro per hour—3.76 Euro for a standard working day. Shopping at their cheapest local markets, the women producing brand-name sportswear in Indonesia need to work 3.75 hours to earn enough to purchase 1.5 kg of uncooked chicken, which for some is all the meat they can afford for a month.[2]

A third result of the new international division of labor is that new specializations have emerged within the core regions of the world-system: high-tech manufacturing and **producer services**. Examples of production services include information services, insurance, and market research that enhance the productivity or efficiency of other firms' activities or that enable them to maintain specialized roles. These producer services industries have themselves become globalized in response to the needs of their most important clients, the global manufacturing corporations. If, say, a law firm or advertising agency wished to keep the business of a major corporation, it had to be able to provide its services in places where that corporation needed them. Thus, advanced business service firms followed their clients along the globalization path in the late 1970s and especially in the 1980s. This meant creating an office network to match clients' needs. After a while, some advanced business service firms used their global office network to win more clients in new markets. By the 1990s, the leading business service firms themselves became global corporations that offer seamless service with offices in key cities around the world.

A New Technology System

The second factor contributing to the globalization of the economy is the new technology system, which is based on a combination of innovations, including solar energy, robotics, microelectronics, biotechnology, and digital telecommunications and information systems. This new technology system has required the geographical reorganization of the core economies. It has also extended the global reach of finance and industry and permitted a more flexible approach to investment and trade. Especially important in this regard have been new and improved technologies in transport and communications—the integration of shipping, railroad, and highway systems through containerization (**Figure 7.20**); the introduction of wide-bodied cargo jets; and the development of fiber-optic networks, communications satellites, and electronic mail and information retrieval systems. Finally, many of these telecommunications technologies have also introduced a wider geographical scope and faster pace to many aspects of political, social, and cultural change, as we shall see in subsequent chapters.

[2]Offside! Oxfam International, "Labour Rights and Sportswear Production in Asia," 2006, p. 2. Accessed April 27, 2011, www.oxfam.org.au/campaigns/labour/06report.

FIGURE 7.20 The impact of containerization on world trade Containerization revolutionized long-distance transport by doing away with the slow, expensive, and unreliable business of loading and unloading ships with manual labor. Before containerization, ships spent one day in port for every day at sea; in the wake of containerization, they spend a day in port for every ten days at sea. By 1965, an international standard for containers had been adopted, making it possible to transfer goods directly from ship to rail to road and allowing for a highly integrated global transport infrastructure. The average container ship today holds 4,000 20-foot containers, but some are able to carry 6,000 or 8,000. Containerization requires a heavy investment in both vessels and dockside handling equipment, however. As a result, container traffic has quickly become concentrated in a few ports that handle high-volume transatlantic and transpacific trade. This photograph shows the Harem container port in Istanbul, Turkey.

Global Consumer Markets

The third factor in globalization has been the growth of consumer markets. Among the more affluent populations of the world, a new and materialistic international culture has taken root, in which people save less, borrow more, defer parenthood, and indulge in affordable luxuries that are marketed as symbols of style and distinctiveness. This culture is easily transmitted through the new telecommunications media, and it has been an important basis for transnational corporations' global branding and marketing of "world products" (German luxury automobiles, Swiss watches, British raincoats, French wines, American soft drinks, Italian shoes and designer clothes, and Japanese consumer electronics, for example). Nine of the ten most trusted brands in India in 2009, for example, were products of European or American transnational corporations: Nokia, Colgate, Lux, Lifebuoy, Dettol, Horlicks, and Pepsodent. Tata Salt, Britannia Foods, and Reliance Mobile were the only Indian brands to make the top-ten list.

This materialistic international culture is reinforced through other aspects of globalization, including the internationalization of television, especially CNN, MTV, Fox Broadcasting, Sky, Star Television, and the syndication of TV movies and light entertainment series. The number of television sets per 1,000 people worldwide doubled between 1980 and 2000, and multimedia industries have been booming. The global market for popular cultural products carried by these media is becoming concentrated, however. At the core of the entertainment industry—film, music, and television—there is a growing dominance of U.S. products, and many countries have seen their homegrown industries wither. Hollywood obtains more than 50 percent of its revenues from overseas, up from just 30 percent in 1980. Movies made in the United States account for about 50 percent of the market in Japan, 70 percent in Europe, and 85 percent in Latin America. Similarly, U.S. television series have become increasingly prominent in the programming of other countries.

The Internationalization of Finance

The fourth factor contributing to today's globalization is the internationalization of finance: the emergence of global banking and globally integrated financial markets. The pivotal moment was a "system shock" to the international economy that occurred in the mid-1970s. World financial markets, swollen with U.S. dollars by the U.S. government's deficit budgeting and by huge currency reserves held by OPEC (Organization of Petroleum Exporting Countries) after it had orchestrated a four-fold increase in the price of crude oil, quickly evolved into a new and sophisticated system of international finance, with new patterns of investment and disinvestment.

Meanwhile, the capacity of computers and information systems to deal very quickly with changing international conditions added a speculative component to the internationalization of finance. International movements of money, bonds, securities, and other financial instruments have now become an end in themselves because they are a potential source of high profits from speculation and manipulation. The global banking and financial network handles trillions of dollars every day—no more than 10 percent of which has anything to do with the traditional world economy of trade in goods and services.

The volume and complexity of international investment and financial trading has created a need for banks and financial institutions that can handle investments on a large scale, across great distances, quickly and efficiently. The nerve centers of the new system are located in just a few places—London, Frankfurt, New York, and Tokyo, in particular. Satellite communications systems and fiber-optic networks made it possible for firms to operate key financial and business services 24 hours a day around the globe (**Figure 7.21**), handling an enormous volume of transactions. Linked to these communications systems, computers permit the recording and coordination of the data. The world's fourth-largest stock market, the National Associated Automated Dealers Quotation System (NASDAQ), has no trading floor at all: Telephone and fiber-optic lines connect its half-million traders worldwide.

FIGURE 7.21 24-hour trading between major financial markets Office hours in the two most important financial centers—New York and London—overlap one another somewhat even though the two cities are situated in broadly separated time zones. While these markets are both closed, Tokyo offices are open. This means that between them the world's three major financial centers span the globe with almost 24-hour trading in currencies, stocks, and other financial instruments.

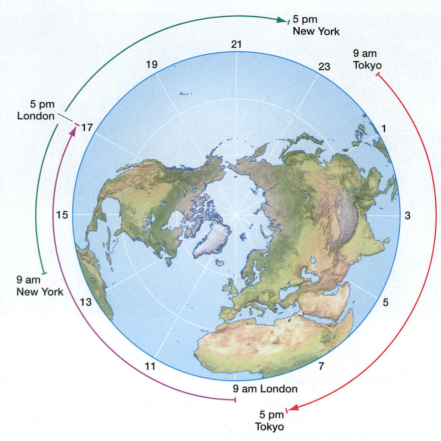

This interconnectedness and complexity contributed to the global financial crisis of October 2008. The failure of several major private financial institutions prompted panic in international financial markets. Suddenly, the world's major economies were thrown into recession, and millions of households in affluent countries had to cope with the loss of jobs, savings, and pension funds. The government of the United States, along with those of other leading economies, intervened to mitigate the recession with hundreds of billions of dollars of support for private financial institutions in an attempt to prop up the international financial system and, with it, their national, regional, and local economies.

How could this have happened? Part of the explanation lies in the steady increase in debt that had been fueling every aspect of the world economy and, in particular, the American economy, since the late 1970s. Consumer spending had been increasingly financed by credit card debt; a housing boom had been financed by an expanded and aggressive mortgage market; and wars in Iraq and Afghanistan had been financed by U.S. government borrowing from overseas. By mid-2008, private debt in America had reached $41 trillion, almost 3 times the country's annual Gross Domestic Product; the external debts of the United States meanwhile had reached $13.7 trillion. All sorts of financial instruments had emerged in the speculative free market climate engendered by neoliberalism: securitization, derivatives, hedge funds, collateralized debt obligations, mortgage-backed securities, and so on. Soon, banks and financial institutions in other countries were joining in, profiting immensely from a global credit binge. Everyone, it seemed, was borrowing from everyone else in an international financial system that had become extremely complex, increasingly leveraged, and decreasingly regulated.

Taking advantage of the relaxed controls on financial institutions resulting from neoliberal policies, mortgage lenders in core countries (and especially the United States) had been selling their mortgages on bond markets and to investment banks in order to fund the soaring demand for housing. Few in the financial services industries fully understood the complexities of the booming mortgage market, and the various risk-assessment agencies were seriously at fault in underestimating the risks associated with these loans. Eventually, when interest rates increased, many households began to default on their monthly payments and, as the bad loans added up, mortgage lenders, in turn, found themselves in financial trouble (**Figure 7.22**).

Because the entire world economy is so interdependent, the problem quickly spread. As credit markets seized up, manufacturers and other businesses found it difficult to get the credit they needed in order to keep going. Understandably, investors big and small were shaken; stock markets collapsed (**Figure 7.23**). Consumer confidence also plummeted, prompting retailers to cut back on orders. The sophisticated flexible production system of global commodity chains meant that the impact was almost instantaneous.

Within days of the U.S. stock market crash in mid-October 2008, workers in Chinese toy factories found themselves unemployed, as did garment workers in Thailand. Meanwhile, in Iceland, the entire economy collapsed almost immediately, a result of its banks having been overextended in their search for profit from the international credit binge. The domino effects of the crisis continued around the world. The U.S. and European governments have led a coordinated attempt to bolster the global financial system and provide a continuing flow of the credit needed to keep the global economy from falling into a deep depression, but in 2010 the Greek economy had to be propped up with a massive US$146 billion loan from the European Union and International Monetary Fund, and a few months later the Irish economy was saved the same way with a loan almost as large.

APPLY YOUR KNOWLEDGE Choose a semiperipheral or peripheral country and determine the ways each of the four factors of globalization discussed in the preceding section of the chapter have affected its economic development process. ■

FIGURE 7.22 Banks in trouble Customers line up in front of an IndyMac Bank branch in Santa Monica, in 2008. The bank reopened after federal seizure during the global economic meltdown.

FIGURE 7.23 Financial meltdown A trader in the Dow Jones Industrial Average stock index futures pit at the Chicago Board of Trade reacts to trading as a monitor above the trading floor broadcasts U.S. President George W. Bush addressing the nation about the financial crisis on October 10, 2008, in Chicago, Illinois.

THE GEOGRAPHY OF ECONOMIC GLOBALIZATION

In this section, we examine some specific impacts of three principal components of the global economy: global assembly lines and supply chains, resulting in large measure from the operations of transnational corporations; the global office, resulting in large measure from the internationalization of banking, finance, and business services; and the pleasure periphery, resulting from the proliferation of international tourism. We also describe a key new trend in economic development: the increasing importance of places as part of consumers' experience.

Global Assembly Lines and Supply Chains

The globalization of the world economy represents the most recent stage in a long process of internationalization. At the heart of this process are private companies that participate not only in international trade but also in production, manufacturing, and/or sales operations in several countries. Almost 80 percent of Ford's workforce, for example, are employed overseas, and foreign sales account for 55 percent of its total revenues. Over 50 percent of IBM's workforce are employed overseas, and 61 percent of its revenues are derived from foreign sales; 82 percent of the workforce of Philips, the Dutch electronics firm, are employed overseas, and 95 percent of its revenues are derived from foreign sales. Many of these transnational corporations have grown large through mergers and acquisitions and their activities span a diverse range of economic activities.

Corporations like theses that consist of several divisions engaged in quite different activities are known as **conglomerate corporations**. Altria (formerly known as Philip Morris), for example, primarily known for its tobacco products (such as Marlboro cigarettes), also controls a large group of assets in the beverage industry (including Miller Brewing) and has extensive interests in real estate, import-export, publishing, and foods (including General Foods, Tobler, Terry's, and Suchard chocolate). Nestlé, the world's largest packaged-food manufacturer, is the largest company in Switzerland but derives less than 2 percent of its revenue from its home country. Its major U.S. product lines and brand names include beverages (Calistoga, Nescafé, Nestea, Perrier, Taster's Choice), chocolate and candy (Butterfinger, Crunch, KitKat), culinary products (Buitoni, Carnation, Libby, Maggi), frozen foods (Lean Cuisine), pet foods (Purina, Fancy Feast, Friskies, Mighty Dog), and drugs and cosmetics (L'Oreal). In addition to its 480 factories in 63 countries around the world, Nestlé operates more than 40 Stouffer hotels.

Transnationals and Globalization

Transnational corporations first began to appear in the nineteenth century, but until the mid-twentieth century there were only a few. Most of these were U.S.- or European-based transnationals concerned with obtaining raw materials, such as oil or minerals, for their domestic manufacturing operations. After World War II, an increasing number of large corporations began to invest in overseas production and manufacturing operations as a means of establishing a foothold in foreign consumer markets. Beginning in the 1970s, the growth of transnational conglomerates increased sharply, not only in the United States but also in Europe, Japan, and even some semiperipheral countries. By 2009, there were about 80,000 transnational corporations in the world. Of these, the top 300 controlled approximately one-quarter of the world's productive assets. Many of the largest transnational corporations are now more powerful, in economic terms, than most sovereign nations. General Motors' economy is larger than Portugal's; Toyota's is larger than Ireland's; and Wal-Mart's annual sales exceed Norway's gross domestic product.

The reason for such growth in the number and scale of transnational conglomerate corporations is that international economic conditions have changed. A recession, triggered by a massive increase in the price of crude oil in 1973, meant that companies everywhere had to reexamine their strategies. At around the same time, technological developments in transport and communications provided larger companies with the flexibility and global reach to exploit the steep differentials in labor costs that exist between core countries and peripheral countries. Meanwhile, these same developments in transport and communications made for intensified international competition, which forced firms to search more intensely for more efficient and profitable global production and marketing strategies. Concurrently, a homogenization of consumer tastes (also facilitated by new developments in communications technologies) has made it possible for companies to more readily cater to global markets.

In effect, the playing field for large-scale businesses of all kinds had been marked out anew. Companies have had to reorganize their operations in a variety of ways, restructuring their activities and redeploying their resources *among different countries, regions, and places*. Local patterns of economic development have been recast again and again as these processes of restructuring, reorganization, and redeployment have played out.

The advantages to manufacturers of a global assembly line are several. First, a standardized global product for a global market allows them to maximize economies of scale. Second, a global assembly line allows production and assembly to take greater advantage of the full range of geographical variations in costs. As noted earlier, basic wages in manufacturing industries, for example, are between 25 and 75 times higher in core countries than in some peripheral

countries. With a global assembly line, labor-intensive work can be done where labor is cheap, raw materials can be processed near their source of supply, final assembly can be done close to major markets, and so on. Third, a global assembly line means that a company is no longer dependent on a single source of supply for a specific component, thus reducing its vulnerability to industrial troubles and other disturbances. Fourth, global sourcing allows transnational conglomerates better access to local markets. For example, Boeing has pursued a strategy of buying a significant number of aircraft components in China and has therefore succeeded in opening the Chinese market to its products.

The automobile industry was among the first to develop a global assembly line. In 1976, Ford introduced the Fiesta, a vehicle designed to sell in Europe, South America, and Asia as well as North America. The Fiesta was assembled in several locations from components manufactured in an even greater number of locations. It became the first of a series of Ford "world cars" that now includes the Focus, the Mondeo, and the Contour. The components of the Ford Escort are made and assembled in 15 countries across three continents. Ford's international subsidiaries, which used to operate independently of the parent company, are now functionally integrated, using supercomputers and video teleconferences.

The other automobile companies followed suit and organized their own global assembly systems (**Figure 7.24**). The global production networks of these companies allow them to process raw materials near sources of supply, to undertake labor-intensive work where labor is cheap, to complete final assembly close to major markets, and to establish multiple sources for components (thus reducing their vulnerability to work stoppages arising from local labor disputes). They employ modular manufacturing for their world cars based on a common underbody platform yet have the flexibility to adapt the interior, trim, body, and ride characteristics to conditions in different countries. Ford, for example, was able to offer the Focus with different engines, transmissions, and other features. Honda has produced three distinct versions of the same car from its Accord world car platform—the bigger, family-oriented Accord for American drivers; the smaller, sportier Accord aimed at young Japanese professionals; and the shorter and narrower Accord, offering the stiff and sporty ride preferred by European drivers.

More than two-thirds of the 60 to 70 million motor vehicles that roll off the production lines each year are made by just ten supergroups of global corporations. In order of size, they are Toyota (including Daihatsu, Hino, Lexus, Scion, Toyota), General Motors (which includes Buick, Cadillac, Chevrolet, Daewoo, GMC, Holden, Opel, Saab, Saturn, and Vauxhall), Ford (incorporating Lincoln, Mercury, Troller, and Volvo Cars), Volkswagen (including Audi, Bentley, Bugatti, Lamborghini, Scania, SEAT, and Škoda), Honda (including Acura), PSA Peugeot Citroën, Nissan (including Infiniti), Renault (including Dacia, and Renault Samsung Motors), and Hyundai.

The global assembly line is constantly being reorganized as transnational corporations seek to take advantage of geographical differences between places and regions and as workers and consumers in specific places and regions react to the consequences of globalization. Nike, the athletic footwear and clothing marketer, provides a good illustration. Nike once relied on its own manufacturing facilities in the United States and the United Kingdom. Today, however, most of its production is subcontracted to suppliers

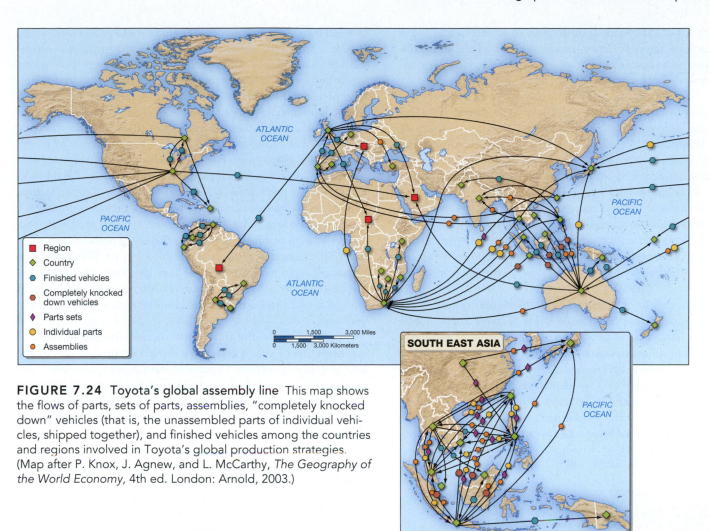

FIGURE 7.24 Toyota's global assembly line This map shows the flows of parts, sets of parts, assemblies, "completely knocked down" vehicles (that is, the unassembled parts of individual vehicles, shipped together), and finished vehicles among the countries and regions involved in Toyota's global production strategies. (Map after P. Knox, J. Agnew, and L. McCarthy, *The Geography of the World Economy*, 4th ed. London: Arnold, 2003.)

in East, South, and Southeast Asia. The geography of this subcontracting evolved over time in response to the changing pattern of labor costs in Asia. The first Asian production of Nike shoes took place in Japan. The company then switched most of its subcontracting to South Korea and Taiwan. As labor costs rose there, Nike's subcontracting spread across more and more peripheral countries—China, Indonesia, Malaysia, and Vietnam—in search of low labor costs. By 2006, Nike subcontractors employed more than 800,000 people in more than 680 different factories. Nike was the largest foreign employer in Vietnam, where its factories accounted for 5 percent of Vietnam's total exports. China, Indonesia, and Thailand were also major components in Nike's expanded global assembly line because of their low wage costs—around $60 per month.

Flexible Production Systems

The strategies of transnational corporations are an important element in the transition from Fordism to Neo-Fordism in much of the world. **Fordism** is named after Henry Ford, the automobile manufacturer who pioneered the principles involved: mass production, based on assembly-line techniques and "scientific" management, together with mass consumption, based on higher wages and sophisticated advertising techniques. In **neo-Fordism** the logic of mass production coupled with mass consumption has been modified by the addition of more flexible production, distribution, and marketing systems. This flexibility is rooted in forms of production that enable manufacturers to shift quickly and efficiently from one level of output to another and, more importantly, from one product configuration to another.

Flexible production systems involve flexibility both within firms and between them. *Within* firms, new technologies now allow a great deal of flexibility. Computerized machine tools, for example, are capable of producing a variety of new products simply by being reprogrammed, often with very little downtime between production runs for different products. Different stages of the production process (sometimes located in different places) are integrated and coordinated through computer-aided design (CAD) and computer-aided manufacturing (CAM) systems. Computer-based information systems monitor retail sales and track wholesale orders, allowing producers to reduce the costs of raw materials stockpiles, parts inventories, and warehousing through sophisticated small-batch, just-in-time production and distribution systems. **Just-in-time production** employs vertical disintegration (see p. 244) within large, formerly functionally integrated firms, such as automobile manufacturers, in which daily and even hourly deliveries of parts and other supplies from smaller (often nonunion) subcontractors and suppliers now arrive "just in time" to maintain "last-minute" and "zero" inventories. The combination of computer-based information systems, CAD/CAM systems, and computerized machine tools has also given firms the flexibility to exploit specialized niches of consumer demand so that economies of scale in production can be applied to upscale but geographically scattered markets.

The Benetton clothing company provides an excellent case-study example of the exploitation of flexible production systems within a single firm. In 1965, the Benetton company began with a single factory near Venice. In 1968, it acquired a single retail store in the Alpine town of Belluno, marking the beginning of a remarkable sequence of corporate expansion. Benetton is now a global organization with over 5,000 retail outlets in more than 120 countries and with its own investment bank and financial services organizations. It achieved this growth by exploiting computers, new communications and transportation systems, flexible outsourcing strategies, and new production-process technologies (such as robotics and CAD/CAM systems) to the fullest possible extent.

Only about 400 of Benetton's employees are located in the company's home base of Treviso, Italy (**Figure 7.25**). From Treviso, Benetton managers coordinate the activities of more than 250 outside suppliers in order to stock its worldwide network of retail outlet franchises. In Treviso, the firm's designers create new shirts and sweaters on CAD terminals, but their designs are produced only for orders already in hand, allowing for the coordination of production with the purchase of raw materials. In factories, rollers linked to a central computer spread and cut layers of cloth

FIGURE 7.25 Headquarters of the Benetton Group, Villa Minelli, Treviso, Italy

in small batches according to the numbers and colors ordered by Benetton stores around the world. Sweaters, gloves, and scarves, knitted in volume in white yarn, are dyed in small batches by machines similarly programmed to respond to sales orders. Completed garments are warehoused briefly (by robots) and shipped out directly (via private package delivery firms) to individual stores to arrive on their shelves within 10 days of manufacture.

Sensitivity to demand, however, is the foundation of Benetton's success. Niche marketing and product differentiation are central to this sensitivity, which requires a high degree of flexibility in exploiting new product lines. Key stores patronized by trendsetting consumers are monitored closely, and many Benetton stores' cash registers operate as point-of-sale terminals so that marketing data are available to company headquarters daily. Another notable feature of the company's operations is the way that different market niches are exploited with the same basic products. In Italy, Benetton products are sold through several different retail chains, each with an image and decor calculated to bring in a different sort of customer.

Between firms, the flexibility inherent in neo-Fordism is achieved through the externalization of certain functions. One way of doing this is to reorganize administrative, managerial, and technical functions into flatter, leaner, and more flexible forms of organization that can make increased use of outside consultants, specialists, and subcontractors. This has led to a degree of vertical disintegration among firms. **Vertical disintegration** involves the evolution from large, functionally integrated firms within a given industry toward networks of specialized firms, subcontractors, and suppliers. Another route to externalization is participation in joint ventures, in the licensing or contracting of technology, and in strategic alliances involving design partnerships, collaborative R&D projects, and the like. **Strategic alliances** are commercial agreements between transnational corporations, usually involving shared technologies, marketing networks, market research, or product development. They are an important contributor to the intensification of economic globalization (**Figure 7.26**).

Maquiladoras and Export-Processing Zones

The governments of many peripheral and semiperipheral countries encourage the type of subcontracting carried out by big transnational corporations. These governments see participation in global assembly lines as a pathway to export-led industrialization. They offer incentives such as tax "holidays" (not having to pay taxes for a specified period) to transnational corporations. In the 1960s, Mexico enacted legislation permitting foreign companies to establish "sister factories"—*maquiladoras*—within 19 kilometers (12 miles) of the border with the United States for the duty-free assembly of products destined for reexport (see **Figures 7.27** and **7.28**). By 2005, more than 3,500 such manufacturing and assembly plants had been established, employing around a million Mexican workers, most of them women, and accounting for more than 30 percent of Mexico's exports. But since 2000, over 500 *maquiladoras* have closed, resulting in the loss of tens of thousands of jobs. The tax breaks that had favored the establishment of the *maquiladoras* are due to expire under the terms of the North American Free Trade Agreement (NAFTA), while lower wage rates and better incentives in other countries—particularly China—have recently proved more attractive to manufacturers.

FIGURE 7.26 Disneyland, Paris The Nestlé food company has a number of strategic alliances, including an alliance since 1990 with the Walt Disney Company in Europe, the United States, Latin America, and other markets that includes making Nestlé the main food company for Disneyland Paris;

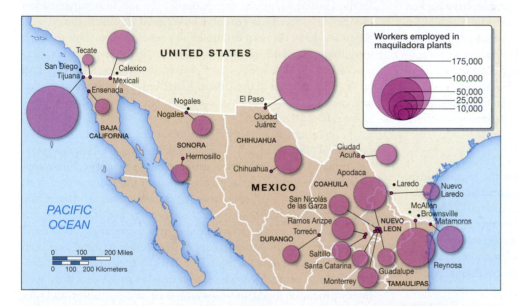

FIGURE 7.27 Principal *maquiladora* centers on the United States–Mexico border Cheap labor and tax breaks for firms manufacturing and assembling goods for reexport have made many Mexican border towns attractive to U.S. companies. Around half a million workers are employed in these *maquiladora* plants, producing electronic products, textiles, furniture, leather goods, toys, and automotive parts. (Adapted from P. Dicken, *Global Shift*, 4th ed. New York: Guilford, 2003; updated per Instituto Nacional de Estadista y Geographia (INEGI), 2011 data, http://dgcnesyp.inegi.org.mx/cgi-win/bdieintsi.exe/NIVR250110009001000370#ARBOL .)

FIGURE 7.28 "Offshore" Manufacturing The production floor at the Flextronics *maquiladora* in Guadalajara, Mexico

Export-processing zones (EPZs)

Export-processing zones (EPZs) are small areas where governments create especially favorable investment and trading conditions in order to attract export-oriented industries. These conditions include minimum levels of bureaucracy, the absence of foreign exchange controls, factory space and warehousing at subsidized rents, low tax rates, and exemption from tariffs and export duties. In 1985, it was estimated that there were a total of 173 EPZs around the world, which together employed 1.8 million workers. In 2006, the International Labor Organization (ILO) estimated that there were 3,500 EPZs, employing about 66 million people.[3] China alone had 164 EPZs, which together employed 40 million workers. The ILO has criticized these "vehicles of globalization" because very few of them have any meaningful links with the domestic economies around them, and most trap large numbers of people in low-wage, low-skill jobs.

In addition to tax incentives and EPZs, many governments also establish policies that ensure cheap and controllable labor. Sometimes countries are pressured to participate in global assembly lines by core countries and by the transnational institutions they support. The United States and the World Bank, for example, have backed regimes that support globalized production and have pushed for austerity programs that help to keep labor cheap in peripheral countries. Countries pursuing export-led industrialization as an economic development strategy do not plan to remain the providers of cheap labor for foreign-based transnational corporations, however. They hope to shift from labor-intensive manufactures to capital-intensive, high-technology goods, following the path of semiperipheral Asian countries like Singapore and South Korea.

Retailing Chains and Global Sourcing

At the other end of the commodity chain from the farms, mines, *maquiladoras*, and factories of the world are the retail outlets and restaurants where the products are sold and consumed. Traditionally, retailing and food services in developed countries have been dominated by small, specialized, independent stores and local cafés, restaurants, pubs, and bars. But the logic of economic rationalization and economies of scale has displaced the traditional pattern of downtown department stores, Main Street shops, corner stores, and local bars with big-box superstores and national and international chains of retail outlets and restaurants. Town centers once filled with a thriving mix of independent and family-owned stores now have "cloned" settings consisting of standardized supermarket retailers, fast-food chains, mobile phone shops, and fashion outlets of global conglomerates. With their cost advantages, these chains have become the economic equivalent of invasive species: voracious, indiscriminate, and often antisocial. Their big, centralized logistical operations have not only put small independent stores out of business but are driving the homogenization of consumption. The retail sector of most towns and cities is now characterized by external control. Decisions about hiring, labor policies, wages, stock, and menus are made in corporate headquarters hundreds of miles or more away.

Fast-food restaurants have become icons of this trend. McDonald's alone has some 31,000 restaurants worldwide and opens new ones at the rate of almost 2,000 each year. It is the largest

purchaser of beef, pork, and potatoes and the largest owner of retail property in the world. In the United States, 40 percent of meals are eaten outside of the home, most of them at fast-food restaurants. One in four adults visits a fast-food restaurant every day. Not surprisingly, the majority of the population is overweight and the frequency of health problems associated with obesity—such as early-onset diabetes and high cholesterol—is rising rapidly. The cost of these problems to personal well-being and to health care systems is already daunting. Meanwhile, fast food's low-paying service sector has become an increasingly significant component of the economy.

Equally significant in terms of local economies, local development patterns, and global supply chains has been the success of "big-box" retail outlets such as Circuit City, Best Buy, Office Max, Home Depot, Target, and Wal-Mart. To many observers, Wal-Mart has come to symbolize the worst characteristics of globalization, including corporate greed, low wages, the decline of small-town mom-and-pop stores, and the proliferation of sweatshops in less developed countries (see Box 7.3, "Geography Matters: Wal-Mart's Economic Landscape").

Supermarket chains have also become particularly influential. In the United Kingdom, for example, the top four supermarket chains—Tesco, Asda (owned by Wal-Mart), Sainsbury, and Safeway—have come to dominate the retail food market through a combination of out-of-town superstores and convenience supermarkets along high streets. As a consequence, they have killed off small general stores in the United Kingdom at the rate of one a day and specialist shops—such as butchers, bakers, and fishmongers—at the rate of 50 per week.

The centralized supply chains of supermarket chains have not only killed off small local businesses but also impacted local farmers. Supermarket chains rely on big suppliers in agribusiness. These suppliers are typically highly subsidized national and transnational firms whose global reach depends heavily on monoculture and extensive husbandry that, in turn, require the extensive use of antibiotics in animals and pesticides, fertilizer, and genetic engineering. As a result, small farmers and fishermen have been squeezed from the market. And with them many traditional local foods have disappeared or are in danger of disappearing. Meanwhile, supermarket shelves are lined with highly processed foods, out-of-season fruit and vegetables, and produce that has traveled a long way and, often, been stored for a while. When the average North American or European family sits down to eat, most of the ingredients have typically traveled at least 1,500 miles from farm, processing, packing, distribution, to consumption.

The external control inherent to centralized chains also brings the prospect of local closures. In Britain, traditional community pubs are increasingly under threat because of this. According to the Campaign for Real Ale, an average of 56 pubs close in Britain every month as the ownership of pubs becomes concentrated in fewer and fewer hands. Corporate chains such as Greene King, Punch Taverns, and Enterprise Inns each own thousands of outlets and are constantly acquiring more in the cause of market penetration and economies of scale. Smaller pubs that do not fit into their business model are sold to a developer to be converted into apartments or restaurants. Following a similar pattern, the number of local bistros in France has fallen to 45,000 from around 225,000 in the 1970s. Many of the surviving local bistros are in danger of having to transform themselves into theme bars, of being swallowed up by large chains, or of going out of business altogether.

[3]International Labor Organization, *ILO database on export processing zones*. Sectoral Activities Programme, Working Paper. Geneva: International Labor Office, 2007.

APPLY YOUR KNOWLEDGE Consider three of the eateries you dine at most often in your community. Are they are locally owned or chains owned by large corporations? If they are chain outlets, research the businesses that were in these locations before them. Were the businesses the chains replaced locally owned? ■

New Geographies of Office Employment

The globalization of production and the growth of transnational corporations have brought about another important change in patterns of local economic development. Banking, finance, and producer services are no longer locally oriented ancillary activities but important global industries in their own right. They have developed some specific spatial tendencies of their own—tendencies that have become important shapers of local economic development processes.

One of the most striking trends has been the geographic decentralization of office employment. This mainly involves "back-office" functions that have been relocated from metropolitan and business-district locations to small-town and suburban locations. Back-office functions such as record-keeping and analytical functions do not require frequent personal contact with clients or business associates. Developments in computing technologies, database access, electronic data interchanges, and telephone call routing technologies are enabling a larger share of back-office work to be relocated to specialized office space in cheaper settings, freeing space in the high-rent locations.

Some prominent examples of back-office decentralization from U.S. metropolitan areas have included the relocation of back-office jobs at American Express from New York to Salt Lake City, Fort Lauderdale, and Phoenix; the relocation of Metropolitan Life's back offices to Greenville, SC., Scranton, PA., and Wichita, KS; the relocation of Hertz's data-entry division to Oklahoma City, Dean Witter's to Dallas, and Avis's to Tulsa; and the relocation of Citibank's MasterCard and Visa divisions to Tampa and Sioux Falls. Some places have actually become specialized back-office locations as a result of such decentralization. Omaha and San Antonio, for example, are centers for a large number of telemarketing firms, and Roanoke, VA., has become something of a mail-order center.

Internationally, this trend has taken the form of offshore back offices. By decentralizing back-office functions to offshore locations, companies save even more in labor costs. Several New York–based life-insurance companies, for example, have established back-office facilities in Ireland, situated conveniently near Ireland's main international airport, Shannon. Insurance claim documents are shipped from New York via Federal Express, are processed in Ireland, and the results are beamed back to New York via satellite or the transatlantic fiber-optic line.

The logical next step is outsourcing. The outsourcing of services is one of the most dynamic sectors of the world economy. Global outsourcing expenditures are expected to grow significantly as small and medium-sized enterprises follow the example of large transnational companies in taking advantage of low wages in semiperipheral and peripheral countries (**Figure 7.29**).

FIGURE 7.29 Outsourcing Assembly workers in Mauritius Island, Republic of Mauritius.

Typically, international outsourcing in service industries involves the work of "routine producers" (who process data by following instructions, perform repetitive tasks, and respond to explicit procedures) rather than "symbolic analysts" (who work with abstract images, are involved with problem-identifying and problem-solving, and make decisions based on critical judgment sharpened by experience). Outsourced services range from simple business-process activities (for example, data entry, word processing, transcription) to more sophisticated, high-value-added activities (for example, architectural drawing, product support, financial analysis, software programming, and human resource services).

India has become one of the most successful exporters of outsourced service activities, ranging from call centers and business-process activities to advanced IT (information technology) services (**Figure 7.30**). More than 150 of the *Fortune* 500 companies, for example, now outsource software development to India. In the Philippines, special electronic "enterprise zones" have been set up with competitive international telephone rates for companies specializing in telemarketing and electronic commerce. Mexico, South Africa, and Malaysia have also become important locations for call centers and business-process activities.

Clusters of Specialized Offices

Decentralization is outweighed, however, by the tendency for a disproportionate share of the new jobs created in banking, finance, and business services to cluster in highly specialized financial districts within major metropolitan areas. The reasons for this localization are to be found in another special case of the geographical agglomeration effects that we discussed earlier in this chapter.

Metropolitan areas such as New York City, London, Paris, Tokyo, and Frankfurt have acquired the kind of infrastructure—specialized office space, financial exchanges, teleports (office parks equipped with satellite Earth stations and linked to local fiber-optic lines), and communications networks—that is essential for delivering services to clients with a national or international scope of

Wal-Mart is the largest company in the world in terms of revenue, with profits of $13.4 billion in 2009. It is also the most highly rationalized, centralized retail chain in the world, designed to run like an assembly line. To some, Wal-Mart represents what happens when the principles of economic geography are implemented in an effort to provide high-quality products at low cost in a global market economy. To others, Wal-Mart illustrates the negative consequences of globalization and corporate power that are consequences of recent trends in the economic geography of the world.

Wal-Mart was founded on a distinctive—and unusual—geographic marketing strategy, which was born in a network of stores in small, isolated towns in the American South. From the firm's headquarters in Bentonville, Arkansas, Wal-Mart's founder, Sam Walton, built a retail empire that reflects the values embedded in the small-town rural South: conservatism, idealized views of family and community, and the principles of hard work, frugality, and competitiveness. Until 1962, Walton's focus was on general merchandise stores, but then it switched to discount stores. Wholesale clubs (Sam's Clubs) were added in 1982, and supercenters (general merchandise/supercenter combinations) in 1988. The company has moved out of the small-town South and into metropolitan markets and international ventures but is still characterized by peripheral store locations, usually a short distance from interstate highways and often not even within the boundaries of an incorporated place (**Figure 7.E**). This is partly due to land costs and tax benefits and partly because of the NIMBYism (NIMBY stands for "Not In My Backyard") of established neighborhoods.

From the start, Wal-Mart's emphasis was on a business model that rests on high-volume turnover through low prices. This required a rapid expansion of stores in order to strengthen the firm's position with suppliers. This, in turn, meant that Wal-Mart quickly developed expertise in logistics—the movement and storage of goods and the management of the entire supply chain, from purchase of raw materials through sale of the final product. Wal-Mart put together its own distribution facilities, fleet of trucks, and satellite communications network in order to maximize the supplier discounts. Another key feature of Wal-Mart's economic landscape is cost control. Pressure is continuously exerted on local governments, employees, and suppliers to get the best deal possible for Wal-Mart. Wal-Mart's labor policies are virulently antiunion; the stores are deliberately understaffed; there is no grievance policy for employees; vendor agreements are tough; suppliers do not give up ownership of the goods until they are sold to the customer; and suppliers are required to drop prices by as much as 5 percent annually.

The company's rapid expansion and tremendous profitability "allowed it a market power unequalled by any of its large corporate competitors, a power that is reshaping the nature of America's and the world's retail industry."[1] Wal-Mart's low prices have reduced consumer inflation and brought many products within

FIGURE 7.E Wal-Mart locations in the Atlanta metropolitan region (Adapted from Figure 2.6 in S. D. Brunn (ed.), *Wal-Mart World*. New York: Routledge, 2006, p. 23.)

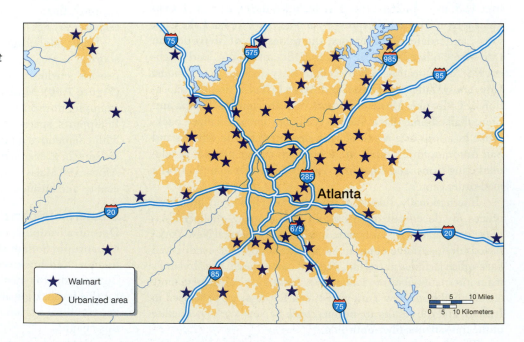

★ Walmart
⬭ Urbanized area

Atlanta

0 5 10 Miles
0 5 10 Kilometers

[1]E. Rosen, "Wal-Mart: The New Retail Colossus." In S. Brunn (ed.), *Wal-Mart World*. New York: Routledge, 2006, p. 92.

FIGURE 7.F Wal-Mart superstore

the range of previously underserved consumers. But there has been a great deal of concern over the impact of Wal-Mart stores on smaller stores and towns. Wal-Mart's predatory pricing has put numerous independent local stores out of business. The result is external control of the local economy. With fewer locally owned businesses, money spent on goods no longer stays within the community but instead is funneled back to corporate headquarters. At the same time, Wal-Mart wages tend to be significantly lower—in the region of 30 percent lower—than in independent, locally owned businesses, so that wages in local labor markets tend to be depressed.

Other concerns include the aesthetic and environmental impact of the big cinderblock boxes and their enormous adjoining parking lots (**Figure 7.F**). Also, although large retailers such as Wal-Mart carry an extensive array of goods, they often stock only a limited variety of any specific product, thereby leaving the consumer with less choice overall. For example, whereas independent local newsagents typically carry a broad range of magazine titles, big-box chain stores typically concentrate on only the titles with the biggest turnover, in order to maximize profit. The same is true of CDs and DVDs. So, not only do corporate chains reduce the range of shops available in small towns, they also reduce the choice of goods readily available.

This lack of choice is compounded, in Wal-Mart's case, by the company's censorship, "protecting" consumers from products that the management deems offensive. This includes refusing to sell CDs with parental warning stickers and obscuring the covers of popular magazines like *Redbook*, *Marie Claire*, *Cosmopolitan*, and *Rolling Stone*. Because of the enormous size of the Wal-Mart network, some magazines willingly send advance copies to corporate headquarters in Bentonville for approval before publication and will even alter cover artwork to avoid losing sales. "Thus, far from being simply a store, Wal-Mart is also

a moral universe external to the community."[2] Wal-Mart advertising campaigns have always emphasized happy staff, customers from cozy families, and support for local communities. Patriotism has been a strong theme. However the firm's Buy-American campaign rollout coincided with Wal-Mart becoming China's sixth-largest trading partner.

The most recent development in Wal-Mart's economic landscape has been internationalization. The company entered markets in Mexico, Puerto Rico, and Canada in the early 1990s, later expanding into Argentina, Brazil, China, El Salvador, Germany, Guatemala, Honduras, Indonesia, Japan, Nicaragua, and South Korea. There were mixed results. More recently, the approach has been to acquire "Wal-Mart ready" chains of stores and then convert their operations to the Wal-Mart model, as with the company's takeover of Asda (the United Kingdom's second-largest supermarket chain) and its 50–50 joint venture with Bharti Enterprises in India.

Wal-Mart's greatest global impact, though, is through its supply chains. Overall, 50 to 60 percent of the goods Wal-Mart imports into the United States each year by ship come from China, and the home office of Wal-Mart Global Procurement is in Shenzhen, China. The company has a network of over two dozen field offices in other countries and oversees the sourcing of products from more than 5,000 factories in 65 countries, including Bangladesh, Brazil, Guatemala, India, Indonesia, Malaysia, the Philippines, Sri Lanka, and Thailand. The reason, of course, is labor costs. But the "natural" cheapness of labor in these countries is not enough for Wal-Mart; the company systematically imposes pressure on suppliers to reduce costs still further. This has led to abuse and violations of labor laws.

Recent evidence from *China Labor Watch*, for example, found that four apparel factories in China that supplied Adidas and Wal-Mart were paying workers just 48 cents an hour—which was not enough to sustain employees unless they worked overtime. A 2007 report by the National Labor Committee, *A Wal-Mart Christmas*,[3] featured Guangzhou Huanya Gift Ltd. Company of Guangdong Province, China, where the 8,000 workers must put in 10- to 12- to 15-hour shifts, seven days a week during the long, eight-month busy season manufacturing Christmas ornaments for Wal-Mart. At least half the workers, some 4,000 people, were working 95 hours, including 55 hours of (mandatory) overtime. The legal minimum wage in Guangzhou is 55 cents an hour, but workers were paid by a piece rate, with some workers earning just 26 cents an hour, half the legal minimum wage.

[2] B. Warf and T. Chapman, "Cathedrals of Consumption: A Political Phenomenology of Wal-Mart." In S. Brunn (ed.), *Wal-Mart World*. New York: Routledge, 2006, p. 165.
[3] National Labor Committee, 2007, "A Wal-Mart Christmas. Brought to You by a Sweatshop in China," http://www.nlcnet.org/article.php?id=498.

FIGURE 7.30 Globalized office work
Workers at a call center in New Delhi, India, where
business is outsourced from western companies.

activity (**Figure 7.31**). These metropolitan areas have also established a comparative advantage both in the mix of specialized firms and expert professionals on hand and in the high-order cultural amenities (available both to high-paid workers and to their out-of-town business visitors). Above all, these metro areas have established themselves as centers of authority, with a critical mass of people in the know about market conditions, trends, and innovations. These people gain one another's trust through frequent face-to-face contact, not just in business settings but also in the informal settings of clubs and office bars. These key cities have become **world city**—places that, in the globalized world economy, are able not only to generate powerful spirals of local economic development but also to act as pivotal points in the reorganization of global space. They are control centers for the flows of information, cultural products, and finance that collectively sustain the economic and cultural globalization of the world (see Chapter 10 for more details).

Offshore Financial Centers

The combination of metropolitan concentration and back-office decentralization fulfills most of the locational needs of the global financial network. There are, however, some needs—secrecy and shelter from taxation and regulation, in particular—that call for a different strategy. The result has been the emergence of a series of **offshore financial centers**: islands and microstates such as the Bahamas, Bahrain, the Cayman Islands, the Cook Islands, Luxembourg, Liechtenstein, and Vanuatu that are specialized nodes in the geography of worldwide financial flows.

The chief attraction of offshore financial centers is simply that they provide low-tax or no-tax settings for savings. They are havens for undeclared income and for hot money. They also provide discreet markets in which to deal currencies, bonds, loans, and other financial instruments without attracting the attention of regulating authorities or competitors. The U.S. Internal Revenue Service estimates that about $400 billion ends up in offshore financial centers each year as a result of tax-evasion schemes. Overall, about 60 percent of all the world's money now resides offshore.

FIGURE 7.31 **Headquarters office space** Office workers outside Lloyds of London, part of a specialized office district that accounts for a quarter of the world market for marine insurance and over a third of the market in aviation risks.

The Pleasure Periphery: Tourism and Economic Development

Many areas of the world, including parts of the world's core regions, do not have much of a primary base (that is, in agriculture, fishing, or mineral extraction), are not currently an important part of the global assembly line, and are not closely tied into the global financial network. For these areas, tourism can offer the otherwise unlikely prospect of economic development. In 2008, international tourism earnings amounted to $944 billion. Tourism is already the world's largest nonagricultural employer, with 1 in every 12 workers worldwide involved in transporting, feeding, housing, guiding, or amusing tourists. The global stock of lodging, restaurant, and transportation facilities is estimated to be worth about $3 trillion.

The globalization of the world economy has been paralleled by a globalization of the tourist industry. In aggregate, there were 922 million international tourist trips in 2008, compared with just 147 million in 1970. The majority of these trips were taken by tourists from the more affluent countries of the world. Spending by German, U.S., U.K., and Chinese tourists accounted for more than 25 percent of total international tourist dollars in 2008. Tourists from France, Italy, Japan, and Canada accounted for another 12 percent. What is most striking, though, is not so much the growth in the number of international tourists as the increased range of international tourism. Thanks largely to cheaper long-distance flights, a significant proportion of tourism is now transcontinental and transoceanic. While Europe (58 percent) and the Americas (18 percent) continue to be the main tourist destinations, visits to countries in Africa, Asia, and the Pacific have grown to account for almost one-quarter of the industry. This has made tourism a central component of economic development in countries with sufficiently exotic wildlife (Kenya), scenery (Nepal, Vietnam), beaches (the Seychelle Islands), shopping (Singapore and Hong Kong), culture (China, India, Japan, and Indonesia), or sex (Thailand).

In addition, "alternative" tourism has been widely advanced as a more sustainable strategy for economic development in peripheral regions (**Figure 7.32**). Alternative tourism emphasizes self-determination, authenticity, preservation of the existing environment, small-scale development, and greater use of local techniques, materials, and architectural styles. Alternative tourism includes ecotourism (bird-watching in Costa Rica; participating in a program to protect endangered sea turtles in Bali; "working" in elephant camps in Thailand), cultural tourism (visiting Machu Picchu, the lost city of the Incas), adventure tourism ("exploring" the Amazon), and industrial tourism (touring the Potteries district of northern England by canal boat).

Ecuador is one country that has fostered alternative tourism. With 6 national parks, 7 nature reserves, and 20 privately protected areas, Ecuador offers great variety for tourists. In a small area straddling the equator, it has some of the world's oldest rain forests and the world's second-highest active volcano—Cotopaxi, at 5,897 meters (19,347 feet). In Ecuador, tourists can see Amazon tribes, spectacular wildlife in the Galapagos Islands, a thriving Andean culture, and a well-preserved legacy of Spanish colonialism. Two-thirds of all organized travel to Ecuador is handled by members of the Ecuadorean Ecotourism Association, an organization sponsored by the private sector as well as the government in an effort to ensure sustainable development through environmental awareness.

FIGURE 7.32 Ecotourism Birdwatchers in Darien National Park, Panama.

Other examples of alternative tourism include guest-house developments in Papua New Guinea, bungalows in French Polynesia, packaged ecotourism in Belize, and "integrated rural tourism" in Senegal. Such developments must be aimed, however, at tourists who are both wealthy and environmentally conscious. This is not, perhaps, a large-enough market on which to pin hopes of significant increases in levels of economic development.

Tourism is by no means confined to the less developed ("unspoiled") peripheral regions. Only 10 percent of Americans, for instance, have passports. Most U.S. tourist dollars are spent in safe and predictable settings where English is spoken—in national parks, specialized resorts, theme parks, big cities, health spas, and renovated historic towns and districts. The growth of tourism and the economic success of places like Baltimore's renovated Harborplace, which attracts about 30 million visitors a year, has meant that few localities exist in America—or anywhere else in the developed world, for that matter—that do not encourage tourism as a central plank of their economic development strategy. It has been estimated that because tourism requires only a basic infrastructure, no heavy plant, and little high-tech equipment, the cost of creating

FIGURE 7.33 Yosemite Summer traffic jam on the road to Yosemite National Park.

one new job in tourism is less than one-fifth the cost of creating a manufacturing job and less than one-fiftieth the cost of creating a high-tech engineering job.

Tourism can provide a basis for economic development, but it is often a mixed blessing. It certainly creates jobs, but they are often seasonal. Dependence on tourism also makes for a high degree of economic vulnerability. Tourism, like other high-end aspects of consumption, depends very much on matters of style and fashion. Some places are sought out by tourists because of their remoteness and their "natural," undeveloped qualities, which are most vulnerable to shifts of style and fashion. Nepal and New Zealand are recent examples of this phenomenon; they are now too "obvious" as destinations and are consequently having to work hard to continue to attract sufficient numbers of tourists. Bhutan, Bolivia, Estonia, Patagonia, and Vietnam have been "discovered" and are coping with their first significant growth in tourism.

Tourism in more exotic tourist destinations is vulnerable in many ways: to political disturbances, natural disasters, outbreaks of disease or food poisoning, and atypical weather. For ski resorts, warm weather represents the equivalent of a harvest failure for an agricultural region. Moreover, although tourism is a multibillion-dollar industry, the financial returns for tourist areas are often not as high as might be expected. The greater part of the price of a package vacation, for example, stays with the organizing company and the airline. Typically, only 40 percent is captured by the tourist region itself. If the package involves a foreign-owned hotel, this number may fall below 25 percent.

The costs and benefits of tourism are not only economic, of course. On the positive side, tourism can help sustain indigenous lifestyles, regional cultures, arts, and crafts, and it can provide incentives for wildlife preservation, environmental protection, and the conservation of historic buildings and sites. On the negative side, tourism can adulterate and debase indigenous cultures and bring unsightly development, pollution, and environmental degradation. In the Caribbean, sewage has poisoned mangrove trees and polluted coastal waters, and boats and divers have damaged coral

reefs. In Kenya, the Maasai Mara National Reserve has been severely degraded by unregulated off-road driving by tour operators.

In the United States, congestion in Yosemite National Park (with over 4 million visitors per year) has become so acute that Park Service rangers frequently have to turn away as many as 1,000 vehicles per day, and park administrators are considering the establishment of a strict advance reservation system (**Figure 7.33**). In the European Alps, where an incredible 40,000 ski runs attract a winter tourist population ten times greater than the resident population, forests have been ripped up, pastures obliterated, rivers diverted, and scenic valleys and mountainsides covered with chalets, cabins, and hotels.

Tourism can also involve exploitative relations that debase traditional lifestyles and regional cultural heritages as they become packaged for outsider consumption. In the process, the behaviors and artifacts that are made available to an international market of outsiders can lose much of their original meaning. Traditional ceremonies that formerly had cultural significance for the performers are now enacted only to be watched and photographed. Artifacts like masks and weapons are manufactured not for their original use but as curios, souvenirs, and ornaments. In the process, indigenous cultures are edited, beautified, and altered to suit outsiders' tastes and expectations.

APPLY YOUR KNOWLEDGE Choose a popular tourist location and list four of the costs and benefits of tourism for this place and its inhabitants. ∎

The Experience Economy and Place Marketing

Tourism is part of what some observers claim is the fourth in a sequence of phases of economic organization. Over the last 200 years, observers argue, there has been a shift from an agrarian economy based on extracting commodities, to an industrial economy based on manufacturing goods, to a service economy based on delivering services, and, now, to an experience economy based on staging experiences. As profitability becomes more challenging in agrarian, industrial, and service enterprises, so businesses must orchestrate memorable events for their customers, with memory itself becoming the "product"—the experience. It is not what is sold that characterizes the experience economy, but rather the way it is sold. Experience thus becomes a competitive advantage for products and services.

James Gilmore, one of the originators of the concept of the experience economy, notes that "Vans has launched a network of Skateparks, where customers pay a $40 annual membership fee, plus $4 to $11 per two-hour skateboarding session. Atlantis, a resort in the Bahamas, charges people who aren't guests just to tour the place! In New Zealand, the country that introduced bungee jumping to the world, people pay to go zorbing, an adventure that involves being strapped into a huge ball and then pushed down a hill."

The experience economy is by no means new. Think, for example, of sports entertainment, arena rock concerts, fancy megamalls with themed entertainment, and visits to museums and galleries, as well as a great deal of traditional tourism. But businesses are increasingly combining commodities, products, and services with compelling experiences. Restaurants organize their services around particular themes (providing marketers with a new term: "eatertainment"). Shops and malls organize shows, events, or expositions ("shoppertainment"). There are now dental practices and general medical practices that double as day spas, while some manufacturers have established flagship stores that have a significant experiential component: Nike Town, SegaWorld, and Warner Studio "Villages," for example. In addition, some businesses are based on selling new kinds of experiences. Ciudad de los niños (Kids City), for example, charges an admission fee to an entire "shopping mall" where children play-shop and play-work. Similarly, American Girl Place offers a combination of attractions specifically targeted at young girls: a doll hair salon, special café, a theater featuring musicals, and shops with dolls and outfits. The American Girl Web site advertises "Café. Theater. Shops. *Memories*."[4]

In the experience economy it is important to capitalize on places because, as we saw in Chapter 6, places have the capacity to arouse distinctive feelings and attachments. At the same time, thanks to economic and cultural globalization, places and regions throughout the world are increasingly seeking to influence the ways in which they are perceived by tourists, businesses, media firms, and consumers. As a result, places are increasingly being reinterpreted, reimagined, designed, packaged, and marketed. Through place marketing, sense of place has become a valuable commodity and culture has become an important economic activity.

Seeking to be competitive within the globalizing economy, many places have sponsored extensive makeovers of themselves, including the creation of pedestrian plazas, cosmopolitan cultural facilities, festivals, and sports and media events—what geographer David Harvey has described as the "carnival masks" and "businessmen's utopias" of global capitalism. An increasing number of places have also set up home pages on the Internet containing maps, information, photographs, guides, and virtual spaces in order to promote themselves in the global marketplace. Not surprisingly, the question of who does the reimagining and cultural packaging can become an important issue for local politics.

Central to place marketing is the deliberate manipulation of culture in an effort to enhance the appeal of places to key groups. These groups include the upper-level management of large corporations, the higher-skilled and better-educated personnel sought by expanding high-technology industries, wealthy tourists, and the organizers of business and professional conferences and other income-generating events. In part, place marketing strategies depend on promoting traditions, lifestyles, and arts that are locally rooted; in part, they depend on being able to tap into globalizing culture through new cultural amenities and specially organized events and exhibitions. Some of the most widely adopted strategies include funding for experiential settings for the arts, investment in public spaces, the re-creation and refurbishment of distinctive settings like waterfronts and historic districts, the expansion and improvement of museums (especially with blockbuster exhibitions of spectacular cultural products that attract large crowds and can be marketed with commercial tie-ins), and the designation and conservation of historic landmarks.

APPLY YOUR KNOWLEDGE Determine if the location you cited in the previous Apply Your Knowledge on page 252 is part of the experience economy. List two ways that manipulation of material and visual culture is used to enhance a sense of place for this location. ∎

Future Geographies

As we have seen, technological breakthroughs and the availability of resources have had a profound influence on past patterns of development, and the same factors will certainly be a strong influence on future economic geographies. The expansion of the world economy and the globalization of industry will undoubtedly boost the overall demand for raw materials of various kinds, and this will spur the development of some previously underexploited but resource-rich regions in Africa, Eurasia, and East Asia.

Raw materials, however, will represent only a fraction of future resource needs. The main issue, by far, will be energy resources. World energy consumption has been increasing steadily over the recent past, and as the periphery is industrialized and its population increases, the global demand for energy will expand rapidly. Basic industrial development tends to be highly energy-intensive. The International Energy Agency, assuming (fairly optimistically) that energy in peripheral countries will be generated in the future as efficiently as it is today in core countries, estimates that developing-country energy consumption will increase dramatically, lifting total world energy demand by almost 50 percent by 2020. Peripheral and semiperipheral countries will then account for more

[4]http://www.americangirl.com/stores/brand_agplace.php

FIGURE 7.34 Energy-consuming goods
An Apple Store in Beijing, China.

than half of world energy consumption. Much of this will be driven by industrialization geared to meet the growing worldwide market for consumer goods, such as private automobiles, air conditioners, refrigerators, televisions, and household appliances (**Figure 7.34**).

Without higher rates of investment in exploration and extraction than at present, production will be slow to meet the escalating demand. As we saw in Box 4.2, many experts believe that current levels of production in fact represent "peak oil" and that by 2020 global oil production may be only 90 percent of its current level. The result might well be a significant increase in energy prices. This would have important geographical ramifications: Companies would need to seriously reconsider their operations and force core-region households into a reevaluation of their residential preferences and commuting behavior, while peripheral-region households would be driven deeper into poverty. If the oil-price crisis of 1973 is anything to go by (after crude oil prices had been quadrupled by the OPEC cartel), the outcome could be a major revision of patterns of industrial location and metropolitan areas. Significantly higher energy costs may change the optimal location for many manufacturers, leading to deindustrialization in some regions and to

new spirals of cumulative causation in others. Higher fuel costs will encourage some people to live nearer to their place of work; others will be able to take advantage of telecommuting to reduce personal transportation costs.

It is also relevant to note that almost all of the increase in oil production over the next 15 or 20 years is likely to come from outside the core economies. This means that the world economy will become increasingly dependent on OPEC governments, which control over 70 percent of all proven oil reserves, most of them in the politically unstable Middle East.

Given what we know about past processes of geographic change and principles of spatial organization, it is clear that changes in transportation technology are also of fundamental importance. Consider, for example, the impact of oceangoing steamers and railroads on the changing geographies of the nineteenth century and the impact of automobiles and trucks on the changing geographies of the twentieth century. Among the most important of the next generation of transportation technologies that will influence future geographies are high-speed rail systems, smart roads, and smart cars.

CONCLUSION

The growth of alternative tourism in Costa Rica, like the growth of the Cayman Islands as an offshore financial center, the emergence of Ireland as a center for back-office activities, and the decline of northern England as a manufacturing region, shows that economic development is not simply a sequential process of modernization and increasing affluence. Economic development involves not only using the latest technology to generate higher incomes but also improving the quality of life through better housing, health care, and social welfare systems and enhancing the physical framework, or infrastructure, on which the economy rests.

Local, regional, and international patterns and processes of economic development are of particular importance to geographers. Levels of

economic development and local processes of economic change affect many aspects of local well-being and so contribute to many aspects of human geography. Economic development is an important place-making process that underpins much of the diversity among regions and nations. At the same time, it is a reflection and a product of variations in natural resources, demographic characteristics, political systems, and social customs.

Economic development is an uneven geographic phenomenon. Regional patterns of economic development are tied to the geographic distribution of resources and to the legacy of the past specializations of places and regions. A general tendency exists toward the creation of regional cores with dependent peripheries. Nevertheless, such patterns are not fixed or static. Changing economic conditions can lead to the modification or reversal of core-periphery patterns, as in the stagnation of once-booming regions like northern England and the spectacular growth of Guangdong Province in Southeast China. Over the long term, core-periphery patterns have most often been modified as a result of the changing locational needs and opportunities of successive technology systems. Today, economic globalization has exposed more places and regions than ever to the ups and downs of episodes of creative destruction—episodes played out ever faster, thanks to the way that technological innovations have shrunk time and space.

At the global scale, the unevenness of economic development takes the form of core-periphery contrasts. Most striking about these contrasts today are the dynamism and pace of change involved in economic development. The global assembly line, the global office, and global tourism are all making places much more interdependent and much faster changing. Parts of Brazil, China, India, Mexico, and South Korea, for example, have developed quickly from rural backwaters into significant industrial regions. Countries like Ecuador and Costa Rica, with few comparative advantages, suddenly find themselves able to earn significant amounts of foreign exchange through the development of ecotourism. This dynamism has, however, brought with it an expanding gap between rich and poor at every spatial scale: international, regional, and local.

Learning Outcomes Revisited

- Scrutinize the nature and degree of unevenness in patterns of economic development at national and international scales.

  At the global scale, unevenness takes the form of core-periphery contrasts. Similar core-periphery contrasts exist at the regional scale. The core regions within the world-system—North America, Europe, and Japan—have the most diversified economies, the most advanced technologies, the highest levels of productivity, and the highest levels of prosperity. Other countries and regions—the periphery and semi-periphery of the world-system—are often referred to as developing or less developed. The average GNI per capita of the ten most prosperous countries in the world is 67 times greater than the average GNI per capita of the ten poorest countries. Overall, more than 80 percent of the world's population live in countries where income differentials are widening.

- Analyze how geographical divisions of labor have evolved with the growth of the world-system and the accompanying variations in economic structure.

  Geographical divisions of labor are national, regional, and locally based economic specializations in primary, secondary, tertiary, or quaternary activities. The relationship between changing regional economic specialization and changing levels of prosperity has prompted the interpretation of economic development in distinctive stages. In reality, however, various pathways exist to development, as well as various processes and outcomes of development. A new international division of labor was initiated in the 1970s as a result of a major wave of corporate globalization. This new international division of labor has resulted in three main changes: the decline of the United States as an industrial producer, the decentralization of manufacturing production from the world's core regions to some semiperipheral and peripheral countries, and the emergence of new specializations in high-tech manufacturing and producer services within the core regions of the world-system.

- Interpret how regional cores of economic development are created, following some initial advantage, through the operation of several basic principles of spatial organization.

  Any significant initial local economic advantage—existing labor markets, consumer markets, frameworks of fixed social capital, and so on—tends to be reinforced through a process of cumulative causation, a spiral buildup of advantages that occurs in specific geographic settings as a result of the development of external economies, agglomeration effects, and localization economies. The agglomeration effects that are associated with various kinds of economic linkages and interdependencies—the cost advantages that accrue to individual firms because of their location among functionally related activities—are particularly important in driving cumulative causation. Spirals of local growth tend to attract people and investment funds from other areas. According to the basic principles of spatial interaction, these flows tend to be strongest from nearby regions and those with the lowest wages, fewest job opportunities, or least attractive investment opportunities.

- Explain how spirals of economic development can be arrested in various ways, including the onset of disinvestment and deindustrialization.

  Core-periphery patterns and relationships can be modified by changes that can slow or modify the spiral of cumulative causation. The main factor is the development of agglomeration diseconomies, the negative economic effects of urbanization, and the local concentration of industry. Spirals of cumulative causation can also be undermined by changes in the relative costs of the factors of land, labor, or capital; by the obsolescence of infrastructure and technology; or by the process of import substitution, whereby goods and services previously imported from core regions come to be replaced by locally made goods and locally provided services. The capital made available from disinvestment in core regions becomes available for investment by entrepreneurs in new ventures based on innovative products and innovative production technologies. This process is often referred to as creative destruction.

- Demonstrate how globalization has resulted in patterns and processes of local and regional economic development that are open to external influences.

The globalization of the world economy involves a new international division of labor in association with the internationalization of finance, the deployment of a new technology system, and the homogenization of consumer markets. This new framework for economic geography has meant that the lives of people in different parts of the world have become increasingly intertwined. Transnational corporations now control a large fraction of the world's productive assets, and the largest of them are more powerful, in economic terms, than most sovereign nations. As these corporations have restructured their activities and redeployed their resources among different countries, regions, and places, they have created many new linkages and interdependencies among places and regions around the world. Even small and medium-sized companies are involved in the myriad global assembly lines and supply chains that characterize the contemporary world economy, linking the fortunes of diverse and often distant local economies.

KEY TERMS

agglomeration diseconomies *(p. 235)*

agglomeration effects *(p.233)*

ancillary industries *(p. 233)*

autarky *(p. 225)*

backwash effects *(p. 234)*

carrying capacity *(p. 220)*

conglomerate corporations *(p. 241)*

creative destruction *(p. 237)*

cumulative causation *(p. 234)*

deindustrialization *(p. 235)*

dependency *(p. 228)*

ecological footprint *(p. 222)*

elasticity of demand *(p. 229)*

export-processing zones (EPZs) *(p. 246)*

external economies *(p. 233)*

flexible production systems *(p. 243)*

Fordism *(p. 243)*

foreign direct investment *(p. 223)*

geographical path dependence *(p. 233)*

gross domestic product (GDP) *(p. 216)*

gross national income (GNI) *(p. 216)*

growth poles *(p. 237)*

import substitution *(p. 230)*

infrastructure (fixed social capital) *(p. 234)*

initial advantage *(p. 233)*

international division of labor *(p. 223)*

just-in-time production *(p. 243)*

localization economies *(p. 233)*

neo-Fordism *(p. 243)*

neoliberal policies *(p. 225)*

newly industrializing countries *(p. 223)*

offshore financial centers *(p. 250)*

primary activities *(p. 223)*

producer services *(p. 238)*

purchasing power parity *(p. 216)*

quaternary activities *(p. 223)*

secondary activities *(p. 223)*

spread effects *(p. 235)*

strategic alliances *(p. 244)*

sustainable development *(p. 222)*

terms of trade *(p. 229)*

tertiary activities *(p. 223)*

trading blocs *(p. 225)*

vertical disintegration *(p. 244)*

world city *(p. 250)*

REVIEW AND DISCUSSION

1. Could your group measure the ecological footprint of your specific place? What data would you need to make an informed statement? List three things in your community that might contribute to sustainable or unsustainable development.

2. Research the businesses in the area connected to your campus. Create a list of the stores. Determine whether or not the stores are locally owned. What is the percentage of locally owned to big-box retail or chain stores? As a comparative study, consider other neighborhoods that are close to yours. Create a list of the locally owned stores and big-box retail or chain stores. How many similarities are there, in terms of businesses? How many differences? What does this say about the economic structures of your campus?

3. After rereading the section on Fair Trade in the chapter, walk with your group to your on-campus food store and create a list of fair trade products. What is the percentage of fair trade products vs. not fair trade products in the store? What does this say about the economic structure of your campus food store?

UNPLUGGED

1. While India's per capita income is well below that of the United States (Figure 7.1), India has more people who earn the equivalent of $70,000 or above a year than the United States does. How can you explain this, and what might be some of the consequences from the point of view of economic geography?

2. Write a short essay (500 words, or two double-spaced typed pages) on any local specialized manufacturing region or office district with which you are familiar. Describe the different kinds of firms that are found there, and suggest the kinds of linkages between them that might be considered examples of agglomeration effects. (*Hint:* You might consider the manufacturing of cars. How many parts are needed to make an entire car? What other firms are needed in the production? Do they come from the same place or are they manufactured in different locations? Are the parts manufactured in one place, region, or country? How would this be an example of agglomeration effects?)

3. Consider the economic structure of the city or town you live in. List two primary, secondary, tertiary, and quaternary activities in your region. How might these activities lead to one part of your city or town being more economically developed than others? Finally, consider where you fit into these economic structures. If you work in a community business, what is your labor used to produce?

Log in to **www.masteringgeography.com** for MapMaster™ interactive maps, geography videos, RSS feeds, flash cards, Web links, an eText version of *Human Geography: Places and Regions in Global Context,* and self-study quizzes to enhance your study of geographies of economic development.

MapMaster™ presents 13 Place Names and 13 Layered Thematic interactive maps to help students practice and master their geographic literacy, spatial reasoning, and critical thinking skills.

8

FOOD AND AGRICULTURE

Learning Outcomes

- Compare and contrast traditional agriculture practices across the globe.

- Describe the three revolutionary phases of agricultural development, from the domestication of plants and animals to the latest developments in biotechnology and industrial innovation.

- Analyze the ways the forces, institutions, and organizational forms of globalization have transformed agriculture.

- Examine the organization of the agro-commodity system from the farm to the retail outlet, including the different economic sectors and corporate forms.

- Scrutinize the ways that agriculture has transformed the environment, including soil erosion, desertification, deforestation, soil and water pollution, and plant and animal species degradation.

- Probe the current issues food-policy experts, national governments, consumers, and agriculturalists face with respect to the availability and quality of food as well as the alternative practices that are emerging to address some of these issues in a world where access to safe, healthy, and nutritious foodstuffs is unevenly distributed.

In Niger, Africa, one in every two children is malnourished and one in every six dies before the age of 5. *Growing a Better Future*, published by Oxfam International, describes the situation of Nigerien people as they face drought, soil depletion, desertification, water scarcity, and predatory food traders. The report sums up the global situation of which Niger is just one part:

> At the start of 2011, there were 925 million hungry people worldwide. By the end of the year, extreme weather and rising food prices may have driven the total back to one billion, where it last peaked in 2008. Why, in a world that produces more than enough food to feed everybody, do so many—one in seven of us—go hungry? (p. 6)

The report argues that the globe is heading for a serious food crisis. One of the factors thought to be adding to it is the increasing amount of cropland once dedicated to food production now being redirected to raising **biofuels,** renewable fuels derived from biological materials that can be regenerated. Replacing even a fraction of fossil fuels with biofuels requires the acquisition and conversion of vast areas of land. National governments around the world have passed laws and are providing inducements for the conversion of tens of millions of hectares to biofuel monocultures like corn, soy, sugarcane, cotton, and grains. Land conversion on this scale often results in evictions of small farmers and poor communities.

Peasants in El Carichi, Ecuador prepare sustainably grown potatoes for market.

In addition to land speculation and the increase in biofuels production, there are other equally significant factors affecting the probability of the global food crisis being in full force by 2030. *Growing a Better Future* argues that the global food system is already reeling from the impacts of climate change, ecological degradation, population growth, rising energy prices, and an increasing demand for meat and dairy products.

The Oxfam report also proposes some workable solutions for repairing the global food system, some of which are already in place. In order to enact these solutions the global community must first agree on a *global governance model* that will make hunger reduction its chief priority. Second, the global community must work towards a new *agricultural future* that puts small-scale farmers at the center of food production. Major gains in productivity, sustainable intensification, and poverty reduction are already being accomplished in some places and can certainly be applied more widely. Finally, the global community must build a new *ecological future*, one that redirects investments and transforms the behaviors of both businesses and consumers toward more sustainable practices and a more even distribution of resources. All of these will require a global agreement on climate change.[1] ∎

[1] Oxfam, *Growing a Better Future: Food Justice in a Resource Constrained World*. Oxfam International: Oxford, England, 2011, http://www.oxfam.org.uk/resources/papers/growing-better-future.html (accessed August 2, 2011).

TRADITIONAL AGRICULTURAL GEOGRAPHY

In this chapter, we examine the geography of agriculture from the global to the household level. We begin by looking at traditional agricultural practices and proceed through the three major revolutions of agricultural change. Much of the chapter is devoted to exploring the ways geographers have investigated the dramatic transformations in agriculture over the last half century as it has become increasingly industrialized through technological, political, social, and economic forces and the effects globalization has had on producing, marketing, delivering, and consuming food. As the example in the chapter-opening discussion of the impending global food crisis makes clear, in order to understand the geography of agriculture, one needs also to take into account a whole host of related factors besides crops and farming.

The study of agriculture has a long tradition in geography. Because of geographers' interest in the relationships between people and land, it is hardly surprising that agriculture has been a primary concern. Geography is committed to viewing the physical and human systems as interactively linked. Such an approach combines an understanding of spatial differentiation and the importance of place and recognition that agriculture practices are affected by local, regional, national, and globally extensive processes. It also provides geographers with a powerful perspective for understanding the dynamics of contemporary agriculture.

One of the most widely recognized and appreciated contributions that geography has made to the study of agriculture is the mapping of the factors that shape it. Geographers map soil, temperature, and terrain, as well as the areal distribution of different types of agriculture and the relationships among and between agriculture and other practices or variables.

Major changes in agriculture worldwide have occurred in the last five decades. Of these, the decline in the number of people employed in farming in both the core and the periphery is perhaps the most dramatic. In addition, the use of chemical, mechanical, and biotechnological innovations and applications has significantly intensified farming practices **(Figure 8.1)**. Agriculture has also become increasingly integrated into wider regional, national, and global economic systems at the same time that it has become more directly linked to other economic sectors, such as manufacturing and finance. The repercussions from these profound changes range from the structure of global finance to the social relations of individual households.

Agricultural Practices

By examining agricultural practices, geographers have sought to understand the myriad ways humans have learned to modify the natural world around them to sustain themselves, their kin, and ultimately the global community. In addition to understanding agricultural systems, geographers are also interested in investigating the lifestyles and cultures of different agricultural communities. They and other social scientists often use the adjective **agrarian** to describe the way of life that is deeply embedded in the demands of agricultural production. *Agrarian* not only defines the culture of distinctive agricultural communities but also refers to the type of tenure (or landholding) system that

FIGURE 8.1 Soybean plantation in Brazil Pictured here is a vast field of genetically modified soybeans. Brazil is the second biggest soybean producer in the world after the United States. Two-thirds of Brazilian soybeans are genetically modified.

determines who has access to land and what kind of cultivation practices are employed there.

Agriculture is a science, an art, and a business directed at the cultivation of crops and the raising of livestock for sustenance and profit. The unique and ingenious methods humans have learned to transform the land through agriculture are an important reflection of the two-way relationship between people and their environments **(Figure 8.2)**. Just as geography shapes our choices and behaviors, so we are able to shape the physical landscape.

While there is no definitive answer as to where agriculture originated, we know that before humans discovered the advantages of agriculture, they procured their food through hunting (including fishing) and gathering. **Hunting and gathering** characterizes activities whereby people feed themselves by killing wild animals and gathering fruits, roots, nuts, and other edible plants. Subsistence agriculture replaced hunting and gathering activities in many parts of the globe when people came to understand that the domestication of animals and plants could enable them to settle in one place

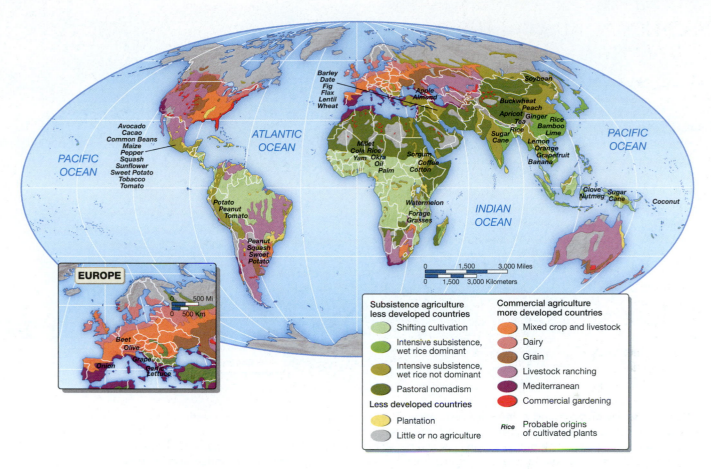

FIGURE 8.2 Global distribution of agriculture, 2005 The global distribution of agricultural practices is illustrated in this map. Notice the differences between core and periphery with respect to commercial versus subsistence agriculture. The periphery, though it does contain commercial agriculture, contains more subsistence agriculture, while the core countries contain virtually none. The origins of cultivated plants can also be seen here as they are spread across both the Old World and the New. (*Source:* Adapted from H. Veregin (ed.), *Goode's World Atlas*, 21st ed. Skokie, IL: Rand McNally, 2005, pp. 38–39.)

over time rather than having to go off frequently in search of edible animals and plants **(Figure 8.3)**. **Subsistence agriculture** is a system in which agriculturalists consume most of that they produce.

During the twentieth century, the dominant agricultural system in the core countries became **commercial agriculture**, a system in which farmers produce crops and animals primarily for sale rather than for direct consumption by themselves and their families. Worldwide, subsistence agriculture is diminishing as increasing numbers of places are incorporated into a globalized economy with a substantial commercial agricultural sector. Still widely practiced in the periphery, however, subsistence activities usually follow one of three dominant forms: shifting cultivation, intensive subsistence agriculture, and pastoralism. Although many people in the periphery rely on these traditional practices to feed themselves, traditional practices are increasingly being abandoned or modified as peasant farmers convert from a subsistence and barter economy to a cash economy.

Shifting Cultivation

In **shifting cultivation**, a form of agriculture usually found in tropical forests, farmers aim to maintain soil fertility by rotating the fields they cultivate. Shifting cultivation contrasts with another method of maintaining soil fertility, **crop rotation**, in which the fields under

cultivation remain the same, but the crops planted are changed to balance the types of nutrients withdrawn from and delivered to the soil.

Shifting cultivation is globally distributed in the tropics—especially in the rain forests of Central and West Africa; the Amazon in South America; and much of Southeast Asia, including Thailand, Burma, Malaysia, and Indonesia—where climate, rainfall, and vegetation combine to produce soils lacking nutrients. The practices involved in shifting cultivation have changed very little over thousands of years **(Figure 8.4)**. Shifting cultivation requires less energy than modern forms of farming, though it can successfully support only low population densities.

The typical agrarian system that supports shifting cultivation is one in which small groups of villagers hold land in common tenure. Through collective agreement or a ruling council, sites are distributed among village families and then cleared for planting by family members. As villages grow, tillable sites must be located farther and farther away. When population growth reaches a critical stage, several families within the village normally split off to establish another village in one of the more remote sites.

Because tropical soils are poor in nutrients, the problem of the rapid depletion of soil fertility through cultivation means that fields are actively planted for less than 5 years. Cultivated plants and heavy tropical rains draw off and wash out the few nutrients

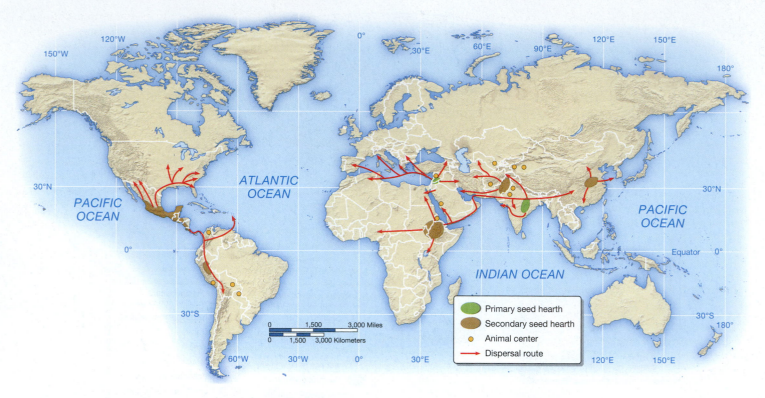

FIGURE 8.3 Areas of plant and animal domestication The origins of plant and animal domestication, however, are not definitively known, and much of what is represented on this map is speculative. Archaeological evidence to date supports the distribution shown here and developed in the mid-twentieth century by Carl Sauer. Primary seed hearths are those places where domestication is believed to have first begun. Secondary seed hearths followed soon after. (*Source:* Adapted from J. M. Rubenstein, *The Cultural Landscape: An Introduction to Human Geography,* 7th ed., Prentice Hall © 2003, p. 319.)

FIGURE 8.4 Shifting cultivation One way to maintain soil fertility is by rotating the fields within which cultivation occurs. Land rotation—also called shifting cultivation—is pictured here as agriculturalists prepare a new field so that the old fields can "rest" and be regenerated. It is also possible to plant certain crops in order to restore soil fertility. That is, a crop that leaches the soil of one kind of nutrient is followed during the next growing season by a dissimilar crop that returns that nutrient to the soil.

that are present in the soil. Once the soil nears exhaustion, a new site is identified and the process of clearing and planting (described in the next paragraph) begins again. It may take over two decades for a once-cleared and cultivated site to become tillable again, after decomposition returns sufficient organic material to the soil.

The typical method for preparing a new site is through **slash-and-burn** agriculture. Existing plants are cropped close to the ground, left to dry for a period, and then ignited. The burning process adds valuable nutrients to the soil, such as potash, which is about the only readily available fertilizer. Once the land is cleared and ready for cultivation, it is known as **swidden** (see Chapter 2). While generally an agricultural practice that is workable when undertaken by small populations on limited portions of land, slash-and-burn is also seen to be ecologically destructive when large numbers of farmers participate, especially in areas with vulnerable and endangered species, such as rain forests **(Figure 8.5)**. Slash-and-burn techniques to plant marijuana, coca, and opium poppies in Colombia since the late 1990s have led to significant devastation in the country's portion of the Amazon rain forest.

Despite its negative impacts on the environment, shifting cultivation can be an elegant response to a fragile landscape. The fallow period, an essential part of the process, is a passive and effective way of restoring plant nutrients to the soil. The burning of stumps and other debris makes the soil more workable, and seeding can proceed with a minimum of effort. Mixing different seeds and seedlings in the same tillage mimics the natural pattern of differing plant heights and types characteristic of the rain forest. It also helps protect the soil from leaching and erosion. Shifting cultivation requires no expensive inputs (except possibly where native seeds are not available) because no manufactured fertilizers, pesticides, herbicides, or heavy equipment—mechanical or otherwise—are necessary. Finally, the characteristically staggered sowing allows for food production throughout the year.

Slash-and-burn agriculture relies largely on human labor, as well as extensive acreage for new plantings because old sites are abandoned frequently when soil fertility is diminished. Although a great deal of labor is involved in cutting and clearing vegetation, once the site is planted, there is little tending of crops until harvest time.

From region to region the kinds of crops grown and their arrangement in the swidden vary depending upon local taste and plant domestication histories. In the warm, humid tropics, tubers—sweet potatoes and yams—predominate, while grains such as corn or rice are more widely planted in the subtropics. The practice of mixing different seeds and seedlings in the same swidden is called **intertillage (Figure 8.6)**. Not only are different plants cultivated, but their planting is usually staggered so that harvesting can continue throughout the year. Staggered planting and harvesting reduces the risk of disasters from crop failure and increases the nutritional balance of the diet.

Shifting cultivation also frequently involves a gender division of labor that may vary from region to region **(Figure 8.7)**.

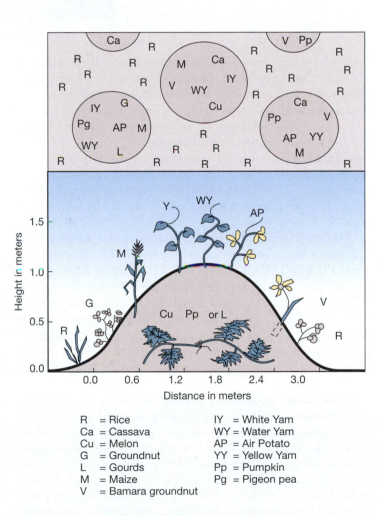

R = Rice	IY = White Yam
Ca = Cassava	WY = Water Yam
Cu = Melon	AP = Air Potato
G = Groundnut	YY = Yellow Yam
L = Gourds	Pp = Pumpkin
M = Maize	Pg = Pigeon pea
V = Bamara groundnut	

FIGURE 8.6 Intertillage Planting different crops together in the same field has many benefits, including the spreading out of food production over the farming season, reduction of disease and pest loss, greater protection from loss of soil moisture, and control of soil erosion. This diagram illustrates what an intertilled site might contain and how the planting is arranged. Hill-planted seeds have tall stalks and a deeper root system; those planted on flat earth tend to be spreading plants that produce large leaves for shading.

FIGURE 8.5 Slash-and-burn, Xishuangbanna, China In the mid-twentieth century, the Xishuangbanna region of China pictured here was a remote tropical rain forest, China's only one. Since then, the government and people of Xishuangbanna have begun to convert the area to rubber tree and sugarcane plantations with slash-and-burn techniques. While the transformations have lifted much of the region out of abject poverty, rapid soil erosion is occurring. Some estimates indicate that Xishuangbanna will be a dust bowl in 30 years.

FIGURE 8.7 Gender division of labor in rice processing in Tamil Nadu, India Rice production is a major source of livelihood for both poor men and women in India. Women in this image are shown threshing the rice while men look on from the rice bales above them. In addition to production activities, in which they participate at a higher level than men, women also process, clean, select, and store the seeds for next year's crop, and are primarily responsible for preparing and cooking rice for household consumption and sometimes for market sale.

For the most part, men are largely responsible for the initial tasks of clearing away vegetation, cutting down trees, and burning the stumps. Women are typically involved with sowing seeds and harvesting crops. Research on shifting cultivation indicates that the actual division of tasks between men and women (and sometimes children) results from traditional cultural practices, as well as the new demands placed upon households by globalization (recall the discussion of Sudanese children in Chapter 6). For instance, many women have found it necessary to complement their subsistence agricultural activities with craft production for local tourist markets.

Although sometimes heralded as an ingenious, well-balanced response to the environmental constraints of the tropics and subtropics, shifting cultivation is not without limitations. Its most obvious limitation is that it can be effective only with small populations. Increasing populations cause cultivation sites to be located farther from villages, with the result that cultivators expend as much energy traveling to sites as they garner from the crops they produce. Indeed, at any one time, it is not unusual for land closest to a village to be entirely fallow or unseeded because the soil is exhausted from previous plantings.

Increasingly, population pressures and ill-considered government policies are undermining the practicality of shifting cultivation, resulting in irreparable damage to the environment in many parts of the world. In Central and South America, for example, national governments have used rural resettlement programs to address urban population pressures. In some cases, relocated individuals not familiar with shifting cultivation techniques have employed them improperly. In others, individuals have been relocated to areas unsuitable for such cultivation practices. In parts of the Brazilian Amazon, for instance, shifting cultivators have acted in concert with cattle grazers, resulting in accelerated environmental degradation.

Although shifting cultivation was likely once practiced throughout the world, population growth and the greater need for increased outputs per acre have led to its replacement by more intensive forms of agriculture.

Intensive Subsistence Agriculture

The second dominant form of subsistence activity is **intensive subsistence agriculture**, a practice involving the effective and efficient use of a small parcel of land to maximize crop yield; a considerable expenditure of human labor and application of fertilizer are also usually involved. Unlike shifting cultivation, intensive subsistence cultivation can often support larger populations. Shifting cultivation is more characteristic of low agricultural densities, whereas intensive subsistence normally reflects high agricultural density. Consequently, intensive subsistence usually occurs in Asia, especially in India, China, and Southeast Asia.

Recall that shifting cultivation involves the application of a relatively limited amount of labor and other resources to cultivation. Conversely, intensive subsistence agriculture involves fairly constant human labor in order to achieve high productivity from a small amount of land. In the face of fierce population pressure and limited arable land, intensive subsistence agriculture reflects the inventive ways in which humans confront environmental constraints and reshape the landscape in the process. In fact, the landscape of intensive subsistence agriculture is often a distinctive one, including raised fields and terraced hillside plots **(Figure 8.8)**.

Intensive subsistence agriculture is able to support large rural populations. In contrast to shifting cultivation, fields are planted year after year as fertilizers and other soil enhancers are applied to maintain soil nutrients. For the most part, the limitations on the size of plots have more to do with the size of the population than geography. In Bangladesh and southern China, for example, where a significant proportion of the population is engaged in intensive subsistence agriculture, land is passed down from generation to generation—usually from fathers to sons—so that each successive generation, if there are multiple male offspring, receives a smaller and smaller share of the family holdings. Yet even with shrinking plot size each family must produce enough to sustain itself.

Under conditions of a growing population and a decreasing amount of arable land, it is critical to plant subsistence crops

FIGURE 8.8 Intensive subsistence agriculture Where usable agricultural land is at a premium, agriculturalists have developed ingenious methods for taking advantage of every square inch of usable terrain. Landscapes like this one—of a terraced rice field in Bali, Indonesia—can be extremely productive when carefully tended and can feed relatively large rural populations.

that produce a high yield per hectare. Different crops fulfill this need in different regional climates. Generally speaking, the crops that dominate intensive subsistence agriculture are rice and other grains.

Rice production predominates in those areas of Asia—South China, Southeast Asia, Bangladesh, and parts of India—where summer rainfall is abundant. In drier climates and places where the winters are too cold for rice production, other sorts of grains—among them wheat, barley, millet, sorghum, corn, and oats—are grown for subsistence. In both situations, the land is intensively used. In fact, it is not uncommon in milder climates for fields to be planted and harvested more than once a year, a practice known as **double cropping**.

Pastoralism

Although not obviously a form of agricultural production, pastoralism is a form of subsistence activity associated with a traditional way of life and agricultural practice. **Pastoralism** involves the breeding and herding of animals to satisfy the human needs for food, shelter, and clothing. Usually practiced in the cold and/or dry climates of savannas (grasslands), deserts, and steppes (lightly wooded, grassy plains), where subsistence agriculture is impracticable, pastoralism can be either sedentary (pastoralists live in settlements and herd animals in nearby pastures) or nomadic (pastoralists travel with their herds over long distances, never settling in any one place for very long). Although forms of commercial pastoralism exist—the regularized herding of animals for profitable meat production, as among Basque Americans in the Basin and Range regions of Utah and Nevada and among the gauchos of the Argentinean grasslands—we are concerned here with pastoralism as a subsistence activity.

Pastoralism is largely confined to parts of North Africa and the savannas of central and southern Africa, the Middle East, and central Asia. Pastoralists generally graze cattle, sheep, goats, and camels, although reindeer are herded in parts of Eurasia. The type of animal herded is related to the culture of the pastoralists, as well

as the animals' adaptability to the regional topography and foraging conditions **(Figure 8.9)**. Nomadism is a form of pastoralism that involves the systematic and continuous movement of groups of herders, their families, and the herds in search of forage. Most pastoralists practice **transhumance**, the movement of herds according to seasonal rhythms: warmer, lowland areas in the winter, and cooler, highland areas in the summer. In addition, women and children in pastoralist groups may be involved with cultivation. They usually split off from the larger group and plant crops at fixed locations in the spring. However, the distinguishing characteristic of pastoralists is that they depend on animals, not crops, for their livelihood.

Like shifting cultivation and intensive subsistence cultivation, pastoralism is not simply a subsistence activity but part of a social system as well. Pastoralist family groups are governed by a leader or chieftain. Groups are divided into units that follow different routes with the herds. The routes are well known; group members are intimately conversant with the landscape, watering places, and opportunities for contact with sedentary groups. Not surprisingly, pastoralism as a subsistence activity is on the decline as more and more pastoralists have become integrated into a global economy that requires more efficient and regularized forms of production. Pastoralists have also been forced off the land by competition from other land uses and the state's need to track citizens for taxation and military reasons.

APPLY YOUR KNOWLEDGE Consult Figure 8.3 showing the distribution of agricultural practices across the globe and locate two additional world maps that show climate and topography. What relationships can you discern between climate and agricultural practice and between topography and agriculture by comparing the three maps? What kind of agriculture happens where and why it happens there? ■

FIGURE 8.9 Pastoralism, Mongolia In this image, sheep forage near the summer settlement of yurts—circular tents of felt or skins on a collapsible framework—at the base of Tsaast Uul mountain in Mongolia, where pastoralism is the main livelihood. Note the dryness of the landscape; pastoralism usually occurs where agriculture is not feasible.

AGRICULTURAL REVOLUTION AND INDUSTRIALIZATION

For a long time human geography textbooks treated the differences in agricultural practices worldwide as systems to be described and cataloged, as we have just done. New conceptual approaches to the agricultural sector, however, have transformed the ways agriculture is viewed. Agriculture has become less a human activity to be described through classification and more a complex component of the global economic system to be explained. While the importance and persistence of traditional agricultural forms are acknowledged, such description must be balanced with an understanding of the ways new commercial practices undermine and otherwise change older forms.

Increasingly, geographers and others have come to see world agricultural practices as having proceeded through "revolutionary" phases, just as manufacturing did. As in manufacturing, practices have not been transformed everywhere at the same time; consequently, some parts of the world are still largely unaffected by certain aspects of agricultural change. By seeing agriculture in this new light, we can recognize that, as in manufacturing, geography and society have changed as the global community has moved from predominantly subsistence to predominantly capital-intensive, market-oriented practices. This history has proceeded in alternating cycles: long periods of very gradual change punctuated by short, explosive periods of radical change, resulting in three distinct revolutionary periods.

The First Agricultural Revolution

The first agricultural revolution is commonly recognized as having been founded on the development of seed agriculture and the use of the plow and draft animals **(Figure 8.10)**. Aspects of this transformation were discussed in Chapter 2. Seed agriculture, which emerged through the domestication of crops such as wheat and rice and animals such as sheep and goats, replaced hunting and gathering as a way of living and sustaining life. Seed agriculture arose during roughly the same period in several regions around the world. The result was a broad belt of cultivated lands across Southwest Asia, from Greece in the west into present-day Turkey and part of Iran in the east, as well as in parts of Central and South America, northern China, northeast India, and East Africa.

The domestication of plants and animals allowed for the rise of settled ways of life. Villages were built, creating types of social, cultural, economic, and political relationships that differed from those that dominated hunter-gatherer societies. On the floodplains along the Tigris, Euphrates, and Nile rivers, important complex civilizations were built upon the fruits of the first agricultural revolution **(Figure 8.11)**. Over time, the knowledge and skill underlying seed agriculture and the domestication of plants diffused outward from these original areas, having a revolutionary impact throughout the globe.

The Second Agricultural Revolution

A great deal of debate exists among historians as to the timing and location of the second agricultural revolution. Though most historians

FIGURE 8.10 Plowing with yoked oxen and camels In many parts of the world, agriculturalists rely on draft animals to prepare land for cultivation. Animals were an important element in the first agricultural revolution. By expanding the amount of energy applied to production, draft animals enabled humans to increase food supplies. Many contemporary traditional farmers view draft animals as their most valuable possessions. Pictured here are camels and oxen pulling the plows of Sikh farmers in Punjab State, India.

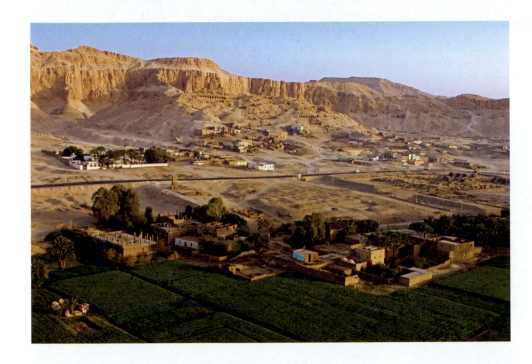

FIGURE 8.11 Agricultural fields along the Nile River, Luxor, Egypt The lush fertile farmland in the foreground is enabled by the silt deposited by the Nile River. At a distance from the river valley, however, dry land not suitable for farming is clearly visible.

agree that it did not occur everywhere at the same time, they disagree over which elements were essential to the fundamental transformation of subsistence agriculture. Important elements included

- dramatic improvements in outputs, such as crop and livestock yields;
- innovations such as the improved yoke for oxen and the replacement of the ox with the horse; and
- new inputs to agricultural production, such as fertilizers and field drainage systems.

The apex of the second agricultural revolution coincided historically and geographically with the Industrial Revolution in England and Western Europe. Although many important changes in agriculture preceded the Industrial Revolution, none had more of an impact than the rise of an industrialized manufacturing sector.

On the eve of the Industrial Revolution—in the middle of the eighteenth century—in Western Europe and England, subsistence peasant agriculture was predominant, though partial integration into a market economy was underway. Many peasants were utilizing a crop-rotation system that, in addition to the application of natural and semiprocessed fertilizers, improved soil productivity and led to increased crop and livestock yields. Additionally, the feudal landholding system—a social and economic system based on peasant service to a lord in exchange for access to land—was breaking down and yielding to a new agrarian system based on an emerging system of private-property relations. Communal lands were being replaced by enclosed, individually owned land or land worked independently by tenants or renters.

Such a situation was a logical response to the demands for food production that emerged from the dramatic social and economic changes accompanying the Industrial Revolution. Perhaps most important of all these changes was the emergence—through the creation of an urban industrial workforce—of a commercial market for food. Many innovations of the Industrial Revolution, such as improvements in transportation technology, had substantial impacts on agriculture. Innovations applied directly to agricultural practices, such as the new types of horse-drawn farm machinery, improved control over—as well as the quantity of—yields.

APPLY YOUR KNOWLEDGE Why was the Industrial Revolution so important to the second agricultural revolution? How did manufacturing technologies change agricultural technologies? Research a technology that debuted in the Industrial Revolution that affected both manufacturing and agricultural practices. ∎

The Third Agricultural Revolution

The third agricultural revolution is fairly recent; it began in the late nineteenth century and gained momentum throughout the twentieth century. Each of the third agricultural revolution's important developmental phases originated in North America. Indeed, the globalization trends framing the discussions in this text are the very same ones that have shaped the third agricultural revolution.

The three phases of the third agricultural revolution are mechanization, chemical farming with synthetic fertilizers, and globally widespread food manufacturing. **Mechanization** is the replacement of human farm labor with machines. Tractors, combines, reapers, pickers, and other forms of motorized machines have, since the 1880s and 1890s, progressively replaced human and animal labor in the United States. In Europe, mechanization did not become widespread until after World War II. **Figure 8.12** shows the global distribution of tractors as a measure of the mechanization of worldwide agriculture.

Chemical farming is the application of synthetic fertilizers to the soil—and herbicides, fungicides, and pesticides to crops—to enhance yields **(Figure 8.13)**. Becoming widespread in the 1950s in the United States, chemical farming diffused to Europe in the 1960s and to peripheral regions of the world in the 1970s. The widespread application of synthetic fertilizers and their impact on the environment is what Rachel Carson wrote about in her highly influential book *Silent Spring* (see Chapter 4).

Food manufacturing also had its origins in late-nineteenth-century North America. **Food manufacturing** adds economic value to agricultural products through a range of treatments—processing, canning, refining, packing, packaging, and so

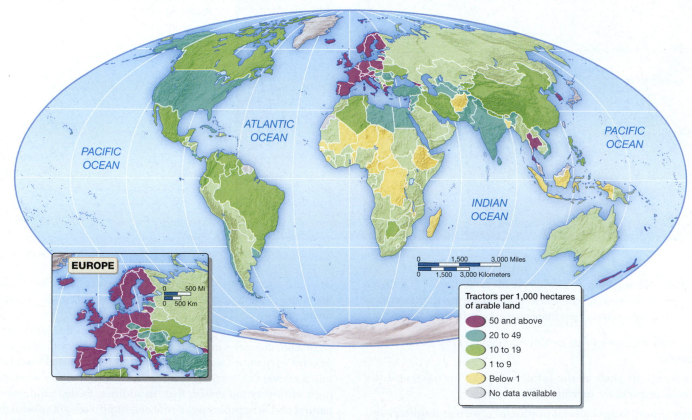

FIGURE 8.12 Tractors per 1,000 hectares Tractor use, a measure of the mechanization of agriculture, is highest in the core countries. Mechanized farming is an expensive undertaking requiring not only machines but the ability to afford fuels and repairs, but it allows for more extensive areas of land to be brought into cultivation.

(*Source:* Reprinted with permission of Prentice Hall, from J. M. Rubenstein, *The Cultural Landscape: An Introduction to Human Geography*, 6th ed., © 1999, p. 341; updated per World Bank, http://data.worldbank.org/indicator/AG.AGR.TRAC.NO)

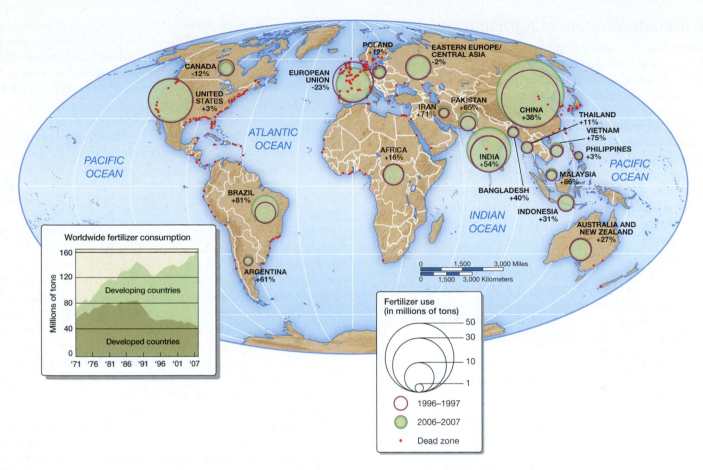

FIGURE 8.13 Worldwide Growth in Fertilizer Use, 2005–2007 As the map indicates, fertilizer use has been growing in peripheral countries faster than in the core, though core countries are still the largest users of fertilizer. One of the biggest problems, and one that is only expected to grow, is increased runoff from fertilizer and resultant dead zones along ocean shores. (*Source:* Adapted from *New York Times,* http://www.nytimes.com/imagepages/2008/04/30/business/20080430_FERTILIZER_GRAPHIC.html, accessed August 17, 2008.)

FIGURE 8.14 Food manufacturing Pictured here is a bread factory in Iowa. The location of the factory allows the wheat grown in the state to be processed close to growing fields and turned into a processed and more valuable product.

on—occurring off the farm and before the products reach the market **(Figure 8.14)**. The first two phases of the third revolution affect inputs to the agricultural production process, whereas the final phase affects agricultural outputs. While the first two are related to the modernization of farming as an economic practice, the third involves a complication of the relationship of farms to firms in the manufacturing sector, which had increasingly expanded into the area of food early in the 1960s. These three developmental phases of the third agricultural revolution constitute the industrialization of agriculture.

APPLY YOUR KNOWLEDGE Take a trip to your local grocery store or open up a kitchen cupboard and read the labels on at least four different products. Identify the various processes involved in making each of the products ready for market. What processing has occurred? Has refining or cooking been involved? What additional ingredients have been added to the product and why (to improve flavor, to extend shelf-life, or for some other reason)? How does the packaging (container, images of the product, colors of the labeling, etc.) enhance the attractiveness of the product? ■

The Industrialization of Agriculture

Advances in science and technology—including mechanical as well as chemical and biological innovations—have driven the industrialization of agriculture over time. As with industrialization more generally, the industrialization of agriculture unfolded as the capitalist economic system became more advanced and widespread. We regard **agricultural industrialization** as the process whereby the farm has moved from being the centerpiece of agricultural production to being one part of an integrated multilevel (or vertically organized) industrial process that includes production, storage, processing, distribution, marketing, and retailing. Experts in the study of agriculture see it as clearly linked to industry and the service sector, thus constituting a complex agro-commodity production system.

Geographers have helped demonstrate the changes leading to the transformation of an agricultural product into an industrial food product. This transformation has been accomplished not only through the indirect and/or direct altering of agricultural outputs (such as tomatoes or wheat) but also through changes in rural economic activities. Agricultural industrialization involves three important developments:

- changes in rural labor activities as machines replace and/or enhance human labor;

- the introduction of innovative inputs—fertilizers and other agrochemicals, hybrid seeds, and biotechnologies—to supplement, alter, or replace biological outputs; and

- the development of industrial substitutes for agricultural products (Nutrasweet instead of sugar, and artificial thickeners instead of cornstarch or flour, for example).

Recall that the industrialization of agriculture has not occurred simultaneously throughout the globe. Changes in the global economic system affect different places in different ways as different states and social groups respond to and shape these changes. For example, the use of fertilizers and high-yielding seeds occurred much earlier in core-region agriculture than in the periphery, where many people still farm without them. In the late 1960s, however, core countries began exporting fertilizers and high-yielding seeds to regions of the periphery (largely in Asia and Mexico) in an attempt to boost agricultural production. In a development known as the **Green Revolution**, they also sent new machines and institutions, all designed to increase global agricultural productivity, as described in Box 8.1, "Window on the World: The Green Revolution and Its Contemporary Challenges."

Economic crises, a decrease in government programs, and reduced trade barriers have slowed the progress of the Green Revolution in many countries. For example, fertilizer use in countries such as Brazil and Mexico has declined in the face of high prices, fewer subsidies, and increased competition from imported corn and wheat, especially from the United States. Many governments have shifted from giving top priority to self-sufficiency in basic grains to encouraging crops that are more competitive in international trade, such as fruit, vegetables, and flowers **(Figure 8.15)**. **Nontraditional agricultural exports (NTAEs)** such as these contrast with traditional exports, such as sugar and coffee. **NTAEs** have become increasingly important in areas of Mexico, Central America, Colombia, and Chile, replacing grain production and traditional exports, such as coffee and cotton. These new crops obtain high prices but also require heavy applications of pesticides and water to meet export-quality standards. They also require fast refrigerated transport to market and are vulnerable to climatic variation and to the vagaries of the international market, including changing tastes for foods and health scares about pesticide or biological contamination.

FIGURE 8.15 Commercial flower production, Argentina Pictured here is a field of allium, flowers that produce a blue, lilac, or purple-colored bloom, on a flower plantation near Mendoza, Argentina. In the background are vineyards. Once grown only in home gardens, alliums are becoming commercialized all over the world, from Argentina to New Zealand.

The Green Revolution and Its Contemporary Challenges

The Green Revolution was an attempt to find ways to feed the world's burgeoning population. In 1943, the Rockefeller Foundation provided funds to a group of U.S. agricultural scientists to set up a research project in Mexico aimed at increasing that country's wheat production through expansion of irrigation infrastructure, modernization of management techniques, distribution of hybridized seeds, synthetic fertilizers, and pesticides. Just 7 years later, scientists distributed the first Green Revolution wheat seeds to Indian farmers. The project was eventually expanded to include research on maize as well. By 1967, Green Revolution scientists were exporting their work to other parts of the world and had added rice to their research agenda (**Figure 8.A**). Norman Borlaug, one of the founders of the Green Revolution, went on to win the Nobel Peace Prize in 1970 for promoting world peace through the elimination of hunger.

The initial focus of the Green Revolution was on the development of seed varieties that would produce higher yields than those traditionally used in the target areas. In developing new, higher-yielding varieties, however, agricultural scientists soon discovered that plants were limited in the amount of nitrogen they could absorb and use. The scientists' solution was to increase nitrogen absorption capacity by delivering nitrogen-based fertilizers in water to plants. This required the building of major water and irrigation projects. Then the scientists discovered that the increased nitrogen and water caused the plants to develop taller stalks. With top-heavy seed heads, the tall stalks fell over easily, reducing the amount of seed that could be harvested. The scientists then came up with dwarf varieties of grains that would

support the heavy seed heads without falling over. Then another problem arose: Short plants in very moist conditions encouraged the growth of diseases and pests. The scientists responded by developing a range of pesticides.

The Green Revolution thus came to constitute a package of inputs: new "miracle seeds," water, fertilizers, and pesticides. Farmers who use all of the inputs—and use them properly—can achieve the yields that scientists produced in their experimental plots, which are two to five times larger than those of traditional crops. In some countries, the resulting yields are high enough to enable export trade, generating important sources of foreign exchange. In addition, the creation of varieties that produce faster-maturing crops has allowed some farmers to plant two or more crops per year on the same land, increasing their individual production—and wealth—considerably.

Thanks to Green Revolution innovations, rice production in Asia grew 66 percent between 1965 and 1985. India, for example, became largely self-sufficient in rice and wheat. Worldwide, Green Revolution seeds and agricultural techniques accounted for almost 90 percent of the increase in world grain output in the 1960s and about 70 percent in the 1970s. In the late 1980s and 1990s, at least 80 percent of the additional production of grains could be attributed to Green Revolution techniques. Thus, although hunger and famine persist, many argue that they would be much worse if the Green Revolution had never occurred (**Figure 8.B**).

The Green Revolution has not been an unqualified success, however. One important reason is that wheat, rice, and maize are

FIGURE 8.A The CIMMYT headquarters The Centro International de Mejorimiento de Maiz y Trigo (CIMMYT) (International Center for the Improvement of Maize and Wheat) in Texcoco, Mexico, is involved in plant breeding and research. High-yield-variety seeds were developed here for the Green Revolution. The center holds the world's premiere collection of corn and wheat germplasm—the genetic components of an organism. Modern, refrigerated storage vaults hold many thousands of varieties.

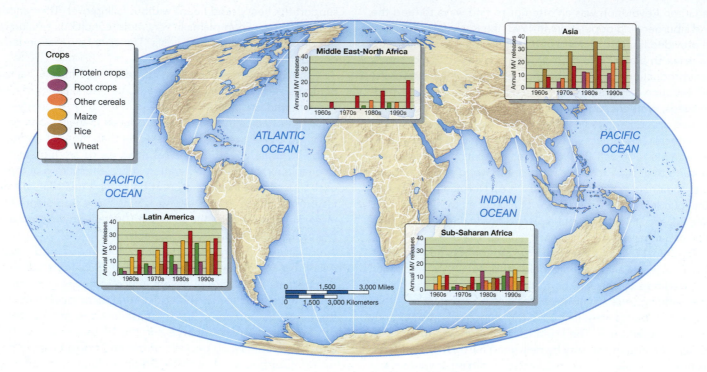

FIGURE 8.B **Effects of the Green Revolution** This map illustrates the increased yields of protein crops (such as beans and peas), root crops, other cereals, maize, rice, and wheat brought about by the Green Revolution in selected countries in Latin America, Asia, sub-Saharan Africa, and the Middle East and North Africa. (Data from: R. E. Evenson and D. Gollin, "Assessing the Impact of the Green Revolution, 1960–2000," *Science, 300,* May 2, 2005, p. 759.)

unsuitable crops in many global regions, and research on more suitable crops, such as sorghum and millet, has lagged behind. In Africa, poor soils and lack of water make progress more difficult to achieve. Another important factor is the vulnerability of the new seed strains to pest and disease infestation, often after only a couple of years of planting. Traditional varieties often have a built-in resistance to the pests and diseases characteristic of an area but genetically engineered varieties often lack such resistance.

A social effect of the Green Revolution technology has been a decreased need for human labor. In southeastern Brazil, machines replaced farm workers, creating significant unemployment. Green Revolution technology and training have also tended to exclude women, who play important roles in traditional food production. In addition, the new agricultural chemicals, especially pesticides, have contributed to ecosystem pollution and worker poisonings, and the more intensive use of irrigation has created salt buildup in soils (*salinization*) and water scarcity (**Figure 8.C**).

Critics also argue that the Green Revolution has magnified social inequities by allowing wealth and power to accrue to a small number of agriculturalists while causing greater poverty and landlessness among poorer segments of the population. In Mexico, a

black market developed in Green Revolution seeds, fertilizers, and pesticides. Poorer farmers, coerced into using them, accrued high debts that they could not begin to repay. Many lost their lands and became migrant laborers or moved to the cities and joined the urban poor.

The new seed varieties sometimes produce grains that are less nutritious, less palatable, or less flavorful. The chemical fertilizers and pesticides that must be used are derived from fossil fuels—mainly oil—and are thus subject to the vagaries of world oil prices. Furthermore, the use of these chemicals, as well as monocropping practices, has produced environmental contamination and soil erosion. Water developments have benefited some regions, but less well-endowed areas have experienced an exaggeration of already existing regional inequities. Worse, pressures to build water projects and to acquire foreign exchange to pay for importation of Green Revolution inputs have increased pressure on countries to grow even more crops for export, often at the expense of production for local consumption.

There are two final criticisms that have raised concern about the overall benefits of the Green Revolution. The first is that it has decreased the production of biomass fuels—wood, crop residues,

FIGURE 8.C Rice paddy, India The introduction of high-yielding, semi-dwarf types of rice, starting in 1962 with the Green Revolution, emphasized the intensive use of fertilizers and pesticides. Rice production increased substantially. However, this achievement was made at a cost to the environment where semi-aquatic organisms, which have always been part of these ecosystems, including wild fish, frogs, shrimps, clams, and snails have disappeared. In addition, to keep soil salinity low, a large quantity of additional water is needed but is seldom available resulting in soil degradation as well as species loss.

and dung—traditionally used in many peripheral areas of the world. The second is that the Green Revolution has contributed to a worldwide loss of genetic diversity by replacing a wide range of local crops and varieties with a narrow range of high-yielding varieties of a small number of crops. Planting single varieties over large areas (monocultures) has made agriculture more vulnerable to disease and pests. Increasing attention is being paid to these limitations. The International Institute of Tropical Agriculture in Ibadan, Nigeria, focuses on foods for humid and subhumid areas, and the International Crops Research Institute for the Semi-Arid Tropics in Hyderabad, India, focuses on researching staples of the Sahel region, such as sorghum, millet, pigeon pea, and groundnut. In both places, the aim is to produce varieties that produce decent yields over good and bad years. They are also working to develop plants that will increase production of biomass in the form of animal fodder and fuel residues, as well as of food, and that give optimal yields when intertilled—a very common practice in Africa. In the Sahel, scientists are working on crops that mature more quickly to compensate for the serious drop in the average length of the rainy season the region has recently experienced.

Despite criticisms, it is clear the global agricultural system has grown spectacularly. And yet, as a recent *New York Times* article states, not spectacularly enough. As described in the chapter opener, the rapid growth in agricultural output that was the hallmark of the twentieth century has declined in the twenty-first to the point where demand is vastly outstripping supply, driven not just by population growth but also by changing food preference

among developing countries like China. With consumption outstripping supply, stockpiles of wheat, rice, soybeans, and grain are also being diminished, creating serious concern among policymakers. Climate change has been identified as one of the most harmful and least easily remediable factors behind lessening food supplies. For instance, the 2003 heat wave in Europe, believed to have been worsened by human-induced global warming, slashed agricultural output in some countries by as much as 30 percent. And a long drought in Australia, also possibly linked to climate change, cut its wheat and rice production. Other contributing events include the prolonged 2009–2010 drought in the Ukraine and 2011 floods in the midwestern United States and in Colombia, South America, which have undermined the global food system.

Although the Green Revolution has come under much justified attack over the years, its main objective of finding innovative new ways to feed the world's peoples has continued. In the process, the world's agricultural system has been expanded into hitherto very remote regions, and important knowledge has been gained about how to conduct science and how to understand the role that science plays in improving agriculture.

Many of the world's leading agricultural researchers believe that it is possible to continue to develop seeds that will be able to withstand drought, flooding, increasing carbon dioxide, and maybe even increasing temperatures (though there is less optimism about the last problem). The twenty-first century poses unforeseen problems for agricultural science, problems that must be solved in order to address the world's growing population.

An example of a nontraditional agricultural export is shrimp, as discussed in Box 8.2, "Geography Matters: The Blue Revolution and Global Shrimp."

One increasingly significant aspect of industrial agricultural transformation is contract farming. **Contract farming** is an agreement between farmers and processing and/or marketing firms for the production, supply, and purchase of agricultural products—from beef, cotton, and flowers to milk, poultry, and vegetables. The legal arrangement requires the firm to provide specified support through, for example, the supply of fertilizer or seeds and the provision of technical advice. The farmer is in turn obliged to produce a specific commodity in quantities and at quality standards determined by the firm. A great deal of agricultural production in the contemporary global system proceeds according to contracts.

Contract farming has been seen by some as a strategy for maintaining the livelihoods of small farmers marginalized by the growth of more corporate forms of farming. For farmers, contractual arrangements can provide access to services and credit as well as new technology. And the price offered for the product can help to reduce risk and uncertainty. On the other hand, there are risks associated with the contract. The most worrisome is the indebtedness that often occurs when farmers are unable to meet the conditions of the contract or when firms fail to honor their agreements.

Biotechnology and Agriculture

In addition to the Blue and the Green revolutions, agriculture has also undergone a **Biorevolution.** The Biorevolution involves the genetic engineering of plants and animals and has the potential to outstrip the productivity increases of the Green Revolution. Ever since the nineteenth century, when Austrian botanist Gregor Mendel identified hereditary traits in plants and French chemist Louis Pasteur explained fermentation, the manipulation and management of biological organisms has been key to the development of agriculture. The central feature of the Biorevolution is **biotechnology,** which is any technique that uses living organisms (or parts of organisms) to improve, make, or modify plants and animals or to develop microorganisms for specific uses. Recombinant DNA techniques, tissue culture, cell fusion, enzyme and fermentation technology, and embryo transfer are some of the most talked-about aspects of biotechnology in agriculture.

A common argument for applying biotechnology to agriculture is the belief that these techniques can help reduce agricultural production costs and serve as a kind of resource management (where certain natural resources are replaced by manufactured ones). Biotechnology has been hailed as a way to address growing concern over the rising costs of cash-crop production, surpluses and spoilage, environmental degradation from chemical fertilizers and overuse, soil depletion, and related challenges now obstructing profitable agricultural production.

Indeed, biotechnology has provided impressive responses to these and other challenges. One particularly spectacular aspect of biotechnology is **biopharming,** in which genes from other life forms (plant, animal, fungal, bacterial, or human) are inserted into host plants. In essence, this is bioengineering plants so that they produce pharmaceuticals. The resulting "pharma crops"—currently in the experimental stage—are expected to include therapeutic proteins; medical and veterinary drugs for treating diarrhea, heart disease, cancer, AIDS, and other illnesses; and vaccines that would be administered by consuming the modified grain or fruit **(Figure 8.16).** Pharma crops are still in the experimental stage and the research is highly confidential. It is estimated that there are over 400 pharma crops currently at the experimental stage, with many of them being anonymously field tested in open-air settings. One of the expected benefits is lowered costs of medication, though it is too soon to tell how this decrease might work in practice. The major disadvantage seen by farmers is that only a small part of the price of the pharma crops is expected to come to them.

While the pharma sector of biotechnology is still in an experimental stage, there are other aspects of biotechnological research

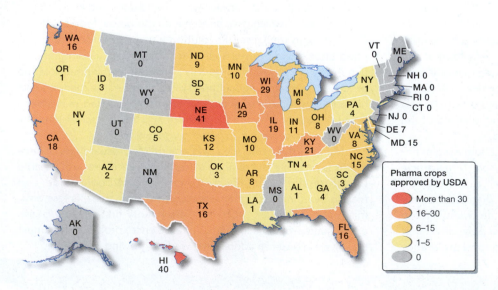

FIGURE 8.16 Biopharma map of the United States This map shows the number of plantings of genetically engineered pharmaceutical and industrial (pharma) crops approved by the U.S. Department of Agriculture (USDA) in individual states. Since 1991, the USDA has approved more than 100 and perhaps as many as 200 applications to grow pharma crops in the United States. The numbers on the map add up to more (395) than the number of applications (100–200) because a single application may include plantings of pharma crops in more than one state.

(*Source:* Pharma crop approvals in the United States map. Copyright © Union of Concerned Scientists, www.ucsusa.org. Reprinted with permission.)

The Blue Revolution and Global Shrimp

By Brian J. Marks

The coastal deltas of the Mississippi and Mekong rivers are vast estuaries where river and sea water mixes in a complex of marshes, swamps, and natural ridges built by river sediment over millennia. These deltas are major shrimp-producing regions for the United States and Vietnam. The similarities and differences between these two shrimping regions speak to the broad forces shaping contemporary global food systems as well as how such forces act on and are shaped by specific places.

In some ways, the Louisiana and Mekong delta shrimp industries could not be more dissimilar. While Louisianians catch shrimp from the sea (**Figure 8.D**), Vietnam relies primarily on **aquaculture**, the growing of aquatic creatures in ponds on shore or in pens suspended in water (**Figure 8.E**). Louisiana's industry involves far fewer people, is oriented to the domestic market, and has a tiny market share; the Mekong Delta's employs many more, is export-oriented, and makes up a much larger share of Vietnam's shrimp harvest. In Louisiana, some 5,700 shrimpers netted 63,000 metric tons from the Gulf of Mexico in 2006. Louisiana's catch makes up more than a third of domestic shrimp production but less than 5% of U.S. shrimp consumption, 90% of which is imported and most of which is farm raised. By contrast, 300,000 shrimp farmers in the Mekong Delta grew 287,000 tons that year, or about 75% of Vietnam's shrimp production, the vast majority of which is exported to core countries.

A commercial shrimp fishery first developed in Louisiana in the late nineteenth century to export sun-dried shrimp to Asia. Filipino and Chinese fishermen were central to the establishment of the shrimp drying industry and taught the practice to others. After World World II, frozen shrimp began to predominate in the market and shrimp prices rose markedly, tapping into the growing affluence of consumers. As people had more disposable income, they went out to eat more often, increasing demand. Louisiana shrimping developed around family ownership and operation of boats. Siblings, in-laws, children, and spouses participated in catching shrimp and pooled financial resources and experience to build boats and set up relatives in the business. Communities of Cajuns, Native Americans, African Americans, Croatians, Canary Islanders (*Isleños*), and, after 1985, Vietnamese and Cambodians have been prominent in domestic shrimping. Shrimpers run everything from 25-foot day boats that work in inshore bays, lakes, and bayous to 100-foot freezer boats that can operate for a month in the deeper waters of the Gulf of Mexico. In recent decades, the industry has seen booms and busts due to changing shrimp prices and costs of production, hurricanes, and Mississippi River floods that disrupt the season, and increasing government regulations put in place to protect endangered sea turtles and other "bycatch" species—other species that may be caught along with the shrimp.

But all these issues did not do as much to cripple the Louisiana shrimp industry, remarkable for its ability to bounce back from such

FIGURE 8.D Louisiana shrimpers These fishermen are sorting wild-caught shrimp harvested near Bayou Lafourche, Louisiana.

FIGURE 8.E Vietnamese shrimp farmer This shrimp farmer is spreading agricultural lime in a shrimp pond in Cà Mau province, the southern tip of the Mekong delta. Shrimp aquaculturalists use lime as a fertilizer and to regulate acidity in the pond water. Most shrimp farms in Vietnam lack the capital to grow shrimp as intensively as this one, which uses paddlewheel aerators to increase dissolved oxygen in the water allowing more shrimp to be grown.

FIGURE 8.F Price trends in Louisiana shrimp and diesel prices, 1998–2009 Shrimp prices fell by some 50% in the 2000s while diesel prices more than doubled, leading to a cost-price squeeze for shrimpers. (*Source*: U.S. Energy Information Agency Short-Term Outlook Real and Nominal Prices, http://www.eia.gov/emeu/steo/realprices/index.cfm and NOAA Fisheries Annual Commercial Landing Statistics, http://www.st.nmfs.noaa.gov/st1/annual_landings.html)

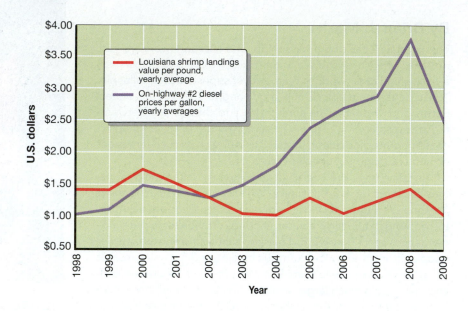

adversity in the past, as the collapse of shrimp prices following the dramatic expansion of shrimp imports in the U.S. market. Between 2000 and 2003, shrimp imports increased 46% and the domestic industry's market share dropped from 17.6% to 11.5%. During the 2000s, Gulf of Mexico shrimp prices fell by 39%; during the same years diesel prices more than doubled (**Figure 8.F**). This created a **cost/price squeeze**, which involves the simultaneous decrease in selling prices and rise in production costs that reduce a business's profit margin. The number of Louisiana shrimpers fell by half over the decade. Those remaining lived at a considerably reduced standard of living, intensified their fishing effort, targeted larger, more expensive shrimp to retail to consumers, diversified into other businesses, or replaced hired deckhands with unpaid family members.

The cost/price squeeze made it more difficult for shrimpers to recover from hurricanes in 2005 and 2008 and the 2010 oil spill because it slashed the income they had to sustain themselves and rebuild from these disasters. While in the short term many Louisiana fishermen have been able to sustain their families, the intergenerational continuity of the industry is being broken as young people seek better and more stable work and their parents discourage them from following their ancestors into shrimping.

The changes in Louisiana are largely the result of the proliferation of shrimp aquaculture, which was responsible for just 5 percent of the global shrimp supply in 1980 but had increased to 52 percent in 2008. A "**Blue Revolution**" in world fisheries introduced larger, more sophisticated vessels into wild fisheries that in turn expanded aquaculture in the late twentieth century. Alongside the

Green Revolution (discussed in Box 8.2), which transformed agriculture in many parts of the periphery, the Blue Revolution has shifted primary-sector activities toward a greater dependence on capitalized inputs—credit, machinery, fuel, feeds, fertilizers, and pesticides—instead of human labor and natural productivity. On the one hand, the Blue and Green revolutions have increased food production in many places, but on the other they have engendered conflict over how the new practices redistribute power and wealth.

Aquaculture has a prominent place in the Blue Revolution. Advances in farming species like carp and tilapia in countries like China have increased the availability of fish for millions of people, yet to date aquaculture has found its greatest economic success in catering to the demand of affluent consumers in the core for products like shrimp and salmon. The major consumers of farmed shrimp are Japan, the United States, and the European Union, while the largest exporters are Thailand, India, Indonesia, Vietnam, and Mexico. Shrimp is big business, the most consumed seafood in the United States and a source of billions of dollars in exports for peripheral countries.

The global shrimp industry is also a major source of controversy; coastal residents in some exporting nations denounce shrimp farming for destroying wetlands; damaging nearby agriculture with salt water; seizing land, water, and access to the sea from communities; and causing violent conflicts between shrimp farmers and their neighbors. The so-called "Pink Gold Rush" of shrimp exports has come with a high social and ecological cost. Globalizing shrimp has also created new economic hazards for shrimp producers themselves.

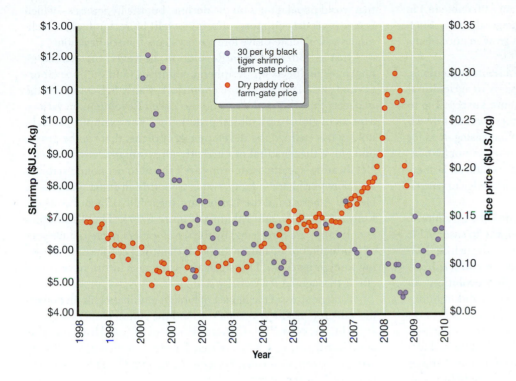

FIGURE 8.G Price trends for shrimp and rice in the Mekong Delta, 1998–2009 When many coastal Mekong Delta farmers abandoned rice for shrimp in 2000, rice was much cheaper and shrimp more expensive than it would be later in the decade. (*Source*: United States Department of Agriculture Vietnam Rice monthly reports, Báo Sóc Trang and Báo Cá Mau (newspapers) weekly shrimp price reports.)

In Vietnam, the modern shrimp industry is much younger than Louisiana's. Most Mekong Delta shrimp farmers grew rice before 2000, after decades of hard work converting saline coastal wetlands into freshwater environments where rice, fruit, and fish were grown for subsistence and domestic markets. By the late 1990s, Vietnam had achieved self-sufficiency in food grains and the national government pursued an export-oriented economic development strategy, allowing farmers to shift away from rice towards more profitable crops. At this time, very low rice prices coincided with exceptionally high shrimp prices (**Figure 8.G**). (Louisiana shrimpers also took on new debt to expand in these years.) This price ratio encouraged farmers to convert some 280,000 hectares of coastal rice land to shrimp, a saltwater crop, in the early 2000s just as global farmed shrimp production surged and prices began their dramatic downward slide. As in Louisiana, family farmers make up the vast majority of shrimp producers in the Mekong Delta, relying on unpaid household labor and limited credit to operate their small farms. Initially, Vietnamese farmers achieved good incomes from shrimp, but disease problems plagued shrimp farming, harming productivity and pushing some out of business. Even many successful farmers' debts grew rapidly to finance intensification.

By the late 2000s, as shrimp prices continued to fall, the cost of fertilizers and chemicals rose with the price of oil. Rice prices skyrocketed, profits from shrimp declined and many producers wanted to go back to rice cultivation. Some areas adopted rice-shrimp farming in rotation, growing shrimp during the saline dry season and rice in the rainy season, but for most farmers in the coastal Mekong Delta the increase of salts in the soil (salinization) and the decay of water control infrastructures meant rice could no longer survive in their fields. Most producers could not reverse course and became effectively "stuck" with shrimp for the foreseeable future even though the price trends that pushed them towards aquaculture had reversed.

For both Louisiana and Mekong Delta shrimp producers, the 2000s have been a decade of profound and often difficult change. Small family producers in both regions sell shrimp to a global market characterized by price volatility that has squeezed their margins while expanding the supply of cheap shrimp for consumers and the major corporations that export, import, and wholesale seafood. While they face common challenges due to globalization, shrimpers in the United States and Vietnam have been affected by the challenges in different ways and have pursued varied coping strategies. A policy for their mutual benefit could include higher and more stable international shrimp prices to reconcile the needs of producers with the operations of the global seafood industry.

that are already fully developed. "Super plants" produce their own fertilizers and pesticides and can be grown on nutrient-lacking soils. Additionally, biotechnologists have been able to clone plants, taking cells of tissues from one plant and using them to grow new plants. A tissue culture no more than 1 cubic cm (.06 cubic in.) in size has the potential to produce millions of identical plants. Such a procedure has decreased the time needed to grow mature plants ready for reproduction.

While such technological innovations can seem miraculous, there is a downside to biotechnological solutions to agricultural problems. For example, cloned plants are more susceptible to disease than natural ones, probably because they have not developed tolerances. This susceptibility leads to an increasing need for chemical treatment. And while industry may reap economic benefits from the development and wide use of tissue cultures, farmers may suffer because they lack the capital or the knowledge to participate in biotechnological applications.

Biotechnology has truly revolutionized traditional agriculture. Its proponents argue that it provides a new pathway to sustainable production. By streamlining the growth process with such innovations as tissue cultures, disease- and pest-resistant and fertilizer-independent plants, optimists believe that the Biorevolution can maximize global agricultural production to keep up with requirements of population and demand. Moreover, the intensification of agriculture, which the Biorevolution (along with the Green and Blue revolutions) enables, has reduced the risk that increased forest resources will be converted into agricultural land. The **Borlaug hypothesis** (named after Norman Borlaug, the force behind the Green Revolution), states that because global food demand is on the rise, restricting crop usage to traditional low-yield methods (such as organic farming) requires either the world population to decrease or the further conversion of forest land into cropland. There is no sign that world population is on the decline, but the hypothesis—which is controversial—proposes that high-yield biotechnologicial techniques will ultimately save forest ecosystems from destruction.

Just as with the Green Revolution, however, biotechnology may have ill effects on peripheral countries (and on poor laborers and small farmers in core countries). **Table 8.1** compares the impacts of the Biorevolution and the Green Revolution on various aspects of global agricultural production. For example, biotechnology has enabled the development of plants that can be grown outside of their natural or currently most suitable environment. Yet location-specific cash crops are critical to economic stability for many peripheral nations—such as cotton in India, bananas in Central America and the Caribbean, sugar in Cuba, and coffee in Kenya, Colombia, and Ethiopia **(Figure 8.17)**. These and other export crops are threatened by the development of alternative sites of production or by multinational agricultural corporations entering foreign markets. Transformations in agriculture have ripple effects throughout the world system.

A tragic story of the impact of foreign multinationals on domestic cash crop producers centers around the cotton farmers in India. In 2011, the New York University School of Law Human Rights and Global Justice Center released a report called *Every Thirty Minutes: Farmer Suicides, Human Rights and the Agrarian Crisis in India*. The report states:

TABLE 8.1 Biorevolution Compared with Green Revolution

Characteristics	Green Revolution	Biorevolution
Crops affected	Wheat, rice, maize	Potentially all crops, including vegetables, fruits, agro-export crops, and specialty crops
Other sectors affected	Pesticides, fertilizer, energy, seeds and irrigation	Pesticides, animal products, pharmaceuticals, processed food products, energy, mining, warfare
Territories affected	Developing countries	All areas, all nations, all locations, including marginal lands
Development of technology and dissemination	Largely public or quasi-public sector, international agricultural research centers (IARCs); R&D millions of dollars	Largely private sector, especially corporations; R&D billions of dollars
Proprietary considerations	Plant breeders' rights and patents generally not relevant	Genes, cells, plants, and animals patentable as well as the techniques used to produce them
Capital costs of research	Relatively low	Relatively high for some techniques; relatively low for others
Access to information	Relatively easy, due to the public funding of research and development	Restricted due to privatization and proprietary considerations
Research skills required	Conventional plant breeding and parallel agricultural sciences	Molecular and cell biology expertise as well as conventional plant-breeding skills
Crop vulnerability	High-yielding varieties relatively uniform; high vulnerability	Tissue culture crop propagation produces exact genetic copies; even more vulnerability
Side effects	Increased monoculture and use of farm chemicals, marginalization of small farmer, ecological degradation. Increased foreign debt due to decrease in biomass fuels and the increasing reliance on costly, usually imported, petroleum	Crop substitution replacing periphery exports; herbicide tolerance; increasing use of chemicals; engineered organisms might affect environment; further marginalization of small farmer

Source: Adapted from M. Kenney and F. Buttel, "Biotechnology: Prospects and Dilemmas for Third-World Development," *Development and Change 16*, 1995, p. 70; and H. Hobbelink, *Biotechnology and the Future of World Agriculture: The Fourth Resource*. London: Red Books, 1991.

FIGURE 8.17 Workers in a coffee plantation For many peripheral countries, the production of cash crops has been a way to boost exports and bring in needed income for the national economy. But transformations in agriculture are making it possible to grow different crops outside of their normal soil, water, and temperature requirements. This makes it difficult for peripheral countries to maintain a foothold in global cash crop production. In Kenya, coffee has for decades been a cash crop grown for export. Luxury exports such as coffee generate some of the capital needed to import staple foods such as wheat.

Economic reforms and the opening of Indian agriculture to the global market over the past two decades have increased costs, while reducing yields and profits for many farmers, to the point of great financial and emotional distress. As a result, smallholder farmers are often trapped in a cycle of debt. During a bad year, money from the sale of the cotton crop might not cover even the initial cost of the inputs, let alone suffice to pay the usurious interest on loans or provide adequate food or necessities for the family. The only way out might be to take on more loans and buy more inputs, which in turn can lead to even greater debt.

In fact, what many of these indebted farmers do out of sheer desperation is commit suicide. In 2009 alone, the most recent year for which official figures are available, 17,638 farmers committed suicide—one farmer every 30 minutes. Over the last 16 years, a quarter of a million Indian farmers have committed suicide, making this the largest wave of recorded suicides in human history. The bottom line is that global agricultural transformations are far from benign and it will require an enormous amount of political cooperation, as underscored in the chapter-opening vignette, to begin to address their profound effect.

APPLY YOUR KNOWLEDGE The industrialization of agriculture has had negative effects on traditional farmers across the globe, including the United States. Referring to the U.S. Census of Agriculture, compare the number of family farms, the mechanization of farming, the application of chemicals to farmland, and the amount of agricultural output for every decade between 1910 and 2010. What observations can you make about these changes and how they have shaped the U.S. food production and distribution system? ■

GLOBAL CHANGE IN FOOD PRODUCTION AND CONSUMPTION

When geographers talk about the globalization of agriculture, they are referring to the incorporation of agriculture into the world economic system of capitalism. A useful way to think about the term **globalized agriculture** is to recognize that, as both an economic sector and a geographically distributed activity, modern agriculture is increasingly dependent on an economy and set of regulatory practices that are global in scope and organization.

Forces of Globalization

Several forces, institutions, and organizational forms play a role in the globalization of agriculture. Technology, economics, and politics have played a central role in propelling national and regional agricultural systems into becoming global in scope. One important way these forces of change have been harnessed is through new global institutions, especially trade and financial organizations. The result is an integrated, globally organized, agro-production system.

The globalization of agriculture has dramatically changed relationships among and within different agricultural production systems. Important outcomes of these changed relationships have been the elimination of some forms of agriculture and the erosion or alteration of some systems as they become integrated into the global economy. Two examples include the current decline of traditional agricultural practices, such as shifting cultivation, and the erosion of a national agricultural system based on family farms (**Figure 8.18**).

Agriculture is one part of a complex and interrelated worldwide economic system. Important changes in the wider

FIGURE 8.18 Family farm Pictured is an idyllic scene of a prosperous U.S. family farm. Since the third agricultural revolution, the number of family farms in the United States and other core countries has declined dramatically as more corporate forms of farming have emerged. In 1920, about one in every three U.S. citizens lived on farms. By 1978, that number had dropped dramatically, to about 1 in 28. This drastic change in the U.S. farm population has caused some commentators to observe that the family farm in the United States is now just a myth because many of them have been bought out by larger corporations or have become incorporated themselves. The local food movement, however, has begun to reinvigorate small-scale farming in the United States and elsewhere.

economy—whether technological, social, political, or otherwise—affect all sectors, including agriculture. National problems in agriculture, such as production surpluses, soil erosion, and food price stability, affect other economic sectors globally, nationally, and locally in different ways. The same is true of global factors—such as the price and availability of oil and other petroleum products critical to commercialized agriculture, the stability of the dollar in the world currency market, and recessions or inflation.

Because of the systemic impact of many problems, integration and coordination of the global economy is needed to anticipate or respond to them. In the last several decades, global and international coordination efforts among states have occurred. These include policies advanced by the World Trade Organization (WTO), as well as the formation of supranational economic organizations such as the European Union (EU) and the Association of Southeast Asian Nations (ASEAN).

It is important to point out, however, that these new forms of cooperation have their opponents. African activists have been involved in frequent and sustained protests against the involvement of European Union companies in biofuel farming there (which we describe in more detail in the material that follows). Similarly, Indian tribal women in the state of Orissa have

been involved for several years in a campaign to prevent foreign and national companies from planting genetically modified (GM) crops. Exhibiting 500 indigenous rice varieties, women argued before the state assembly that the government would put at risk its rich array of rice species if genetically modified organisms entered the region. These and many other protests across the globe demonstrate that the global transformation of agriculture faces resistance.

At the same time that supranational organizations and coordination efforts have been addressing global food problems, states continue to be essential in mediating crises at national levels. By changing public policy, states attempt to regulate agro-industries in order to maintain production, consumption, and corporate profits. One way that governments try to maintain the profitability of the agricultural sector while keeping food prices affordable is through direct and indirect subsidies to agricultural producers. For example, the U.S. government subsidizes agriculture in a number of ways. One is by paying farmers not to grow certain crops that are expected to be in excess supply. Another is by buying up surplus supplies and guaranteeing a fixed price for them.

Billions of dollars are paid out each year in agricultural subsidies, the effects of which are complex and global in impact. Government efforts, while perhaps stabilizing agricultural production in the short term, can lead to problems within the larger national and international agricultural system. For instance, guaranteeing a fixed price for surplus food can act as a disincentive for producers to lower their production, so the problem of overproduction continues. Once in possession of the surplus, governments must find ways to redistribute it. The U.S. government often sells or donates its surplus to foreign governments, where the "dumping" of cheap foodstuffs may undermine the local price structure for food and reduce economic incentives for local farmers to farm.

Many reasons exist for state intervention in agriculture. Economic interests can be both internally and externally driven. At the internal level, governments routinely intervene in one economic sector or another in order, for instance, to correct wider problems of inflation or depression. In the 1930s, the U.S. government attempted to address the problem of economic depression through policies intended to reduce overproduction. States also intervene in the agricultural sector with respect to consumers' interests. Because of subsidies to farm income, the real cost of food may be quite high, but many states in both the developed and less developed world also subsidize the price of food in the marketplace. Such policies are meant to keep the workforce well-fed and healthy, as well as to avoid problems of civil unrest should food prices exceed the general population's ability to pay. Nineteenth-century bread riots—a response to the high cost of flour and bread—were a common occurrence in Europe. More recently, the toppling of Haiti's government in 2009 followed extensive protests by the population over high food prices.

For years, bread subsidies were the norm across the Middle East, used as a way of placating the population. But in 2011 in Tunisia civil unrest was predicated in part on the high cost of food because national grain subsidies had not kept pace with soaring

FIGURE 8.19 Demonstrators in Yemen holding up bread In the spring of 2011 the protesters in Yemen used bread as a way to express their discontent with their government. On the bread is written the word "Leave" in Arabic demanding the resignation of Yemeni President Ali Abdullah Saleh, in Sanaa, Yemen.

prices. It is important to keep in mind that the revolution in Tunisia is more than just about bread. Basic human rights are the most pressing demands across the Middle East from Tunisia to Yemen. When a government fails to provide those rights for the majority of its citizens, instead using handouts or subsidies as a substitute for democratic or economic reforms, bread becomes a powerful symbol of all these citizens lack. In Yemen, demonstrators baked loaves of bread that spelled out the command "leave" in Arabic. The message they were sending is that the very commodity that the regime has used to ensure obedience has now become a symbol and source of defiance **(Figure 8.19)**.[2]

APPLY YOUR KNOWLEDGE Research agricultural subsidies in the United States. Which growers receive the most subsidies? Once you have established this fact, compare the U.S. data with two other countries—one from the periphery and one from the core. How much in government subsidies do farmers in other countries receive and how does it compare with subsidies in the United States. What impacts might this have on the agriculture industry of periphery nations? ∎

[2]Annia Ciezaldo, "Let Them Eat Bread: How Food Subsidies Prevent (and Provoke) Revolutions in the Middle East," *Foreign Affairs* (online), March 23, 2011, accessed June 6, 2011, http://www.foreignaffairs.com/articles/67672/annia-ciezadlo/let-them-eat-bread.

The Organization of the Agro-Food System

Although the changes that have occurred in agriculture worldwide are complex, certain elements that serve as important indicators of change can help us understand them. Geographers and other scholars interested in contemporary agriculture have noted three prominent and interconnected forces that signal a dramatic departure from previous forms of agricultural practice: agribusiness, food chains, and integration of agriculture with the manufacturing, service, finance, and trade sectors.

The concept of agribusiness has received a good deal of attention in the last three decades, and in the popular mind it has come to be associated with large corporations, such as ConAgra or DelMonte. Our definition of agribusiness departs from this conceptualization. Although multi- and transnational corporations (TNCs) are certainly involved in agribusiness, a **agribusiness** is a system rather than a kind of corporate entity. It is a set of economic and political relationships that organizes food production from the development of seeds to the retailing and consumption of the agricultural product. Defining agribusiness as a system does not mean that corporations are not critically important to the food production process. In the core economies, the transnational corporation is the dominant player, operating at numerous strategically important stages of the food production process. TNCs have become dominant for a number of reasons, but mostly because of their ability to negotiate the complexities of production and distribution in many geographical locations. That capability requires special knowledge of national, regional, and local regulations and pricing factors.

The concept of a food supply chain (a special type of commodity chain; see Chapter 2) is a way to understand the organizational structure of agribusiness as a complex political and economic system of inputs, processing and manufacturing, and outputs. A **food supply chain** is composed of five central and connected sectors (inputs, production, processing, distribution, and consumption) with four contextual elements acting as external mediating forces (the state, international trade, the physical environment, and credit and finance). **Figure 8.20** illustrates these linkages and relationships, including how state farm policies shape inputs, prices, farm structure, and even the physical environment.

The food supply chain concept illustrates the network of connections among producers and consumers and regions and places. Consider for example, the linkages that connect cattle production in the Amazon and Mexico, the processing of canned beef along the United States–Mexico border, the availability of frozen hamburger patties in core grocery stores, and the construction of McDonald's restaurants in Moscow **(Figure 8.21)**. Because of complex food chains such as this, it is now common to find that traditional agricultural practices in peripheral regions have been displaced by expensive, capital-intensive practices.

That agriculture is not an independent or unique economic activity is not a particularly new realization. Beginning with the second agricultural revolution, agriculture began slowly but inexorably to be transformed by industrial practices. What is different about the current state of the food system is the way in which farming has become just one stage of a complex and multidimensional economic process. This process is as much about distribution and marketing—key elements of the service sector—as it is about growing and processing agricultural products.

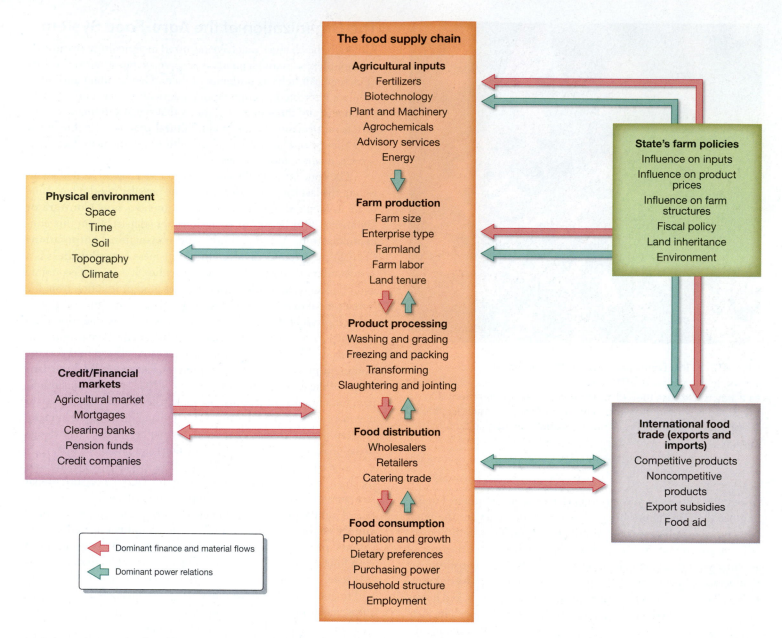

FIGURE 8.20 The food supply chain The production of food has been transformed by industrialization into a complex system that comprises distinctly separate and hierarchically organized sectors. Mediating forces (the state, the structure and processes of international trade, credit and finance arrangements, and the physical environment) influence how the system operates at all scales of social and geographical resolution. (*Source:* After I. Bowler (ed.), *The Geography of Agriculture in Developed Market Economies.* New York: J. Wiley & Sons, 1992, p. 12.)

Food Regimes and Alternative Food Movements

A **food regime** is a specific set of links that exist between food production and consumption. Like the agricultural revolutions already described, food regimes emerge during key historical periods, when different cultural, political, and economic forces are in operation. In contrast to a food chain, which describes the network within which specific food items are produced, manufactured, and marketed, a food regime indicates the ways a particular type of food item is dominant during a specific time period. Although hundreds of food chains may be in operation at any one time, agricul-

tural researchers believe that only one food regime dominates each particular period.

During the decades surrounding the turn of the nineteenth century, an independent system of nation-states emerged and colonization expanded (see Chapters 2 and 9). At the same time, the industrialization of agriculture began. These two forces of political and economic change were critical to the fostering of the first food regime, in which colonies became important sources of exportable foodstuffs by supplying the industrializing European states with cheap food in the form of wheat and meat. The expansion of the colonial agriculture sectors, however, created a crisis in production. The crisis stemmed from the higher cost efficiency of colonial

FIGURE 8.21 World's largest cattle feedlot With over 120,000 head, this feedlot near Greeley, Colorado, is a subsidiary of the food giant ConAgra.

food production, which undercut the prices of domestically produced food, put domestic agricultural workers out of work, and forced members of the agricultural sector in Europe to improve cost efficiency. Industrialized agriculture helped to drive down operating costs and restabilized the sector (reducing even more the need for farm workers). This movement toward the integration of agriculture and industry is also referred to as agro-industrialization.

A wheat and livestock food regime characterized global agriculture until the 1960s; researchers believe that a fresh fruit and vegetable regime is currently dominant. This new pattern of food consumption and production has been called the "postmodern diet." The more perishable agro-commodities of fresh fruits and vegetables have become central to diets of people in the core. Integrated networks of food chains, using integrated networks of refrigeration systems, deliver fresh fruits and vegetables from all over the world to Western Europe, North America, and Japan. Echoing the former food networks that characterized nineteenth-century imperialism, peripheral production systems supply core consumers with fresh, often exotic and off-season, produce. Consumers in the core regions have come to expect the full range of fruits and vegetables to be available year-round, and unusual and exotic produce has become increasingly popular.

Alongside the emergence of a core-oriented food regime of fresh fruits and vegetables, it is important to note an additional aspect of food production practices that has been taking hold in core regions over the last 25 years and accelerating especially over the last 10. One goal of sustainability in agriculture has been increased commitment to organic crops—both proteins (such as beef, eggs, soy) and produce (fruits and vegetables). **Organic farming** describes farming or animal husbandry that occurs without the use of commercial fertilizers, synthetic pesticides, or growth hormones. It is important to point out that organic food production is not the primary mode of practice, but rather that it has become a growing force alongside the dominant **conventional farming** (an approach that uses chemicals in the form of plant protectants

and fertilizers and intensive, hormone-based practices to breed and raise animals) **(Table 8.2)**. We discuss issues of food safety in more detail later in the chapter, but safety issues as well as overall health concerns and an evolving commitment to food as a gastronomic experience among mostly white, middle-class households is at the center of a number of alternative movements that are driving organically produced food as well as the local food and slow food movements.

Local food is usually also organically grown and its designation as local means that it is produced within a fairly limited distance from where it is consumed. Most understandings of local food set a 100-mile radius as a border **(Figure 8.22)**. Thus, if one is a "locavore," the food one consumes should be produced on farms that are no further than 100 miles from the point of distribution. Local food movements have resulted in the proliferation of communities of individuals who have joined together to support the growth of new farms within the local "food shed" (the area within the 100-mile radius). These individuals have been behind a movement known as Community Supported Agriculture, or CSA. The CSA operates by collecting subscriptions and then paying a farmer up front to grow food locally for CSA members for a season or a specified time frame. The farmer is then to purchase seeds, hire workers, cultivate produce and livestock, and deliver the harvest without relying on interest-bearing loans. The CSA members, in return for their investment in the farm, receive weekly shares of produce (and sometimes meats, eggs and cheese, flowers, and milk) that reflect the season and local growing conditions.

Thus, while a CSA in Tucson, Arizona, might receive verdolagas (a green that is native to the region) in addition to squash, snow peas, beets, and carrots, they are unlikely to ever receive pineapples or celery because neither can be grown naturally in the desert. CSA farmers are expected to grow their food organically and members accept what the farmers grow at the same time that they assume the financial risks that are part of the vagaries of any agricultural enterprise. CSA farms are usually small, independent, and

TABLE 8.2 Conventional vs. Alternative Food

Conventional	Alternative
Cheap food to most of the Western world, where the amount of money spent on food in the household budget has steadily decreased.	Logic of sustainable development, not mass production.
Quantity over quality where agriculture is part of consumption; food is a commodity like any other.	The idea of (real or idealized) "quality" predominates—expressed in such parameters as farmers' markets, CSAs, free-range, organic, GMO-free, integrated pest-management schemes, authenticity, traceability.
Fast, convenient foods: for nutritionists, the year 2000 was significant since it was the first time the number of overweight people matched the number of undernourished people—at 1.1 billion.	Slow, pleasurable food and consumers with an ethical consciousness; a new moral economy surfaces where issues of provenance take a central role and consumers demand knowledge of where their food comes from, whether it contains additives, and whether pesticides were used in production.
"Hard Power"—retailer-led supply chains, emasculation of primary producers as supermarkets drive down costs to below that of even production.	"Soft Power"—food sells itself through ethical persuasion and the new "moral economy."
Corporate capital, combined with science and technology enable large-scale food processing and a reliance on large retailers—at the expense of small, independent farms.	Social, as opposed to corporate capital. Direct farmer-to-consumer interaction, relational, trust-based, sense of "community," and attempts to restore small farm enterprises.
Food is "liberated" from nature; technological fixes predominate.	Food is relocated in specific ecological niches and the externalized costs of the conventional system—such as soil degradation, water pollution, animal welfare issues, and health care costs—are addressed.
Logic of deterritorialization—where geographies are a result of the intensification and stretching of the links and networks between production and consumption.	Logic of reterritorialization—where geographies are shaped by the resurgence of local and regional practices as central forces in new articulations of production and consumption.
Geographic specialization, monoculture, and spatial homogenization in North America and Europe produce a large quantity of standardized products.	The "local" as a set of social networks and cultural contexts that are harnessed in place—which provide the foundation for development, innovation, and economic relations.
Placeless landscape of food production (U.K., U.S.) versus place-based landscapes (France, Italy)—where region-specific foods are much more prevalent and provenance is understood as a selling point.	Transparent food chain, traceable, where producers attempt to regain power against the conventional price-driven "race to the bottom" and consumers' knowledge of provenance helps to produce local networks and promotion of social capital.

FIGURE 8.22 **Local food advertisement** Large chain grocery stores in core countries like the United States—such as Shaw's, a New England based supermarket pictured here—are becoming increasingly involved in marketing produce and even processed foods generated in the local area. Much of this increase is the result of consumer demand.

labor-intensive. The CSA movement is seen to be helping to restore the family farm to the national landscapes from which they have been disappearing for 30 years as more corporate forms of farming have taken hold. The CSA movement originated in Japan but it has spread to other core regions, including North America and Europe. In addition to the emergence of CSAs, it is also important to note that urban agriculture is also on the rise. **Urban agriculture** is the establishment or performance of agricultural practices in or near an urban or city-like setting (see Box 8.3 "Visualizing Geography: The Growth of Urban Agriculture").

Europe has contributed its own focus to alternative food production and consumption through the slow food movement. **Slow food** (see Chapter 6), as the name suggests, is an attempt to resist fast food by preserving the cultural cuisine and the associated food and farming of an ecoregion. The movement was started in Italy in the mid-1980s and there are now over 85,000 chapters (known as *convivia*) worldwide.

Although slow food, local food and CSAs, and organic and ecologically sustainable agricultural practices are movements that signal a shift in food production and consumption, they in no way are challenging the dominance of more conventionally produced, distributed, marketed, and consumed food. Moreover, as a number of critics of the movements have pointed out, these alternative practices are largely organized and promoted by white, middle-class members of core regions and exclude, often simply through cost and associated accessibility, poor people of color. The latter are denied these purportedly healthier eating opportunities because they lack the information, income, and proximity to access them. In core countries, the result is that poor people turn to cheap, easily accessible food, also known as fast food.

APPLY YOUR KNOWLEDGE Identify the sites in your region where locally produced food is available, including grocery stores as well as farmers' markets. Create a map of where these sources are located. Is the distribution of outlets even across the city or are there places where no local food is available? Do any particular characteristics of the population correspond to availability or lack of availability? ■

Fast Food

Fast food was born in the United States as a product of the post–World War II economic boom and the social, political, and cultural transformations that occurred in its wake. The concept of **fast food**—edibles that can be prepared and served very quickly in packaged form in a restaurant—actually preceded the name for it. What made fast food *fast*, was the adoption of industrial organizational principles applied to food preparation in the form of the Speedee Service System. The system, invented by Richard and Maurice McDonald in their San Bernardino, California, McDonald Brothers Burger Bar Drive-In in 1948, revolutionized the restaurant business. To preclude the labor and material costs of standard restaurant food preparation, each Burger Bar worker was assigned one task in an assembly-line operation—cooking the burger, placing it on the bun, or packaging it for take-away, for instance—and the product was standardized so that the same condiments—ketchup,

onions, mustard, and two pickles—were added to each patty. As Eric Schlosser writes in his *New York Times* best-selling book *Fast Food Nation*, once Raymond Kroc came along in 1954 and convinced the McDonald brothers to agree to franchise (licensing of trademarks—the golden arches—and methods of doing business), the fast-food concept began to expand all over the United States, so that by 1951, *Merriam-Webster's Dictionary* included it as a new word; and by 1964, there were 657 McDonald franchises. In time, other enterprising individuals seized the fast-food idea and opened hundreds of In-N-Out Burger, Burger King, Chik-Fil-A, and Subway restaurants.

Today, fast food is so ubiquitous in the United States—and is becoming increasingly so throughout the world—that it is taken for granted by many consumers. To appreciate just how omnipresent fast food is, consider the following statistics published by the National Restaurant Association. In the United States in 2010 there were 50,000 fast-food locations nationwide. Serving over 70 billion meals or snacks, their annual sales totaled $558,000 billion. Over 13 million workers are employed in the fast-food industry, making it one of the largest private-sector employers in the country. Every day, 1 in 3.5 Americans visits a fast-food establishment. French fries are the most eaten vegetable in the country and McDonald's feeds more than 47 million people a day—more than the entire population of Spain.

Health practitioners, alternative food activists, journalists, and others have turned a critical eye on fast food in order to expose some of its dietary, labor-related, and ecological shortcomings. The concerns about the effects of fast food on the nation's health have risen as our consumption of (or, as some health experts would argue, addiction to) fast food increases and our levels of routine exercise decrease **(Figure 8.23)**. Worries focus on the increased incidence of obesity (an amount of fatty body mass that is a danger to health) and diabetes (a disorder that affects the human body's use of food for energy), particularly among children. Since the late 1970s, the rate of obesity and diabetes among U.S. children has doubled. In a 2005 study published in the prestigious medical journal *Lancet*, which followed children's eating habits for 15 years, researchers found that study subjects who consumed fast food two or more times a week gained approximately 10 more pounds over the 15-year period and had twice as great an increase in insulin resistance than subjects who consumed fast food less than once per week. More recently, another *Lancet* article reported that by 2010, more than 34 percent of children in North American and eastern Mediterranean World Health Organization regions, 38 percent in Europe, 22 percent in the western Pacific, and 22 percent in southeast Asia are overweight or obese. These numbers, though certainly alarming, are not surprising given how heavily the fast-food industry advertises to children, a large percentage of whom can recognize the golden arches of McDonald's before they know their own name.

Diseases related to the increased consumption of fast food and many prepared, processed foods—energy-dense, nutrient-poor foods with high levels of sugar and saturated fats—as well as reduced physical activity, have also led to adult obesity rates that have risen threefold or more since 1980 across North America, the United Kingdom, Eastern Europe, the Middle East, the Pacific Islands, Australasia, and China. Moreover, the obesity epidemic is often increasing faster in rapidly developing peripheral countries than in the core. The explanation for this epidemic reflects the impacts of

Although most people think of agriculture as a rural activity, urban agriculture made possible the emergence of the world's first cities. Until recently, however, urban agriculture was largely ignored in the development of urban economic policies, apparently because produce generated from it was seen as belonging to the informal sector of the local economy and not significant in terms of income-generating potential. Most definitions of **urban agriculture** focus on the establishment or performance of agricultural practices in or near an urban or city like no keep setting or on the urban fringes, where it's known as **peri-urban agriculture**. In countries like China, official policies have long recognized and even fostered urban and peri-urban agricultural practices. In many core countries, however, particularly since the Industrial Revolution, urban agriculture has been officially discouraged or made difficult as arable land has been used for real estate development or seriously degraded through industrial processes.

Whether encouraged or discouraged by official policy, urban residents across the globe are increasing their participation in growing crops and raising livestock, for reasons ranging from food security to income production to taste and health concerns. A recent publication by the Food and Agricultural Organization of the United Nations has estimated that 800 million people are actively engaged in urban agriculture around the world. And 200 million of these produce goods for market sale. One-half of the vegetables consumed in Havana, Cuba, are grown in the city's farms and gardens. Singapore has 10,000 urban farmers who produce 80 percent of the city's poultry and 25 percent of its vegetables. Currently, 14 percent of Londoners and 44 percent of Vancouver's residents grow some food in their gardens. Most official discussions of urban agriculture indicate that it is primarily engaged in by poor households where people often go hungry for days, although not necessarily the poorest of people participate since those individuals often lack access to land (**Figure 8.H**).

In the United States, up to 30 percent of agricultural production occurs within metropolitan areas. Throughout the core, urban agriculture is often seen as a leisure activity that helps supplement the routine purchase of commercial foodstuffs. Increasingly, however, urban agriculture, in the form of community and school gardens has become a way to improve the diets of low-income adults and children by enabling them to have easy access to fresh, healthy fruits and vegetables and, in some cases, poultry and even beef, pork, and fish (through aquaculture). Detroit, for instance, has experienced a dramatic growth in urban agriculture and provides an interesting case of how poverty in rich countries like the United States is being partly addressed by providing low-income individuals with the skills and the resources they need to grow, and sometimes sell, their own fresh produce.

One organization that emerged in Detroit, a global organization that has gone on to develop community gardens throughout most of the major cities in the United States as well as some areas abroad,

is Urban Farming™. Starting with three gardens in 2005, Urban Farming™ now has over 50,000 community, residential and partner gardens that are a part of the Urban Farming Global Food Chain®, drawing communities together to plant wholesome, fresh food in cities like Los Angeles, Phoenix, Chicago, Denver, New York, and Cincinnati. The mission of Urban Farming™ is "To create an abundance of food for people in need by planting, supporting and encouraging the establishment of gardens on unused land and space while increasing diversity, raising awareness for health and wellness, inspiring and educating youth, adults and seniors to create an economically sustainable system to uplift communities around the globe."[1]

One of the hallmarks of the non-profit organization, Urban Farming™ is that the food from the Urban Farming Community Gardens™ is free for the community. There are many local organizations in most major cities that are involved in reclaiming unused land for urban food production. Within Detroit, a consortium of organizations began in 2003, the Garden Resource Program, which helps to support 875 urban gardens and farms in the metropolitan area (**Figure 8.I**). Organizations supporting urban gardening exist in most U.S. large cities and many towns as well. Gardening has been taken up in K–12 schools. In New York City, where open space is at a premium, organizations such as schools and cooperatives have transformed their rooftops into gardens. Urban Farming™ installed an Edible Wall™ in Harlem in 2009 at an after school facility.

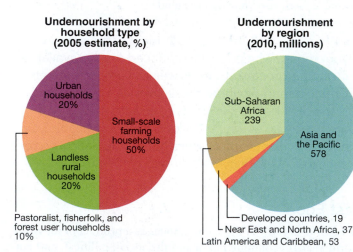

Undernourishment by household type (2005 estimate, %)

- Urban households 20%
- Small-scale farming households 50%
- Landless rural households 20%
- Pastoralist, fisherfolk, and forest user households 10%

Undernourishment by region (2010, millions)

- Sub-Saharan Africa 239
- Asia and the Pacific 578
- Developed countries, 19
- Near East and North Africa, 37
- Latin America and Caribbean, 53

FIGURE 8.H Where the hungry people are Most of the hungry people are in rural areas in developing countries. But as this graph also makes clear, one-fifth of urban populations contain undernourished individuals. Urban hunger is as real as rural hunger, despite the availability of food in urban markets. (Source: Figure 17 on page 32 from *Growing a Better Future: Food Justice in a Resource-Constrained World*, 2011, is reproduced with the permission of Oxfam GB, www.oxfam.org.uk. *Sources:* UN Millennium Project, http://www.unmilleniumproject.org/reports/tf_hunger.htm; FAO http://www.fao.org/hunger/en/ and http://www.fao.org/economic/ess/ess-data/ess-fs/ess-fadata/en/)

[1]urbanfarming.org Web site: http://www.urbanfarming.org/, accessed June 6, 2011.

(a)

FIGURE 8.I Urban gardening movement in Detroit, Michigan Pictured left here are members of a Detroit neighborhood who've converted land close to their homes into a community garden. Pictured below is an Urban Farming Community Garden™.

(b)

FIGURE 8.J Areal photo of how Detroit is growing greener This image shows metropolitan Detroit with a a close-up view of its downtown area. In the view of the entire city, north is up. The Canadian city of Windsor can be seen across the river from Detroit. The shortwave infrared (TM band 5), infrared (TM band 4), and visible green (TM band 2) channels are displayed in the images as red, green, and blue respectively. In this combination, barren and/or recently cultivated land appears red to pink, vegetation appears green, and water is dark blue. Also shown to the left are logos of some of Detroit's many active food organizations.

Teachers use the gardens not only to inform their students about soil quality, crop production, irrigation systems, and other aspects of successful gardening, they also use the gardens to teach about food security, food cultures, the importance of cooperation, and the satisfaction of seeing a challenging goal come to fruition. And gardening has also been taken up on university campuses where students interested in sustainable practices have convinced university officials to give over land to university community gardens.

Throughout the United States and other core countries, urban agriculture is experiencing a resurgence, drawing in a wide range of individuals, organizations, and businesses. Urban Farming™ was featured on the Ellen Degeneres Show in 2010 and has been on the forefront of the urban agriculture movement which has helped fellow gardeners around the world (**Figure 8.J**). With First Lady Michelle Obama planting a vegetable garden on the White House lawn and political figures like Gavin Newsome, former mayor of San Francisco, turning the front lawn of city hall into a vegetable garden in 2008 and worked to bring an Urban Farming Community Garden™ to San Francisco, it's clear that urban agriculture in the United States has left the province of a few isolated garden activists and is becoming more broadly popular, enabling ordinary people to experience sustainable farming practices firsthand.

FIGURE 8.K Urban rooftop gardens in China A green rooftop of a residential building in Chongqing, southwest China, is shown here. Many apartment dwellers here in Chongqing make full use of the space on the rooftops where gardens include green plants, flowers, and vegetables. With inflation squeezing the income of Chinese people, many residents grow vegetables to combat rising commodity prices.

Although urban agriculture is a growing movement in core countries, it is in the periphery where it more often is the sole means of economic and personal survival. As wage cuts, inflation, job loss, civil strife, and natural disasters become more frequent, urban agriculture in the periphery has become a way to address greater food insecurity. This can occur in a number of ways. At the household level, urban agriculture can provide income through the sale of produce at markets. It also provides family members with more nutritionally rich foods and a more diverse diet. Further, urban agriculture can help to stabilize household food consumption against temporary shortages due to layoffs or loss of family member's income due to illness.

As we discuss in Chapter 10, urban populations throughout the world are growing more than twice as fast as rural populations. According to the UN Center for Human Settlements, half of the world's population now lives in cities. And as development experts seek ways of maintaining economic growth without destroying the environment, urban agriculture has increasingly drawn their attention as a way of making cities sustainable. Proponents of urban agriculture contend that it should not be understood as an alternative to conventional agriculture, but rather as a supplementary branch of modern agricultural systems. For most development experts, an ideal urban agricultural system would incorporate various elements of modern, sustainable agriculture

based on reusable, self-contained waste and nutrient cycles through resource conservation and management based on non-chemical fertilizers and pest-management techniques.

It is important to recognize that urban agriculture cannot solve the world's food-security problems. For example, small urban gardens will not replace agribusiness as the primary players in the global food system. Moreover, there are legitimate health concerns surrounding urban agriculture, particularly in terms of recycling urban waste into agricultural inputs. In arid parts of the world, where water is scarce, the use of wastewater from domestic or commercial uses seems an obvious solution to the irrigation needs of urban agriculture. In some parts of the world wastewater is effectively treated and used for secondary applications, yet in others it is not treated and can easily carry disease, which can be spread into the food system when applied to crops. Clearly, policies and practices need to be developed that carefully address the health implications of urban agriculture in the different settings in which it is being practiced.

While health concerns about urban agricultural practices should not be taken lightly, there is evidence that they are far outweighed by the current and potential benefits of urban agriculture. Particularly in developing countries and poorer inner-city neighborhoods throughout the world, urban agriculture can be crucial to a family's survival and is certainly a boost to its overall health.

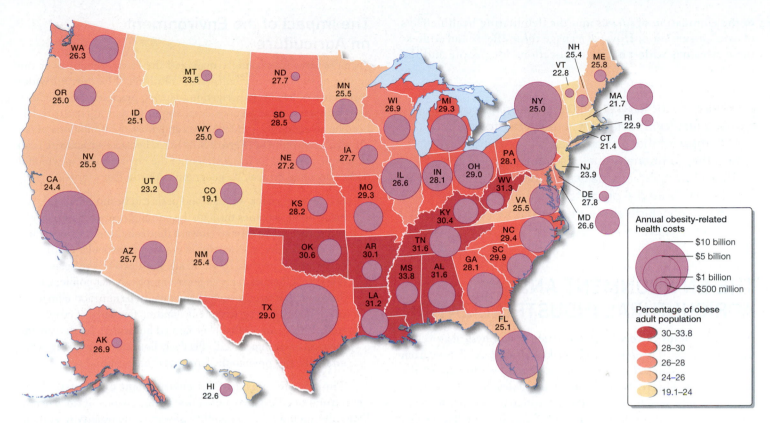

FIGURE 8.23 Economic costs of obesity in the United States, 2009 This map shows the percentage by state of individuals whose Body Mass Index (BMI) is 30 or higher. (BMI measures percentage of body fat based on weight and height; a BMI of 30 or above indicates obesity.) It also provides economic impact data that address health costs, absenteeism, lost employment opportunities, and even the higher cost of gasoline that airlines expend transporting obese people. (*Source:* Data from ODC Behavioral Risk Factor Surveillance System; US Census Bureau, Population Division, December 2009.)

economic growth, modernization, and globalization on daily nutritional practices. The World Health Organization explains:

> As incomes rise and populations become more urban, diets high in complex carbohydrates give way to more varied diets with a higher proportion of fats, saturated fats, and sugars. At the same time, large shifts towards less physically demanding work have been observed worldwide. Moves towards less physical activity are also found in the increasing use of automated transport, technology in the home, and more passive leisure pursuits.[3]

Obesity is of concern to the international health community because it poses a major risk for serious diet-related chronic diseases, including type 2 diabetes, cardiovascular problems, hypertension and stroke, and certain forms of cancer. The consequences range from increased risk of premature death to grave chronic conditions that reduce the overall quality of life.

In addition to its negative impact on human health, the rapacious growth of fast-food production and distribution processes is dramatically affecting Earth's resources, especially forests and farmland. The Beyond Beef Campaign, an international coalition of environment, food safety, and health activist organizations, lists the following facts about beef production and its environmental impacts:

One quarter-pound of hamburger beef imported from Latin America requires the clearing of 6 square yards of rain forest and the destruction of 165 pounds of living matter, including 20 to 30 different plant species, 100 insect species, and dozens of bird, mammal, and reptile species.

Cattle degrade the land by stripping vegetation and compacting the earth. Each pound of feedlot steak costs about 35 pounds of eroded topsoil.

Nearly half of the total amount of water used annually in the United States goes to grow feed and provide drinking water for cattle and other livestock. Producing a pound of grain-fed steak requires almost 1,200 gallons of water.

Cattle produce nearly 1 billion tons of organic waste each year. The average feedlot steer produces more than 47 pounds of manure every 24 hours.

Much of the carbon dioxide released into the atmosphere is directly attributable to beef production: burning forests to make way for cattle pasture and massive tracts of agricultural waste from cattle feed crops.

U.S. cattle production has caused a significant loss of biodiversity on both public and private lands. More plant species in the United States have been eliminated or threatened by livestock grazing than by any other cause.[4]

The impacts of fast-food production and consumption on people and places are complex, from the convenience for busy consumers and the provision of play spaces for children

[3]Source: World Health Organization, accessed June 9, 2011, http://www.who.int/dietphysicalactivity/publications/facts/obesity/en.

[4]Source: Beyond Beef, accessed June 6. 2011, http://www.mcspotlight.org/media/reports/beyond.html.

to the elimination of forests and the debilitating health effects of overconsumption. How we manage these effects and address the causes has wide-ranging implications for people and the planet.

> **APPLY YOUR KNOWLEDGE** Determine how many fast-food restaurants are in your town. Pick three of these and research their menu and the nutritional value of several popular items. This information is usually available on the Internet. Note the prices for these items. What can you conclude about the nutritional value of each item and its price? ■

THE ENVIRONMENT AND AGRICULTURAL INDUSTRIALIZATION

Agriculture always involves the interaction of biophysical as well as human systems. This relationship makes agriculture distinct from forms of economic activity that do not depend so directly on the environment. This relationship also necessitates that communities determine how best to manage the environment in order to facilitate the continued production of food. Because the relationships between the human system of agriculture and the biophysical system of the environment are highly interactive, it is important to look at the ways they shape each other.

The Impact of the Environment on Agriculture

Farmers have increasingly managed the environment over the course of the three agricultural revolutions. The current widespread use of fertilizers, irrigation systems, pesticides, herbicides, and industrial greenhouses even suggests that agriculture has become an economic practice that can ignore the limitations of the physical environment **(Figure 8.24)**. Yet because agriculture is an economic activity, its management of the environment in which it occurs becomes critical. As geographer Martin Parry writes:

> Soil, terrain, water, weather, and pests can be modified and many of the activities through the farming year, such as tillage and spraying, are directed toward this. But these activities must be cost-effective; the benefits of growing a particular crop, or increasing its yield by fertilizing, must exceed the costs of doing so. Often such practices are simply not economic, with the result that factors such as soil quality, terrain, and climate continue to affect agriculture by limiting the range of crops and animals that can profitably be farmed be farmed. In this way the physical environment still effectively limits the range of agricultural activities open to the farmer at each location.[5]

Though the impact of the environment on industrialized agricultural practices may not at first seem obvious, it does occur. Examples include the availability of water, the quality of soil, air temperature, and length of the growing season, among other factors. Of course, on the other hand, there are also many readily observable contemporary and historical examples of the ways

FIGURE 8.24 Modern irrigation system Modern irrigation systems are electronically programmed to deliver different amounts of water at different times of the day or days of the week. This figure shows a computer screen with a program that allows farmers to calculate the quantity of water used to irrigate the lands on the outskirts of Valencia, Spain. The Moorish invaders that once ruled Spain brought with them an irrigation system that helped turn arid land into verdant fields. More than 1,000 years later it is still largely in use, and Spain remains one of Europe's breadbaskets. But after years of chronic drought coupled with vastly increased water use, not to mention climatic change, farm groups have realized it's time to go digital.

[5]M. Parry, "Agriculture as a Resource System," in I. Bowler (ed.), *The Geography of Agriculture in Development Market Economics*. Harlow, England: Longman Scientific and Technical, 1992, p. 208.

that agriculture destroys, depletes, or degrades the environmental resources on which its existence and profitability depend, as we discuss in the material that follows.

The Impact of Agriculture on the Environment

As discussed in Chapter 4, one of the earliest treatises on the impact of chemical pesticides on the environment was Rachel Carson's *Silent Spring*, which identified the detrimental impacts of synthetic chemical pesticides—especially DDT—on the health of human and animal populations **(Figure 8.25)**. Although the publication of the book and the environmental awareness that it generated led to a ban on the use of many pesticides in most industrialized nations, chemical companies continued to produce and market these types of products in peripheral countries. Some of these pesticides, aimed at combating malaria and other insect-borne diseases, are applied to crops that are sold in the markets of developed countries. Thus, a kind of "circle of poison" has been set into motion, encompassing the entire global agricultural system.

Among the most pressing issues facing agricultural producers today are soil degradation and denudation, which are occurring at rates more than a thousand times the natural rates. Although we in the United States tend to dismiss soil problems such as erosion as an artifact of the 1930s Dust Bowl, the effects of agriculture on worldwide soil resources are dramatic, as **Table 8.3** illustrates. Most forms of agriculture tend to increase soil degradation. Although severe problems of soil degradation persist in the United States—which has a federal agency devoted exclusively to managing soil conservation—more severe problems are occurring in peripheral countries.

FIGURE 8.25 Impact of pesticides on pollinators Pollination is the transfer of pollen from one flower to another and is critical to fruit and seed production. Pollination is frequently provided by insects seeking nectar, pollen or other floral rewards. The U.S. Department of Agriculture, however, reports that the country is facing a pollination crisis as both wild and managed pollinators are disappearing at alarming rates as a result pesticide poisoning, as well as habitat loss, diseases and pests.

TABLE 8.3   Global Soil Degradation

Region	Degrading area (km²)	mi²	% Territory	% Global degrading area	% Total population	Affected people
Asia	9,128,498	3,524,533	25.39	0.890	23.58	1,070,737,071
Africa	6,596,641	2,546,977	26.65	0.441	28.31	230,604,253
Europe	656,007	253,286	12.88	0.089	9.75	48,457,913
South America	4,719,162	1,822,079	24.86	0.457	22.90	149,905,245
Oceania	2,364,959	913,116	34.68	1.224	26.29	5,494,554
North America	3,968,971	1,532,428	20.25	9.755	14.24	36,654,152
World Total (land, excluding inland water body)	35,058,104	1,35,360,010	23.54	100.000	23.89	1,537,679,148

Source: Adapted from Z. G. Bai, D. L. Dent, L. Olsson, and M. E. Schaepman, *Global Assessment of Land Degradation and Improvement 1: Identification by Remote Sensing.* Report 2008/01, FAO/ISRIC, Rome and Wageningen, 2008, p. 24, Table 1.

FIGURE 8.26 Desertification, Gansu, China A woman walks along the leading edge of the Kumtag Desert as it threatens to engulf her onion farm. The sand is advancing at the rate of up to 4 meters (13 feet) a year.

The problem of soil degradation and loss is particularly critical in the humid tropical areas of the globe—especially in South America and Asia. Arguably the most critical hotspot with respect to soil and related environmental impacts is the moist Brazilian *cerrado,* or grassland, which is being converted to soybean production in order to feed the growing biofuels industry (which we discuss in more detail in the next section). In addition to soil erosion and degradation, the conversion of the *cerrado* is also threatening the species richness of birds, fishes, reptiles, amphibians, and insects. Some local animal and plant species face extinction.

The quantity and quality of soil worldwide are important determining factors for the quantity and quality of food that can be produced **(Figure 8.26)**. The loss of topsoil worldwide is a critical problem because topsoil is a fixed resource that cannot be readily replaced. It takes, on average, 100 to 500 years to generate 10 millimeters (½ inch) of topsoil, and it is estimated that nearly 50,000 million metric tons (55,000 million tons) of topsoil are lost each year to erosion.

The nature-society relationship previously discussed in Chapter 4 is very much at the heart of agricultural practices. Yet as agriculture has industrialized, its impacts on the environment have multiplied and in some parts of the globe are at crisis stage. In some regions the agricultural system leads to overproduction of foodstuffs, but in others the quantity and quality of water and soil severely limit the ability of a region's people to feed themselves.

APPLY YOUR KNOWLEDGE Choose three common commercial crops (such as coffee, cotton, citrus, lettuce, potatoes, or peanuts) that are grown in the United States. Identify the areas in the country where these three crops are grown and consult soil, temperature, and rainfall maps to learn what the local environmental conditions are that enable these crops to be grown in those areas. Next, identify what effect crop cultivation has on that same environment. Which of the crops has the most negative impact on the local environment? Which has the least? ■

EMERGING PROBLEMS AND OPPORTUNITIES IN THE GLOBAL FOOD SYSTEM

In this final section, we examine two problematic issues in the world food system, as well as an encouraging new prospect. These cases certainly do not illuminate the myriad challenges and possibilities facing food producers and policymakers today, but they do provide a sense of the broad range of issues.

Food and Health

We have spent most of this chapter describing the ways food is cultivated, processed, engineered, marketed, financed, and consumed throughout the world. What we have yet to do is talk about access to this most essential of resources. Although there is more than enough food to feed all the people who inhabit Earth, access to food is uneven, and many millions of individuals in both the core and the periphery have had their lives shortened or harmed because war, poverty, or natural disaster has prevented them from securing adequate nutrition. In fact, hunger is very likely the most pressing problem facing the world today **(Figure 8.27)**.

Hunger can be chronic or acute. Chronic hunger is nutritional deprivation that occurs over a sustained period of time: months or even years. Acute hunger is short-term and is often related to catastrophic events—personal or systemic. Chronic hunger, also known as **undernutrition**, is the inadequate intake of one or more nutrients and/or of calories. Undernutrition can occur in individuals of all ages, but its effects on children are dramatic, leading to stunted growth, inadequate brain development, and a host of other serious physical ailments.

Globally, 24,000 people a day die from the complications brought about by undernutrition. Children are the most frequent victims. Children who are poorly nourished suffer up to 160 days of illness each year, and poor nutrition plays a role in at least half of the 10.9 million child deaths each year. Moreover, undernutrition magnifies the effect of every disease, including measles and malaria.

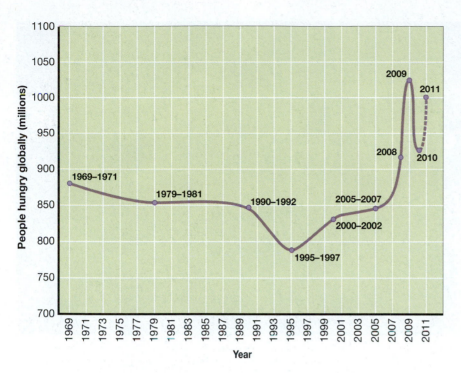

FIGURE 8.27 Number of hungry people worldwide Almost one in seven people worldwide is chronically undernourished. After steadily declining, global hunger began to rise again in the 1990s. The number of hungry people peaked in 2009 at 1 billion and then dropped in 2010. The number is, however, expected to rise again in 2011. Most hungry people live in rural areas and are small-scale producers.

(*Source:* Figure 16 on page 31 is adapted from *Growing a Better Future: Food Justice in a Resource-Constrained World,* 2011, with the permission of Oxfam GB, www.oxfam.org.uk. *Sources:* FAO, http://www.fao.org/hunger and Financial Times, http://cachef.ft.com/cms/s/0/68b31de6-329e-11e)-97ca-00144feabdc),s01=2.html.)

Malnutrition can also be caused by diseases, such as the diseases that cause diarrhea, by reducing the body's ability to convert food into usable nutrients. **Malnutrition** is the condition that develops when the body does not get the right amount of the vitamins, minerals, and other nutrients it needs to maintain healthy tissues and organ function. A person with malnutrition can be either under- or overnourished.

Perhaps the most widely publicized examples of acute hunger are famines, especially those that have occurred in parts of the periphery over the last few decades. **Famine** is acute starvation associated with a sharp increase in mortality. The most widely publicized famines of the late twentieth century occurred in Bangladesh in 1974 and Ethiopia in 1984–1985. The causes of these two famines (and the other twentieth-century famines that preceded them) were complex. The crisis of starving people so often publicized by the news media is usually just the final stage of a process that has been unfolding for a far longer period, sometimes years or even decades. Experts who study famine argue that there are at least two critical factors behind long-standing vulnerability to famine. The first has to do with a population's command over food resources in terms of their livelihood. The second has to do with a trigger mechanism, which may be a natural phenomenon like drought or a human-made situation such as civil war.

People who study famine and other forms of hunger have come to conceptualize nutritional vulnerability in terms of the notion of food security. **Food security** for a person, a household, or even a country is assured access to enough food at all times to ensure active and healthy lives. And while famine is a dramatic reminder of the precarious nature of food security, it is important to appreciate that chronic hunger resulting from food insecurity is a far more widespread and devastating problem than famine, which tends to be shorter in duration and more contained geographically. Related to food security is the concept of **food sovereignty,** the right of peoples, communities, and countries to define their own agricultural, labor, fishing, food, and land policies that are ecologically, socially, economically, and culturally appropriate to their unique circumstances.

Today in the United States, where food is abundant and overeating is a national problem, 10 percent of the population is undernourished or experiences food-security problems at one time or another each year. In the periphery, where food availability is more limited than in core countries like the United States, undernutrition is far more pervasive. Furthermore, in some parts of the world, among some social classes, there are higher levels of undernutrition among women and girls than among men and boys. This is largely because different cultural and social norms favor men and boys, who eat first, leaving the leftovers for females, or who eat certain high-status foods, such as proteins like meat or fish, that women are not allowed to eat.

As mentioned in the chapter-opening content, one of the factors thought to be adding to the most recent food emergency is the increasing amount of cropland around the globe now being redirected to raising biofuels. Land grabs for biofuels are happening across Asia, Latin America, and Africa and often involve violence. Some 150,000 families in Argentina and 90,000 families in Paraguay have been displaced by soy production. In Tanzania, the government has handed over 9,000 hectares of land on which over 11,000 people depend for their livelihoods to the U.K. firm Sun Biofuels PLC for a jatropha (a shrubby plant that produces seeds that are 40 percent oil) plantation (**Figure 8.28**). A recent report by the World Bank states that biofuels production has forced

FIGURE 8.28 Jatropha plantation Workers are shown here in a field of jatropha plants, a source of bio-diesel, at the International Crops Research Institute for the Semi-Arid Tropics, in Hyderabad, India. Bio-diesel contains no petroleum, but it can be blended with petroleum diesel to create a Bio-diesel blend or can be used in its pure form for fuel.

global food prices up by 75 percent. It is our view that the price of food is a complex variable affected by causes that include the emergence and growth of biofuels production but is also related to rising demand among increasingly more affluent populations, commodities speculation, rapidly escalating energy prices, and poor harvests.

While it is important to understand the complex causes and repercussions of hunger and undernutrition, perhaps the most relevant point to take away from this discussion is that it is a problem that can be solved. Neither short-term nor long-term hunger is due to inadequate supply, so the solution must lie in improving access to supplies. This could occur, under a radical scenario, through a massive redistribution that would give all the world's people access to the same amount of food resources. Barring such a dramatic restructuring of the current economic system, the solution to the problem of hunger lies in improving access to livelihoods that pay well enough that adequate nutrition becomes a human right and not dependent upon the vagaries of economic or natural systems.

At the same time, hunger is a serious problem for peripheral populations worldwide. For both core and peripheral populations, issues of food safety have also grown in significance. Over the last several years, "food scares" caused by tainted foodstuffs have been routinely in the headlines: California spinach tainted with *E. coli* bacteria, *salmonella* in canned tomatoes from Mexico and in hummus in the United Kingdom, toxic chemicals routinely added to pickled vegetables in China, *E. coli* tainted cucumbers in Germany, and potentially cancer-causing dioxins (which entered grazing pastures through illegally dumped toxic waste) in buffalo milk used in the production of Italian mozzarella.

The impact of food scares is widespread, affecting not only those who have become ill or died but also farmers and grocers as well as the credibility of government agencies meant to protect consumers from tainted and toxic foods. Mostly, the recent apparent escalation in food scares reminds consumers of the vulnerability of the global food system.

Genetically Modified Organisms

A **genetically modified organism**, or **GMO** as it is more commonly known, is any organism that has had its DNA modified in a laboratory rather than through cross-pollination or other forms of evolution. Examples of GMOs include a bell pepper with DNA from a fish added to make it more drought-tolerant, a potato that releases its own pesticide, and a soybean that has been genetically engineered to resist fungus.

Genetic modification has both critics and supporters. Proponents argue that it allows great advances in agriculture (for instance, plants that are more resistant to certain diseases or water shortages), as well as other beneficial creations, such as the petroleum-eating bacteria that can help clean up oil spills. But opponents worry that genetically modified organisms may have unexpected and irreversible effects on human health and the environment and result in maturation problems in children or in mutant plant and animal species (**Figure 8.29**).

In the United States, genetic modification is permitted, on the principle that there is no evidence yet that it is dangerous. GMO foods are fairly common in the United States, and estimates of their market saturation vary widely. It is not easy to identify GMOs in the grocery store, as there are no labeling requirements. While the U.S. food-safety establishment maintains that GMOs are safe until proved otherwise, countries in Europe have taken the opposite position: genetic modification has not been proved safe, so they will not accept genetically modified food from the United States or any other country. The World Trade Organization has determined that not allowing GMO food into a country creates an unnecessary obstacle to international trade.

GMOs represent perhaps the most highly technological way to date in which nature and society come together in the global food system. At present, so little is known about the impacts of GMOs on human health, the environment, or even the wider global economic system that it is difficult to sort the costs from

FIGURE 8.29 Protest over GMOs, Paris, France Food activist and leader of the *Confederation paysanne*, a French agricultural union that supports sustainable agriculture, Jose Bove (center) leads thousands of GMO activists in protest in the streets of Paris close to the Assemblée Nationale where ministers were voting on GMO laws in spring 2008.

the benefits of their increasing incorporation into global food production. What is clear is that genetic modification is not just a passing fad and the debate cannot be reduced to a simple question of "good" vs. "bad." GMOs are neither entirely evil nor entirely good; certain applications may be widely beneficial, while others may not. Regulatory structures are crucial to protecting human health and the environment, as well as spreading whatever benefits may accrue from GMOs beyond the core and into the periphery. But regulatory structures are not easily implemented.

Protests against GMO regulatory structures have been most effective in Europe and parts of Asia, Africa, and Latin America as well as in Canada and Mexico. In these regions, national governments are devising regulations to control or publicize the entry of GMO commodities into the food system and are requiring more research in order to better understand the long-term effects of GMO consumption on humans as well as the impact of GMOs on the food chain. Widespread protests against GMO foods have also been organized in the United States, but the government has not responded with support for these movements. Congress has passed domestic legislation making the labeling of GMOs voluntary, and at the international level the American government has threatened action against countries who are arguing for the mandatory labeling of GMO foods traded in the global marketplace. It is important to

be aware that the position taken by the United States is pro-trade, rather than strictly pro-GMO. The U.S. government is mostly concerned with the economic impact GMO labeling is likely to have and is reluctant to put the trade of U.S. agricultural products at risk. This strategy is followed because the United States is the largest exporter of agricultural goods in the world as well as the largest producer of GMO foods.

Global debate and activism—both popular and governmental—over GMOs is still in its early stages. Some of the most vocal opponents of GMOs are concerned that engineered food is destined for consumption by poor people in the periphery, while "real" food, produced by artisans and organic growers, will only be available to rich people in the core and little attention will be paid to the importance of reducing world hunger or access to a safe diet as a global human right.

APPLY YOUR KNOWLEDGE   Search the Internet for news stories on recent protests against GMO food. What are the protesters' concerns and how do they go about pressing their demands? Write a paragraph stating whether or not you believe their concerns are justified. Provide scientific evidence to support your argument. ∎

Future Geographies

We began this chapter with a story about the impending global food crisis that is predicted to be full-blown by the year 2030. Not surprisingly, the crisis is expected to affect poor people in developing countries where food resources are already scarce for political, climatological, and financial reasons. There is very little

disagreement among policymakers and scholars that this food crisis will occur unless the world's governments take a different attitude and approach to food production (**Figure 8.30**). An effective approach requires global cooperation, the support of small farmers across the globe, and an appreciation for how sustainable

FIGURE 8.30 Where governments are investing As this graphic shows, investment in food is one of the smallest expenditures that governments make. Fossil-fuel consumption subsidies are the greatest expenditure. There needs to be a reversal with respect to these choices or we face a future where more than 7 billion people will go hungry. (*Source:* Figure 24 on page 60 is adapted from *Growing a Better Future: Food Justice in a Resource-Constrained World*, 2011, with the permission of Oxfam GB, www.oxfam.org.uk. *Sources:* Clean Energy Progress Report, OECD/IEA 2011; IEA (2010) World Energy Outlook; Government contributions to WFP in 2009 were $3.47 billion; http://www.wfp.org/about/donors/wfp-donors; OECD Producer Support Estimates – 2009 estimate of $252, 522 million; OECD DAC5 Official Bilateral Commitments by Sector (total for all donors, 2009. Includes agriculture, forestry & fishing).

$3.5 billion
Contributions to WFP (World Food Program)

$9.8 billion
ODA (official development assistance) for agriculture

$20 billion
Biofuels subsidies

$57 billion
Worldwide subsidies for renewable energy

$252 billion
Industrialized countries' agricultural support

$312 billion
Worldwide subsidies for fossil fuels (consumption only)

agricultural practices are and a recognition of the impacts of climate change on growing seasons and farmlands. There is reason to be optimistic, as policymakers are already responding to the anticipated problem and small farmers are beginning to receive resources and financial support from governments and nonprofit organizations. But whether these efforts will be adequate is still unclear.

Other trends that are currently unfolding suggest that issues like obesity in the United States and elsewhere may begin to decline as more and more core consumers express their desire for the increased availability of healthy foods in restaurants as well as in grocery stores and markets. There are signs for optimism in the growing alternative food movements that exist not only in wealthy countries but also in India and South America, where food is becoming so expensive as to be beyond the reach of millions. Sustainable, community gardening will not be able to feed the world, but it may be able to provide healthy produce to those who cannot afford to buy their food on the open market.

The global production systems seem poised to continue to produce vast quantities of grain, enough to feed the world. But feeding the world will also require the recognition that not all people have the resources to afford the increasing cost of food on the world market. Issues of redistribution of food also need to be confronted. Food revolutions such as biotechnology are likely to continue to expand, but the results of new approaches are not likely to be equally affordable across the socioeconomic spectrum.

The future of the global food system is being shaped at this very moment in food science laboratories, in corporate boardrooms, on the street in organized protests, and in settlements throughout the world. The biggest issues food-policy experts, national governments, consumers, and agriculturalists face concern the availability and quality of food in a world where access to safe, healthy, and nutritious foodstuffs is unevenly distributed. For the periphery, the most pressing concern is adequate food supplies to feed growing populations. For the core, concerns about food quality abound in a system that is increasingly industrialized and biologically engineered.

CONCLUSION

Agriculture has become a highly complex, globally integrated system. While traditional forms of agricultural practices, such as subsistence farming, continue to exist, they are overshadowed by the global industrialization of agriculture. This industrialization includes not only mechanization and chemical applications but also the linking of the agricultural sector to the manufacturing, service, and finance sectors of the economy. In addition, states have become important players in the regulation and support of agriculture at all levels, from the local to the global.

The dramatic changes that have occurred in agriculture affect different places and different social groups. Households in both the core and the periphery have strained to adjust to these changes, often disrupting existing patterns of authority and access to resources. Just as people have been affected by the transformations in global agriculture, so have the land, air, and water.

The geography of agriculture today is a far cry from what it was 100 or even 50 years ago. As the globalization of the economy has accelerated in the last few decades, so has the globalization of agriculture. The changes in global agriculture do not necessarily mean increased prosperity in the core, nor are the implications of these changes simple. For example, the production of oranges in Florida is directly influenced by the newer Brazilian orange industry. Both industries affect the prices of oranges in the marketplaces of Europe and Asia. Additionally, other forces, such as social reactions to genetically engineered foods, agricultural research, trade concerns, and a host of other factors, have repercussions throughout the world food system.

Learning Outcomes Revisited

- Compare and contrast traditional agriculture practices across the globe.

  Although traditional agricultural practices, including subsistence farming, shifting cultivation, and pastoralism, no longer dominate agricultural practices on a global scale, they are still engaged in throughout the world, in some cases alongside more mechanized forms. Globalization, in addition to restructuring entire national farming systems, has also transformed farming households in the core, periphery, and semi-periphery. Thus, traditional forms of agriculture are waning as more and more places are irresistibly drawn into a globalized economy supported by a strong commercial agricultural sector.

- Describe the three revolutionary phases of agricultural development, from the domestication of plants and animals to the latest developments in biotechnology and industrial innovation.

  The three revolutionary phases have not occurred simultaneously throughout the globe, but have been adopted and adapted to differing degrees, based on levels of development, culture, and physical geography. The first phase involved the domestication of seeds and animals. The second revolved around the improvement of outputs and innovations for making farming more efficient, such as an improved yoke for oxen. The third was based on inputs to production such as fertilizers and field drainage systems.

- Analyze the ways that the forces, institutions, and organizational forms of globalization have transformed agriculture.

  Two of the most important forces behind agricultural transformation are multinational and transnational corporations and states. Institutions like the World Trade Organization as well as regional associations like the European Union have also been important influences. And the organization of agriculture itself has experienced significant changes as it has moved from a family-oriented business model to a corporate undertaking that stretches across national boundaries.

- Examine the organization of the agro-commodity system from the farm to the retail outlet, including the different economic sectors and corporate forms.

  The farm is no longer the central piece in the chain of agricultural organization, but one of several important components that include seed and fertilizer manufacturers, food processors, food distributors, and consumers. The organizational structure of agriculture is composed of five central and connected sectors (inputs, production, processing, distribution, and consumption) with four contextual elements acting as external mediating forces (the state, international trade, the physical environment, and credit and finance).

- Scrutinize the ways that agriculture has transformed the environment, including soil erosion, desertification, deforestation, soil and water pollution, and plant and animal species degradation.

  While most of the core countries have instituted legislation to address some of the environmental problems associated with agriculture, these problems exist throughout the global system to greater and lesser degrees. As agriculture has industrialized, its impacts on the environment have multiplied and spread so that some parts of the globe are at crisis stage. In some regions, the agricultural system has led to overproduction of foodstuffs, but in others the quantity and quality of food production is severely limited by physical constraints and environmental degradation. The challenge of the twenty-first century is to work toward a more sustainable relationship between humans and the environment, especially with respect to food production.

■ Probe the current issues food-policy experts, national governments, consumers, and agriculturalists face with respect to the availability and quality of food as well as the alternative practices that are emerging to address some of these issues in a world where access to safe, healthy, and nutritious foodstuffs is unevenly distributed.

Genetic modification is one way of improving productivity, though it does not address issues of access to food. Opportunities for the world's poor—who are increasingly residing in urban settings—to grow their own food is another way. Another way to improve food availability and quality is to recognize access to food as a human right and work toward more even distribution. Other responses include promoting more sustainable farming practices and supporting small farmers' efforts to produce efficiently and effectively.

KEY TERMS

agrarian *(p. 260)*

agribusiness *(p. 281)*

agricultural industrialization *(p. 270)*

agriculture *(p. 260)*

aquaculture *(p. 275)*

biofuels *(p. 258)*

biopharming *(p. 274)*

Biorevolution *(p. 274)*

biotechnology *(p. 274)*

Blue Revolution *(p. 276)*

Borlaug hypothesis *(p. 278)*

chemical farming *(p. 268)*

commercial agriculture *(p. 261)*

contract farming *(p. 274)*

conventional farming *(p. 283)*

cost/price squeeze *(p. 276)*

crop rotation *(p. 261)*

double cropping *(p. 265)*

famine *(p. 293)*

fast food *(p. 285)*

food supply chain *(p. 281)*

food manufacturing *(p. 268)*

food regime *(p. 282)*

food security *(p. 293)*

food sovereignty *(p. 293)*

globalized agriculture *(p. 279)*

GMO (genetically modified organism) *(p. 294)*

Green Revolution *(p. 270)*

hunting and gathering *(p. 260)*

intensive subsistence agriculture *(p. 264)*

intertillage *(p. 263)*

local food *(p. 283)*

malnutrition *(p. 293)*

mechanization *(p. 268)*

nontraditional agricultural exports (NTAEs) *(p. 270)*

organic farming *(p. 283)*

pastoralism *(p. 265)*

shifting cultivation *(p. 261)*

slash-and-burn *(p. 262)*

slow food *(p. 285)*

subsistence agriculture *(p. 261)*

swidden *(p. 263)*

transhumance *(p. 265)*

undernutrition *(p. 292)*

urban and peri-urban agriculture *(p. 286)*

REVIEW AND DISCUSSION

1. Search the term "Green Revolution" on the Internet. From what you find, create a time line of the Green Revolution identifying its most significant milestones as well as the countries it has most greatly affected. Make a list of the pros and cons of the Green Revolution and decide where you stand with respect to the controversy and why.

2. Visit your local supermarket and identify seven different types of produce. Determine what company produced them and where and how they were grown. You may need to ask the produce manager where they come from but the tiny labels (with the PLU codes) will tell you whether they are commercially grown, organic, or GMO crops. Once

you have determined the source of the produce, calculate how far it traveled (food miles) to reach your town.

3. Explore further the impact of the environment on agriculture by researching the projections that are being made about climate change and agricultural change. How will climate change affect the traditional areas of the globe where staples (such as rice, wheat, potatoes) are grown? How will it affect the areas where more popular fruits (such as oranges, apples, and wine grapes) are grown? Speculate on what you think the political and economic effects of the changing geography of global agriculture will be.

UNPLUGGED

1. The Food and Agriculture Organization (FAO) has published a range of yearbooks containing statistical data on many aspects of global food production since the mid-1950s. Using these *State of Food and Agricultural Production* yearbooks, compare the changes that have occurred in agricultural production between the core and the periphery since the middle of the last century. You can use just two yearbooks for this exercise, or you may want to use several to get a better sense of when and where the most significant changes have occurred. Once you have identified where the changes have been most significant, craft an explanation as to why these changes may have occurred.

2. The U.S. Department of Agriculture (USDA) also provides statistics on food and agricultural production, though, of course, data is limited to the United States. Contained in volumes called *Agricultural Statistics* are a range of important variables, from what is being grown where, to who is working on farms, and what kinds of subsidies the government is providing. Drawing on the USDA's *Agricultural Statistics,* examine the changing patterns of federal subsidies. Using a map of the United States, show which states have received subsidies from the 1940s (immediately following the Great Depression) to the present day. Have subsidies increased for some parts of the country and not others? Have subsidies increased or decreased overall for

the entire country? Which farm sectors and, therefore, which regions have most heavily benefited from federal agricultural subsidies? Why? Summarize and explain your findings.

3. Odds are your breakfast is the result of the activities of a whole chain of producers, processors, distributors, and retailers whose interactions provide insights into both the globalization of food production and the industrialization of agriculture. Consider the various foods you consume in your typical breakfast and describe not only where (and by whom) they were produced—grown and processed—how they were transported (by whom) from the processing site—but also where and by whom they were retailed. Summarize how the various components of your meal illustrate both the globalization and the industrialization of agriculture.

Log in to **www.masteringgeography.com** for MapMaster™ interactive maps, geography videos, RSS feeds, flash cards, Web links, an eText version of *Human Geography: Places and Regions in Global Context,* and self-study quizzes to enhance your study of food and agriculture.

MapMaster™ presents 13 Place Name and 13 Layered Thematic interactive maps to help students practice and master their geographic literacy, spatial reasoning, and critical thinking skills.

9

POLITICAL GEOGRAPHIES

In January 2011, the people of Sudan voted on whether the ten states of the southern part of the country, which has been autonomously governed since 2005, would separate from the north and become an independent country. The referendum was the result of the longest civil war in African history. Flaring first from 1955 until 1972, the war was reignited in 1983 and ended in 2005 when the Comprehensive Peace Agreement was reached. Roughly 2 million people died as a result of the war; another 4 million were displaced.

Sudan's boundaries were determined by the British, who ruled (in an awkward partnership with Egypt) from 1899 until 1955. The North, where the capital Khartoum is located, is populated by some 22 million people, most of them Arabic-speaking Muslims (residents also use a non-Arabic mother tongue). It is more economically developed and contains most of the country's urban centers. The South, which contains about 6 million people, is largely rural with a subsistence-based economy. Southern Sudanese people practice indigenous religions, though some of them are Christians. The South is home to many more tribal groups, and many more distinct languages are spoken there. The North is largely a desert region, whereas the South contains both deserts and tropical areas.

The animosity between North and South has been characterized in several different ways: as a racial conflict between Arabs and Africans and as a religious conflict between Muslims

Citizens of the new state of South Sudan celebrating on independence day.

and Christians and animists. But it is also the case that the kingdoms and powerful rulers of the North have exploited the southern peoples for centuries. Britain added to this tension by treating the two parts of the country separately during the colonial period and continuing to exploit the underdeveloped south to the advantage of the north.

The January 2011 referendum, in which nearly 98 percent of eligible voters participated, provided a landslide of support for secession. Africa's newest nation gained formal independence on July 9, 2011. Unfortunately, the path to full independence continues to be difficult because of many unresolved issues. Among these are a North-South border that has never been clearly defined. The border region dissects water basins and differing tribal areas, creating an arbitrary divide that may separate peoples and their livelihoods. Some areas of the border remain contested by the two governments as well as local communities and tribes who fail to recognize any demarcations whatsoever. Northern air and ground attacks upon Abeyi, the border state that contains rich oil reserves, also continue, as no agreement has been made on how to share the oil, Sudan's most valuable natural resource. Finally, tribal differences in the South stand in the way of popular political accord. A more unified population will be critical for the new state—called South Sudan—to succeed. ■

THE DEVELOPMENT OF POLITICAL GEOGRAPHY

This chapter explores how new maps and geopolitical arrangements continue to emerge around the globe at the same time that established boundaries persist. Exploration, imperialism, colonization, decolonization, and the Cold War between East and West are powerful forces that have created and transformed national boundaries. Much of the political strife that currently grips the globe involves local or regional responses to the impacts of globalization of the economy. The complex relationships between politics and geography are two-way. In addition, political geography is not just about global or international relationships. It is also about the many other geographic and political divisions that stretch from the globe to the neighborhood and to the individual body.

Political geography is a long-established subfield in the wider discipline of geography. The ancient Greek philosopher Aristotle is often considered the first political geographer. His model of the state is based upon factors such as climate, terrain, and the relationship between population and territory. Other important political geographers have promoted theories of the state that incorporate elements of the landscape and the physical environment as well as the population characteristics of regions. From about the fourteenth through the nineteenth century, scholars interested in political geography theorized that the state operated cyclically and organically. They believed that states consolidated and fragmented based on complex relationships among and between factors such as population size and composition, agricultural productivity, land area, and the role of the city.

Political geography at the turn of the nineteenth century was influenced by two important traditions within the wider discipline of geography: the people-land tradition and environmental determinism. Different theorists placed more or less emphasis on each of these traditions in their own political geographic formulations. The traditions' effects are evident in the factors deemed important to state growth and change. Why these factors were identified as central undoubtedly had much to do with the widespread influence of Charles Darwin on intellectual and social life. His theory of competition inspired political geographers to conceptualize the state as a kind of biological organism that grew and contracted in response to external factors and forces. It was also during the late nineteenth century that foreign policy as a focus of state activity began to be studied. This new field came to be called *geopolitics*.

The Geopolitical Model of the State

Geopolitics is the state's power to control space or territory and shape international political relations. Within the discipline of geography, geopolitical theory originated with Friedrich Ratzel (1844–1904), a German trained in biology and chemistry. Ratzel was greatly influenced by the theories embodied in *social Darwinism* that emerged in the mid-nineteenth century. His model portrays the state as behaving like a biological organism, with its growth and change seen as "natural" and inevitable. Ratzel's views continue to influence theorizing about the state today. What has been most enduring about his conceptualization is the conviction that geopolitics stems from the interactions of power and territory.

Although it has evolved since Ratzel first introduced the concept, geopolitical theory has become one of the cornerstones of contemporary political geography and state foreign policy more generally. And, although the organic view of the state has been abandoned, the twin features of power and territory still lie at the heart of political geography. In fact, the changes that have occurred in Africa, Europe, and the former Soviet Union over the last 25 years suggest that Ratzel's most important insights about geopolitics are still valid.

Figure 9.1 illustrates Ratzel's conceptualization of the interaction of power and territory through the changing face of Europe from the end of World War I to the present. The fluidity of the maps reflects the unstable relationship between power and territory, especially some states' failure to achieve stability. The most recent map of Europe portrays the precarious nation-state boundaries in the post–Cold War period. Estonia, Latvia, and Lithuania have regained their sovereign status. Czechoslovakia has dissolved into Czechia and Slovakia. Yugoslavia has dissolved into seven states, but not without much civil strife and loss of life. The former Soviet Union is now the Commonwealth of Independent States, with Russia the largest and most powerful. And still the map is not fully settled.

The tensions within the former Soviet Union, for instance, are especially high as new republics such as Georgia and Ukraine move politically closer to Europe and the United States and further away from Russia. These emerging Western allegiances signal a seismic geopolitical shift that is prompting retaliation by Russia, particularly against Georgia. Tensions in that region began to mount when Mikhail Saakashvili was elected Georgian president in 2004. He promised to reunite the country by bringing back into the national fold South Ossetia and Abkhazia, two breakaway regions containing large ethnic minorities currently occupied by Russia (**Figure 9.2**). The result has been an escalation of tensions between Russia and the West, especially the European Union, with Cold War-like rhetoric issuing from both sides, illustrating the centrality of territorial boundaries to the operations of the state.

Boundaries and Frontiers

Boundaries enable territoriality to be defined and enforced and allow conflict and competition to be managed and channeled. The creation of boundaries is, therefore, an important element in making territories. Boundaries are normally *inclusionary*. That is, they are constructed to regulate and control specific sets of people and resources within them. Encompassed within a clearly defined territory, all sorts of activity can be controlled and regulated—from birth to death. The delimited area over which a state exercises control and which is recognized by other states is **territory**. Such an area may include both land and water, and even airspace.

Boundaries can also be *exclusionary*. They are designed to control people and resources outside them. National boundaries, for example, can control the flow of immigrants or imported goods. Municipal boundaries may separate different tax structures or even

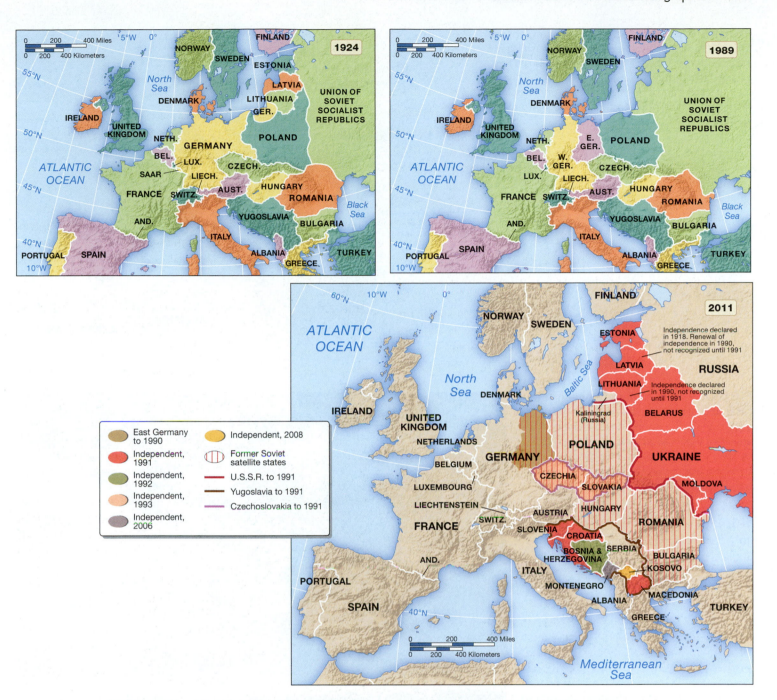

FIGURE 9.1 The changing map of Europe: 1924, 1989, 2011 The boundaries of the European states have undergone dramatic changes since World War I. The changing map of Europe illustrates the instability of international politics and the dynamism in the geography of political boundaries.

access to alcohol; land-use zoning boundaries can regulate access to upscale neighborhoods; field boundaries can regulate access to pasture, and so on (**Figure 9.3**).

One aspect of boundaries that is of increasing interest to geographers is how they are organized and policed. Since 9/11, for example, the relevance of policing borders with respect to international travel has grown dramatically. States across the world, led by front-runners like Israel, the United Kingdom, and the United States, have been instituting a wide range of practices to secure borders by electronic screening. By way of passports that contain RFID chips (radio frequency identification), as well as retinal scanning and related biometrics (electronic technologies for recognizing an individual based on one or more physical or behavioral elements), state agencies are increasingly able to know more and more personal information about travelers. These practices have the effect of creating virtual (technologically generated) boundaries that fortify the territorial ones.

Once established, boundaries tend to reinforce spatial exclusion as well as differentiation. The outcomes result partly from different sets of rules, both formal and informal, that apply within

FIGURE 9.2 Disputed territories in Georgia
South Ossetia and Abkhazia are both pushing for independence from Georgia. While Georgia has allowed the regions a large measure of autonomy, they are still part of its national territory. Russia has recognized their independence, though no other country has.

FIGURE 9.3 Boundary between rural and urban places Some boundaries signal differences in settlement activities that may actually be governed by land-use regulations. The division between agricultural activities and suburban living is clearly shown in this image.

different territories, and partly because boundaries often restrict contact between people and foster the development of stereotypes about "others." This restricted contact, in turn, reinforces the role of boundaries in regulating and controlling conflict and competition between territorial groups.

Boundaries can be established in many ways and with differing degrees of permeability. At one extreme are informal, implied boundaries that are set by markers and symbols but never delineated on maps or in legal documents. Good examples are the "turf" of a city gang, the "territory" of an organized crime "family," or the range of a pastoral tribe. At the other extreme are formal boundaries established in international law, delimited

on maps, demarcated on the ground, fortified, and aggressively defended against the movement not only of people but also of goods, money, and even ideas. An example of this sort of boundary is the one between the United States and Mexico (**Figure 9.4**). There are also formal boundaries that have some degree of permeability. The boundaries between the states of the European Union, for example, have become quite permeable, and people and goods from member states can now move freely between them with no customs or passport controls.

Impermeability does not necessarily mean immutability, however. The boundary between East and West Germany, part of the Iron Curtain for more than 40 years, was aggressively

FIGURE 9.4 Boundary between the United States and Mexico The U.S.–Mexico border is exclusionary. It is heavily patrolled and lined with barbed-wire chain-link fences along the highly urbanized parts. Aerial surveillance is also extensive along the border. This photo shows the U.S.–Mexico border along the Tijuana River estuary, with southern California on the left and Mexico on the right.

FIGURE 9.5 Berlin Wall The boundary between East and West Germany was virtually impermeable for more than 40 years. This photograph is of the scene on November 12, 1989, when Berliners tore the wall down in celebration of the reunification of Germany.

defended, yet it was removed in 1989 when Germany was reunified (**Figure 9.5**). Similarly, as mentioned previously, the boundaries of the former Soviet Union have been dramatically redrawn since 1989, allowing states like Lithuania and Estonia to reappear.

Boundaries are important elements of geopolitics and of the geography of domestic politics.

APPLY YOUR KNOWLEDGE Using Google Earth or aerial photographs, compare and contrast the international boundaries between the United States and Canada and between the United States and Mexico. How does the U.S.-Mexican boundary differ from the U.S.-Canadian boundary? How are they similar? Speculate as to why these differences and similarities exist. ■

Frontier Regions

Frontier regions occur where boundaries are very weakly developed. They involve zones of underdeveloped territoriality, areas that are distinctive for their marginality rather than for their belonging. In the nineteenth century, vast frontier regions still existed—major geographic realms that had not yet been conquered, explored, and settled (such as Australia, the American West, the Canadian North, and sub-Saharan Africa). All of these are now largely settled, with boundaries set at a range of jurisdictional levels from individual land ownership to local and national governmental borders. Only Antarctica, virtually unsettled, exists today as a frontier region in this strict sense of the term.

There remain, nevertheless, many regions that are still somewhat marginal in that they have not been fully settled or do not have a recognized economic potential, even though their national political boundaries and sovereignty are clear-cut. Such regions—the Sahara Desert, for example—often span national boundaries simply because they are inhospitable, inaccessible, and (at the moment) economically unimportant. Political boundaries are drawn through them because they represent the line of least territorial resistance. These frontier areas may exist at the local level as well. Not fully integrated into the realm of any one sociocultural group or economic activity, these spaces are often transitional in nature, with a relatively rapidly changing pattern of land use and an equally rapidly changing profile of residents.

Boundary Formation

Generally speaking, formal boundaries tend first to follow natural barriers, such as rivers, mountain ranges, and oceans. Good examples of countries with important mountain-range boundaries include France and Spain (the Pyrenees); Italy and France, Switzerland, and Austria (the Alps); and India and Nepal and China (the Himalayas). Chile, though, provides the ultimate example: a cartographic freak, restricted by the Andes to a very long and relatively thin strip along the Pacific coast. Examples of countries with boundaries formed by rivers include China and North Korea (the Yalu Tumen), Laos and Thailand (the Mekong), and Zambia and Zimbabwe (the Zambezi). Similarly, major lakes divide Canada and the eastern United States (along the Great Lakes), France and Switzerland (Lake Geneva), and Kenya and Uganda (Lake Victoria).

Where no natural features occur, formal boundaries tend to be fixed along the easiest and most practical cartographic device: a straight line. Examples include the boundaries between Egypt, Sudan, and Libya, between Syria and Iraq, and between the western United States and Canada. Straight-line boundaries are also characteristic of formal boundaries established through colonization, which is the outcome of a particular form of territoriality. The reason, once again, is practicality. Straight lines are easy to survey and even easier to delimit on maps of territory that remains to be fully charted, claimed, and settled. Straight-line boundaries were established, for example, in many parts of Africa during European colonization in the nineteenth century.

In detail, however, formal boundaries often detour from straight lines and natural barriers in order to accommodate special needs and claims. Colombia's border, for instance, was established to contain the source of the Orinoco River; the Democratic Republic of Congo's border was established to provide a corridor of access to the Atlantic Ocean; and Sudan's north central border detoured to include the settlement Wadi Halfa (**Figure 9.6**).

After primary divisions have been established, internal boundaries tend to evolve as smaller, secondary territories are demarcated. In general, the higher the population density, the smaller these secondary units tend to be. Their configuration is generally similar to that of larger units, following physical features, accommodating special needs, and assuming straight lines where there are no appropriate natural features or where colonization has made straight boundaries expedient. This last reason explains the rectilinear pattern of administrative boundaries in the United States to the west of the Mississippi.

Territories delimited by formal boundaries—national states, states, counties, municipalities, special districts, and so on—are known as *de jure* spaces or regions. *De jure* means "legally recognized." Historically, the word referred to a loose patchwork of territories (with few formally defined or delimited boundaries). More recently it has evolved to describe nested hierarchies (**Figure 9.7**) and overlapping systems of legally recognized territories.

FIGURE 9.6 Borders between Egypt and Libya and Sudan The triangular area along the Red Sea on the eastern end of the border of Sudan and Egypt, called "Administrative Body," once belonged to Egypt but is now in Sudan. That area was ceded to Sudan by Egypt in an attempt to induce it to join the now disbanded United Arab Republics (UAR). Sudan never joined the UAR but retained the land, which is a point of recurring controversy and sometimes outright conflict between the two countries.

These *de jure* territories are often used as the basic units of analysis in human geography, largely because they are both convenient and significant units of analysis. They are often, in fact, the only areal units for which reliable data are available. They are also important in their own right because of their status as units of governance or administration. A lot of regional analysis and nearly all attempts at regionalization, therefore, are based on a framework of *de jure* spaces.

GEOPOLITICS AND THE WORLD ORDER

There is, arguably, no other concept to which political geographers devote more attention than the state. The state is one of the most powerful institutions—if not the most powerful—implicated in the process of globalization. The state effectively regulates, supports, and legitimates the globalization of the economy.

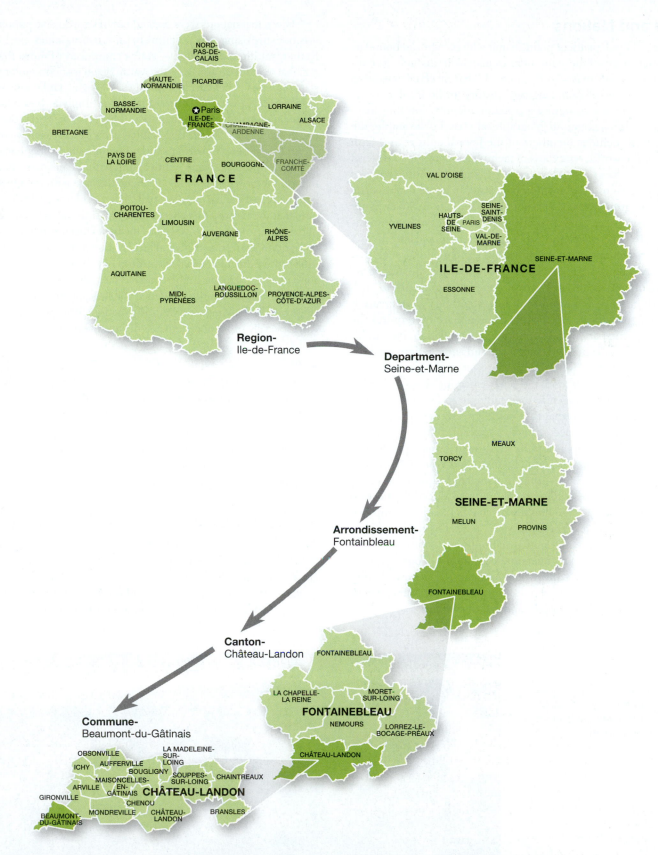

FIGURE 9.7 Nested hierarchy of *de jure* territories *De jure* territories are constructed at various spatial scales. Administrative and governmental territories are often "nested," with one set of territories fitting within the larger framework of another, as in this example of region, department, arrondissement, canton, and commune in France.

States and Nations

The **state** is an independent political unit with recognized boundaries, although some of these boundaries may be in dispute. In contrast to a state, a **nation** is a group of people sharing certain elements of culture, such as religion, language, history, or political identity. Members of a nation recognize a common identity, but they need not reside within a common geographical area. For example, the Jewish nation includes members of the Jewish culture and faith throughout the world, regardless of their places of origin. The term **nation-state** refers to an ideal form consisting of a homogeneous group of people governed by their own state. In a true nation-state, no significant group exists that is not part of the nation. **Sovereignty** is the exercise of state power over people and territory; that power is recognized by other states and codified by international law.

Citizenship is a category of belonging to a nation-state that includes civil, political, and social rights. Following the overthrow or decline of monarchies in Europe in the late eighteenth to mid-nineteenth centuries, a number of new republics were created. Republican government, as distinct from monarchy, requires the democratic participation and support of its population. Monarchical political power is derived from force and subjugation; republican political power derives from the support of the governed. By creating a sense of nationhood, the newly emerging states of Europe were attempting to homogenize their multiple and sometimes conflicting constituencies so that they could govern with the population's active cooperation according to a sense of common purpose.

Modern citizenship as a political category was a product of the popular revolutions—from the English Civil War to the American War of Independence—that transformed monarchies into republics. In the process, these revolutions produced the need to reimagine the socially and culturally diverse population occupying the territory of the state. Where once they were considered subjects with no need for a unified identity, nationhood required of these same people a sense of an "imagined community," one that rose above divisions of class, culture, and ethnicity. This new identity was called *citizen*, and citizenship came to be based on a framework of civil, political, and social rights and responsibilities.

Given that nations were created out of very diverse populations, it is not surprising that no entirely pure nation-states exist today. Rather, multinational states—states composed of more than one regional or ethnic group—are the norm. Spain is such a multinational state (composed of Catalans, Basques, Gallegos, and Castilians), as are France, Kenya, the United States, and Bolivia. Indeed, a very limited number of nation-states are uni-national; Iceland is one of the few that are. Since World War I, it has become increasingly common for groups of people sharing an identity different from the majority, yet living within the same political unit with the majority, to agitate to form their own state. This has been the case with the Québécois in Canada and the Basques in Spain. It is out of this desire for autonomy that the term *nationalism* emerges. **Nationalism** is the feeling of belonging to a nation, as well as the belief that a nation has a natural right to determine its own affairs. Nationalism can accommodate itself to very different social and cultural movements, from the white supremacy movements in the United States and Europe (**Figure 9.8**) to the movements for independence in Estonia, Latvia, and Lithuania (**Figure 9.9**) The impact of minority nationalism on the world map was especially pronounced during the twentieth century.

> **APPLY YOUR KNOWLEDGE** Nationalism as a political movement is on the rise around the globe. Identify two national movements in places other than the United States—one that is left-wing (politically progressive) and one that is right-wing (politically conservative)—and trace their origins and current missions. How do these two movements differ? In what ways are they the same? ■

Russia's State and National Transformation

The history and the present status of the former Soviet Union clearly illustrates the tensions among and between states, nations, and nationalism. Both enduring nationalism and the desire for sovereignty are evident in the history of the Russian Empire. Russia's strategies to bind the 100-plus nationalities (non-Russian ethnic

FIGURE 9.8 English Defense League, 2011 Pictured are members of the right-wing EDL in a march in the city of Luton. The group opposes the spread of Islamism in England.

FIGURE 9.9 Independent states of the former Soviet Union For the most part, the administrative structure of the USSR has remained in place. The differences are that autonomous regions and republics now have more than nominal local control, and the former federal republics have become independent states (pictured here in various colors), as have the popular democracies of Eastern and Central Europe. Despite, or perhaps because of, recent democratic reforms, nationalist movements continue to plague the consolidation of the state in autonomous republics like Chechnya, located on the map in the green area just northeast of Georgia. (*Source:* Reprinted with permission from Prentice Hall, from J. M. Rubenstein, *The Cultural Landscape: An Introduction to Human Geography*, 5th ed. © 1996, p. 318.)

peoples) into a unified Russian state were often punitive and not at all successful. Non-Russian nations were simply expected to conform to Russian cultural norms. Those that did not were more or less persecuted. The result was opposition and, among many if not most of the nationalities, sometimes rebellion and refusal to bow to Russian cultural dominance.

Such was the legacy that Vladimir Lenin and the Bolsheviks inherited from the Russian Empire following the overthrow of the aristocracy in 1917. The solution to the "national problem" orchestrated by Lenin was *recognition* of the many nationalities through the newly formed Union of Soviet Socialist Republics (USSR). Lenin believed that a *federal system*, with *federal units* delimited according to the geographic extent of ethnonational communities, would ensure political equality among at least the major nations in the new state. This political arrangement recognized the different nationalities and provided them a measure of independence. Federation was also a way of bringing reluctant areas of the former Russian Empire into the Soviet fold. A **federal state** allocates

power to units of local government within the country. The United States is a federal state with its system of state, county, and city/ town government. A federal state can be contrasted with a **unitary state**, in which power is concentrated in the central government. The Russian state under the czar was a unitary state.

Although a federal system theoretically remained in place when General Secretary of the Communist Party Mikhail Gorbachev came to power in 1985, at that time the USSR actually operated as a unitary state with power concentrated in Moscow. Gorbachev's goal was a massive restructuring of the Soviet economy through radical economic and governmental reforms (*perestroika*) and the direct democratic participation of the republics in shaping those reforms through open discussions, freer dissemination of information, and independent elections (*glasnost*). Effectively, Gorbachev lifted the restrictions that had been placed on the legal formation of national identity.

By 1988, grassroots national movements were emerging, first in the Baltic republics and later in Transcaucasia, Ukraine, and Central Asia. By 1991, the relatively peaceful breakup of the Soviet Union was under way, and new states had emerged to claim their independence, facilitated by the federated structure that had existed under the USSR. **Figure 9.9** is a recent map of the former USSR, including all the newly independent states now in the Commonwealth of Independent States (CIS) and the Baltic states of Estonia, Latvia, and Lithuania, which chose not to be members of the CIS. The CIS is a **confederation**, a group of states united for a common purpose. The newly independent states that chose confederation did so mostly for economic (and to a lesser extent, for military) purposes. A similar case can be found in the Confederate States of America, the 11 southern states that seceded from the United States between 1860 and 1861 to achieve economic and political solidarity. This secession led ultimately to the Civil War, a bloody conflict that caused a massive loss of American lives for both the Union and the Confederacy.

With the fall of communism in Eastern Europe in 1991, regions such as the Balkans (the mountainous isthmus of land between the Danube River and the plains of northern Greece that includes Albania, Bulgaria, continental Greece, southeast Romania, European Turkey, and most of the territories formerly organized as Yugoslavia) also experienced national movements resulting in the redrawing of political boundaries. And while the breakup of the former USSR had been primarily peaceful (with significant exceptions such as Georgia and Chechnya), the recent redrawing of national boundaries in the Balkans has resulted in bitter and widespread ethnic conflict. The region is situated at a geopolitical crossroads where East meets West, Islam meets Christendom, the Ottoman Empire met the Austro-Hungarian Empire, and communism once confronted capitalist democracy. Conflict has been characteristic of the region for centuries. In the twentieth century, for instance, ethnic tensions spilled over the boundaries of the Balkan region, eventually pulling all of Europe (and later the United States as well as other, more far-flung countries) into World War I in 1914–1915.

The more recent conflicts have been concentrated in the region organized in 1918 as the Kingdom of Serbs, Croats, and Slovenes following the collapse of the Austro-Hungarian Empire. In 1929, this region was renamed Yugoslavia. After 1946, the country included six republics: Slovenia, Croatia, Bosnia-Herzegovina, Macedonia, Serbia, and Montenegro. The first four declared independence in 1991, while Serbia and Montenegro formed a new Yugoslavian state in 1992.

The most horrifying conflict in the region occurred during the late 1990s in Kosovo, a region within Serbia that lost its autonomy in 1989 when Serbian nationalist leader Slobodan Milosevic placed it under military occupation (**Figure 9.10**). A year later Kosovo's parliament was abolished and its political leaders fled. Milosevic orchestrated numerous attempts to rid the region of ethnic Albanians, and in 1995 Serb refugees from Croatia began flooding into Kosovo. Fueled by the racist rhetoric and military support of Milosevic, a civil war between Serbs and ethnic Albanians eventually erupted. By late 1999, eight hundred thousand ethnic Albanians had fled Kosovo and tens of thousands had been massacred. An 11-week air war by NATO helped to bring the conflict to a halt, although atrocities continued as ethnic Albanians retaliated against Serbs. An international war tribunal was set up in The Hague to try Milosevic and others on human rights violations, including the possible charge of genocide. Milosevic acted in his own defense at the trial; he was found in his cell in 2006, dead of apparent heart failure, before a verdict was reached.

Theories and Practices of States

The definition of the state provided in the previous section is a static one. The state, through its institutions—such as the military or the educational system—can act to protect national territory and harmonize the interests of its people. Therefore, the state is also a *set of institutions* for the protection and maintenance of society. A state is not only a place, a bounded territory, it is also an active entity that operates through the rules and regulations of its various institutions, from social-service agencies to governing bodies to the courts, in order to shape national populations.

An especially influential state theorist whose work has inspired a large number of political geographers is Louis Althusser. Althusser developed an approach that identifies the operations of

FIGURE 9.10 Kosovo, 1999 Children run through the NATO bomb-damaged neighborhood of Pristina, the capital and largest city of the Republic of Kosovo.

the state in two ways. The first is to view the state as an ideological force operating through the institutions of the schools, the media, the family, and religion to produce citizens who conform to state expectations. The second, and complementary, way is to view the state as repressive. It uses force, through different institutions like the courts, army, and the police, to compel citizens to comply with its rules. These two related aspects of a model for understanding the state have been employed by political geographers in a number of ways, but perhaps the most common is to explore how the spaces of these various institutions, such as the school or the family, are produced and operate to do the work of shaping citizens.

A theorist who has also been a significant influence on political geography is Michel Foucault. Foucault pushed Althusser's ideas further by formulating a way for understanding just *how* the various institutions of the state operate to do the work of shaping citizens. Foucault explored the way that power, knowledge, and discourse operate in concert to produce particular kinds of subjects: citizens, women, soldiers, terrorists. For Foucault, **discourse** is an institutionalized way of thinking. For example, an army turns civilians into soldiers through the conjunction of power (in the form of both force and cooperation) and knowledge (accumulated insights about training, discipline, warfare) that come together as discourse, the institutionalized way of thinking that produces rules, identities, practices, exclusions, and a range of other elements that make large numbers of individuals function as an effective collective.

A third state theorist important to geographers is Giles Deleuze, who sees the state not as a set of institutions—the courts, the legislature, the military, etc.—but as a force. This force is greater than formal institutions at the same time that it works through them. For Deleuze, the state is not a thing but a principle that operates through power and authority. Rather than seeing the state as having been created during a particular period of human history and then expanding its power over time, he sees the state as having always existed in different forms, even before the emergence of the institutions by which we most clearly recognize it now. Deleuze believes that the state is best thought of as a *machine*, with its purpose being to regulate and dominate. This machine of the state operates through mundane practices that produce a population willing to submit to political sovereignty. In short, the state is a machine that produces a population willing to submit to the power of a ruler whether that ruler be a police officer, a forest ranger, a teacher, a president.

Geographers have incorporated these and other theories and models of the state to explain the unfolding and impact of different kinds of political spaces, from the organization and influence of the classroom to the development and effects of international laws. For political geographers, theories and models of geopolitics have been their most prominent contributions to understanding the role and behavior of the state. The revolutionary fervor that spread across North Africa and the Middle East in 2011 is an illustration of geopolitics and the relationship of a national population to its state (see Box 9.1, "Window on the World: The Arab Spring").

APPLY YOUR KNOWLEDGE Based on what you have leaned about Foucault's theory of discourse and Deleuze's conception of the state, write a paragraph describing the Arab Spring from each of these theorist's perspectives. ∎

Imperialism and Colonialism

Geopolitics may involve the extension of power by one group over another. Two ways this may occur are through the related processes of imperialism and colonialism. Recall from Chapter 2, imperialism is the extension of state authority over the political and economic life of other territories. Over the last 500 years, imperialism has resulted in the political or economic domination of strong core states over the weaker states of the periphery. Imperialism does not necessarily imply formal governmental control over the dominated area; it may also involve a process by which some countries pressure the independent governments of other countries to behave in certain ways. This pressure can take many forms, such as military threat, economic sanctions, or cultural domination. Imperialism always involves some form of *authoritative control* of one state by another.

As discussed in Chapter 2, the process of imperialism begins with exploration (**Figure 9.11**), often prompted by the state's perception that there is a scarcity or lack of a critical natural resource. It culminates in development via colonization or the exploitation of indigenous people and resources, or both. In the first phases of imperialism, the core exploits the periphery for raw materials. As the periphery becomes developed, colonization may occur and economies based on money transactions—or "cash economies"—may be introduced where none previously existed. The periphery may also become a market for the manufactured goods of the core. Eventually, though not always, the periphery—because of the availability of cheap labor, land, and other inputs to production—can become a new arena for large-scale capital investment. Some peripheral countries improve their status and become semiperipheral or even core countries. **Figure 9.12** is a map of the colonies created by European imperialism in Africa.

Colonialism is a form of imperialism. It involves the formal establishment and maintenance of rule by a sovereign power over a foreign population through the establishment of settlements. The colony does not have any independent standing within the world system and instead is considered an adjunct of the colonizing power. From the fifteenth to the early twentieth centuries, colonialism constituted an important component of core expansion. Between 1500 and 1900, the primary colonizing states were Britain, Portugal, Spain, the Netherlands, and France. **Figure 9.13** illustrates the colonization of South America, largely by Spain and Portugal.

Important states more recently involved in colonization and imperialist wars include the United States and Japan. Although colonial penetration often results in political dominance by the colonizer, such is not always the case. For example, Britain may have succeeded in setting up British colonial communities in China, but it never succeeded in imposing British administrative or legal structures in any widespread way. And at the end of the colonial era a few former colonies, such as the United States, Canada, and Australia, became core states themselves. Others, such as Rwanda, Bolivia, and Cambodia, remain firmly within the periphery. Some former colonies, such as Mexico and Brazil, have come close to the core but have not fully attained core status and are categorized as being within the semiperiphery.

The Effects of Colonialism

An example of colonialism is the extension of British rule in India, which began with the establishment of the East India Trading

While many political observers would consider the pivotal moment of the Arab Spring—that period of revolutionary fervor that telegraphed across the Middle East and North Africa in Spring 2011—to have happened in Tunisia, others would point first to Iran. Despite the fact that Iran is not Arab, and its "Green Revolution" (also known as the "Persian Awakening") occurred in June 2009, the protests that gathered momentum in Tehran and swept across the country (and to major cities throughout the world), can be seen as at least a powerful inspiration for what happened in Tunisia and ultimately diffused eastward, westward, and to the south.

Millions of protesters took to the streets in the Iranian Green Revolution charging that votes were manipulated and the election was rigged, keeping President Mahmoud Ahmadinejad illegally in power. In response to the protests, police and the *Basij*, a paramilitary group, used batons, pepper spray, sticks, and, in some cases, firearms against both peaceful demonstrators and rioters. The government of Iran reports that 36 people died as a result of the police response to the protests; supporters of opposition candidate, Mir-Hossein Mousavi, allege that there were 72 deaths. The most high-profile killing was of a young Iranian woman, Neda Agha-Soltan, who was shot by the *Basij*. Amateur videos of the killing spread virally across the Internet after being posted on Facebook and YouTube and incited further protests. Since the protests began, Iranian authorities have closed universities in Tehran, blocked Web sites, blocked cell phone transmissions and text messaging, and banned all rallies.

Tunisia's was the next uprising to occur in the region and there is no doubt that its "Jasmine Revolution" served to embolden other national protests in the Spring of 2011. Sustained public anger there, which forced President Zine al-Abidine Ben Ali out of office after nearly a quarter century, was ignited by the suicide of a young, unemployed man, Mohamed Bouazizi, who set fire to himself on December 17 after officials prevented him from selling fruits and vegetables on the street. The United Nations estimates that 219 people have died in the Tunisian protests. A key reform that has been enacted in response to the protests is the dismantling of the political police as well as the state security force, which appears to have been responsible for extensive human rights abuses.

From Tunisia, the desire for accountability in government and the end of corruption, police brutality, restrictions on free speech, poverty, rising food prices, as well as anger about the personal enrichment of the political elite have fueled protests in Morocco, Algeria, Jordan, Saudi Arabia, Bahrain, Oman, Egypt, Libya, Yemen, and Syria (**Figure 9.A(a)–(j)**). In all of these places the protests have been largely and deliberately peaceful as individuals have attempted to move their leaders by a firm

FIGURE 9.A(a) Benghazi, Libya Shown here are local residents of Benghazi who have joined a protest against dictator Muammar Gaddfi. The protests turned into civil war there.

FIGURE 9.A(b) Casablanca, Morocco In the city of Casablanca, the economic center of the country, demonstrators assembled through much of the summer calling for reforms in the Arab world's oldest reigning monarchy.

commitment to nonviolence and open dialogue. In Egypt, the world watched in awe as President Hosni Mubarak, in power for three decades, resigned in response to the demands of millions of protestors as well as pressure from the international community. The transitional government in Egypt estimates that at least

846 people were killed during the uprising and more than 6,400 people were injured. Instrumental in sparking the protests was the video blogging of 26-year-old Asmaa Mahfouz. She urged the Egyptian people to join her on January 25 in Tahrir Square to bring down Mubarak's regime. Her posts went viral and the event attracted 80,000 attendees. The first of many trials against members of the Mubarak regime commenced in May, with more scheduled. Mubarak and his two sons have been imprisoned in Cairo and are to be tried for the deaths of the antigovernment protesters.

In response to thousands of protesters gathering across Morocco's cities in February 2010, King Mohammed VI, a direct descendent of the Prophet Mohammed and a member of the dynasty that has been ruling Morocco for some 350 years, promised "comprehensive constitutional reform." In Morocco—rather different from its neighbors in that the country has a successful economy, an elected parliament, and a reformist monarchy—the protesters are pushing for reform that will "restore dignity and end graft." Some observers believe that Morocco may yet experience the more massive protests besetting other North African states if real reforms are not enacted. Behind the facade of a relatively prosperous and democratic country is a brewing problem of a growing impoverished youthful population with little chance of employment and a government elite living obscenely lavish lives.

Like Morocco, Algeria is a country that on the surface seems generally prosperous, based on its sizeable oil and gas reserves. Protesters there, who have been assembling since January, have been demanding that the Algerian constitution be changed to limit presidential terms. While protestors here have not reached the levels of indignation, nor the overall numbers, of neighboring countries' unrest, there is fear among the ruling elites that the situation is a highly volatile one. The protests appear to have been triggered by high food prices. In February, President Bouteflika lifted the country's 19-year state of emergency, and in April, he promised to create a constitutional commission that would strengthen democracy. New social and economic programs are also being proposed to address poverty.

Saudi Arabia has experienced no sizeable protests, probably for several reasons. First, opposition movements are banned there; second, the country's vast oil reserves make it one of the wealthiest of the Arab states; third, the country is deeply conservative with a population that supports King Abdullah. Protests in Saudi Arabia have been small and have involved minority Shia Muslims who have gathered in support of Shia Muslim protesters in Bahrain. In Bahrain, 30 people have died in protests demanding that King Hamad take action to increase

FIGURE 9.A(c) Qatif, Saudi Arabia Shiite protesters wearing masks chant slogans demanding democratic reforms.

FIGURE 9.A(d) Amman, Jordan While strict controls on protesting have been exercised here, some risked arrest in order to broadcast their demands for more democratic government.

political freedoms and eliminate job discrimination that favors the governing Sunni Muslim minority. As small groups of demonstrators gathered in the center of the capital city Manama, King Hamad swiftly imposed a state of emergency and cleared the protesters' camps in a brutal show of force aided by Saudi and United Arab Emirates soldiers. The state of emergency was lifted on June 1, 2011, with no promises made to address the demands of the protesters.

In both Jordan and Oman, protests have been relatively limited but the demands of the protestors are similar to those of all of the other countries involved in the Arab Spring movement: job creation, controls on food prices, an end to government corruption, and more democratic government. Youthful populations across the Middle East and North Africa are usually the key foundational elements of the protests.

We have saved Libya, Yemen, and Syria for last because it is in these three countries where revolutions have turned into armed insurrection. This turn to aggression is uncharacteristic of the larger movement in the region, where nonviolent protest was the explicit objective. Why protests in these three have become violent has much to do with the entrenched intransigence of the leaders and the deep frustration and righteous anger of the protesters.

Libyans have risen up against their ruler, Colonel Muammar Gaddafi, one of the most autocratic and longest-ruling dictators in the Middle East and North Africa. Citizens have been demanding a return to the 1952 constitution and a multiparty democracy, a reduction in unemployment and poverty, freedom from domestic surveillance, and the right to gather peacefully to express dissent. The protests began peacefully on February 15, 2011, in the city of Benghazi, the second-largest city in Libya. Shockingly, the protests were immediately met with violence as Gaddafi ordered the police to fire on the protesters, killing 500–700 of them. In response, the protests grew into an armed uprising that spread across the country, pitting government forces against civilians. An opposition government—the National Transitional Council—was established in Benghazi on February 27 to act as the "political face of the revolution" with the ultimate goal of overthrowing Gaddafi and holding democratic elections. As the fighting escalated, the rebels received international support from a NATO-led coalition. Several thousand people have been killed so far in the conflict. On October 21, 2011 Colonel Gaddifi was himself killed by Libyan rebel forces. Several days later he was later buried in an unmarked grave in the desert to prevent it from becoming a loyalist shrine or a target for his opponents.

The situations in Syria and Yemen have also turned violent. Yemen is the Arab world's poorest nation. Almost half of its 23 million people live on $2 a day or less, one-third suffer from chronic hunger. The presence of al-Qaeda in Yemen as well as separatist challenges to government authority have made the country an increasingly unstable place. And the protests that began on February 11, 2011, against the rule of President Ali Abdullah Saleh are in many ways unsurprising. Government corruption is rampant, unemployment is high, and the country's

FIGURE 9.A(e) Algiers, Algeria Physicians were part of the protests that spread in Algeria against the government arguing for improvements to working conditions and access to healthcare.

FIGURE 9.A(f) Manama, Bahrain Protesters here formed a seven kilometer (more than four miles) human chain to register their anti-government sentiments.

median age, at 17.9 years, is the youngest in the region. What has been shocking is the level of violence waged by security forces against unarmed protesters wearing pink ties and scarves as a sign of solidarity with the peaceful Jasmine Revolution in Tunisia. With President Saleh out of the country for the treatment of injuries suffered in an assassination attempt and the protests continuing across the country, Yemen is on the brink of becoming a failed state.

The Syrian government, too, has met peaceful protest with an armed response. Popular unrest came late to Syria; the first protest broke out in mid-March. Since then, rights activists have estimated that 1,200 Syrians have died as demonstrations have spread across the country and the government has branded them "an armed insurrection." Government forces have deployed tanks, shelled residential areas, and used snipers on rooftops to kill protesters. The mutilated and tortured body of 13-year-old Hamza al-Khatib has become a symbol of the Syrian uprising. The killing of children—and there have been many others in addition to Hamza al-Khatib—has galvanized the anger among the protestors.

Throughout the region, unrest continues to unfold. Extreme violence characterizes the situations in Libya, Yemen, and Syria. Peaceful protests are regularly organized in the remainder of the region. Some places are facing the possibility of new elections while others are receiving promises of increased food aid and access to education, among other things. As of Summer 2011, the Arab Spring continues.

Excerpted from: Middle East Protests: country by country, BBC World News, June 27, 2011, http://www.bbc.co.uk/news/world-12482311, accessed July 3, 2011; Stephen Blight and Sheil Pulham, The Path of Protest, guardian.co.uk, June 8, 2011, http://www.guardian.co.uk/world/interactive/2011/mar/22/middle-east-protest-interactive-timeline, accessed July 3, 2011.

FIGURE 9.A(g) Hurghada, Egypt Protests occurred throughout Egypt and were largely peaceful but insistent.

FIGURE 9.A(h) Sana'a, Yemen At a Friday prayer ceremony, people also staged anti-government protests, as shown here.

FIGURE 9.A(i) Muscat, Oman While anti-government protests were protesting the lack of a constitution in this monarchical state, pro-government demonstrators, pictured here, were defending their leader Sultan Qaboos.

FIGURE 9.A(j) Syrian refugees in Hatay, Turkey As the Syrian government began to use violence against protesters, many Syrians began leaving the country in fear ful of the more sustained attacks by the army that followed. Here Syrian found refuge in Turkey and are protesting against Syrian President Bashar al-Asaad.

FIGURE 9.11 Principal steps in the process of exploration This diagram illustrates the main elements in the process of exploration, beginning with a need in the home country. Geographers have figured prominently in the process of exploration by identifying areas to be explored as well as actually traveling to those places and cataloging resources and people. Nineteenth-century geography textbooks are records of these explorations and the ways geographers conceptualized the worlds they encountered. Exploration is one step in the process of imperialism; colonization is another. (Adapted from J. D. Overton, "A Theory of Exploration," *Journal of Historical Geography*, 7, 1981, p. 57.)

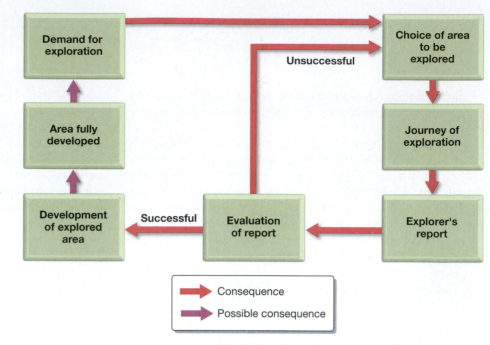

FIGURE 9.12 European colonies in Africa, 1496–1912 Lying within easy reach of Europe, Africa was the most likely continent for early European expansion. The Belgian, Italian, French, German, and Portuguese states all laid claim to various parts of Africa and in some cases went to war to protect those claims. The partitioning of the African continent by the imperial powers created a crazy quilt that crosscut preexisting affiliations and alliances among the African peoples. (Adapted from *Harper Atlas of World History*. New York: HarperCollins, 1992, p. 139.)

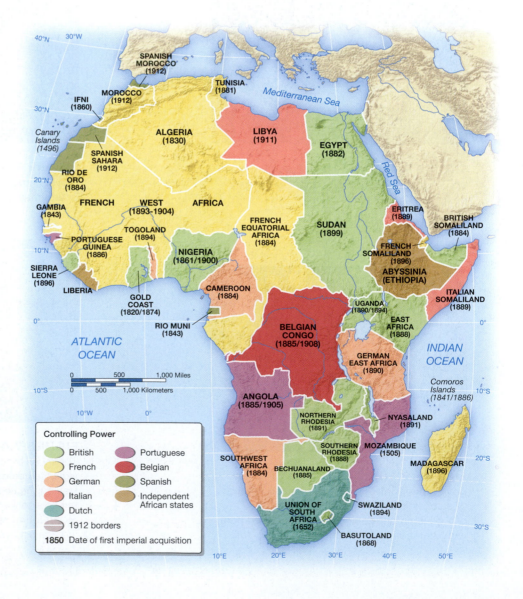

FIGURE 9.13 Colonization in South America and the Caribbean, 1496–1667 The Spanish and Portuguese dominated the colonization and settlement of South America. The Dutch, French, and English were a minor and largely tentative presence. African colonization focused upon gaining new subjects and the acquisition of additional territories; South American colonization yielded rich commodity and mineral returns. (Adapted from *Rand McNally Atlas of World History.* Skokie, IL: Rand McNally, 1992, p. 85.)

treated Arabs. This same argument can be applied to the British in India or to other Western powers with respect to their colonies: The colonized are seen as inferior and in need of disciplining in the eyes of the superior and enlightened colonizer. The colonizer has a moral obligation to colonize, according to this logic.

The postcolonial history of the Indian subcontinent has included partition and repartition, as well as the eruption of regional and ethnic conflicts. In 1947, Pakistan split off from India and became a separate Muslim state. In 1971, Bangladesh, previously part of Pakistan, declared its independence. Regional conflicts include radical Sikh movements for independence in the states of Kashmir and Punjab. Ethnic conflicts include decades of physical violence between Muslims and Hindus over religious beliefs and the privileging of Hindus over Muslims in the national culture and economy. It would be misleading, however, to attribute all of India's current strife to colonialism. The Hindu caste system, which distinguishes social classes based on heredity and plays a significant role in political conflict, preceded British colonization and persists to this day.

Company in the mid-eighteenth century. The British government gave the company the power to establish forts and settlements, as well as to maintain an army. The company soon established settlements—including factories—in Mumbai (formerly Bombay), Chennai (formerly Madras), and Calcutta. What began as a small trading and manufacturing operation burgeoned over time into a major military, administrative, and economic presence by the British government and did not end until Indian independence in 1947. During that 200-year period, Indians were brutalized and killed and their society transformed by British influence **(Figure 9.14)**. That influence permeated nearly every institution and practice of daily life—from language and judicial procedure to railroad construction and cultural identity.

The reasons Britain was able to be so callous in its colonial practices are complex. Theorist Edward Said has proposed the concept of Orientalism to explain them, at least in part. For Said, **Orientalism** is a discourse (as described earlier in the chapter) that positions the West as culturally superior to the East. Said developed the concept to describe the way the West has both historically and contemporarily

Since the turn of the nineteenth century, the effects of colonialism have continued to be felt as peoples all over the globe struggle for political and economic independence. The 1994 civil war in Rwanda is a sobering example of the ill effects of colonialism. As in India, where an estimated 1 million Hindus and Muslims died in a civil war when the British pulled out, the exit of Belgium from Rwanda left colonially created tribal rivalries unresolved and seething. Although the Germans were the first to colonize Rwanda, the Belgians, who arrived after World War I, established political dominance among the Tutsi by allowing them special access to education and the bureaucracy.

Previously, a complementary relationship had existed between the Tutsi, who were cattle herders, and the more numerous Hutu, who were agriculturists. In effect, colonialism introduced difference into an existing political and social structure that had operated more or less peacefully for centuries. In 1959, the Hutus rebelled and the Belgians abandoned their Tutsi favorites to side with the Hutus. In 1962, the Belgians ceded independence

FIGURE 9.14 Indians in the United Kingdom
The British presence in India affected culture, politics, the economy, and the layout of cities, as well as numerous other aspects of everyday life. Indian society absorbed and remolded many British political and cultural practices so that contemporary Indian government, for example, embodies both British and Indian ideals and practices. Importantly, Indian culture has come to influence British culture even more directly through immigrants who've migrated to the United Kingdom. For example, the most popular dish in England is curry.

FIGURE 9.15 Refugees returning to Rwanda Fleeing civil unrest in their own country, Rwandans from the Hutu tribe increasingly sought refuge in the Democratic Republic of Congo (formerly Zaire) when the Tutsi-led government assumed power in 1994. Two-and-a-half years later, over half a million Rwandan refugees in the Democratic Republic of Congo occupied some of the largest refugee camps in the world. In late 1996, they began streaming back into Rwanda when the Tutsi-led government urged them to come help rebuild the country. Faced with two difficult alternatives—dire conditions in the camps or possible violence in Rwanda—many refugees chose to go home. Tens of thousands of Rwandans jammed the road between eastern Democratic Republic of Congo and Rwanda for over three days.

to Rwanda, leaving behind a volatile political situation that has erupted periodically ever since, most tragically in 1994's civil war. After a year of violence in which over half a million Tutsis were killed, the Hutus were driven across the border to the Democratic Republic of Congo (DRC) and a new Tutsi-led Rwandan government was formed. The Hutu refugees gathered in UN camps that gradually came to be controlled by armed extremists, who transformed them into virtual military bases and used them to attack the Tutsis in Rwanda (**Figure 9.15**). When Rwanda's Tutsi-led military, with Uganda's support, invaded the DRC to break up the camps, over a million refugees were released. Refugees have since dispersed across Central Africa.

To help address the culture of impunity that angered so many Rwandans, the UN established a truth commission, known

as "Gacaca Court System," based on Rwandan traditional justice. Although its work has had some important moral impacts, the commission's aim—enabling Tutsis and Hutus to work together to rebuild their society and economy—has not been realized. Hundreds of thousands of children have become orphans as their parents either died in the atrocities or their mothers were infected with HIV/AIDS through rape and have since died. The country spends more on debt repayments to international banks than it does on education and health. Seventy percent of all households in Rwanda live below the national poverty line. Half the adult population can neither read nor write and one in three children does not attend school. In short, the economic situation in Rwanda is worse than it was before the 1994 ethnic cleansing, which doesn't bode well for the country's long-term political stability.

On the Internet, find three different recent news articles from reputable sources about a political issue in the Middle East. Consider the articles from the perspective of Orientalism. How do the articles depict the West and the Middle East? Is the West depicted as being culturally superior to the Middle East? If so, show how that is accomplished. If not, explain how the journalists avoid adopting an Orientalist explanation. ■

The North/South Divide and Decolonization

The colonization of Africa, South America, parts of the Pacific, Asia, and smaller territories scattered throughout the Southern Hemisphere resulted in a political geographic division of the world into North and South, known as the **North/South divide**. In the North are the imperialist states of Europe, the United States, Russia, and Japan. In the South are colonized states. Though the equator is often used as a dividing line, some so-called Southern territories, such as Australia and New Zealand, actually are part of the North in an economic sense.

The crucial point is that a relation of dependence was set up between countries in the South (the periphery) and those in the North (the core) that began with colonialism and persists even today. Only a few peripheral countries have become prosperous and economically competitive since achieving political autonomy, though that is beginning to change, as this book addresses in Chapter 7. Political independence is markedly different from economic independence, and the South remains very much oriented to the economic demands of the North.

The reacquisition by colonized peoples of control over their own territory is known as **decolonization**. In many cases, sovereign statehood has been achievable only through armed conflict. From the Revolutionary War in the United States to the twentieth-century decolonization of Africa, the world map created by the colonizing powers has repeatedly been redrawn. Many former colonies achieved independence after World War I. Deeply desirous of averting wars like the one that had just ended, the victors (excluding the United States, which entered a period of isolationism following the war) established the League of Nations. One of the first international organizations ever formed, the League of Nations had a goal of international peace and security. An **international organization** is one that includes two or more states seeking political and/or economic cooperation with each other. **Figure 9.16** shows the member countries of the League.

Within the League, a system was designed to assess the possibilities for independence of colonies and to ensure that the process

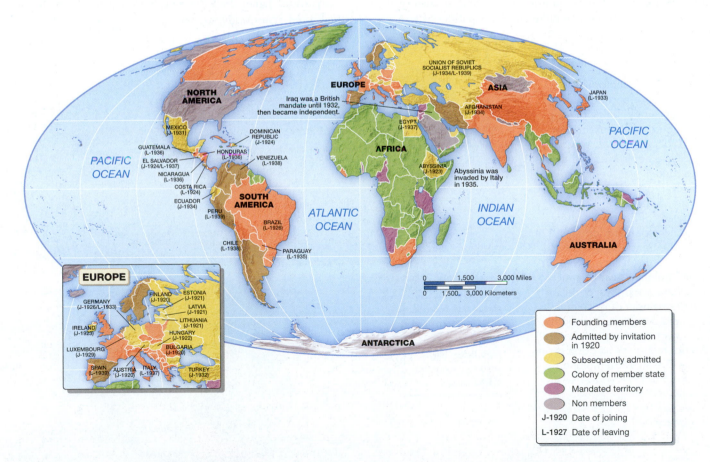

FIGURE 9.16 Countries participating in the League of Nations Woodrow Wilson was a central figure in the creation of the League of Nations, so his inability to convince the U.S. Senate to join was a major blow to the effectiveness of this, the first international organization of the twentieth century. Britain and France played important roles in the League but were never able to secure arms limitations and security agreements among the membership. Perhaps the League's greatest success before it was dissolved in 1946 was pressing for the decolonization of Africa.

occurred in an orderly fashion. Known as the *colonial mandate system*, it had some success in overseeing the dismantling of numerous colonial administrations. **Figures 9.17**, **9.18**, and **9.19** illustrate decolonization during the twentieth century in Africa, Asia, and the South Pacific and during the nineteenth century in South and Central America. (Although the League of Nations proved effective in settling minor international disputes, it was unable to prevent aggression by major powers and was dissolved in 1946. It did, however, serve as the model for the United Nations.)

Decolonization does not necessarily constitute an end to domination within the world system, however. Even though a former colony may exhibit all the manifestations of independence, including its own national flag, governmental structure, currency, educational system, and so on, its economy and social structures may continue to be dramatically shaped in a variety of ways by core states. Participation in foreign aid, trade, and investments from core countries subjects the periphery to relations that are little different from those they experienced as colonial subjects. In former British colony Kenya, for example, core countries' provision of foreign aid monies, development expertise, and educational opportunities to selected individuals has created a class of native civil servants that is in many ways more strongly connected to core processes and networks than those operating within Kenya. This relatively small group of men and women, often foreign educated, emerged as the first capitalist middle class in Kenyan history and their children routinely enter middle-class occupations as well.

Commercial relations also enable core countries to exert important influence over peripheral, formerly colonized, countries. For example, as discussed in more detail in Chapter 8, *contract farming* has become a central mechanism around which agricultural production in the periphery is organized for core consumption. Conditions of production of specified agricultural commodities are dictated by core countries to growers in the periphery. For example, Japanese and EU firms issue contracts that set the conditions of production for the Thai broiler (chicken) industry; Chiquita Brands, Inc., a U.S. firm, issues contracts for Honduran banana production. In this way, a core country can invoke a new form of colonialism in places it never formally colonized. As explained in Chapter 2, this new form, known as *neocolonialism*, is the domination of peripheral states by core states not by direct political intervention (as in colonialism), but by economic and cultural influence and control.

Exploration and to a lesser extent, colonization are still occurring in Antarctica (**Figure 9.20**). This ice-covered landmass provides an unusual example of twentieth-century imperialism in which strong states exerted power in an area where no people and, therefore, no indigenous state power existed. At present, while no one country exclusively "owns" the continent, 15 countries lay claim to territory and/or have established research stations there: Argentina, Australia, Belgium, Brazil, Chile, China, France, Germany, India, Japan, New Zealand, Norway, the United Kingdom, the United States, and Uruguay.

FIGURE 9.17 Decolonization of Africa, before and after 1960 Britain, France, and Belgium—the dominant European presences in African colonization—were also the first to divest themselves of their colonies. Britain was the first colonial power to grant independence. France granted independence to its African colonies soon after. In the French-speaking former colonies, the transition to independence occurred largely without civil strife. Belgium's withdrawal as well as the withdrawal of Britain from the remainder of its colonial holdings did not go at all smoothly, with civil wars breaking out. Portugal did not relinquish its possession of Guinea Bissau, Mozambique, or Angola until 1974. (Adapted from *The Harper Atlas of World History, Revised Edition,* Librairie Hachette, p. 285. Copyright © 1992 by HarperCollins Publishers, Inc.)

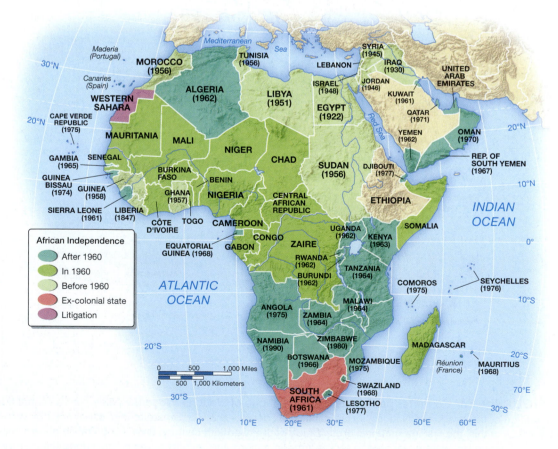

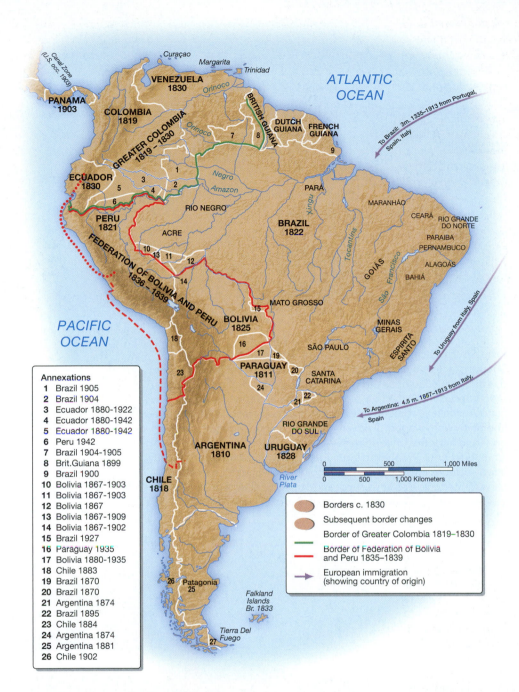

FIGURE 9.18 Independent South America, nineteenth century Independence came much earlier to South America than to Africa (it had also been colonized much earlier than Africa). What was most influential in the independence movements there was the presence of local Spanish and Portuguese elites. This colonial ruling class became frustrated with edicts and tax demands from the home country and eventually waged wars—not unlike the U.S. Revolutionary War—against Spain and Portugal for independence. (Adapted from *Rand McNally Atlas of World History*. Skokie, IL: Rand McNally, 1992, p. 113.)

Annexations

1 Brazil 1905
2 Brazil 1904
3 Ecuador 1880-1922
4 Ecuador 1880-1942
5 Ecuador 1880-1942
6 Peru 1942
7 Brazil 1904-1905
8 Brit.Guiana 1899
9 Brazil 1900
10 Bolivia 1867-1903
11 Bolivia 1867-1903
12 Bolivia 1867
13 Bolivia 1867-1909
14 Bolivia 1867-1902
15 Brazil 1927
16 Paraguay 1935
17 Bolivia 1880-1935
18 Chile 1883
19 Brazil 1870
20 Brazil 1870
21 Argentina 1874
22 Brazil 1895
23 Chile 1884
24 Argentina 1874
25 Argentina 1881
26 Chile 1902

Borders c. 1830
Subsequent border changes
Border of Greater Colombia 1819–1830
Border of Federation of Bolivia and Peru 1835–1839
European immigration (showing country of origin)

Heartland Theory

Because imperialism and colonialism shaped the world political map, it is helpful to understand one of the theories that drove them. By the end of the nineteenth century, numerous formal empires were well established and imperialist ideologies were dominant. To justify the strategic value of colonialism and explain the dynamic processes and possibilities behind the new world map created by imperialism, Halford Mackinder (1861–1947) developed a theory. Mackinder was a professor of geography at Oxford University and director of the London School of Economics. He later went on to serve as a member of Parliament from 1910 to 1922 and as chairman of the Imperial Shipping Committee from 1920 to 1945. Given his background in geography, economics,

and government, it is not surprising that his theory highlighted the importance of geography to world political and economic stability and conflict.

Mackinder believed that Eurasia was the most likely base from which a successful campaign for world conquest could be launched. He considered its closed heartland to be the "geographical pivot," the location central to establishing global control. Mackinder premised his model on the conviction that the age of maritime exploration, beginning with Columbus, was drawing to a close. He theorized that land transportation technology, especially railways, would reinstate land-based power, and sea prowess would no longer be as essential to political dominance. Eurasia, which had been politically powerful in earlier centuries, would rise again because it

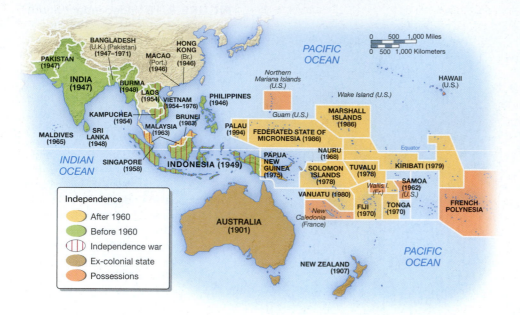

FIGURE 9.19 Independence in Asia and the South Pacific, before and after 1960 Decolonization and independence are not uniform phenomena. Different factors influence the shape that independence takes. The form of colonial domination that was imposed is as much a factor as the composition and level of political organization that existed in an area before colonization occurred. Some former colonies gain independence without wars of liberation; in Asia, these include India and Australia. In other places, the former colonizers were prepared to surrender their colonies only after wars of liberation were waged. This was the case in Indochina, where the domino theory influenced the French colonizers—and later the United States—to go to war against the Vietcong. The war for independence in Vietnam lasted from 1954 until 1976, exacting huge costs from all involved. Mostly, decolonization and political independence forced societies into a nation-state mold for which they had little, if any, preparation. It is little wonder, then, that few former colonies have succeeded in competing effectively in a world economy. (Adapted from *The Harper Atlas of World History, Revised Edition,* Librairie Hachette, p. 283. Copyright © 1992 by HarperCollins Publishers, Inc.)

FIGURE 9.20 Territorial divisions of Antarctica Even the uninhabitable terrain of Antarctica has become a site for competition among states. The radial lines delineating the various claims bear no relationship to the physical geography of Antarctica; rather, they are cartographic devices designed to formalize and legitimate colonial designs on the region. The overlay provides an indication of the mineral wealth of the continent and the drive behind much of the territorial claims. (*Source:* Reprinted with permission from Prentice Hall, from J. M. Rubenstein, *The Cultural Landscape: An Introduction to Human Geography,* 5th ed., 1996, p. 294; and http://www.coolantarctica.com/Antarctica%20fact%20file/science/threats_mining_oil.htm, accessed August 30, 2011).

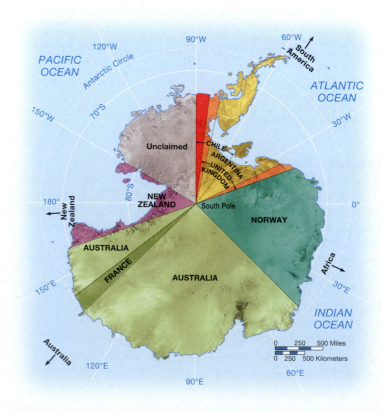

was adjacent to the borders of so many important countries and it was strategically buttressed by an inner and outer crescent of landmasses (**Figure 9.21**).

When Mackinder presented his geostrategic theory in 1904, Russia controlled a large portion of the Eurasian landmass protected from British sea power. In an address to the British Royal Geographical Society, Mackinder suggested that the "empire of the world" would be in sight if one power, or combination of powers, came to control the heartland. He believed that Germany allied with Russia and China organized by Japan were alliances to be feared. Mackinder's theory was a product of the age of imperialism. To understand why Britain adopted this theory, it is important to remember that antagonism was increasing among the core European states, leading to World War I a decade later.

Mackinder's geopolitical theory has certain resonance today. An example is the aggressive nationalism of Vladimir Putin, the current prime minister of Russia. As mentioned earlier in this chapter, the breakup of the Soviet Union created extensive regional realignments in the former confederation. Putin's response to the Western leanings of Estonia and Ukraine and Russia's invasion of Chechnya (a region that is part of Russia) and more recently Georgia have some observers suggesting that a new round of Russian empire-building is underway. Although Putin is not likely to attempt a further expansion into Central Europe, he could undermine those countries by capturing their next-door neighbor, Ukraine. And expanding into Georgia would give Russia access to the Mediterranean, the transit route between the Caspian Sea and the Black Sea. Many political geographers today believe that Mackinder's theories

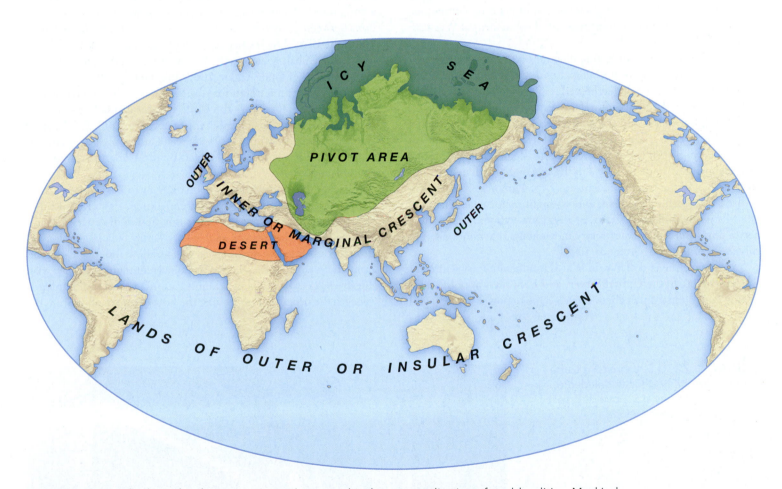

FIGURE 9.21 The heartland A quintessential geographical conceptualization of world politics, Mackinder's heartland theory has formed the basis for important geopolitical strategies throughout the decades since the theory's inception. While the pivot area (the heartland) of Eurasia is wholly continental, the outer crescent is mostly oceanic and the inner crescent part continental and part oceanic. It is interesting to compare the Mercator map projection, which Mackinder used to promote his theory, with the Dymaxion projection used in several maps in this text (See Figures 2.9, 2.12, and 2.14 for example). The Mercator projection decreases the size of the northern and southern oceans, which are vast and significant natural barriers. The spatial distortions inherent in the Mercator projection overemphasize the importance of Asia and the splitting of North and South America so that they appear on both sides of the map. The Dymaxion projection, as a northern polar representation, deemphasizes the centrality of any one landmass but exaggerates distances between continents. Mackinder's worldview map provides a good example of how cartographic representations can be employed to support ideological arguments. (Adapted from M. I. Glassner and H. DeBrij, *Systematic Political Geography*, 3rd ed. New York: J. Wiley & Sons, 1980, p. 291.)

are outdated, but Prime Minister Putin's recent aggressions and animosity toward what he regards as the "traitorous" former republics and breakaway regions underscore the current relevance of his geostrategic formulations.

The East/West Divide

In addition to a North/South divide based on imperialism and colonialism, the world order of states can also be viewed as dividing along an East/West split. The **East/West divide** refers to the gulf between communist and noncommunist countries, respectively. The East/West divide has played a significant role in global politics since at least the end of World War II in 1945 and, perhaps more accurately, since the Russian Revolution in 1917. By the second decade of the twentieth century, the major world powers were backing away from colonization. Still, many were reluctant to accelerate decolonization for fear that independent countries in Africa and elsewhere would choose communist political and economic systems instead of some form of Western-style capitalism.

Cuba provides an interesting illustration of an East/West tension that persists despite the official end of the Cold War. (The Cold War was the state of heightened military and political tension as well as economic competition between the former Soviet Union and its satellite states and the United States and its allies, lasting from 1946 to 1991). Although Cuba did not become independent from Spain until 1902, U.S. interest in the island dates back to the establishment of trade relations in the late eighteenth century. In the first half of the twentieth century, U.S. economic imperialism replaced Spanish colonialism and wrought major changes in the island's global linkages. Cuba, at that time, was experiencing a series of reform and revolutionary movements that culminated in the rise of Fidel Castro in the 1950s.

Following the end of World War II, anticommunist sentiment was at its peak in the United States, and U.S. fears about Cuba's "going communist" intensified. Castro came to power by 1959 and the United States began training Cuban exiles in Central America for an attack on Cuba. This training proceeded despite Castro's publicly declaring his alliance with the West in the Cold War during a 1959 visit to the United States. In 1961, an unsuccessful attempt by U.S.-trained Cuban exiles to invade Cuba at The Bay of Pigs and overthrow Castro further undermined Cuba's relations with the United States and paved the way for improved Cuban-Soviet relations.

Since that time, the U.S. opposition to communism in Cuba has been economic, not military. Having failed to overthrow Castro militarily, the United States fell back upon a 1960 embargo designed to destabilize the Cuban economy. The intent of the embargo, which is still in place, was to create such popular dissatisfaction with economic conditions that Cubans themselves would eventually overthrow Castro. The nearly 50-year-old embargo forbids economic trade between the United States and Cuba across a wide range of activities and products. However, a thaw in relations between the two countries began about a decade ago when, in 2000, the United States began selling food and agricultural products to the Cuban government. Today, the United States is Cuba's main supplier of food. And in 2003, President Bush established a Cabinet-level Commission for Assistance to a Free Cuba, whose mission is to explore ways the United States can help hasten and ease a democratic transition there. In February 2008, Fidel Castro announced that he would step down as Cuba's president and his brother Raul has since been elected by the Cuban parliament to that position.

In late 2010, President Obama began to chip away further at the embargo by lifting travel, remittance, postal, and business restrictions with respect to the interaction of Americans and Cubans. These changes will enable U.S. family members unlimited visits to Cuban family members as well as allow Americans to send as much money to their relatives in Cuba (remittances) as they wish (**Figure 9.22**). The Obama administration has also said it intends to begin the process of improving the flow of information from the

FIGURE 9.22 U.S. tourists in Cuba
Tourists visit the Bodeguita del Medio, a bar frequented by U.S. novelist Ernest Hemingway in Havana, Cuba.

United States to Cuba by enabling telecommunications networks between the two countries.

Domino Theory

Although the United States was unable to stop Castro's rise to power in Cuba, the end of World War II marked the rise of the United States to a dominant position among countries of the core. Following the war, the tension that arose between East and West translated into an American foreign policy pitched in opposition to the former Soviet Union. Domino theory underlay that foreign policy, which included economic, political, and military objectives directed at preventing Soviet world domination. **Domino theory** held that if one country in a region chose, or was forced to accept, a communist political and economic system, neighboring countries would fall to communism as well, just as one falling domino in a line of dominos causes all the others to fall. The means of preventing the domino-like spread of communism was often military aggression.

Domino theory first took root in 1947, when the postwar United States feared communism would spread from Greece to Turkey to Western Europe. It culminated in U.S. wars in Korea, Vietnam, Nicaragua, and El Salvador. However, preventing the domino effect was based not just on military aggression. Cooperation was also emphasized. For example, the international military alliance NATO (North Atlantic Treaty Organization) was established in 1949 with the stated purpose of *safeguarding* the West against Soviet aggression. After World War II, the core countries set up a variety of foreign aid, trade, and banking organizations to open foreign markets and bring peripheral countries into the global capitalist economic system. The strategy not only improved productivity in the core countries but also was seen as a way of strengthening the position of the West in its confrontation with the East.

The Vietnam War and its aftermath were probably the most serious global manifestation of Cold War competition and wrought terrible social and environmental damage on Southeast Asia as well as serious impacts on U.S. domestic and international politics. More than a million Vietnamese died, together with 58,000 Americans. The U.S. forces sprayed 2 million hectares (5 million acres) of Vietnam with defoliants, such as Agent Orange, that poisoned ecosystems and caused irreparable damage to human health **(Figure 9.23)**. Neighboring Cambodia and Laos were also bombed with napalm and defoliated to disrupt communist supply lines and camps. The media images of destruction, the loss of U.S. lives, and the financial cost of the war created considerable opposition in the United States, including protests on college campuses and marches on Washington.

The United States began to withdraw its forces in the early 1970s and left South Vietnam altogether as the Vietcong approached Saigon in 1973; in 1975, Vietnam was unified under communist rule. Two million people left South Vietnam fearing repression after unification, many (the so-called Vietnamese boat people) sailing away in small, fragile boats. The communist government confiscated farms and factories from owners to create state- and worker-owned enterprises, resettled hill tribes into intensive agricultural zones, and moved 1 million people into new economic development regions. But U.S.-led economic sanctions from 1973 to 1993 limited the potential for exports and restricted

FIGURE 9.23 Agent Orange in the Vietnam War Approximately 20 million gallons of herbicides, given the code name Agent Orange, were used in Vietnam between 1962 and 1971 to defoliate the jungle in order to deny cover to enemy forces. The dense jungle created significant challenges for the U.S. military in fighting the Vietcong. The toxins contained in Agent Orange resulted in serious health problems for U.S. service personnel exposed to the spray as well as for Vietnamese civilian and military personnel on the ground. These include diabetes, chloracne, Hodgkin's disease, multiple myeloma, non-Hodgkin's lymphoma, prostate cancer, respiratory cancer, soft tissue sarcoma, and neuropathic problems as well as birth defects in the children of those exposed.

some critical imports, such as medicines. At present, some sanctions are being lifted and trade and diplomatic relations with the Socialist Republic of Vietnam have improved dramatically since the mid-1990s.

In Cambodia, Khmer Rouge communist revolutionaries overcame the U.S.-backed military government in 1975 and instituted a cruel regime under the leadership of Pol Pot. The Khmer Rouge suspended formal education, emptied the cities, and set out to eliminate the rich and educated and to isolate the country from the world, renaming it Kampuchea. It is estimated that nearly 2 million people—approximately 30 percent of Cambodia's population—died in the brutal death march out of the capital, Phnom Penh, in 1975–1976 and in the "killing fields" of the countryside where civilians were tortured and executed. Conflict in Cambodia did not end even with the 1991 UN peace settlement and the return of the monarch, King Sihanouk, because the Khmer Rouge retook power in a coup in 1997. Today, the Khmer Rouge have again been defeated, and the Cambodian people, with the help of the international community, are building a democratic society as they recover from the emotional, physical, and economic brutality of the recent past.

The New World Order and Terrorism

With the fall of the Berlin Wall in 1989 and the opening up of former socialist and communist countries, such as China and Russia, to Western-style capitalist economic development, the Cold War is widely regarded as over. In March 1991, President George H. W. Bush made a speech referring to a "new world order" following the collapse of the Soviet Union. Francis Fukuyama, author of *The End of History and the Last Man*, helped further establish the term in the popular imagination. The **new world order** assumes that with the triumph of capitalism over communism, the United States is the world's only superpower and therefore its policing force. With the political, economic, and cultural dominance of the United States comes the worldwide promotion of liberal democracy and of a global economy predicated on transnational corporate growth through organizations like the World Bank and the World Trade Organization.

However, the move toward liberal, Western-style democracies and the capitalist consumption practices necessary to the success of the new global economy have created instability in some parts of the world. This instability is especially problematic where the Cold War struggle between the United States and the Soviet Union was once waged, in countries that once appeared ripe for succumbing to communism. The recent history of Afghanistan provides a telling illustration of this instability and its geopolitical implications (see Box 9.2, "Geography Matters: From the Cold War to the New World Order and the War on Terror").

With the emergence of a new world order, radical forms of warfare and political practices replaced more conventional ones. The attacks on the World Trade Center and the Pentagon on September 11, 2001, and the resulting war on terrorism are prime examples of the political and economic restructuring impulses of the new world order. The terrorist attacks and the subsequent response by the United States and other governments make clear that terrorism is the pivotal factor in current global geopolitics.

Terrorism is a complicated concept whose definition very much depends on social and historical context. A very simple definition is that **terrorism** is the threat or use of force to bring about political change. It is most commonly understood as actions by individuals or groups of individuals against civilian populations to undermine state practices or institutions. But the state can also be an agent of terrorism. Terrorism involves violent acts directed against society—whether by antigovernment actors, governments themselves, angry mobs or militants, or even psychotic individuals—and it will always mean different things to different people.

The term *terrorism* was first used during the French Revolution (1789–1795) to describe the new revolutionary government's repression of its people during the "Reign of Terror." Fifty years later, the term was used to describe revolutionaries who violently opposed existing governments. As the nineteenth century came to a close, the definition of "terrorism" was expanded to apply to militant worker and nationalist political organizations. By the mid-twentieth century, some used the term to describe many left-wing groups, as well as subnationalists (minority groups within the nation-state), or radical ethnic groups. In the 1980s, the violent activity of hate groups—anti-gay, racist, neo-Nazi organizations—in the United States was described by some as terrorist. At the same time, terrorism internationally was identified as a brand of ethnic or subnational warfare sponsored by rogue regimes.

Ethnic and subnational terrorism affects many countries today, including Russia, Uzbekistan, India, China, Colombia, the Philippines, Israel, and Rwanda. In Sri Lanka, the Tamil minority rebelled in the early 1980s against the oppressive Sinhalese majority. Originally a political movement, eventually the Liberation Tigers of Tamil Eelam launched a brutal campaign of terror that left over 65,000 people dead. Another ethnic or subnational violent struggle involves the Russian region of Chechnya, where the population has demanded independence. For the Muslim Chechens, Russia is acting as a terrorist state by employing military force to keep them from gaining independence. Chechens believe they have little in common politically, historically, or culturally with Russia.

Faith-based Terrorism

While subnational resistance organizations using terrorist tactics continue to operate throughout the world, the most widely recognized terrorism of the new century has religious roots. The September 11 attacks have helped to bring the realities of religious terrorism sharply into public focus. The connection between religion and terrorism is nothing new as terrorism has been perpetrated by religious fanatics for more than 2,000 years. Indeed, words like *zealot*, *assassin*, and *thug* all stem from fundamentalist religious movements of previous eras. And while the links between Muslims and terrorism in the world today are especially strong and geographically widespread, it is critical to understand that Muslim terrorism is not the only form of terrorism and that domestic terrorism sparked by fundamentalist Christian organizations has taken the lives of hundreds of innocent people and continues to be a threat in the United States and elsewhere.

A chilling example of domestic terrorism in the United States is the bombing of the Alfred P. Murrah Federal Office Building in Oklahoma City in April 1995, when 168 people were killed **(Figure 9.24)**. In June 1997, Timothy McVeigh, a U.S. Army veteran and Christian white supremacist, was convicted in federal court of perpetrating the attack. McVeigh was executed by lethal injection in June 2001, and his accomplice, Terry L. Nichols, was sentenced to life in prison. Both McVeigh and Nichols were connected to the Christian Identity movement through a group that came to be known as the Michigan Militia Corps, a paramilitary survivalist organization. The Christian Identity movement is just one of several religious extremist groups in the United States, including other forms of white supremacy movements, apocalyptic cults, and Black Hebrew Israelism. The Christian Identity movement is based on a belief in the superiority of whiteness as ordained by God.

Another, more recent example of faith-based terrorism is the twin bombings that occurred in Norway on July 24, 2011. In this case, a young Christian man, fearing that Muslims were polluting

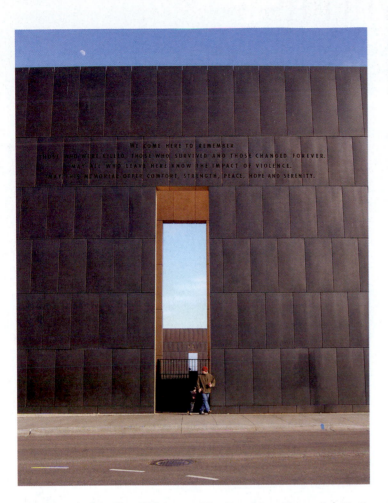

FIGURE 9.24 The Oklahoma City Bombing Memorial Until the attacks of September 11, 2001, the Oklahoma City bombing was the worst terrorist incident—domestic or foreign—to be perpetrated on U.S. soil. The blast was caused by a deadly combination of fuel oil and fertilizer placed in a van parked outside the building and detonated remotely.

FIGURE 9.25 Oslo, Norway terrorist bombing An Oslo neighborhood where bombs set off by Anders Behring Breivik in summer 2011 destroyed buildings.

the Norwegian nation, is suspected of having detonated two bombs killing 93 people, most of them young people **(Figure 9.25)**. Terrorism with roots in religious fervor is widespread and has occurred in both rich and poor parts of the globe.

The War on Terror in Iraq

The United States responded to the terrorist attacks of September 11, 2001, by declaring a global war against terrorism and identifying first Afghanistan and then Iraq as the greatest threats to U.S. security. Although the evidence of involvement in the 9/11 attacks by Iraq and its leader, Saddam Hussein, was highly questionable, on March 19, 2003, after amassing over 200,000 U.S. troops in the Persian Gulf region, President George W. Bush ordered the bombing of the city of Baghdad, Iraq. The declaration of war and invasion occurred without the explicit authorization of the UN Security Council, and some legal authorities take the view that the action violated the UN Charter. Some of the staunchest U.S. allies (Germany, France, and Canada) as well as Russia opposed the attack and hundreds of thousands of antiwar protestors repeatedly took to the streets throughout the world

for the weeks and months preceding and following the onset of war launched by coalition forces of the United Kingdom and the United States.

The motivation for the war, as expressed by British Prime Minister Tony Blair and George W. Bush, was that Iraq had stockpiled "weapons of mass destruction"—chemical and biological weapons capable of massive human destruction. In the days leading up to the war, UN weapons inspector Hans Blix and his team were unable to locate any weapons despite an intensive search of the country. President Bush, however, proceeded to justify a dramatically stepped-up "war on terrorism" (following the invasion in Afghanistan) on the grounds that "neutralizing" Iraq's leader, Saddam Hussein, was necessary to global security.

Despite the fact that "major combat" has ended, peace has not returned to Iraq. Iraq continues to host violent conflict between U.S. and Iraqi soldiers and forces described by the occupiers as insurgents. Tactics in use include mortars, suicide bombers, roadside bombs, small-arms fire, and rocket-propelled grenades, as well as sabotage against the oil infrastructure of the country. As of Summer 2011, the total number of deaths of U.S. soldiers as a direct result of the Iraq invasion reached 4,459, with somewhere between 100,000 and 110,000 combined Iraqi civilian and military deaths **(Figure 9.26)**. An especially distressing and long-term aspect of the Iraq war, as well as other U.S. wars, is its impact on returning veterans' mental health; the Department of Veteran Affairs director has stated that there are about 1,000 suicide attempts per month among the nation's 25 million veterans (from all U.S. wars). Besides the mental health impacts, there are, of course, also very significant physical ones, including lost limbs and serious brain and organ injuries.

Afghanistan, known in ancient times as Gandhar, was once famous for its wealth, art, and culture (**Figure 9.B**). Its trading centers were important links on ancient trading routes between Central Asia and South Asia, and their wealth soon attracted invaders. Alexander the Great swept into Afghanistan—then a part of the Persian Empire—in 329 B.C.E. This invasion paved the way for a cultural awakening and the emergence of the Gandhar school of art, known for its amalgamation of Indian and Greek styles.

Afghanistan is dominated by the rugged Hindu Kush Mountains, which sweep from the east to the west, sinking into the desert near the northwestern city of Herat. Tens of thousands of square kilometers of the Hindu Kush form an intricate and seemingly endless maze of valleys and ravines. Jagged scree-strewn mountains and rugged valleys and caves provide ideal territory in which to fight a guerrilla war against invaders or occupying forces. The problems of topography are compounded by the weather. By late October swirling snow descends on the mountains, sealing off many of the passes, valleys, and high plateaus and making troop movement almost impossible until late spring.

Despite the inhospitable terrain and weather, Afghanistan has attracted one invader after another from the Arabs to the Persians to the Turkic Ghaznavids to the Mongol invasion, led by Genghis Khan. Late in the eighteenth century, Afghanistan's geopolitical significance increased still more. For the eastward-expanding Russian empire, Afghanistan represented the last barrier to a thrust toward the rich plains of India. For the British, who were establishing a hold on India, Afghanistan represented a bastion against Russian expansion. Both the Russians and the British desperately wanted to control Afghanistan and so began the Great Game—as the struggle between the two imperial powers for control of Afghanistan was called. The British were able to block the Russians but were not able to establish territorial control. After three wars with the stubborn Afghans, the British finally granted Afghanistan independence in 1921. A brief period of Afghan independence followed, during which Mohammad Zahir Shah, who reigned from 1933 to 1973, established a relatively liberal constitution. Five years later, a bloody coup led by the People's Democratic Party of Afghanistan (PDPA) imposed a Marxist-style reform program that ran counter to deeply rooted Islamic traditions in the country. As a result, Afghan opposition to the Marxist government emerged almost immediately.

The Soviet Union moved quickly to take advantage of the 1978 coup, signing a bilateral treaty of friendship and cooperation with Afghanistan that included a military assistance program. Before long, as the opposition insurgency intensified, the PDPA

FIGURE 9.B Afghanistan and Pakistan As this map illustrates, Afghanistan is a landlocked, mountainous country sharing borders with six other countries. When war occurred here in the twentieth century, many of the people fled to these neighboring countries for safety, including Pakistan, where al-Qaeda and the Taliban have established strongholds.

regime's survival was wholly dependent upon Soviet muscle. In December 1979, faced with a rapidly deteriorating security situation, the Soviet Union sent a large airborne force and 120,000 ground troops to Kabul under the pretext of a field exercise but was unable to establish authority outside Kabul. An overwhelming majority of Afghans opposed the communist regime, and Islamic freedom fighters (*mujahideen*) made it almost impossible for the regime to maintain control outside major urban centers. Poorly armed at first, the *mujahideen* began receiving substantial assistance in the form of weapons and training from the United States, Pakistan, and Saudi Arabia in 1984. The guerrilla war ended in 1989 with the withdrawal of the Soviet Union, but as the victorious *mujahideen* entered Kabul to assume control over the city and the central government, a new round of fighting began between the various militia groups that had coexisted uneasily during the Soviet occupation.

With the demise of their common enemy, the militias' ethnic, clan, religious, and personality differences surfaced, and civil war ensued. Fighting in Kabul and in northern provinces caused thousands of civilian deaths and created new waves of displaced persons and refugees, hundreds of thousands of whom trekked across the mountains to Pakistan for sanctuary. Eventually, the hard-line Islamist faction of the *mujahideen*—the Taliban—gained control of Kabul and most of Afghanistan. By 2001, the only remaining resistance to the Taliban regime was the Northern Alliance, a loose coalition of minority ethnic groups, including the Tajiks, who controlled between 5 and 10 percent of Afghan territory.

The Taliban regime not only imposed harsh religious laws and barbaric social practices on the Afghan population but also harbored an entirely new geopolitical force with worldwide implications: Osama bin Laden and his al-Qaeda terrorist network, who were responsible for numerous attacks on the West, including the September 11, 2001, attacks on the Pentagon and the World Trade Center. The United States, with support from the United Kingdom, Australia, Canada, and the Northern Alliance, invaded Afghanistan in October 2001 as part of its war on terrorism campaign. The military campaign was named Operation Enduring Freedom and its aim was to capture Osama bin Laden, who had been a resident

there since 1996; destroy the al-Qaeda network; and overthrow the Taliban regime. Bombing raids over Afghanistan officially ceased in 2002 and Osama bin Laden was assassinated on May 1, 2011, in Pakistan.

After 10 years of fighting, increased troop deployment to Afghanistan following the formal end of the war in Iraq, and the elimination of Osama bin Laden, President Obama, after adding 30,000 new troops in 2009 (bringing the total to 100,000), has promised a scheduled drawdown of 33,000 U.S. troops in Summer 2012. Now costing $2 billion a week (71 percent of the Pentagon's budget in fiscal 2011) and having resulted in the loss of 1,500 American lives, the war in Afghanistan is becoming increasingly unpopular. Troops are being routinely killed by roadside bombs, small-arms fire and rocket attacks, and in unspecified combat operations. The biggest concern for defense officials is that the sheer levels of violence are higher than they ever have been, with 2010 being the deadliest year on record.

With U.S. troops on the ground in Afghanistan, al-Qaeda moved operations to neighboring Pakistan, where Osama bin Laden was killed, without the knowledge of the Pakistani government. For years there had been speculation that bin Laden was sequestered in Pakistan's remote and semiautonomous tribal areas in the northwestern part of the country, the Federally Administered Tribal Areas (FATA) (**Figure 9.C**). But Abbottabad, the town in which he was found and killed,

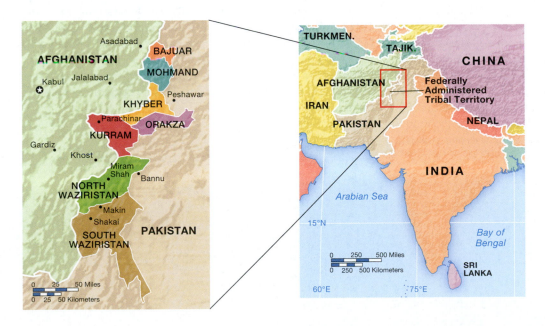

FIGURE 9.C Federally Administered Tribal Areas, Pakistan FATA is a tribal region in northwestern Pakistan. Jurisdictionally, the region consists of tribal areas and frontier regions. Historically, the FATA acted as the "Great Game" buffer zone between Russia and British-controlled India. (*Source:* Source for Pakistan internal boundaries, Geocart; source for Federally Administered Tribal Areas, CIA.)

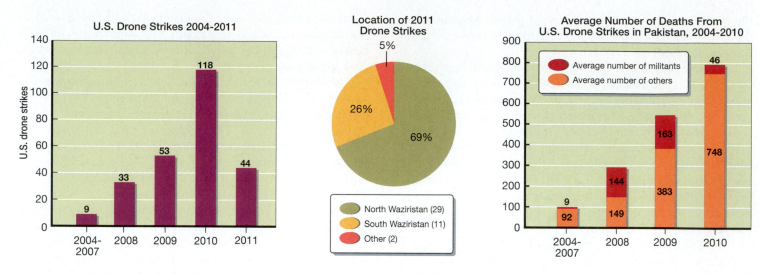

FIGURE 9.D U.S. drone strikes, 2004–2011 The New America Foundation study shows that the 257 reported drone strikes in northwest Pakistan have killed approximately 1,575 to 2,486 individuals, of whom 1,282 to 2,015 were described as militants in reliable press accounts. The report estimates that the true nonmilitant fatality rate since 2004 is approximately 20 percent. (*Source:* Reprinted by permission of New American Foundation: National Security Studies Program.).

is just a short distance from Islamabad. It is also the location of the country's main military training institution, located several hundred miles from Waziristan—an area of FATA on the Afghanistan border—where the CIA drone strike campaign has been concentrated.

Significantly, the United States, while not sending troops to Pakistan, has been deploying drones (unmanned aircraft) in Pakistan's tribal areas against suspected al-Qaeda and Taliban targets since 2004. The Pakistani government simultaneously privately supports and publicly condemns the drone attacks. A highly regarded recent study by the New America Foundation has shown that there have been 257 reported U.S. drone strikes there. Presently, the U.S. military has close to 7,000 unmanned

aircraft, with 39 combat-air patrols flying over Iraq and Afghanistan constantly. This number is expected to rise to 65 a day by 2013. These strikes have been implicated in the deaths of up to 2,486 individuals (**Figure 9.D**). Tense relations now exist between Pakistan and the United States following the killing of bin Laden. The U.S. covert assassination operation humiliated the Pakistani government—whom many regional specialists believe knew where bin Laden was all along—and it has retaliated by cutting back on counterterrorism cooperation with the United States.

New America Foundation, "The Year of the Drone," 2011, available at: www. Newamerica.net, accessed June 30, 2011.

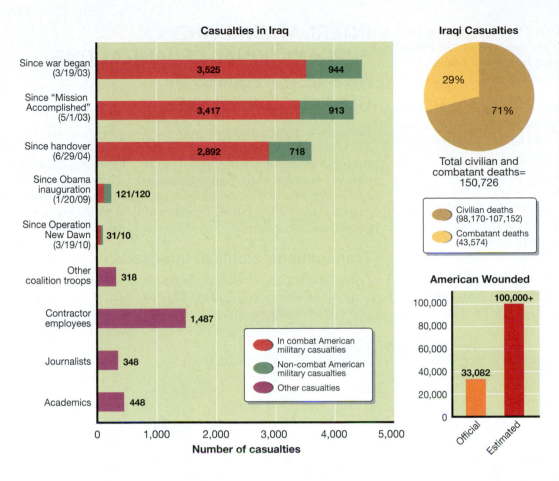

Casualties in Iraq

	In combat American military casualties	Non-combat American military casualties
Since war began (3/19/03)	3,525	944
Since "Mission Accomplished" (5/1/03)	3,417	913
Since handover (6/29/04)	2,892	718
Since Obama inauguration (1/20/09)	121/120	
Since Operation New Dawn (3/19/10)	31/10	
Other coalition troops	318	
Contractor employees	1,487	
Journalists	348	
Academics	448	

Legend:
- In combat American military casualties
- Non-combat American military casualties
- Other casualties

Iraqi Casualties

29% / 71%

Total civilian and combatant deaths= 150,726

- Civilian deaths (98,170–107,152)
- Combatant deaths (43,574)

American Wounded

Official: 33,082
Estimated: 100,000+

FIGURE 9.26 Casualties from the war in Iraq, 2011 This graphic includes both the wounded and dead among military and civilian personnel participating in the U.S. war effort in Iraq. More than a third of those who died have come from six states: California, Florida, Pennsylvania, Ohio, New York, and Texas; more than half were between the ages of 18 and 24. (Data source: http://antiwar.com/casualties)

In early 2004, the 9/11 Commission (more formally known as the National Commission on Terrorist Attacks Upon the United States) concluded that there was no credible evidence that Saddam Hussein, the captured and executed former president of Iraq had assisted the terrorist organization al-Qaeda to prepare for or carry out the 9/11 attacks. There is also general agreement among U.S. intelligence and military personnel that Iraq had most likely destroyed its programs for biological and chemical weapons production before the UN team began its inspection.

In 2007, President Bush ordered, over congressional opposition, a surge of yet more troops to Iraq with the intention of providing security to the city of Baghdad and the Al Anbar Province. The surge met with some success, however patchy and unstable, particularly in Al Anbar, where Sunni tribes are cooperating with the U.S. Army to fight al-Qaeda terrorism. And Sunni tribes around the country have formed "Awakening Councils" paid for by the United States that cooperate with U.S. troops. These have also led to an overall decrease in sectarian violence. In 2010, with 144,000 troops in Iraq, President Obama began a troop drawdown there. As a result, by August 2010 there were 50,000 troops still operating in the country. Remaining troops are mostly involved in "advise and assist" activities, supporting and training Iraqi security forces, protecting American personnel and facilities, and mounting counterterrorism operations. Unfortunately, sectarian violence still occurs on a regular basis throughout much of the country, especially in Baghdad. Moreover, the civilian humanitarian situation is among the most critical in the world. Millions of Iraqis have insufficient access to clean water, sanitation, and health care.

Although the war in Afghanistan and its implications for Pakistan are addressed in Geography Matters Box 9.2, it is also important to point out here the implications of the assassination of Osama bin Laden that occurred on May 1, 2011, in Abbottabad, Pakistan, which is seen by many as a symbolic end to the war on terror. Bin Laden was the founder of al-Qaeda, the jihadist organization responsible for the September 11, 2001, attacks on the United States. A member of a wealthy Saudi family, bin Laden had been in hiding for a decade before he was killed in a covert operation ordered by President Obama and carried out by U.S. Navy SEALs and CIA operatives. Bin Laden believed that the United States and its political, economic, and cultural values persecuted and oppressed Muslims around the world and the only way to end this violence was with violence. Bin Laden's fundamental approach was to seduce leading economic powers, such as the United States and the former Soviet Union, into endless war in Muslim countries. At the same time, al-Qaeda continued to recruit new jihadists to the organization's cause, perpetuating an unconventional but highly costly conflict. Bin Laden believed that endless war would lead to the economic collapse of the United States.

For the United States, the costs of the war are indeed still mounting. The U.S. Treasury estimates the direct cost at $845 billion. Nobel-prize winning economist Joseph Stiglitz and his coauthor Linda Bilmes argue that the true cost of the war is closer to $3 trillion when the indirect costs are also accounted for, including interest on the debt raised to fund the war, the rising cost of oil, health care costs for returning veterans, and the cost of replacing destroyed military hardware and degraded operational capacity.[1]

With Osama bin Laden dead, the war on terror has not actually ceased. In fact, the war still goes on in Afghanistan and, more covertly, in Pakistan. U.S. Secretary of Defense Robert Gates, speaking in Afghanistan in June 2011, said that it is too soon to tell when the war on terror will be over.

APPLY YOUR KNOWLEDGE Conduct an Internet search on the Iraq war and how it has affected your city and state. How many soldiers from your area have been killed? In economic terms, how much has the war cost your city and state? (To determine the economic impact of the war, consult the National Priorities Project database.) ■

INTERNATIONAL AND SUPRANATIONAL ORGANIZATIONS AND NEW REGIMES OF GLOBAL GOVERNANCE

Just as states are key players in political geography, international and supranational organizations have become important participants in the world system in the last century. These organizations have become increasingly important means of achieving goals that could otherwise be blocked by international boundaries. These goals include the freer flow of goods and information and more cooperative management of shared resources, such as water.

Transnational Political Integration

Perhaps the best-known international organization operating today is the United Nations (**Figure 9.27**). The period since World War II has seen the rise and growth not only of large international

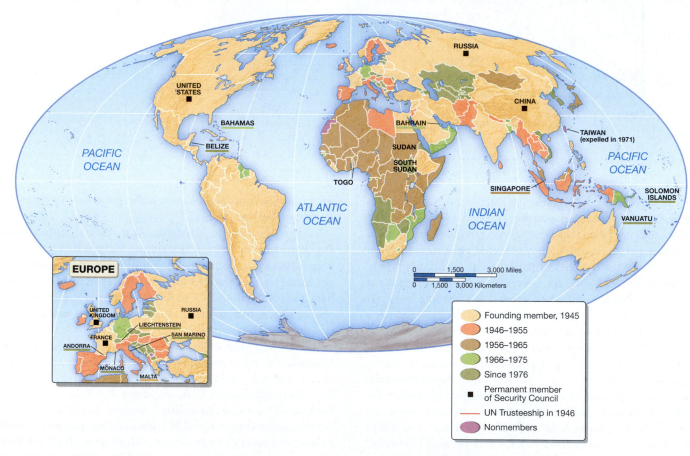

FIGURE 9.27 UN member countries Following World War II and the demise of the League of Nations, a renewed effort was made to establish an international organization aimed at instituting a system of international peace and security. The UN Charter was approved by the U.S. Senate in July 1945, raising hopes for a more long-lived organization than the ineffective League of Nations. Located in New York City, the United Nations is composed of a Security Council, which includes permanent members, the United States, Britain, China, France, and Russia, and a General Assembly, which includes all those countries identified on the map. At the same time that the United Nations was set up, the United States lobbied for the creation of the International Monetary Fund (IMF) and the World Bank. The U.S. government had believed that World War II resulted from the collapse of world trade and financial dislocation caused by the Great Depression. The task of the IMF and the World Bank is to provide loans to stabilize currencies and enhance economic growth and trade.

[1] J. Stiglitz and L. Bilmes, *The Three Trillion Dollar War*. New York: W.W. Norton, 2008.

FIGURE 9.28 Membership in the European Union The goal of the European Union is to increase economic integration and cooperation among the member states. The EU was established in 1992, when the Maastricht Treaty was ratified by the 12 members of the European Economic Community, or EEC (Belgium, Denmark, France, Germany, Greece, Ireland, Italy, Luxembourg, Netherlands, Portugal, Spain, and the United Kingdom), which was created in 1967. Upon ratification of the treaty, the countries of the EEC became members of the EU, and the EEC became the policymaking body of the EU. The treaty established European citizenship for citizens of each member state, enhanced EEC customs and immigration agreements, and allowed for the establishment of a common currency, the euro, which is currently in circulation among all of the original 12 members except Denmark and the United Kingdom. Newer members, including Finland, Austria, Slovakia, Latvia, Lithuania, and Malta, also use the euro in addition to their own currencies. (*Source:* Updated per Europa: Gateway to the European Union, http://europa.eu/abc/european_countries/index_en.htm.)

organizations but also of new regional arrangements. These arrangements vary from local ones, such as the Swiss-French cooperative management of Basel-Mulhouse airport, to more extensive ones, such as the North American Free Trade Agreement (NAFTA), which joins Canada, the United States, and Mexico into a single trade region. Regional organizations and arrangements now address a wide array of issues, including the management of international watersheds and river basins, such as the Great Lakes of North America and the Danube and Rhine rivers in Europe. They also oversee the maintenance of health and sanitation standards and coordinate regional planning and tourism. Such regional arrangements seek to overcome the barriers to the rational solution of shared problems posed by international boundaries. They also provide larger arenas for the pursuit of political, economic, social, and cultural objectives.

A **supranational organization** is a collection of individual states with a common goal that may be economic and/or political in nature. By organizing and regulating designated operations of the individual member states, these organizations diminish, to some extent, individual state sovereignty in favor of the collective interests of the entire membership. The European Union (EU) is perhaps the best example of a supranational organization.

At the end of World War II, European leaders realized that Europe's fragmented state system was insufficient to the demands and levels of competition coalescing within the world political and economic system. They responded by creating an entity that would preserve important features of state sovereignty and identity. They also intended to create a more efficient intra-European marketing system and an entity more competitive in global transactions. **Figure 9.28** shows the original member countries of the European

Economic Community, which evolved into the EU in 1992, the existing members of the EU, and those countries that are currently candidates for admission. The EU holds elections, has its own parliament and court system, and decides whether and when to allow new members to join. Generally speaking, the EU aims to create a common geographical space within Europe in which goods, services, people, and information move freely and in which a single monetary currency prevails. Whether an EU foreign policy will ever be accomplished remains to be seen, but a common European currency, the euro, has been in circulation since 2002.

Importantly, the most recent global recession has wreaked havoc on the euro and EU monetary stability. Ireland, Portugal, and Greece have huge budget deficits that have put their own economies as well as the monetary union at great risk. Spain has also showed signs of insolvency, with some EU finance personnel anticipating that Italy may soon follow. Indeed, the single currency adopted by most European nations is now in danger of looking more like a trap for individual countries that, by giving up their own currency, also give up economic flexibility.

Globalization, Transnational Governance, and the State

As we have already noted, globalization has been as much about restructuring geoeconomics as it has been about reshaping geopolitics. In fact, some globalization scholars believe that the impact of globalization on politics has been so profound that globalization is leading to the diminution of the powers of the modern state, if not its ultimate disappearance. These scholars believe that because the modern state is organized around a bounded territory and because globalization is creating a new economic space that is transnational, the state is increasingly incapable of responding to the needs of the new transnational economy. Although we do not subscribe to this position, we do recognize that the state is undergoing dramatic changes that are restructuring its role with respect to both local, domestic concerns as well as global, transnational ones.

In the twentieth century, from the end of World War II until 1989, when the Berlin Wall was dismantled, world politics was organized around two superpowers. The capitalist West rallied around the United States, and the communist East around the Soviet Union. But with the fall of the Berlin Wall signaling the "end of communism," the bipolar world order came to an end. The concept of the new world order, organized around global capitalism, emerged and has increasingly solidified around a new set of political powers and institutions, which have recast the role of the state.

The increasing importance of trade-facilitating organizations such as the EU, NAFTA, the Association of Southeast Asian Nations (ASEAN), the Organization of Petroleum Exporting Countries (OPEC), and the World Trade Organization (WTO) is the most telling indicator that the world, besides being transformed into one global economic space, is also experiencing global geopolitical transformations. These organizations are unique in modern history, as they aim to treat the world and different regional clusters as seamless trading areas free of the rules that ordinarily regulate national economies.

The state must now contend with a whole new set of processes and other important political actors on the international stage as well as within its own territory. For instance, the transnational financial network that was established in the 1980s is far beyond the control of any one state, even a very powerful state like the United States, to regulate effectively. In fact, what the increasing importance of transnational flows and connections—from flows of capital to flows of migrants—indicates is that the state is less a container of political or economic power and more a site of flows and connections.

The increasing importance of flows and connections means that contemporary globalization has made possible a steadily shrinking world. In short, politics can move beyond the confines of the state into the global political arena, where rapid communications enable complex supporting networks to be developed and deployed, facilitating interaction and decision making. One indication of the increasingly global nature of politics outside of formal political institutions is the rise in environmental organizations whose purview and membership are global, as discussed in Chapter 4.

What is interesting about the institutionalization of global politics is that it has been less involved with the traditional preoccupations of relations between states and military security issues and more involved with issues of economic, ecological, and social security. The massive growth in flows of trade, foreign direct investment, financial commodities, tourism, migration, crime, drugs, cultural products, and ideas has been accompanied by the emergence and expansion of global and regional institutions to manage and regulate these flows. The modern state has been drawn increasingly into this complex of global, regional, and multilateral systems of governance. And as the state has been drawn into these new activities, it has shed or deemphasized some of its previous responsibilities, such as maintenance of social welfare.

The involvement of the state in new global activities, the growth of supranational and regional institutions and organizations, the critical significance of transnational corporations to global capital, and the proliferation of transnational social movements and professional organizations are captured by the term **international regime**. The term reflects the fact that the arena of contemporary politics is now international, so much so that even city governments and local interests groups—from sister-city organizations to car clubs—are making connections and conducting their activities both within and beyond the boundaries of their own states. An example of this is the human rights movement that has gained ascendancy over the last four to five decades.

Human Rights

Human rights, including the rights to justice, freedom, and equality, are considered by most societies to belong automatically to all people. Until World War II, safeguarding human rights was the provenance of states whose rules and regulations legislated the proper treatment of its citizens, from prisoners to schoolchildren. Since the late 1940s and 1950s, nearly all states have come to accept the importance of a comprehensive political and legal framework that focuses on human rights and that allows an international organization to intervene in the operations of a sovereign state that is in violation of the International Bill of Human Rights adopted by the United Nations in 1948.

In 1998, the United Nations realized another step in the protection of human rights by adopting a treaty to establish a permanent International Criminal Court (ICC). In establishing the court, Kofi Anan, then secretary-general of the United Nations, stated: "Our hope is that, by punishing the guilty, the ICC will bring some comfort to the surviving victims and to the communities that have been targeted. More important, we hope it will deter future war criminals, and bring nearer the day when no ruler, no State, no junta and no army anywhere will be able to abuse human rights with impunity." By forming the treaty, the United Nations aimed to create a permanent mechanism to bring to justice the perpetrators of such crimes as genocide, ethnic cleansing, and sexual slavery and put an end to the impunity so often enjoyed by those in positions of power. The court has the mandate to try individuals, not states, for crimes committed in the present (after July 2002). What makes the ICC unique is the principle of complementarity, which means that the court can only exercise its jurisdiction when a national court is unable or unwilling to genuinely do so itself. Thus, the importance of the international scale has increased over the last 50 years of the twentieth century and appears likely to continue to do so into the twenty-first century. It is important to note, however, that not all states support the ICC. Seven UN members voted against the treaty to establish the ICC. These were the United States, China, Iraq, Israel, Libya, Qatar, and Yemen.

An example of global human rights legislation is the UN Declaration on Rights of Indigenous People, approved in 2007 when 143 countries voted for it; 11 countries abstained; and 4—Australia, Canada, New Zealand, and the United States, all countries with very significant numbers of indigenous people—voted against it. The declaration emphasizes the rights of indigenous peoples to maintain and strengthen their own institutions, cultures, and traditions and development in keeping with their own needs and aspirations. It also prohibits discrimination against them and promotes their full and effective participation in all matters that concern them. The four "no" voting countries expressed concern about the vague language of the declaration and the impact this might have on existing, already settled treaties.

Children's Rights

An additional and often overlooked aspect of human rights is **children's rights**. In 1989, the United Nations adopted the Convention on the Rights of the Child, and despite the convention's nearly universal ratification (only the United States and Somalia have not ratified it), many of the most basic rights of children are still not being protected **(Figure 9.29)**. The convention promises children around the world the fundamental right to life, liberty, education, and health care. Among the fundamental safeguards the convention provides are protection of children in armed conflict; protection from discrimination; protection from torture or cruel, inhuman, or degrading treatment or punishment; protection within the justice system; and protection from economic exploitation. And yet, despite the existence of the convention, around the world street children are killed or tortured by police, recruited or kidnapped to serve as soldiers in military forces or labor under extremely difficult conditions, forced into prostitution, and imprisoned. Refugee children, often separated from their families, are vulnerable to exploitation, sexual abuse, and domestic violence. Keeping the promises made in the Convention on the Rights of the Child is one of the biggest challenges of the twenty-first century.

The emergence of human rights as a globally relevant issue has occurred as groups and organizations, both governmental and nongovernmental, have been able to debate and discuss issues that concern all people everywhere and can do so at the international level through conferences, e-mail, listservs, and direct action at international events. The phenomenon of different people and groups across the world in common cause is known as global civil society. **Global civil society** is composed of the broad range of institutions that operate between the private market and the state.

In the next section, we move from the international level to national, regional, and local levels in an effort to show that political geography occurs at all levels of political organization and that each facilitates significant insights about the politics of geography and the geography of politics.

APPLY YOUR KNOWLEDGE Research arguments made by the United States about why it has not ratified the UN convention on the Rights of the Child. Do you find the arguments compelling? Explain in a paragraph your reaction to the arguments and whether or not you support them and why. ∎

The Two-Way Street of Politics and Geography

Political geography can be viewed according to two contrasting orientations. The first orientation sees it as the *politics of geography*. This perspective emphasizes that *geography*—the areal distribution/differentiation of people and objects in space—has a very real and measurable impact on politics. Regionalism and sectionalism, discussed later in this section, illustrate how geography shapes politics. This politics-of-geography orientation is also a reminder that politics occurs at all levels of the human experience, from the international order down to the neighborhood, household, and individual body.

The second orientation sees political geography as the *geography of politics*. This approach analyzes how *politics*—the tactics or operations of the state—shapes geography. Mackinder's heartland theory and the domino theory, discussed earlier in this chapter, attempt to explain how the geography of politics works at the international level. In the heartland theory, the state expands into new territory in order to relieve population pressures. In the domino theory, as communism seeks new members, it expands geographically to incorporate new territories. As an example of the geography-of-politics orientation, consider Palestine and Israel. An examination of a series of maps of Palestine/Israel since 1923 reveals how the changing geography of this area is a response to changing international, national, regional, and local politics (see Box 9.3, "Visualizing Geography: The Israel-Palestine Conflict").

FIGURE 9.29 UN Convention on the Rights of Children This child-friendly poster outlines the rights of children around the world to a dignified, safe, and supportive environment.

The Israel-Palestine Conflict

By Jeff Wilson

The Israel-Palestine conflict is a modern territorial dispute with its roots in post–World War I colonial politics. Under the auspices of the League of Nations, the British government assumed control of Palestine and its population in 1923, marking the beginning of the British Mandate of Palestine, which included the areas of Transjordan and Palestine. Within the borders of Palestine mandate, the conflict began to take shape. **Figure 9.E** maps out how subsequent conflicts have created territorial changes over time.

Subsequent to the mandate period, three particular events stand out as defining moments for the geopolitical shape of the former territory of Palestine and contemporary Israel: the 1947 United Nations (UN) Partition Plan, the Palestine War of 1948, and the Six-Day War in 1967. For more details on these events, refer

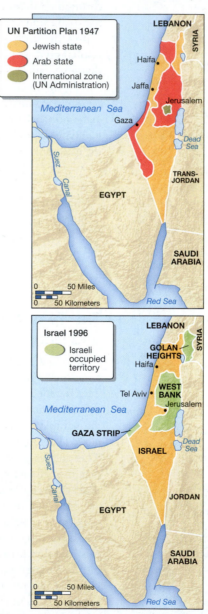
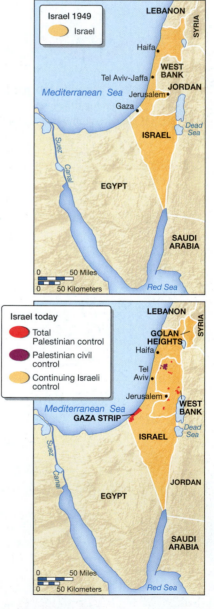

FIGURE 9.E Changing geography of Israel/Palestine, 1923–2011 Since the creation of Israel out of much of what had been Palestine in 1947, the geography of the region has undergone significant transformation. A series of wars between Israelis and Arabs and a number of political decisions by Israel have produced the changing geographies we see here. (*Source: Adapted with permission from Prentice Hall, from J. M. Rubenstein, The Cultural Landscape: An Introduction to Human Geography, 5th ed., 1996, p. 233.*)

to the time line of the conflict in **Figure 9.F**. At its very core, the conflict is between the Israelis and the Palestinians. On the one side, efforts by the Jewish community to establish an exclusively Jewish state in Palestine have largely succeeded, as is clear by the creation of the modern state of Israel. On the other side, the Palestinian community continues to struggle against the effects of displacement and colonization and remains committed to the founding of an independent Palestinian state. The conflict is essentially about land. Currently, Israel controls over 75 percent of the former Palestine; this includes a military and settler occupation of the West Bank. Israel also controls land, sea, and flight access to the Gaza Strip. Despite this external control, Palestinians remain firmly engaged in a national project of establishing their own independent state in the remaining 25 percent of the land. This struggle is inseparable from Palestinian efforts to resist the Israeli military occupation of the West Bank and Gaza Strip.

People of the Middle East consistently rank a resolution to the Palestinian-Israeli Conflict as one of the most pressing regional concerns. The Palestinians' passionate desire for self-rule coupled with Israel's security needs make political negotiations at the governmental level complex, protracted, and emotionally loaded. General discourse surrounding the conflict is often framed solely from the vantage point of state politics and policies. However, state and international institutions are not the only political actors seeking a resolution to the conflict. Many nongovernmental organizations (NGOs) have formed—made up of Palestinians, Israelis, and international volunteers—to advocate for Palestinian rights, self-rule, and a lasting security for all involved.

These local groups often shape the political landscape at the national and international levels through grassroots organizing. They underscore the need for expanding the discourse surrounding the conflict, one that too often constructs the Palestinians simply as "terrorists" and the Israelis as "victims" and confines politics to the decisions of heads of states. Such constructions ignore the complex reality of all groups and severely handicap effective political action. In this box we provide snapshots of four organizations, both Israeli and Palestinian, that, from their unique standpoints, are addressing the ramifications of the conflict. Each organization is addressing aspects of the occupation (human rights violations, land confiscation, abuse, etc.) using nonviolent means.

The International Solidarity Movement (ISM) is a Palestinian-led organization working in the West Bank and Gaza Strip to resist the Israeli military occupation. ISM emerged in the summer of 2001 out of a convergence of a handful of Palestinian and Israeli activists, along with the members from the Bayt Sahur-based Center for Rapprochement Between People.[1] ISM employs nonviolent direct action methods and principles as its method for engaging with the occupation, which currently presides over and administers the territories. It views the military occupation as the root cause of all other violence between Israelis and Palestinians. In the view of ISM, a just solution to the conflict would require an end to the military occupation, which will bring peace to both groups of people. Crucial to ISM's work is the participation of international volunteers, who come from all over the world to join Palestinians and Israelis in local acts of resistance. These volunteers provide personal witness to the everyday conditions of Palestinians' lives under occupation. ISM has found the presence of international volunteers at demonstrations and protests often curbs various types of violence perpetrated by the Israeli military or police.[2]

The village of Bi'lin just west of Ramallah is a central location for one of ISM's most successful and long-lasting campaigns. The Palestinian villagers and international volunteers have held weekly nonviolent protests against the Israeli "security" fence or wall for the past 7 years. The barrier, which the Israeli government claims is necessary to stop suicide bombers, severed access to at least 50 percent of Palestinian Bi'lin farmland. Due in large part to ISM's consistent civil resistance, the Israeli Supreme Court ordered the military to reroute the fence in 2007 because the barrier could not be justified on security grounds. Nearly 4 years after the ruling, the fence was finally rerouted in the summer of 2011.[3]

B'Tselem, an Israeli human rights organization, was formed in 1989. The organization was put together by a group of distinguished Israeli academics, journalists, attorneys, and members of the Israeli legislative branch, the Knesset. B'Tselem reports on human rights conditions within the Occupied Territories. By issuing reports, holding press conferences, and actively engaging with the Knesset, B'Tselem seeks to ensure that the Israeli government abides by both domestic and international laws that govern the Occupied Territories. They also aim to inform Israeli society about the human rights conditions in the West Bank and Gaza Strip. B'Tselem reports on, amongst other issues, the number of Palestinians killed by Israeli security forces.[4] Under the heading of "Use of firearms" on the organization's Web site, for example, it is reported that 4,927 Palestinians have been killed in the Gaza Strip and West Bank since 2000, 970 of whom were under the age of 18. This number does not include the 1,389 Palestinians killed in the Israeli's 2009 military campaign, "Operation Cast Lead," in the Gaza Strip.[5] B'Tselem also reports on land expropriation, settlement activities, the separation barrier, issues related to East Jerusalem, and other human rights abuses.

[1]Charmaine Seitz, "ISM At the Crossroads: The Evolution of the International Solidarity Movement," *Journal of Palestine Studies*, 32 (4, Summer) 2003, pp. 50–67, http://www.jstor.org.proxy.lib.wayne.edu/stable/10.1525, accessed March 5, 2011.

[2]"About ISM," International Solidarity Movement, http://palsolidarity.org/about-ism/, accessed March 20, 2011.

[3]Isabel Kershner, "Israeli Court Orders Barrier Rerouted," New York Times, Sept. 5, 2007, http://www.nytimes.com/2007/09/05/world/middleeast/05mideast. html?ref=middleeast, accessed June 25, 2011.

[4]"About B'Tselem," B'Tselem, http://www.btselem.org/about_btselem, accessed March 21, 2011.

[5]"Use of Firearms," B'Tselem, http://www.btselem.org/topic/firearms, accessed March 21, 2011.

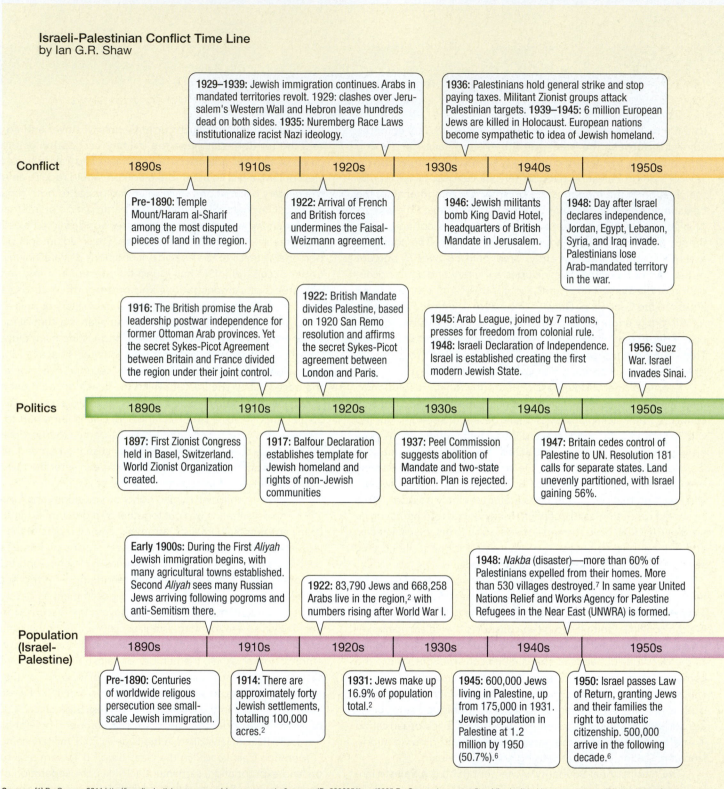

Israeli-Palestinian Conflict Time Line
by Ian G.R. Shaw

Conflict

1929–1939: Jewish immigration continues. Arabs in mandated territories revolt. 1929: clashes over Jerusalem's Western Wall and Hebron leave hundreds dead on both sides. 1935: Nuremberg Race Laws institutionalize racist Nazi ideology.

1936: Palestinians hold general strike and stop paying taxes. Militant Zionist groups attack Palestinian targets. 1939–1945: 6 million European Jews are killed in Holocaust. European nations become sympathetic to idea of Jewish homeland.

| 1890s | 1910s | 1920s | 1930s | 1940s | 1950s |

Pre-1890: Temple Mount/Haram al-Sharif among the most disputed pieces of land in the region.

1922: Arrival of French and British forces undermines the Faisal-Weizmann agreement.

1946: Jewish militants bomb King David Hotel, headquarters of British Mandate in Jerusalem.

1948: Day after Israel declares independence, Jordan, Egypt, Lebanon, Syria, and Iraq invade. Palestinians lose Arab-mandated territory in the war.

1916: The British promise the Arab leadership postwar independence for former Ottoman Arab provinces. Yet the secret Sykes-Picot Agreement between Britain and France divided the region under their joint control.

1922: British Mandate divides Palestine, based on 1920 San Remo resolution and affirms the secret Sykes-Picot agreement between London and Paris.

1945: Arab League, joined by 7 nations, presses for freedom from colonial rule.
1948: Israeli Declaration of Independence. Israel is established creating the first modern Jewish State.

1956: Suez War. Israel invades Sinai.

Politics

| 1890s | 1910s | 1920s | 1930s | 1940s | 1950s |

1897: First Zionist Congress held in Basel, Switzerland. World Zionist Organization created.

1917: Balfour Declaration establishes template for Jewish homeland and rights of non-Jewish communities

1937: Peel Commission suggests abolition of Mandate and two-state partition. Plan is rejected.

1947: Britain cedes control of Palestine to UN. Resolution 181 calls for separate states. Land unevenly partitioned, with Israel gaining 56%.

Early 1900s: During the First *Aliyah* Jewish immigration begins, with many agricultural towns established. Second *Aliyah* sees many Russian Jews arriving following pogroms and anti-Semitism there.

1922: 83,790 Jews and 668,258 Arabs live in the region,[2] with numbers rising after World War I.

1948: *Nakba* (disaster)—more than 60% of Palestinians expelled from their homes. More than 530 villages destroyed.[7] In same year United Nations Relief and Works Agency for Palestine Refugees in the Near East (UNWRA) is formed.

Population (Israel-Palestine)

| 1890s | 1910s | 1920s | 1930s | 1940s | 1950s |

Pre-1890: Centuries of worldwide religous persecution see small-scale Jewish immigration.

1914: There are approximately forty Jewish settlements, totalling 100,000 acres.[2]

1931: Jews make up 16.9% of population total.[2]

1945: 600,000 Jews living in Palestine, up from 175,000 in 1931. Jewish population in Palestine at 1.2 million by 1950 (50.7%).[6]

1950: Israel passes Law of Return, granting Jews and their families the right to automatic citizenship. 500,000 arrive in the following decade.[6]

Sources: [1] ProCon.org 2011 http://israelipalestinian.procon.org/view.resource.php?resourceID=000635#israel2005 ProCon.org is a nonprofit public charity with no government affiliation. It contains an amalgam of population and statistical data on deaths for both Israelis and Palestinians, including multiple sources (e.g., UN, Israeli Ministry of Foreign Affairs, and so on). [2] ProCon.org http://israeli palestinian.procon.org/viewresource.asp?resourceID=000636 [3] BBC News http://news.bbc.co.uk/1/shared/spl/hi/middle_east/03/v3_ip_timeline/html/1967.stm [4] Palestinian Centre for Human Rights 2009 http://www.pchrgaza.org/files/PressR/English/2008/36-2009.html [5] BBC News 2009 http://news.bbc.co.uk/1/hi/world/middle_east/7838618.stm [6] Council on Foreign Relations 2009 http://www.cfr.org/publication/15268/ [7] The Electronic Intifada 2007 http://electronicintifada.net/bytopic/197.shtml [8] UNWRA 2008 http://www.un.org/unrwa/publications/pdf/rr_countryandarea.pdf [9] Congressional Research Service 2008 http://www.un.org/unrwa/publications/pdf/rr_countryandarea.pdf. All accessed June 29, 2009.

FIGURE 9.F Time line of Israeli/Palestinian conflict and change since 1890 The history of modern Israel/Palestine is one of enduring conflict as Israel increasingly occupies Palestinian territory and as Palestinians continue to fight to keep their homelands. This time line, prepared by geographer Ian Shaw for use in this textbook, shows the different conflicts as well as the larger political and population issues around which these conflicts have been played out. (*Source:* Ian Shaw.)

Timeline 1 (orange band)

Above:

1967: Arab–Israeli "6 Day" War between Israel and neighboring Egypt, Jordan, and Syria. Results in Israeli control of West Bank, Gaza Strip, Golan Heights, and Sinai Peninsula.

1982: Israel invades Lebanon to wipe out Palestinian guerrilla bases on northern border. Israeli forces push all the way to Beirut. Allied Lebanese Phalangists kill hundreds of Palestinian refugees.

2008–9: Israel launches air strikes against Hamas and invades Gaza. Israeli Defense Force (IDF) continues its actions despite UN Security Council Resolution 1860.

2006: Israel–Lebanon War lasts over a month. Hamas offers *hudna* (truce), then and again in **2009.**

Band: 1960s | 1970s | 1980s | 1990s | 2000s

Below:

1972: Munich Olympics, 11 members of Israeli team murdered by Black September Organization.

1981: Israel annexes Golan Heights.

1987–1993: First Palestinian "Intifada" composed of widespread civil disobedience, demonstrations, and strikes.

2000: Second Intifada follows Sharon's visit to Temple Mount. Widespread use of suicide bombings against innocent Israelis precipitates barrier construction.

2010: Israeli Commandos board a flotilla of ships carrying aid for Gaza and kill nine activists.

Timeline 2 (green band)

Above:

1967: UN Security Council Resolution 242 calls for Israeli armed forces to withdraw from occupied territory.

1974: UN resolution 3236 affirms the inalienable rights of the Palestinians, and the right to national independence.

1993: Oslo Accords and Declaration of Principles signed. Oslo II signed in **1995.** Mid–1990s saw intense period of Arab-Israeli peacemaking.

2005: Israel withdraws from Gaza. Some settlers accept compensation, others forcibly moved by the IDF.

2011: Palestinian rovals Fatah and Hamas sign reconciliation pact.

Band: 1960s | 1970s | 1980s | 1990s | 2000s

Below:

1964: Palestinian Liberation Organization (PLO) founded, vowing to resist Zionism.

1978: Camp David Accords, brokered by Pres. Jimmy Carter. Followed by **2000** negotiations with Pres. Bill Clinton.

1987: Hamas Founded. **1988:** Hamas Covenant calls on Muslims to liberate Palestine through violent means.

1994: Palestinian Authority formed.

2003: Roadmap for two-state solution (over one-state) is endorsed by U.S., UN, EU, and Russia.

Timeline 3 (purple band)

Above:

1967–present: Palestinian Refugees. 500,000 Palestinians displaced from the West Bank and Gaza Strip immediately after 1967.[3] 1991 Gulf War saw 350,000 displaced from Kuwait. Today, over 260,000 internally displaced Palestinians live in Israel, and over 4.5 million[8] refugees are registered with UNWRA, living in camps in the West Bank, Gaza, Lebanon, Syria, and Jordan.

2005: Jewish population about 5.3 million and Arabs 5.1 million in Israel and Occupied Territories.

2007: Israel receives $2.5 billion dollars of U.S. government aid. Since 1949 this brings the total U.S. aid to Israel to over $101 billion.[9]

Band: 1960s | 1970s | 1980s | 1990s | 2000s

Below:

1984: Operation Moses airlifts some 8,000 Ethiopian Jews from famine-struck Sudan.

1987–2000: Conflict results in 1,551 Palestinian deaths and 422 Israeli deaths. Between **2000–2006:** 4,269 Palestinian and 1,017 Israeli causalities.

2008–2009: War in Gaza kills 1,417 Palestinians including 926 civilians.[4] Over 50,000 homeless and 400,000 without running water. 13 Israeli deaths.[5]

Formed in 2004, Breaking the Silence is an organization of Israeli military veterans who served in the Occupied Territories from September 2000 onward. They strive to give voice to combatants and provide a space for them to speak about what they witnessed and engaged in while serving. The group collects and publishes the testimonies of willing veterans and holds lectures in Israel and around the world. In Israel, military service is required of both women and men who have turned 18 years of age. In light of the large numbers of citizens who serve in the military, Breaking the Silence attempts to fill a gap between what soldiers experience on the ground and the general lack of dialogue in Israeli society about the effects highly charged military situations have on young soldiers. Since its inception, Breaking the Silence has collected the testimonies of over 700 soldiers, many of which have been translated and are available in English.[6]

By sailing humanitarian aid into Gazan ports, the Free Gaza Movement endeavors to end the Israeli and Egyptian blockade of the Gaza Strip, which was intensified in 2007 after the Palestinian group Hamas came to power. The blockade significantly limits the volume of imports for the 1.5 million residents of Gaza, down to a quarter of what they received prior to 2005.[7] Items that have been banned include cement for construction, light bulbs, musical instruments, and even tea and coffee. This ban has led to a grave humanitarian situation in which 80 percent of the population depends on UN aid[8] and 61 percent are considered food-insecure persons.[9] In an attempt to break the blockade, using nonviolent direct action and civil resistance, the Free Gaza Movement sailed on eight voyages from Cyprus to the Gaza Strip between August 2008 and the summer of 2010. Their cargo consists of several tons of aid, as well as international volunteers. Their work has garnered several reactions from Israel. In May of 2010, one of many ships voyaging to Gaza, the *Mavi Marmara*, was boarded by the Israeli military. During this operation, the Israeli military killed nine of the passengers and 24 others were seriously injured by live ammunition.

These snapshots represent the engagement of a diverse grouping of people committed to resolving the conflict through direct, nonviolent means. Their grassroots organizing offers concrete examples of what is needed for all the people who are affected by the conflict, no matter what nation they are from, to live in a secure and just world.

Jeff Wilson is an independent scholar living in Tucson, Arizona.

[6]"Organization," Breaking the Silence: Israeli soldiers talk about the occupied territories, http://www.breakingthesilence.org.il/about/organization, accessed June 21, 2011.

[7]"Guide: Gaza under blockade: On 17 June, Israel announced that it will 'liberalise' the blockade of the Gaza Strip for civilian goods," BBC News, July 6, 2010, http://news.bbc.co.uk/2/hi/middle_east/7545636.stm,, accessed June 27, 2011.

[8]United Nations Relief and Works Agency, "Thousands missing out on education in Gaza," July 6, 2010, http://www.unrwa.org/etemplate.php?id=731, accessed June 27, 2011.

[9]World Food Program and Food and Agriculture Organization, "Occupied Palestinian Territory: Food Security and Vulnerability Analysis Report," December 2009, http://home.wfp.org/stellent/groups/public/documents/ena/wfp213663.pdf, accessed June 27, 2011.

The Politics of Geography

Territory is often regarded as a space to which a particular group attaches its identity. Related to this concept of territory is the notion of **self-determination**, which refers to the right of a group with a distinctive politico-territorial identity to determine its own destiny, at least in part, through the control of its own territory.

Regionalism and Sectionalism

Different groups with different identities—religious or ethnic—sometimes coexist within the same state boundaries. At times, discordance between legal and political boundaries and the distribution of populations with distinct identities leads to movements to claim or reclaim particular territories. These movements, whether conflictual or peaceful, are known as *regional movements.* **Regionalism** is a feeling of collective identity based on a population's politico-territorial identification within a state or across state boundaries.

Regionalism often involves ethnic groups who seek autonomy from an interventionist state and the development of political power. The Basque provinces of northern Spain and southern France have sought autonomy from those states for most of the twentieth century. The Basque people are one of the oldest European peoples, with a distinctive culture and language. Since the 1950s, agitation for political independence has included—especially for the Basques in Spain—terrorist acts. The Spanish move to parliamentary democracy and the granting of autonomy to the Basque provinces did not squelch the Basque thirst for self-determination. On the French side, the Basque separatist movement is neither as violent nor as active as the movement in Spain.

Another separatist movement is the EZLN (Ejército Zapatista de Liberación Nacional), the Zapatista Army of Liberation, which practices a different type of revolutionary autonomy. Composed largely of indigenous peasants from Chiapas, one of the poorest states in Mexico, the EZLN opposes the Mexican federal state and its embrace of corporate globalization, arguing that it oppresses them by denying the peasant way of life. A key element of the Zapatista ideology is their aspiration to practice politics in a truly participatory way, from the "bottom-up," by guaranteeing the right of indigenous peoples to form and govern traditionally their own municipalities. In doing so, the Zapatistas reject official authorities and elect their own at the same time that they refuse federal government involvement and control. The Zapatistas have been building this autonomous form of governing since 1994 with dramatically beneficial results for community health, education, housing, and general welfare.

We need only look at the long list of territorially based conflicts that have emerged in the post–Cold War world to realize the extent to which territorially based ethnicity remains a potent force in the politics of geography. For example, the Kurds continue to fight for their own state separate from Turkey and Iraq **(Figure 9.30)**. A significant proportion of Québec's French-speaking population, already accorded substantial autonomy, continues to advocate complete independence from Canada. In the 1996 plebiscite on the issue, the separatists were only very narrowly defeated. Consider also the former Yugoslavia, whose geography has fractured along ethnic and religious lines **(Figure 9.31)**. Regionalism also underlies efforts to sever Scotland from the United Kingdom.

Regionalism may also be based on economics. For example, in a nonbinding referendum put before the California voters in 1993, drawing on sentiment over a century old, California's northern and mostly rural counties voted to separate from the southern, more urban counties. The desire for separation was based on the belief that political representation in the state legislature economically and politically advantaged the south over the north. A similar movement emerged in Arizona in 2011, where members of the Democratic Party in Pima County, part of southern Arizona, have

FIGURE 9.30 Kurdish independence protest Kurdish protesters take part in a demonstration as they march on a bridge, in 2011 in Istanbul after Turkey's electoral board barred prominent Kurdish candidates from standing in upcoming elections. The Peace and Democracy Party (BDP), Turkey's main Kurdish political movement, urged an extraordinary parliamentary session to resolve the problem.

FIGURE 9.31 Map of the former Yugoslavia The former Yugoslavia now consists of seven nations: Slovenia, Croatia, Bosnia and Herzegovina, the new Yugoslavia (made up of Serbia and Montenegro), Kosovo and Macedonia. For the most part, the boundaries of the Yugoslav states were laid out only in the twentieth century, across segments of the Austro-Hungarian and Ottoman empires that had acquired a complex mixture of ethnic groups. The history of these boundaries has also been the history of ethnic conflict revolving around claims to territory, as well as religious intolerance. As this map shows, with the exception of Slovenia, the new states are home to a mix of nationalities. (Redrawn with permission from Prentice Hall, from J. M. Rubenstein, *The Cultural Landscape: An Introduction to Human Geography*, 6th ed., © 1999, p. 260.)

advocated secession from the rest of the state because of its conservative politics (**Figure 9.32**). Although many additional steps would have to be taken—such as state senate approval, approval by the governor and by the voters, and finally by the U.S. Congress—before separation could occur, the Arizona and California movements represent an interesting example of *political and economic* regionalism.

Although the two sometimes coexist, **sectionalism**, an extreme devotion to local interests and customs, should not be confused with regionalism. Sectionalism has been identified as an overarching explanation for the U.S. Civil War. It was an attachment to the institution of slavery and to the political and economic way of life slavery made possible that prompted the southern states to secede from the Union. Although the Civil War was waged around the real issue of permitting or prohibiting slavery, it also involved disagreement over the power of the states. The Union went to war to ensure that sectional interests would not take priority over the unity of the whole and states' rights would not undermine the power of the federal government. As **Figure 9.33** illustrates, the election of Abraham Lincoln to the presidency in 1860 reflected the sectionalism that dominated the country: He received no support from slave states.

Suburbs vs. Cities, and Rural vs. Urban

Sectionalism persists today, but in different forms. In the United States one of the most apparent manifestations of sectionalism

may be found in the politics of differentiation between suburbs and cities. Two examples are the "taxpayer revolts" (expressed in the passage of Proposition 13 in California and Proposition 2½ in Massachusetts) and growth-control movements. The former were voter-approved mandates to cut property taxes and limit government spending, while the latter are government-enforced caps on population growth approved by residents to limit the density and extent of growth. Both types of actions have been primarily pursued by suburban U.S. voters.

Arising out of the taxpayer revolts, the growth-control movements of the 1980s drew their support from suburban home-owning constituencies. Arguing for the maintenance of open space, low density, large lots, and community control over land-use planning and development, the growth-control movements have attempted to keep the city out of the suburb and retain a "country feel." Critics of growth control complain that it discriminates against newcomers to the community. Related to growth-control movements is *NIMBYism* ("Not In My Backyard"). NIMBYism is action by neighborhood residents against the introduction of unwanted land use, ranging from group homes for developmentally disabled adults or battered women's shelters to low-income housing. NIMBY supporters see these land uses as a threat to the composition and quality of the neighborhood and the market value of single-family residences.

The politics of geography, in terms of sectionalism, also finds strong focus today in rural versus urban politics. In France, for example, attitudes about birth control (and birth rates themselves) are significantly different between the urbanized north of the country and the more rural south. Throughout the EU, farmers have fought the removal of farm subsidies and tariff arrangements advocated by urban-based policymakers because the existing arrangements have long protected agricultural productivity. The dispute pits the politics of local farmers against an international organization. In England and Wales, a rural-versus-urban conflict over foxhunting persists. In the countryside, foxhunting is largely seen as a ritual of rural life. In the early 2000s, a largely urban constituency of animal welfare proponents opposed foxhunting and pressured the government to act to protect the foxes. The result was the Hunting Act of 2004 that made it illegal to hunt a mammal using a dog. The act pitted a group called the Countryside Alliance, which continues to seek repeal of the act, against animal rights activists, mostly headquartered in large British cities, and the London-based Parliament.

Competition also exists among and between cities, as well as among and between states. The most ubiquitous form of this competition revolves around the desire by local and state authorities to attract corporate investment (as discussed in Chapter 7). Often corporations play the jurisdictions against each other in attempts to obtain the most attractive investment packages. At other times, cities and states compete to induce the government to locate government facilities within their jurisdictions.

APPLY YOUR KNOWLEDGE Use the Internet to identify a self-determination movement somewhere in the world. Collect information on who is involved in the movement, what its aims are, how long the movement has been active, and any other information you feel is pertinent. Describe the role territory and power play in the aims of the group you have identified. ■

FIGURE 9.32 Free Baja Arizona, 2011 This map shows the line of demarcation that would separate the proposed 51st state of Baja Arizona from the current state of Arizona. While secession is highly unlikely, Baja Arizona provides another example of the relationship between geography and politics. (*Source:* "America's 51st state…movement starts to free Baja, Arizona," by Hugh Holub, *Tucson Citizen.com*, February 21, 2011. Reprinted by permission of the author, Hugh Holub.)

FIGURE 9.33 The 1860 presidential election The U.S. presidential election of 1860 graphically illustrates the role sectionalism plays in determining who gets votes from which geographical regions. In a four-way race, Abraham Lincoln failed to win the support of any of the slave states. (Adapted from *Presidential Elections Since 1789*, 4th ed. Washington, DC: Congressional Quarterly Inc., 1987.)

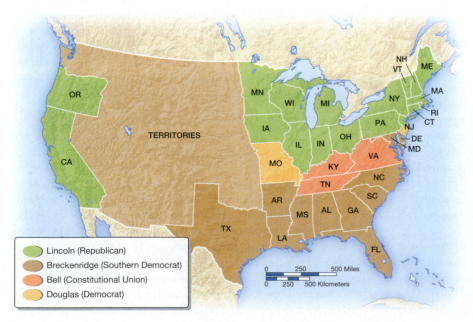

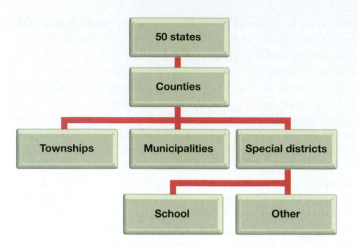

FIGURE 9.34 U.S. geographical basis of representation
This diagram is a breakdown of the types of voter districts at each level of political representation. Each type of district is territorially defined, creating a complicated and overlapping pattern of political units.

The Geography of Politics and Geographical Systems of Representation

An obvious way to show how politics shapes geography is to demonstrate how systems of political representation are geographically anchored. For instance, the United States has a political system in which democratic rule and territorial organization are linked by the concept of territorial representation.

Democratic rule describes a system in which public policies and officials are directly chosen by popular vote. **Territorial organization** is a system of government formally structured by area, not by social groups. Thus, voters vote for officials and policies that will represent them and affect them *where they live*. The territorial bases of the U.S. system of representation are illustrated in **Figure 9.34**. The United States is a federation of 50 states, which are subdivided into over 3,000 counties or parishes. Counties and parishes are further broken down into municipalities, townships, and special districts, which include school districts, water districts, library districts, and others.

The electoral divisions established for choosing elected officials in the United States range from precincts and wards to congressional districts and states. State power is applied within geographical units, and state representatives are chosen from geographical units. The bottom line is that in the United States—as in many other representative democracies—politics is geography. People and their interests gain representation in government through the location of their interests in particular places and through their relative ability to capture political control of *geographically based political units*.

For example, election of the president involves a popular vote carried out at the precinct level but totaled at the state level. Thus, even though particular precincts, cities, or counties may vote for one candidate over another, if the majority of votes at the state level supports the opposing candidate, then that person is declared the winner in that state. The president is not elected by the popular vote, but by the electoral college. The electoral college is composed of a specified number of delegates allocated to each state based on that state's population as of the most recent official census. Thus, it is the state-level voting tally that drives the process. A candidate may win the countrywide popular vote but lose in the electoral college if that candidate fails to win enough states to secure the required majority of the electoral votes. This is what happened to candidate Al Gore in the 2000 presidential elections. Furthermore, as was evidenced by the 1992 election of Bill Clinton, a candidate may win with considerably less than 50 percent of the popular vote if a third-party candidate (in this case, H. Ross Perot) siphons off enough of the popular vote to prevent the opposing candidate (in this case, George H. W. Bush) from winning a sufficient number of electoral votes. The geographical implications of the U.S. presidential voting arrangement are crucial to candidates' campaign strategies. To be a winner requires concentrating time and energy on capturing a majority of votes in some of the nation's most populous states, as Barack Obama did in 2008.

In particular, Obama sought out young voters in the populous states (and elsewhere), many of whom were voting in their first presidential election. His campaign targeted 18 to 29 years olds, that group of people who rely most heavily on Internet media for information, including political information about the election. Obama's campaign managers were savvy enough to recognize that the reason younger voters tended to ignore politicians was because politicians tended to ignore the issues which most concerned them. Young voters were key to Obama's successful campaign and the absence of social media campaigning among directed at young people was a gaping absence for losing candidate John McCain **(Figure 9.35)**.

Other systems of representation exist throughout the world. Many electoral systems are based on representing special constituencies in the legislative branch of government. In Pakistan, for example, there are four seats for Christians, four seats for Hindus and people belonging to the scheduled castes, and one seat each for the Sikh, Buddhist, and Parsi communities. Systems of representation are very much tied to the history of a country, with some very sensitive to the way that history and geography (who lives where) come together. These systems are both a product of and an important influence on the political culture of a country.

Reapportionment and Redistricting

The U.S. Constitution determines the allocation of legislators among states, guaranteeing each state a representative system of government. For U.S. presidential and senatorial elections, candidates are elected at large within each state, not on the basis of electoral districts. U.S. representatives, however, are elected from congressional districts of roughly equal population size. This is also the case for state senators, representatives, and, often, other elected officials, from city council members to school board members. It is the responsibility of each state's legislature to create the districts that will elect most federal and state representatives. Other levels of government—from counties to special districts—also establish their own electoral districts. The result is that representatives are elected

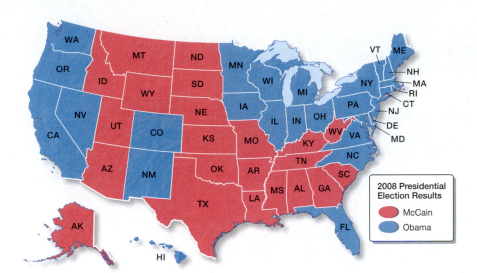

FIGURE 9.35 The vote for president, 2008
The map shows that Obama was able to win eight of the ten most populous states (which include in order California, Texas, New York, Florida, Illinois, Pennsylvania, Ohio, Michigan, Georgia and North Carolina) in the country helping him to secure the election. (*Source:* NPR, http://www.npr.org/news/specials/election2008/2008-election-map.html#/president?view=race08.)

at any number of levels of government in a collection of districts that is complicated, extensive, and by no means systematic.

Problems of the proper "fit" between political representation and territory emerge when population changes. Because most forms of representation are based on population, it often becomes necessary to change electoral district boundaries to distribute the total population more evenly among districts. **Reapportionment** is the process of allocating electoral seats to geographical areas. **Redistricting** is the defining and redefining of territorial district boundaries. For example, the number of congressional representatives in the United States as a whole is fixed at 435. These 435 seats must be reapportioned in accordance with population change every 10 years. (Recall from Chapter 3 that the federal government is required to count the U.S. population every 10 years. One of the chief reasons for this is to maintain the proper match between population and representation.) Both reapportionment and redistricting are political, geographical, and statistical exercises. As geographer Richard Morrill writes:

> The process is *political* in that the design and approval of systems of districts is usually done by bodies of elected representatives; the balance of power between groups and areas is often involved; identification of citizens with a traditional electoral territory is altered; and the incumbency of individuals is usually at stake. . . . The process is *geographic* in that areas must be allocated to districts and boundaries drawn (or territories partitioned into districts); communities of interest, which may have arisen in part from pre-existing systems of districts, may well be affected; restructuring of basic electoral geography is altered; and accessibility of voters to their representatives or centers of decision-making may be changed. . . . Redistricting is also *statistical* or mathematical in that there is a requirement for reasonably current and accurate data on population and its characteristics and, sometimes, of property and its valuation.[2]

R. Morrill, *Political Redistricting and Geographic Theory.* Washington, DC: Association of American Geographers, 1981, p. 1.

The results of the 2010 U.S. Census mean that redistricting activities are again underway as population growth and geographical shifts will result in the redrawing of district boundaries in order to maintain equal representation on the basis of population so that each person's vote is equally weighted.

Gerrymandering

The purpose of redistricting is to ensure the equal probability of representation among all groups. Redistricting for partisan purposes is known as **gerrymandering**. Gerrymandering involves boundaries of districts being redrawn to advantage a particular political party or candidate or to prevent or ensure a loss of power of a particular subpopulation (like African Americans). The term immortalizes Governor Elbridge Gerry of Massachusetts, who signed into law a bill designed to maximize the election of Republican-Democrats over Federalists in the election of 1812. As the result of this bill, an electoral district north of Boston was redrawn so that most Federalist votes were contained there. With the majority of Federalist votes restricted to a small number of districts, their voting power was reduced. Although Federalists won the most votes overall, Republican-Democrats took 29 of 40 seats because they won the most districts.

Gerrymandering persists today. Although the Federal Voting Rights Act intended that redistricting enhance minority representation in Congress, a fine line exists between "enhancement" and creating a district solely to ensure that a minority person is elected. **Figure 9.36** shows North Carolina's reengineered Twelfth District, the constitutionality of which was upheld by a Supreme Court decision in 2001.

APPLY YOUR KNOWLEDGE Use the Internet to forge a basic understanding of how the U.S. electoral college works. Identify what you feel are the advantages and disadvantages of the system. Establish a position on whether the United States should retain the current electoral college or change it and indicate why you have taken that position, making sure that you address the *geographical issues* involved. ■

FIGURE 9.36 North Carolina's Twelfth Congressional District The drawing of electoral district boundaries remains a politically volatile exercise. A 1996 case before the Supreme Court concerned the redrawn Twelfth Congressional District of North Carolina, whose shape (in red on the map) was as contorted as a salamander. Accusations of gerrymandering circulated around the drawing of the district, which was intended to consolidate African American voting strength. Although this redistricting was immediately challenged legally, the Supreme Court upheld its constitutionality in a 2001 decision.

Future Geographies

The end of the Cold War, the widespread availability of telecommunications, the transnational reorganization of industry and finance, and the liberalization of trade all suggest that a new global political and economic order is already underway. With the rapid growth of India and China and the prominence of the European Union, it would seem that a multipolar world is emerging, one that includes the United States but may not necessarily be so wholly dominated by the United States. This new, multipolar world might come close to what President George H. W. Bush heralded as a "new world order." And while the prospects for political stability in core countries are likely to remain strong, for peripheral ones the future looks far less bright.

The prospect for increasing ethnic rivalry and conflict is high. For instance, despite the recent popular vote for the creation of Southern Sudan, conflict there between Muslim majority and Christian and animist minorities is still ongoing. Internal conflicts such as this risk escalation into regional conflicts and possibly even the failure of states. Weak states can become hosts to violent drug cartels (as in Mexico) and terrorist groups (as in Afghanistan). The most feared form of terrorism in the twenty-first century is **bioterrorism**, the deliberate use of microorganisms or toxins from living organisms to induce death or disease **(Table 9.1)**. Biological and chemical agents that could be used for bioterrorist purposes range from anthrax to West Nile virus and when deployed could result in the widespread loss of life across human and animal populations as well as crops.

TABLE 9.1 Place Vulnerability Index for Select U.S. Cities

Atlanta, GA 1.683	Grand Rapids, MI −0.636	Philadelphia, PA 1.737
Austin, TX −0.109	Helena, MT −1.729	Phoenix/Glendale/Mesa, AZ −0.904
Baton Rouge, LA 3.016	Honolulu, HI −0.237	Providence, RI −0.280
Boise, ID 1.696	Houston, TX 1.844	Rochester, NY −0.263
Boston, MA 0.604	Indianapolis, IN −0.379	Salt Lake City/Ogden, UT 0.190
Buffalo, NY 0.279	Juneau, AK −2.421	San Diego, CA −0.519
Charlotte, NC 1.046	Lancaster, PA −0.115	San Francisco, CA 0.050
Chicago, IL 1.404	Los Angeles, CA 0.421	Seattle/Tacoma, WA −0.315
Columbus, OH −0.036	Memphis, TN 0.913	St. Louis, MO 0.944
Colorado Springs, CO −1.262	Milwaukee, WI 0.361	Tampa/St. Petersburg, FL 1.150
Concord, NH −1.484	Minneapolis/St. Paul, MN 0.340	Topeka, KS −0.716
Dallas/Fort Worth, TX 0.447	New Orleans, LA 3.119	Trenton, NJ 0.410
Denver, CO 0.171	New York, NY/Newark, NJ 2.154	Tucson, AZ −0.768
Detroit/Warren, MI 0.188	Norfolk, VA 2.326	Washington, DC 1.978
El Paso, TX 0.189	Oklahoma City, OK 0.049	Worcester, MA −0.854

Source: "Benchmark Analysis for Quantifying Urban Vulnerability to Terrorist Incidents," Walter W. Piegorsch, Susan Cutter, Frank Hardisty, Risk Analysis, vol. 27, issue 6, pp. 1411–1425.

CONCLUSION

The globalization of the economy has been largely facilitated by states extending their spheres of influence and paving the way for the smooth functioning of markets and industries. Political geography is as much about what happens at the global level as it is about what happens at other levels of spatial resolution, from the region to the neighborhood to the household and the individual.

Theories of the state have been one of geography's most important contributions to understanding politics. Ratzel's emphasis on the relationship between power and territory and Mackinder's model of the geographical pivot remind us that space and territory shape the actions of states in both dramatic and mundane ways. Time and space shape politics, and events distant in time and space—such as colonialism—continue to have impacts long afterwards. For example, the civil war in Northern Ireland, instigated by English colonial practices now centuries old, has only recently shown signs of ceasing. The impacts of English colonization have also been felt in countries throughout the Northern Hemisphere, as well as by neighbors living unhappily for several generations side by side in cities like Belfast and Boston.

Continuing strife also characterizes the enduring North/South divide that pits core countries against peripheral, mostly formerly colonial, countries. Perhaps the most surprising political/geographical transformation of this century has been the near dissolution of the East/West divide. Although it is too soon to tell whether communism has truly been superseded by capitalism, the distinctions between them are certainly more blurred than they once were.

Perhaps the most significant aspect of contemporary globalization is the emergence of a new world order and a resulting focus on terrorism as well as the growth of transnational institutions of governance. Both forces are reshaping not only governing structures and economic processes (creating new layers of rules, regulations, and policies, as well as new ways of political interaction among and between nation-states) but also the practices of everyday life (as citizens deal with increased personal security measures and the transformation of human rights).

The pairing of the terms *politics* and *geography* serves to remind us that politics is clearly geographical at the same time that geography is unavoidably political. The divisions of area into states, counties, cities, towns, and special districts mean that where we live shapes our politics, and vice versa. Geography is politics, just as politics is geography. And geographical systems of representation, as well as identity politics based on regional histories, confirm this interactive relationship.

Learning Outcomes Revisited

- Express the geopolitical model of the state and explain how it links geography and state practices with respect to the key issues of power and territory.

  The ancient Greek philosopher Aristotle is often considered the first political geographer because his model of the state is based upon factors such as climate, terrain, and the relationship between population and territory. Other important political geographers have promoted theories of the state that incorporated elements of the landscape and the physical environment, as well as population characteristics of regions. Later scholars theorized that the state operated cyclically and organically. Twentieth-century theorists such as Foucault, Althusser, and Deleuze have shifted focus away from viewing the state as a set of institutions; they are more concerned with how state power is assembled and deployed.

- Compare and contrast the ways that different contemporary theorists—from Deleuze to Althusser—approach the state as a political and geographical entity.

  The state is also a set of institutions for the protection and maintenance of society. A state is not only a place, a bounded territory, it is also an active entity that operates through the rules and regulations of its various institutions. State theorist Louis Althusser views the state as both ideological and repressive. Michel Foucault, another state theorist, has explored the ways that power, knowledge, and discourse operate to produce particular kinds of state subjects. Giles Deleuze sees the state as a force that is greater than the formal institutions that constitute it. Deleuze believes that the state is best thought of as a *machine* whose purpose is to regulate and dominate.

- Interpret how imperialism, colonialism, heartland theory, domino theory, the end of the Cold War, and the emergence of the new world order are key examples of ways geography has influenced politics and politics has influenced geography.

  Geopolitics may involve the extension of power by one group over another. There are many different manifestations of this phenomenon. Imperialism and colonialism involve occupation and control by one state over another. Heartland theory recognizes that a central location is pivotal to political and geographical control, whereas domino theory reflects the significance of proximity in the extension of power and control. During the Cold War blocks of the global political system—capitalist versus communist—were in direct and indirect conflict. The current new world order is a manifestation of the decline of those old conflicts and the emergence of new ones.

■ Demonstrate how the growth and proliferation of international and supranational organizations created the foundation for the emergence of global forms of governance.

Just as states are key players in political geography, so too have international and supranational organizations become important participants in the world system in the last century. These organizations have become increasingly important means of achieving goals such as the freer flow of goods and information and more cooperative management of shared resources, such as water.

■ Recognize how events of international political significance are usually the result of East/West and North/South divisions, whereas national and local political issues emerge out of tensions related to regionalism and sectionalism.

Capitalist colonialism and imperialism were key factors in producing global state divisions around capitalism versus communism (East/West) and rich versus poor (North/South). More recently, local divisions have emerged that reflect differences in ethnicity, political orientations, and economic commitments, among others. These are expressed through regionalism and sectionalism.

■ Describe the difference between the politics of geography and the geography of politics as manifestations of the two-way relationship between politics and geography.

Political geography can be viewed through two contrasting lenses. The first orientation sees it as the *politics of geography*. This perspective emphasizes that *geography*—the areal distribution/differentiation of people and objects in space—has a very real and measurable impact on politics. The second orientation sees political geography as the *geography of politics*. This approach analyzes how *politics*—the tactics or operations of the state—shapes geography.

KEY TERMS

bioterrorism *(p. 346)*

children's rights *(p. 335)*

citizenship *(p.308)*

confederation *(p. 310)*

decolonization *(p. 319)*

discourse *(p. 311)*

democratic rule *(p.344)*

domino theory *(p.325)*

East/West divide *(p.324)*

federal state *(p.309)*

geopolitics *(p.302)*

gerrymandering *(p.345)*

global civil society *(p.335)*

human rights *(p.334)*

international organization *(p.319)*

international regime *(p.334)*

nation *(p.308)*

nation-state *(p.308)*

nationalism *(p.308)*

new world order *(p.326)*

North/South divide *(p.319)*

Orientalism *(p.317)*

reapportionment *(p.345)*

redistricting *(p.345)*

regionalism *(p.341)*

sectionalism *(p.342)*

self-determination *(p.341)*

sovereignty *(p.308)*

state *(p.308)*

supranational organization *(p.333)*

territorial organization *(p.344)*

territory *(p.302)*

terrorism *(p.326)*

unitary state *(p.309)*

REVIEW AND DISCUSSION

1. Research organizations that assist refugees in your town or city. Are refugees who come to your area typically from a specific location? If so, once you have identified the state or region a majority of the refugees come from, list four geopolitical factors that might have led to the displacement of these people. To do so, research the specific history of a region and map out the refugees' journey.

2. As a group, pick an international organization and research its efforts. For example, you might research the International Bill of Human Rights, the UN Declaration on the Rights of Indigenous People, or the International Criminal Court. Outline the history of the international

regime and the impact it has had on international politics. Once you have done this, compare and contrast these international regimes with your own local government and laws. List three similarities and three differences.

3. Conduct an Internet search to determine if redistricting has occurred in your town or city. Create a time line demonstrating the frequency of local redistricting. Also determine the actors primarily responsible for redistricting and the arguments they employ. Finally, determine the political consequences of redistricting. Explain in what ways it has (or has not) changed your local political landscape.

UNPLUGGED

1. International boundaries are a prominent feature of the political geography of the contemporary world. In this exercise, you will explore the impact of a boundary on nationalist attitudes and behaviors. You will need to use *The New York Times Index* to complete this assignment. Using the United States–Mexico border as your key word, describe the range of issues that derive from this juxtaposition of two very different nations. Concentrate on a 5-year period, indicating which issues grew in importance, which issues declined, and which issues continued to have a consistent news profile throughout the period.

2. National elections usually tell a story about the ways regional ideas and attitudes shape the political agenda. In the 2000 and 2004 U.S. presidential elections, pollsters considered religion an important issue, with the religious right playing a crucial role in reelecting George W. Bush. Using national election results data available through the Government Documents Division of your college or university library, describe the political geography of fundamentalist or evangelical Christians. Did these Christians vote for Bush in all regions of the country? If not, which ones did not, and what might explain the regional distribution of this powerful voting block? What might the 2000 and 2004 polling data potentially indicate for 2012 election results?

3. U.S. presidential elections provide a snapshot of the changing political geography of the country. Compare the 1992 results with those for the year 2008 using maps that show the voting results. What are the most significant differences between the two maps? What are some reasons for these differences? If you are able to get maps that are disaggregated by race, ethnicity, or gender, what further explanations can you offer for the differences in the maps based on these additional variables?

4. Compare two maps of Europe (not including Russia and the former Soviet Union), one from 1930 and one from the present. How do issues of ethnicity, religion, and political-economic system (communist, capitalist) help to explain the changes in boundaries that have occurred? Identify any areas on the map that you feel may be the sites of future border changes and explain why.

Log in to www.masteringgeography.com for MapMaster™ interactive maps, geography videos, RSS feeds, flash cards, Web links, an eText version of *Diversity Amid Globalization*, and self-study quizzes to enhance your study of political geographies.

MapMaster™ presents 13 Place Name and 13 Layered Thematic interactive maps to help students practice and master their geographic literacy, spatial reasoning, and critical thinking skills.

10 URBANIZATION

Learning Outcomes

- Explain how the urban areas of the world are the linchpins of human geographies at the local, regional, and global scales.

- Describe how the earliest towns and cities developed independently in the various hearth areas of the first agricultural revolution.

- Scrutinize how the expansion of trade around the world, associated with colonialism and imperialism, established numerous gateway cities.

- Assess why and how the Industrial Revolution generated new kinds of cities—and many more of them.

- Interpret how a small number of "world cities," most of them located within the core regions of the world-system, have come to occupy key roles in the organization of global economics and culture.

- Compare and contrast the differences in trends and projections between the world's core regions and peripheral regions.

Mostly off the world's radar, on a dusty plain in West Africa, is a city of 1.6 million people. Bisected by the River Niger, its two halves—with about 800,000 inhabitants each—are linked by only two bridges. The pressure of movement is so strong that every morning one of these bridges is dedicated exclusively to incoming traffic: minibuses, bicycles, motorbikes, pedestrians, and occasionally private cars. In the evenings, leaving the center involves joining an exodus of people toward the minibus depots as the process reverses. Green vans loaded with passengers file out to residential neighborhoods as far as 20 kilometers (12.5 miles) away.

This is Bamako, Mali. It contracts into its center every morning and breathes out again in the evening. With each breath Bamako grows bigger. It happens to be one of the fastest-growing cities in the world. Natural population growth is supplemented by migration from the countryside and other Malian cities. Its population in 2008 was 50 percent larger than in 1998, resulting in a city that is now roughly the same size as Budapest, Dubai, or Warsaw. It has 10 times more inhabitants than the next biggest Malian city and accommodates 70 percent of the country's industrial establishments. New neighborhoods—*quartiers*—formerly villages, become consolidated with the rest of the city, as it grows to the south, east, and west. Some people are now moving out of Bamako into surrounding neighborhoods in search of cheaper land and some tranquility, but they remain within reach of the city because it provides their livelihoods.

Cars, motorcycles and pedestrians crowd the busy streets of Bamako, Mali, West Africa.

Despite its industriousness, Bamako is one of the sleepier cities in West Africa. Many of the manufactured staples travel 1,184 kilometers (736 miles) by road from one of the region's metropolises, Abidjan in Côte d'Ivoire, which has more than twice Bamako's population. And Abidjan in turn seems small beside Lagos, Nigeria, where activity is so concentrated that its residents speak of living in a pressure cooker.

In Lagos, some families rent rooms to sleep for 6 hours and then turn them over to another family that takes their place. Shopping does not necessarily require travel: Goods are brought on foot and cart to drivers stuck in Lagos's interminable traffic jams. To some, like the authors of Lagos's 1980 master plan written when the city had just 2.5 million residents, the continuing growth of the city is "undisciplined."

What can possibly be so attractive about living in Lagos that, despite its congestion and crime, it continues to draw migrants? The short answer: economic density. Lagos is not the most economically dense city in the world, nor even the most densely populated. Those distinctions belong to Central London and Mumbai, respectively. Even so, Nigeria's economic future and Lagos's growth are as inextricably tied as Britain's economy is to London's growth. No country has developed without the growth of its cities. As countries become richer, economic activity becomes more densely packed into towns, cities, and metropolises. ■

URBAN GEOGRAPHY AND URBANIZATION

Urbanization is one of the most important geographic phenomena in today's world. The United Nations Center for Human Settlements (UNCHS) notes that the growth of cities and the urbanization of rural areas are now irreversible because of the global shift to technological-, industrial-, and service-based economies. The proportion of the world's population living in urban settlements is growing at a rapid rate, and the world's economic, social, cultural, and political processes are increasingly being played out within and between the world's systems of towns and cities. In this chapter we describe the extent and pattern of urbanization across the world, explaining its causes and the resultant changes wrought in people and places.

Urbanization and Changing Human Geographies

From small market towns and fishing ports to megacities of millions of people, the urban areas of the world are the linchpins of human geographies. They have always been a crucial element in spatial organization and the evolution of societies, but today they are more important than ever. Between 1980 and 2010, the number of city dwellers worldwide rose by 1.7 billion. Cities now account for over half the world's population. Much of the developed world has become almost completely urbanized (**Figure 10.1**), and in many peripheral and semiperipheral regions the current rate of urbanization is without precedent (**Figure 10.2**).

Many of the largest cities in the periphery are growing at annual rates of between 4 and 7 percent. At the higher rate their populations will double in 10 years; at the lower rate they will double in 17 years. The *doubling time* of a city's population is the time needed for it to double in size, at current growth rates. To put the situation in numerical terms, metropolitan areas like Mexico City and São Paulo are adding half a million persons to their population each year, nearly 10,000 every week, even taking into account losses from deaths and out-migration. It took London 190 years to grow from half a million to 10 million. It took New York 140 years. By contrast, Mexico City, São Paulo, Buenos Aires, Kolkata (Calcutta), Rio de Janeiro, Seoul, and Mumbai (Bombay) all took less than 75 years to grow from half a million to 10 million inhabitants. Urbanization on this scale is a remarkable geographical phenomenon—one of the most important processes shaping the world's landscapes.

Towns and cities are centers of cultural innovation, social transformation, and political change. They can also be engines of economic development. The gross domestic product of large cities like

FIGURE 10.1 Percentage of population living in urban settlements, 2009 The lowest levels of urbanization—less than 25 percent—are found in Central Africa and South and Southeast Asia. Most of the core countries are highly urbanized, with between 65 and 95 percent of their populations living in urban settlements.

(*Source:* Data from the World Bank: http://data.worldbank.org/indicator/SP.URB.TOTL.IN.ZS.)

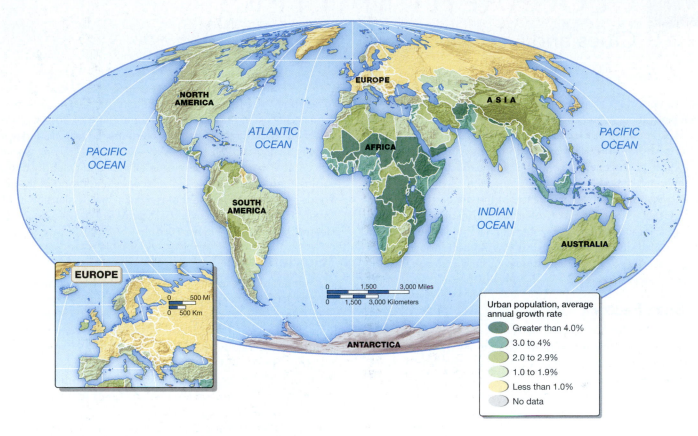

FIGURE 10.2 Rates of growth in urbanization, 2000–2010 This map shows the annual average growth rate between 2000 and 2010 in the proportion of people in each country living in urban settlements. Core countries, already highly urbanized, grew quite slowly. The urban populations of peripheral countries such as Angola, Afghanistan, Burundi, Liberia, Rwanda, and Somalia, on the other hand, grew by more than 5 percent each year, creating tremendous pressure on their cities' capacity to provide jobs, housing, and public services. (*Source:* Data from United Nations Human Settlements Programme, *Planning Sustainable Cities. Global Report on Human Settlements 2009*, pp. 238–241; updated data from World Bank 2009 data, http://data.worldbank.org/indicator/SP.URB.TOTL.IN.ZS.)

London, Los Angeles, Mexico City, and Paris is roughly equivalent to that of entire countries like Australia and Sweden. Although they often pose social and environmental problems, towns and cities are essential elements in human economic and social organization. Experts on urbanization point to four fundamental aspects of the role of towns and cities in human economic and social organization:

- *The mobilizing function of urban settlement.* Urban settings, with their physical infrastructure and their large and diverse populations, are places where entrepreneurs can get things done. Cities provide efficient and effective environments for organizing labor, capital, and raw materials and for distributing finished products. In developing countries, urban areas produce as much as 60 percent of total gross domestic product with just one-third of the population.

- *The decision-making capacity of urban settlement.* Because urban settings bring together the decision-making machinery of public and private institutions and organizations, political and economic power resides in cities and towns.

- *The generative functions of urban settlement.* The concentration of people in urban settings makes for much greater interaction and competition, which facilitates the generation of innovation, knowledge, and information (see Box 10.1, "Geography Matters: Cities and Civilization").

- *The transformative capacity of urban settlement.* The size, density, and variety of urban populations tend to have a liberating effect on people, allowing them to escape the rigidities of traditional, rural society and to participate in a variety of lifestyles and behaviors.

APPLY YOUR KNOWLEDGE Provide two examples of how the transformative capacity of urban settlements can be liberating for some people. ■

Studying Urbanization

Urban geography is concerned with the development of towns and cities around the world, with particular reference to the similarities and differences both *among* and *within* urban places. For urban geographers, some of the most important questions include: What attributes make towns and cities distinctive? How did these distinctive identities evolve? What are the relationships and interdependencies between particular sets of towns and cities? What are the relationships between cities and their surrounding territories? Do significant regularities exist in the spatial organization of land use within cities, in the patterning of neighborhood populations, or in the layout and landscapes of particular kinds of cities?

Cities are important arenas of cultural production, seedbeds of cultural innovation, centers of fashion and the creation of taste. As Jane Jacobs pointed out long ago in her famous book *The Death and Life of Great American Cities*[1], the density and diversity of urban populations generate serendipity, unexpected encounters and "new combinations" that lead to innovation. Geographers Ash Amin and Stephen Graham[2] suggest that the generative capacity of cities rests on the multiplexity of urban life, involving four main dimensions: intense face-to-face interactions; the development of dense, specialized quarters and districts; the heterogeneity and cultural hybridity of urban populations; and the concentration of institutional assets in cities.

The Buzz Factor

Personal communications devices and social networking sites have enhanced the generative capacity of contemporary cities, but density, diversity, and face-to-face interactions are still key. They are especially important in facilitating the "creative buzz" associated with the most vibrant and creative city districts. Planning professor Elizabeth Currid has likened it to the dynamics of Andy Warhol's famous "Factory" in Manhattan, a center of social interaction among artists, photographers, actors, and others, as well as a workplace. As Currid notes, job opportunities, along with professional

knowledge, are often heavily reliant on buzz, social contacts, and acquaintances. In Manhattan, the comingling of artists, artisans, designers, photographers, actors, students, educators, and writers in cafés, restaurants, and clubs and—for some—gallery openings, fashion after-parties, music release events, and celebrity birthday parties contributes to a hip, cool "scene," a blurring of the social worlds of work and lifestyle that is a distinctive dimension of Manhattan districts like the Lower East Side, SoHo, Chelsea, and the Meatpacking District (**Figure 10.A**).

"Golden Ages"

In some cities, these elements of generative capacity are supercharged by other factors, creating distinctive "golden ages" of innovation and creativity: Athens 500–400 B.C.E, for example; Rome 50 B.C.E. to 100 C.E.; Florence 1400–1500, Vienna 1780–1910 (**Figure 10.B**), London 1825–1900, Paris 1870–1910 (**Figure 10.C**), and Berlin 1918–1933. Geographer Peter Hall[3] has written about these and other cities, noting that all of them led their respective states or empires, which made them magnets for the immigration of talent. As they drew talent from the far corners of the empires they controlled, it made them cosmopolitan. They were all also wealthy trading cities. Out of trade came new ways of economic organization, and out of those came new forms of production.

FIGURE 10.A The buzz factor Evening crowds in the Meatpacking District of Manhattan, New York City.

[1]Jacobs, J., *The Death and Life of Great American Cities.* New York: Random House, 1961.
[2]Amin, A. and S. Graham, "The Ordinary City," *Transactions of the Institute of British Geographers, New Series*, 22.4, 1997, pp. 411–429.
[3]Hall, P., *Cities in Civilization.* London: Weidenfeld and Nicolson, 1998.

FIGURE 10.B Vienna The coffee house was the classic setting for meetings of the artists and intelligentsia of Vienna's golden age. Shown here is the Cafe Landt-mann, where people continue to gather.

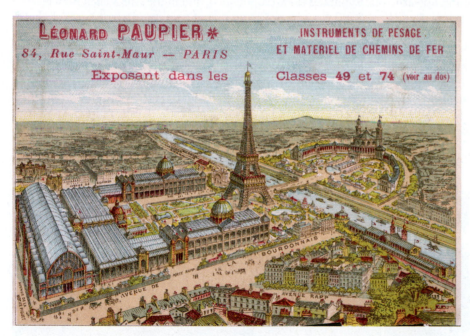

FIGURE 10.C Paris The golden age of Paris was celebrated in a series of World Expositions. This retail advertisement shows a bird's-eye view of the Paris Exposition of 1889.

Perhaps the most important factor, though, was that all were in the process of rapid economic and social transformation. As a result, they were in a state of uneasy tension between conservative forces and values—aristocratic, hierarchical, religious, conformist—and radical values that were the exact opposite: open, rational, skeptical. These radical values were articulated more often than not by creative people who felt themselves outsiders because they were young or provincial or even foreign, or because they did not belong to the established order of power and prestige. In such circumstances, the ferment of new movements in the arts, new philosophies, new political ideals, and new cultural practices lead to "golden ages," with wealthy individuals and well-funded institutions providing patronage for the avant-garde.

Urban geographers also want to know about the causes of the patterns and regularities they find. How, for example, do specialized urban subdistricts evolve? Why does urban growth occur in a particular region at a particular time? And why does urban growth exhibit a distinctive physical form during a certain period? In pursuing such questions, urban geographers have learned that the answers are ultimately to be found in the wider context of economic, social, cultural, and political life. Towns and cities must be viewed as part of the economies and societies that maintain them.

Urbanization, therefore, is not simply the population growth of towns and cities. It also involves many other changes, both quantitative and qualitative. Geographers conceptualize these changes in several different ways. One of the most important of these is examining the attributes and dynamics of urban systems. An **urban system**, or city system, is any interdependent set of urban settlements within a given region. For example, we can speak of the French urban system, the African urban system, or even the global urban system. As urbanization takes place, urban systems reflect the increasing numbers of people living in ever-larger towns and cities. They also reflect other important changes, such as changes in the relative size of cities, changes in their functional relationships with one another, and changes in their employment base and population composition.

Another important aspect of change associated with urbanization processes concerns urban form. **Urban form** refers to the physical structure and organization of cities in their land use, layout, and built environment. As urbanization takes place, not only do towns and cities grow bigger physically, extending upward and outward, but they also become reorganized, redeveloped, and redesigned in response to changing circumstances.

These changes, in turn, are closely related to a third aspect of change: transformations in patterns of urban ecology. **Urban ecology** is the social and demographic composition of city districts and neighborhoods. Urbanization not only brings more people to cities, it also brings a greater variety of people. As different social, economic, demographic, and racial subgroups become sorted into different territories, distinctive urban ecologies emerge. As new subgroups arrive or old ones leave, these ecologies change.

A fourth aspect of change associated with urbanization concerns people's attitudes and behavior. New forms of social interaction and new ways of life are brought about by the liberating and transformative effects of urban environments. These changes have given rise to the concept of urbanism, which refers to the distinctive nature of social and cultural organization in particular urban settings. **Urbanism** describes the way of life fostered by urban settings, in which the number, physical density, and variety of people often result in distinctive attitudes, values, and patterns of behavior. Geographers are interested in urbanism because of the ways in which it varies both within and between cities.

URBAN ORIGINS

It is important to put the geographic study of towns and cities in historical context. After all, many of the world's cities are the product of long periods of development. We can only understand a city, old or young, if we know something about the reasons behind its growth, its rate of growth, and the processes that have contributed to this growth.

In broad terms, the earliest urbanization developed independently in the various hearth areas of the first agricultural revolution (see Chapter 2). The very first region of independent urbanism was in the Middle East, in the valleys of the Tigris and Euphrates (in Mesopotamia), and in the Nile Valley from around 3500 B.C.E. (see Chapter 4). Together, these intensively cultivated river valleys formed the so-called Fertile Crescent. In Mesopotamia, the growth in size of some of the agricultural villages located on the rich alluvial soils of the river floodplains formed the basis for the large rival city-states of the Sumerian empire. They included Ur (in present-day Iraq), the capital from about 2300 to 2180 B.C.E., as well as Eridu, Uruk, and Erbil (Ancient Arbela—**Figure 10.3**). These fortified city-states contained tens

FIGURE 10.3 Erbil Erbil (Ancient Arbela) in northeast Iraq is located atop a *tell*, a mound representing the remains of generations of sun-dried mud-brick buildings, visible as a hill rising high above the surrounding plain. The 100-foot-high Erbil tell is believed to represent 6,000 years of continuous occupation.

of thousands of inhabitants; social stratification, with religious, political, and military classes; innovative technologies, including massive irrigation projects; and extensive trade connections. By 1885 B.C.E., the Sumerian city-states had been taken over by the Babylonians and then the Neo-Babylonians, who governed the region from their capital city, Babylon.

In Egypt, which became a unified state as early as 3100 B.C.E., large irrigation projects controlled the Nile's waters for agricultural and other uses, supporting a series of capital cities that included Thebes, Akhetaten (Tell el-Amarna), and Tanis. Internal peace in Egypt meant that there was no need for massive investments in these cities' defensive fortifications. Also, each pharaoh was free to locate a new capital at any site he selected for his tomb, and after his death the city was usually abandoned to the priests.

By 2500 B.C.E., cities had appeared in the Indus Valley, and by 1800 B.C.E., they were established in northern China. Other areas of independent urbanism include Mesoamerica (from around 100 B.C.E.) and Andean America (from around 800 C.E.). Meanwhile, the original Middle Eastern urban hearth continued to produce successive generations of urbanized world-empires, including those of Greece, Rome, and Byzantium.

Explanations of these first transitions from subsistence minisystems to city-based world-empires differ. The classical archaeological interpretation emphasizes the availability of an agricultural surplus large enough to allow the emergence of specialized, nonagricultural workers. Some urbanization, however, may have resulted from the pressure of population growth. This pressure, it is thought, disturbed the balance between population and resources, causing some people to move to marginal areas. Finding themselves in a region where agricultural conditions were unfavorable, these people either had to devise new techniques of food production and storage or establish a new form of economy based on services such as trade, religion, or defense. Any such economy would have required concentrations of people in urban settlements.

Most experts agree that changes in social organization were an important precondition for urbanization. Specifically, urbanization required the emergence of groups who were able to exact tributes, impose taxes, and control labor power, usually through some form of religious persuasion or military coercion. Once established, this elite group provided the stimulus for urban development by using its wealth to build palaces, arenas, and monuments to display its power and status. This activity not only created the basis for the physical core of ancient cities but also required an increased degree of specialization in nonagricultural activities—construction, crafts, administration, the priesthood, soldiery, and so on—which could be organized effectively only in an urban setting. By 1000 C.E., city-based world-empires had emerged in Europe, the Middle East, and China, including a dozen major cities with populations of 100,000 or more (**Figure 10.4**).

The urbanized economies of world-empires were a precarious phenomenon, however, and many of them lapsed into ruralism before being revived or recolonized. In a number of cases, the decline of world-empires was a result of demographic setbacks associated with wars or epidemics. Such disasters left too few people to maintain the social and economic infrastructure necessary for urbanization. This lack of labor power seems to have been a major contributing factor to the eventual collapse of the Mesopotamian empire, and it may also have contributed to the abandonment of much of the Mayan empire more than 500 years before the arrival of the Spanish. Similarly, the population of the Roman empire began to decline in the second century C.E., giving rise to labor shortages, abandoned fields, and depopulated towns and allowing the infiltration of "barbarian" settlers and tribes from the German lands of East-Central Europe.

The Roots of European Urban Expansion

In Europe the urban system introduced by the Greeks and reestablished by the Romans almost collapsed during the Dark Ages of the early medieval period (476–1000 C.E.). During this time, feudalism gave rise to a fragmented landscape of inflexible and inward-looking world-empires. Feudalism was a rigid, rurally oriented form of economic and social organization based on the communal chiefdoms of Germanic tribes that had invaded the disintegrating Roman empire. From this unlikely beginning an elaborate urban system developed, its largest centers eventually growing into what would become the nodal centers of a global world-system.

Early medieval Europe, divided into a patchwork of feudal kingdoms and estates, was mostly rural. Each feudal estate was more or less self-sufficient regarding foodstuffs, and each kingdom or principality was more or less self-sufficient regarding the raw materials needed to craft simple products. Most regions, however, did support at least a few small towns. The existence of these towns depended mainly on their role:

- *Ecclesiastical or university centers*—Examples include St. Andrews in Scotland; Canterbury, Cambridge, and Coventry in England; Rheims and Chartres in France (**Figure 10.5**); Liège in Belgium; Bremen in Germany; Trondheim in Norway; and Lund in Sweden.

- *Defensive strongholds*—Examples include the hilltop towns of central Italy, such as Foligno, Montecompatri, and Urbino (**Figure 10.6**); the bastide, or fortress, towns of southwestern France, such as Aigues-Mortes and Montauban; and gateway towns such as Bellinzona, Switzerland.

- *Administrative centers (for the upper tiers of the feudal hierarchy)*—Examples include Cologne (**Figure 10.7**), Mainz, and Magdeburg in Germany; Falkland in Scotland; Winchester in England; and Toulouse in France.

From the eleventh century onward, however, the feudal system faltered and disintegrated in the face of successive demographic, economic, and political crises. These crises arose from steady population growth in conjunction with only modest technological improvements and limited amounts of cultivable land. To bolster their incomes and raise armies against one another, the feudal nobility began to levy increasingly higher taxes. Peasants were consequently obliged to sell more of their produce for cash on the market. As a result, a more extensive money economy developed, along with the beginnings of a pattern of trade in basic

Córdoba (population 450,000)
The largest and most prosperous city of the time, Córdoba was at the cultural forefront in A.D. 1000, renowned for its architecture, craftwork, and dedication to learning.

Seville (population 90,000)
One of the wealthiest and most cultured cities in the Muslim state of Andalusia, Seville excelled in science and the arts.

Constantinople (population 300,000)
Located at a strategic crossroads between Europe and Asia, Constantinople was the center of the Byzantine Empire and a major trading hub.

Rayy (population 100,000)
Known for its superior silks and ceramics, the city was described at the time as stunningly beautiful.

Isfahan (population 100,000)
Located high atop a fertile plain, Isfahan was a producer of grains and silk and was well-known for its metalwork and rugs.

Neyshabur (population 125,000)
One of Persia's most progressive cities, Neyshabur also served as a major source of turquoise.

Kaifeng (population 400,000)
Situated near the Yellow River, this Song dynasty capital benefited from its proximity to the empire's industrial center and canal network.

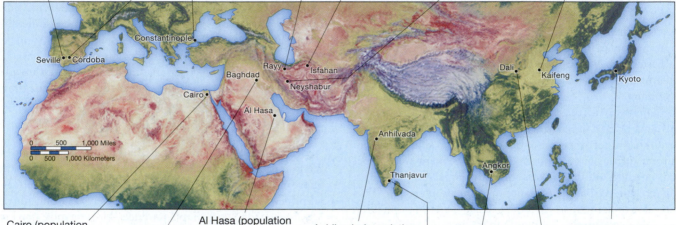

Cairo (population 135,000)
Capital of the Fatimid dynasty, Cairo was known for its many libraries and colleges.

Baghdad (population 125,000)
The capital of the Abassid caliphate, Baghdad was known in 1000 C.E. as the intellectual center of the world. Persian influence pervaded the city's architecture, literature, and court life.

Al Hasa (population 110,000)
Al Hasa was the center of the Qarmatian movement, a radical arm of the Shiite Muslim sect that advocated widespread social equality.

Anhilvada (population 100,000)
The size and location of Anhilvada, like many Indian cities, were subject to changes in the path and flow of nearby rivers.

Angkor (population 200,000)
This Khmer capital was the political center of Southeast Asia and the main market for rice produced by the Khmer empire's high-yield irrigation system.

Thanjavur (population 90,000)
Thanjavur was the capital of India's Chola dynasty. There King Rajaraja built a massive stone temple dedicated to the god Shiva.

Kyoto (population 175,000)
Japan's capital since the late eighth century, Kyoto was a religious and cultural center. It was also renowned for its silk works.

Dali (population 90,000)
Dali peaked in A.D. 986, but the fine marble that was widely sought for buildings and sculptures is still quarried there today.

FIGURE 10.4 Major cities in 1000 C.E. The most important cities in 1000 C.E. were the seats of world-empires—the Islamic caliphates, the Byzantine empire, the Chinese empire, and Indian kingdoms—that had developed well-established civilizations with urban systems based on regional trade and protected by strong military rule. (*Source:* Data from T. Chandler, *Four Thousand Years of Urban Growth: A Historical Census.* Washington, D.C.: Worldwatch Institute, 1987; "The Year 1000," *U.S. News & World Report,* August 16, 1999, pp. 66–70.)

FIGURE 10.5 Chartres, France Chartres was an important ecclesiastical center. The cathedral, built in the?thirteenth century, is widely considered to be the finest gothic cathedral in France.

(a)

(b)

agricultural produce and craft manufactures. Some long-distance trade even began in luxury goods, such as spices, furs, silks, fruit, and wine. Towns began to increase in size and vitality on the basis of this trade.

The regional specializations and trading patterns that emerged provided the foundations for a new phase of urbanization based on merchant capitalism (**Figure 10.8**). Beginning with networks established by the merchants of Venice, Pisa, Genoa, and Florence (in northern Italy) and the trading partners of the Hanseatic League (a federation of city-states around the North Sea and Baltic coasts), a trading system of immense complexity soon spanned Europe from Bergen to Athens and from Lisbon to Vienna. By 1400, long-distance trading was well established, based not on the luxury goods of the pioneer merchants, but on bulky staples such as grains, wine, salt, wool, cloth, and metals. Milan, Genoa, Venice, and Bruges had all grown to populations of 100,000 or more. Paris was the dominant European city, with

FIGURE 10.6   (a) Aigues-Mortes, France The walled medieval town of Aigues-Mortes in southern France is one of the best-preserved examples of thirteenth-century military architecture. The town of rectilinear streets is surrounded by a wall with five towers and ten fortified gates. Aigues-Mortes was originally built as a port, with access to the sea via a canal dug through the ponds and marshes of the Camargue.

(b) **Urbino, Italy** An important strategic center in the thirteenth century with a classic hilltop defensive site, Urbino became an important artistic center during the Renaissance.

FIGURE 10.7 Cologne In the late 1400s, when this woodcut was made, Cologne had a population of less than 25,000 but was already a vital commercial and manufacturing center, with an important cathedral and a university that was already more than 100 years old.

a population of about 275,000. Europe stood poised to extend its grasp on a global scale.

Between the fifteenth and seventeenth centuries, a series of changes occurred that transformed not only the cities and city systems of Europe but the entire world economy. Merchant capitalism increased in scale and sophistication; economic and social reorganization was stimulated by the Protestant Reformation and the scientific revolution. Meanwhile, aggressive overseas colonization made Europeans the leaders, persuaders, and shapers of the rest of the world's economies and societies.

Spanish and Portuguese colonists were the first to extend the European urban system into the world's peripheral regions. They established the basis of a Latin American urban system in just 60 years, between 1520 and 1580. Spanish colonists founded their cities on the sites of Indian cities (in Oaxaca and Mexico City, Mexico; Cajamarca and Cuzco, Peru; and Quito, Ecuador) or in regions of dense indigenous populations (in Puebla and Guadalajara, Mexico; and Arequipa and Lima, Peru). These colonial towns were established mainly as administrative and military centers from which the Spanish Crown could occupy and exploit the New World. Portuguese colonists, in contrast, situated their cities—Recife, Salvador, São Paulo, and Rio de Janeiro—with commercial rather than administrative considerations in mind. They, too, were motivated by exploitation, but their strategy was to establish colonial towns in locations best suited to organizing the collection and export of the products of their mines and plantations.

In Europe, reorganization during the Renaissance saw the centralization of political power and the formation of national states, the beginnings of industrialization, and the funneling of plunder and produce from distant colonies. In this new context, the port cities of the North Sea and Atlantic coasts enjoyed a decisive locational advantage. By 1700, London had grown to 500,000 people, while Lisbon and Amsterdam had each grown to about 175,000. The cities of continental and Mediterranean Europe expanded at

a more modest rate. By 1700, Venice had added only 30,000 to its 1400 population of 110,000, and Milan's population did not grow at all between 1400 and 1700.

The most important aspect of urbanization during this period, however, was the establishment of gateway cities around the rest of the world (**Figure 10.9**). **Gateway cities** serve as a link between one country or region and others because of their physical situation. These control centers command entrance to, and exit from, their particular country or region. European powers founded or developed thousands of towns in other parts of the world as they extended their trading networks and established their colonies. The great majority of these urban enclaves were ports. Protected by fortifications and European naval power, they began as trading posts and colonial administrative centers. Before long, they developed manufacturing of their own to supply the pioneers' needs, along with more extensive commercial and financial services.

As colonies were developed and trading networks expanded, some of these ports grew rapidly, acting as gateways for colonial expansion into continental interiors. Into their harbors came waves of European settlers; through their docks were funneled the produce of continental interiors. Rio de Janeiro (Brazil) grew on the basis of gold mining; Accra (Ghana) on cocoa; Buenos Aires (Argentina) on mutton, wool, and cereals; Kolkata (India, formerly Calcutta) on jute, cotton, and textiles; São Paulo (Brazil) on coffee; and so on. As these cities grew into major population centers, they became important markets for imported European manufactures, adding even more to their functions as gateways for international transport and trade.

APPLY YOUR KNOWLEDGE From an atlas map, create a list of probable gateway cities along the Atlantic seaboard of North America. Use the Internet to check on the early histories of these cities, noting their principal imports and exports. ■

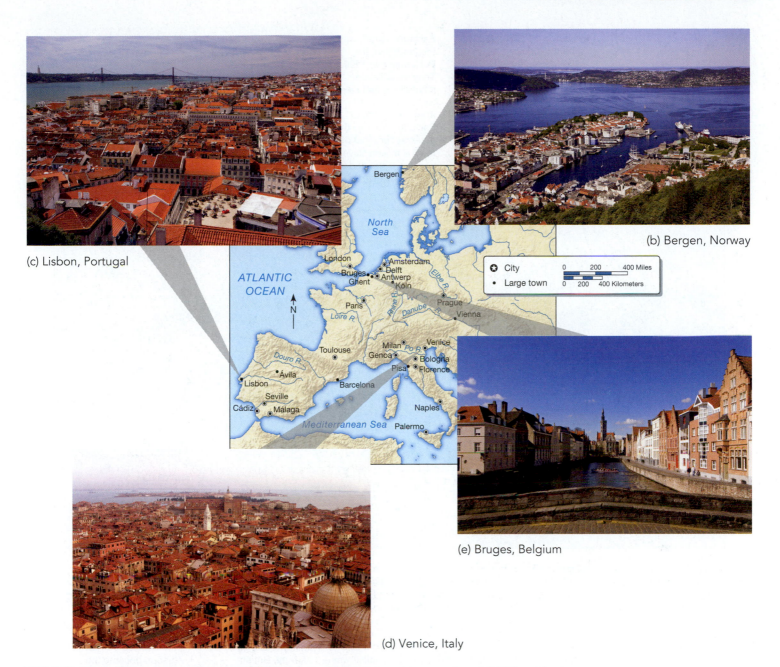

(c) Lisbon, Portugal

(b) Bergen, Norway

(e) Bruges, Belgium

(d) Venice, Italy

FIGURE 10.8 The towns and cities of Europe, ca. 1350 Cities with more than 10,000 residents were uncommon in medieval Europe except in northern Italy and Flanders. In the cities of Florence and Delft, the spread of cloth production and the growth of trade permitted relatively intense urbanization. Elsewhere, large size was associated with administrative, religious, educational, and economic functions. By 1350, many of the bigger towns (for example, Barcelona, Cologne, Prague) supported universities as well as a variety of religious institutions. Most urban systems, reflecting the economic and political realities of the time, were relatively small. (b) Bergen, Norway (c) Lisbon, Portugal (d) Venice, Italy (e) Bruges, Belgium (*Source:* (a) Adapted from P. M. Hohenberg and L. H. Lees, *The Making of Urban Europe 1000–1950.* Cambridge, MA: Harvard University Press, 1995.)

Industrialization and Urbanization

It was not until the late eighteenth century, however, that urbanization began to become an important dimension of the world-system in its own right. In 1800, less than 5 percent of the world's 980 million people lived in towns and cities. By 1950, however, 16 percent of the world's population was urban, and more than 900 cities of 100,000 or more existed around the world. The Industrial Revolution and European imperialism had created unprecedented concentrations of humanity that were intimately linked in networks and hierarchies of interdependence.

Cities were synonymous with industrialization. Industrial economies could be organized only through the large pools of labor; the transportation networks; the physical infrastructure of factories, warehouses, stores, and offices; and the consumer markets provided by cities. As industrialization spread throughout Europe in the first half of the nineteenth century and then to other parts of

New York, at first a modest Dutch fur-trading port, became the gateway for millions of European immigrants and for a large volume of U.S. agricultural and manufacturing exports.

Boston first flourished as the principal colony of the Massachusetts Bay Company, exporting furs and fish and importing slaves from West Africa, hardwoods from central America, molasses from the Caribbean, manufactured goods from Europe, and tea (via Europe) from South Asia.

Salvador, Brazil, was the landfall of the Portuguese in 1500. They established plantations that were worked by slave labor from West Africa. Salvador became the gateway for most of the 3.5 million slaves who were shipped to Brazil between 1526 and 1870.

Guangzhou was the first Chinese port to be in regular contact with European traders—first Portuguese in the sixteenth century and then British in the seventeenth century.

Nagasaki was the only port that feudal Japanese leaders allowed open to European traders, and for more than 200 years Dutch merchants held a monopoly of the import-export business through the city.

Havana was founded and developed by the Spanish in 1515 because of its excellent harbor. It was used as the assembly point for annual convoys returning to Spain.

Panama City, founded by the Spanish in 1519, became the gateway for gold and silver on its way by galleon to Spain.

Cape Town was founded in 1652 as a provisioning station for ships of the Dutch East India Company. Later, under British rule, it developed into an import-export gateway for South Africa.

Mombasa (in present-day Kenya) was already a significant Arab trading port when Vasco da Gama visited it in 1498 on his first voyage to India. The Portuguese used it as a trading station until it was recaptured by the Arabs in 1698. It did not become an important gateway port until it fell under British Imperial rule in the nineteenth century, when railroad development opened up the interior of Kenya, along with Rwanda, Uganda, and northern Tanzania.

Colombo's strategic situation on trade routes saw it occupied successively by the Portuguese, the Dutch, and the British. It became an important gateway after the British constructed an artificial harbor to handle the exports from tea plantations in Ceylon (now Sri Lanka).

Sydney, Australia, was not settled until the late eighteenth century, and even then many of the settlers were convicts who had been forcibly transported from Britain. It soon became the gateway for agricultural and mineral exports (mostly to Britain) and for imports of manufactured goods and European immigrants.

FIGURE 10.9 Gateway cities in the evolving world-system periphery Many of the world's most important cities grew to prominence as gateway cities because they commanded routeways into and out of developing colonies. Gateway cities are control centers that command entrance to and exit from their particular country or region.

the world, urbanization increased at a faster pace. The higher wages and greater variety of opportunities in urban labor markets attracted migrants from surrounding areas. The countryside began to empty. In Europe the *demographic transition* caused a rapid growth in population as death rates dropped dramatically (see Chapter 3). This growth in population provided a massive increase in the labor supply throughout the nineteenth century, further boosting the rate of urbanization, not only within Europe itself but also in Australia, Canada, New Zealand, South Africa, and the United States as emigration spread industrialization and urbanization to the frontiers of the world-system.

Transport networks and urban systems

Within the world's core regions, the transformation of urban systems hinged on successive innovations in transport technology that opened up agrarian interiors and intensified intercity and interregional trading networks. The first phase of this transformation was based on an old technology: the canal (**Figure 10.10**). Merchant trade and the beginnings of industrialization in both Britain and France were underpinned by extensive navigation systems that joined one river system to another. By 1790, France had just over 1,000 kilometers (620 miles) of canals and canalized rivers; Britain had nearly 3,600 kilometers (2,230 miles). The Industrial Revolution provided both the need and the capital for a spate of additional canal building that began to integrate emerging industrial towns. Market towns and hill towns that were not connected to canal systems were isolated from commerce and immediately fell behind.

In the United States, the opening of the Erie Canal in 1825 enabled New York, a colonial gateway port, to reorient itself toward the nation's growing interior. The Erie Canal was so profitable that it set off a "canal fever" that resulted in the construction of some 2,000 kilometers (1,240 miles) of navigable waterways in the next 25 years. But the scale of the United States was so great that a network of canals was a viable proposition only in more densely settled regions. The effective colonization of the interior and the growth of American cities did not take place until the development of steam-powered transportation—first riverboats and then railroads. The first steamboats, developed in the early 1800s, offered the possibility of opening up the vast interior by way of the Mississippi and its tributaries. The heyday of the river steamboat was between 1830 and 1850. During this period, river ports such as New Orleans, St. Louis, Cincinnati, and Louisville grew rapidly, extending the frontier of industrialization and modernization.

By 1860, the railroads had taken over as the dominant mode of transportation, further extending the frontier of settlement and industrialization and intensifying the growth of existing cities. The railroad originated in Britain, where George Stephenson engineered the world's first commercial railroad, a 20-kilometer (12.4-mile) line between Stockton and Darlington that was opened in 1825. The *Rocket*, the first locomotive for commercial passenger trains, was designed mainly by Stephenson's son Robert for the Liverpool & Manchester line, which opened 4 years later. The economic success of this line sparked two railroad-building booms that eventually created a highly integrated urban system and allowed Britain's manufacturing industry to flourish.

In other core countries, where sufficient capital existed to license (or copy) the locomotive technology and install the track, railroad systems led to the first full stage of urban system integration. The first passenger-train services averaged little more than

FIGURE 10.10 Canal systems The canal systems that opened up the interiors of Europe and North America in the eighteenth century were initially dependent on horsepower. Early examples carried horse-drawn barges. Later, barges were able to utilize steam- and oil-powered engines, extending the importance of larger canals and navigable rivers. Barges are still an important mode of transport for bulky goods. This photograph shows part of the Burgundy canal in France.

20 to 35 kilometers per hour (15–20 mph), but locomotive technology changed rapidly and made it easier, faster, and cheaper to conquer vast distances. While the railroads integrated the economies of entire countries and allowed vast territories to be colonized, they also brought some important regional and local restructuring and differentiation. In the United States, for example, the railroads led to the mushrooming of Chicago as the focal point for railroads. This extended the Manufacturing Belt's dominance over the West and South (**Figure 10.11**). The reorientation of the nation's transportation system effectively ended the role of the South's cotton regions as outliers of the British trading system. Instead, they became outliers of the U.S. Manufacturing Belt, supplying factories in New England and the Mid-Atlantic Piedmont. The phenomenal growth of New Orleans, which had thrived on cotton exports, came to an abrupt end.

In the twentieth century, the internal combustion engine powered further rounds of urban system development and integration. The development of trucks in the 1910s and 1920s suddenly released factories from the need to locate near railroads, canals, and waterfronts. Trucking allowed goods to be moved farther, faster, and cheaper than before. As a result, trucking made it feasible to locate factories on inexpensive land on city fringes and in smaller towns and peripheral regions where labor was cheaper. It also increased the market area of individual factories and reduced the need for large product inventories.

APPLY YOUR KNOWLEDGE How have changing transportation technologies affected the history of the town or city in which you live? ∎

(a)

(b)

FIGURE 10.11 Railroads (a) The Illinois Central Railroad issued a promotional poster in the 1860s celebrating the role it envisaged for itself. It hoped that its trunk line from Chicago to New Orleans would significantly affect the settling of the American interior. Note the deliberate juxtaposition of the high technology of the railroad locomotive and the telegraph line, contrasting with the obsolescent technologies of the stagecoach (inset top right) and barge (inset bottom right). Even the oceangoing ship is shown with masts and sails to diminish its status in comparison to the railroad locomotive. (b) The poster issued 50 years later by the same entity, on the other hand, shows the railroad locomotive in the company of several important new technologies: aircraft, automobiles and trucks, and electric streetlights.

Industrial cities

The shock city of nineteenth-century European industrialization was Manchester, England, which grew from a small town of 15,000 in 1750 to a city of 70,000 in 1801; a metropolis of 500,000 in 1861; and a world city of 2.3 million by 1911. A **shock city** is seen at the time as the embodiment of surprising and disturbing changes in economic, social, and cultural life. Manchester's phenomenal growth was based on new textile manufacturing technologies. The city's first cotton mill was built in the early 1780s, and by 1830 there were 99 cotton-spinning mills. The opening of the Suez Canal in 1869 halved the travel time between Britain and India. It ruined the Indian domestic cotton textile industry, but it allowed India to export its raw cotton to Manchester. Around the same time, British colonialists established cotton plantations in Egypt and Uganda, providing another source of supply. As Manchester grew, it spilled out into the surrounding countryside, bringing its characteristic landscape of red-brick terrace housing and "Dark Satanic Mills" with their tall brick chimneys (**Figure 10.12**).

When industrialization took hold in North America, the shock city was Chicago, which grew from under 30,000 in 1850 to 500,000 in 1880, 1.7 million in 1900, and 3.3 million in 1930 (**Figure 10.13**). When Chicago was first incorporated as a city in 1837, its population was only 4,200. Its growth followed the arrival of the railroads, which made the city a major transportation hub. By the 1860s, lake vessels were carrying iron ore from the Upper Michigan ranges to Chicago's blast furnaces, and railroads were hauling cattle, hogs, and sheep to the city for

slaughtering and packing. The city's prime geographic situation also made it the nation's major lumber-distributing center by the 1880s.

Manchester and Chicago were archetypal forms of an entirely new kind of city—the *industrial city*—whose fundamental reason for existence was not, as it was for earlier generations of cities, to fulfill military, political, ecclesiastical, or trading functions. Rather, it existed simply to assemble raw materials and to fabricate, assemble, and distribute manufactured goods. Both Manchester and Chicago had to cope with unprecedented rates of growth and the unprecedented economic, social, and political problems that were a consequence of their growth. Both were also world cities, cities in which a disproportionate part of the world's most important business—economic, political, and cultural—is conducted. Such cities experience growth largely as a result of their role as key nodes in the world economy.

During the Industrial Revolution and for much of the twentieth century, a close and positive relationship existed between rural and urban development in the core regions of the world (**Figure 10.14**). The appropriation of new land for agriculture, together with mechanization and the innovative techniques that urbanization allowed, resulted in increased agricultural productivity. This extra productivity released rural labor to work in the growing manufacturing sector in towns and cities. At the same time, it provided the additional produce needed to feed growing urban populations. The whole process was further reinforced by the capacity of urban labor forces to produce agricultural tools,

(a) Manchester's first cotton mill was built in the early 1780s, and by 1830 there were 99 cotton-spinning mills. As the city grew, it spilled out into the surrounding countryside, bringing its characteristic landscape of red-brick terrace housing and "Dark Satanic Mills" with their tall brick chimneys.

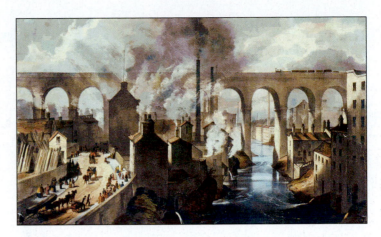

(b) Railway viaducts, like this one in Stockport, just outside Manchester, brought rail transportation to Manchester early in the nineteenth century and helped to make the city a major transportation hub.

(c) Canals were at the heart of the development of industrial Manchester. They enabled coal and raw materials to be carried right to the heart of the city and finished goods to be transported away easily. The Bridgewater Canal, one of the earliest modern canals, ran into Manchester from the Duke of Bridgewater's coal mine at Worsley, about 16 kilometers away. Later, other canals radiated outwards from the area and the Ship Canal brought seagoing vessels close to the city center.

FIGURE 10.12 Manchester, England The "shock city" of the nineteenth century.

FIGURE 10.13 Growth of Chicago In 1870, when Manchester was already a thriving metropolis, Chicago was at the beginning of a period of explosive growth. A year later 9 square kilometers (4 sq. mi.) of the city, including the business district, were destroyed by fire. Chicago was rebuilt rapidly, with prosperous industrialists taking the opportunity to build impressive new structures in the downtown area. The city's economic and social elite colonized the Lake Michigan shore. Heavy industry, warehouses, and rail yards crowded the banks of the Chicago River, stretching northwestward from the city center. To the south of the city were the Union Stockyards and a pocket of heavy industry where the Calumet River met Lake Michigan. The surrounding neighborhoods of working families' homes spread rapidly outward as wave after wave of immigrants arrived in the city.

FIGURE 10.14 The urbanization process in the world's core regions Urbanization was stimulated by advances in farm productivity that (1) provided the extra food to support the increased numbers of townspeople and (2) made many farmers and farm laborers redundant, prompting them to migrate to cities. Labor displaced in this way ended up consuming food rather than producing it, but this was more than compensated for by increased agricultural productivity and increased capacity of enlarged urban labor forces to produce agricultural tools, machinery, fertilizers, and so on that contributed further to agricultural productivity.

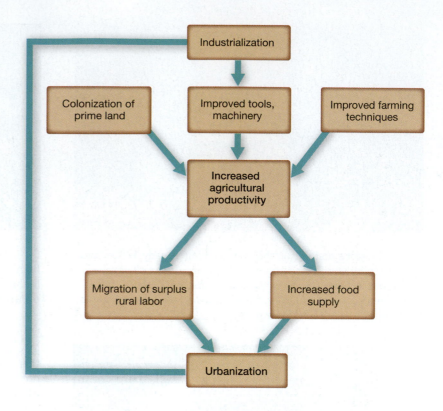

machinery, fertilizer, and other products that made for still greater increases in agricultural productivity. This kind of urbanization is a special case of *cumulative causation* (Chapter 7), in which a spiral buildup of advantages is enjoyed by particular places as a result of the development of external economies, agglomeration effects, and localization economies.

APPLY YOUR KNOWLEDGE Conduct an Internet search to find a current example of what you consider to be a shock city. What are the surprising/disturbing changes in the economic, social, and cultural life of the city that make it shocking? (*Hint:* You might want to consider cities in the Middle East.) ■

Imperialism and Peripheral Urbanization

The industrialization of the core economies was highly dependent on the exploitation of peripheral regions. Inevitably, the new international division of labor that resulted from this relationship also had a significant impact on patterns and processes of urbanization in the periphery. European imperialism led to the creation of new gateway cities in peripheral countries and, as Europeans raced to establish economic and political control over continental interiors, these colonial cities were established as centers of administration, political control, and commerce.

Colonial cities are those that were deliberately established or developed as administrative or commercial centers by colonial or imperial powers. Geographers often distinguish between two types of colonial city. The pure colonial city was usually established, or "planted," by colonial administrations in a location where no significant urban settlement had previously existed. Such cities were laid

out expressly to fulfill colonial functions, with ceremonial spaces, offices, and depots for colonial traders, plantation representatives, and government officials; barracks for soldiers; and housing for colonists. Subsequently, as these cities grew, they added housing and commercial space for local peoples drawn by the opportunity to obtain jobs as servants, clerks, or porters. Examples of pure colonial cities are the original settlements of Mumbai (Bombay), Kolkata (Calcutta), Ho Chi Minh City (Saigon), Hong Kong, Jakarta, Manila, and Nairobi.

In the other type of colonial city, colonial functions were grafted onto an existing settlement, taking advantage of a good site and a ready supply of labor. Examples include Delhi, Mexico City, Shanghai, and Tunis. In these cities the colonial imprint is most visible in and around the city center in the formal squares and public spaces, the layout of avenues, and the presence of colonial architecture and monuments. This architecture includes churches, city halls, and railway stations (**Figure 10.15**); the palaces of governors and archbishops; and the houses of wealthy traders, colonial administrators, and landowners.

The colonial legacy can also be read in the building and planning regulations of many colonial cities. Often, colonial planning regulations were copied from those that had been established in the colonizing country. Because these regulations were based on Western concepts, many turned out to be inappropriate to colonial settings. Most colonial building codes, for example, are based on Western models of family and work, with a small family living in a residential area that is some distance from the adults' places of work. This is at odds with the needs of large, extended families whose members are involved with a busy domestic economy and with family businesses that are traditionally integrated with the residential setting. Colonial planning, with its gridiron street layouts, zoning regulations that do not allow for a mixture of land uses, and building

FIGURE 10.15 Colonial architecture and urban design Cities in the periphery of the world-system have grown very rapidly since the colonial era, but the legacy of the era can still be seen in the architecture, monuments, and urban design of the period. This photograph shows part of the High Court building in Mumbai.

codes designed for European climates, ignored the specific needs of local communities and misunderstood their cultural preferences.

URBAN SYSTEMS

Every town and city is part of the interlocking urban systems that link regional-, national-, and international-scale human geographies in a complex web of interdependence. These urban systems organize space through hierarchies of cities of different sizes and functions. Many of these hierarchical urban systems exhibit common attributes and features, particularly in the relative size and spacing of individual towns and cities.

Towns and cities function as market centers and this results in a hierarchical system of central places. A **central place** is a settlement where certain types of products and services are available to consumers. **Central place theory** seeks to explain the tendency for central places to be organized in hierarchical systems, analyzing the relative size and geographic spacing of towns and cities as a function of consumer behavior. A fundamental tenet of central place theory is that the smallest settlements in an urban system provide only those goods and services that meet everyday needs (bakery and dairy products and groceries, for example) and that these small settlements are situated relatively close to one another because consumers, assumed to be spread throughout the countryside, are not prepared to travel far for such items. On the other hand, people are willing to travel farther for more expensive, less frequently purchased items. This means that the larger the settlement, with a broader variety of more specialized goods and services, the farther it will be from others of a similar size.

Although consumer behavior certainly helps to explain some aspects of urban systems, there are relatively few regions today where the functions of towns and cities are dominated by local markets and shopping. Nevertheless, the urban systems of most regions do exhibit a clear hierarchical structure. This is partly a legacy of past eras, when towns and cities did function mainly as

market centers for surrounding agricultural areas. **Figure 10.16** shows a typical example: the Spanish urban system, with smaller towns and cities functioning interdependently with successively larger ones. Note that the whole system is dominated by one or two metropolitan areas whose linkages are national in scope.

Urban systems also exhibit clear *functional* differences within such hierarchies. This is yet another reflection of the interdependence of places. The geographical division of labor resulting from such processes of economic development (Chapter 7) means that many medium- and larger-size cities perform specialized economic functions and so acquire distinctive characters. Thus the industrial era produced steel towns (for example, Pittsburgh, Pennsylvania; Sheffield, England), textile towns (for example, Lowell, Massachusetts; Manchester, England), and auto-manufacturing towns (for example, Detroit, Michigan; Turin, Italy; Toyota City, Japan).

Some towns and cities, of course, do evolve as general-purpose urban centers, providing an evenly balanced range of functions for their own particular sphere of influence. **Figure 10.17** shows the urban system in the United States. In America, the top tier of cities consists of centers of global importance (including Chicago, New York, and Los Angeles) that provide high-order functions to an international marketplace. The second tier consists of general-purpose cities with diverse functions but only regional importance (including Atlanta, Miami, and Boston), and the third and fourth tiers consist of more specialized centers of subregional and local importance.

City-Size Distributions, Primacy, and Centrality

The functional interdependency between places within urban systems tends to result in a distinctive relationship between the population size of cities and their rank within the overall hierarchy. This relationship is known as the **rank-size rule**, which describes a certain statistical regularity in the city-size distributions of countries and regions. The relationship is such that the

FIGURE 10.16 The Spanish urban system Note how the smaller cities tend to be linked to middle-order cities, while these in turn are linked to regional metropolises, which are linked to the national metropolises, Madrid and Barcelona. These linkages represent the major flows of capital, information, and goods within the Spanish urban system. (Source: Adapted from L. Bourne, R. Sinclair, M. Ferrer, and A. d'Entremont (eds.), *The Changing Geography of Urban Systems.* Navarra, Spain: Department of Human Geography, Universidad de Navarra, 1989, fig. 2, p. 46.)

FIGURE 10.17 Functional specialization within an urban system Different cities tend to specialize in particular kinds of economic activities, and some provide a much greater range of functions than others. This map shows a functional classification of the U.S. urban system. The top tier of the system consists of "world cities" that provide high-order functions to a global marketplace. Among these, New York stands alone in the U.S. as the dominant metropolis. The next tier consists of cities with diverse functions but only regional importance (for example, Atlanta, Minneapolis). The third tier consists of more specialized centers of business, government, and producer services (Austin, Texas; Albany, New York; and Hartford, Connecticut). The fourth tier consists of still more specialized cities of various kinds: manufacturing centers (Buffalo, New York; Chattanooga, Tennessee), mining/industrial centers (Charleston, West Virginia; Duluth, Minnesota), industrial/military centers (Newport News, Virginia; San Diego, California), and resort/retirement centers (Las Vegas, Nevada; Orlando, Florida).

(Source: Reprinted with permission of Prentice Hall from P. L. Knox, *Urbanization,* © 1994, p. 64.)

*n*th largest city in a country or region is $1/n$ the size of the largest city in that country or region. Thus, if the largest city in a particular system has a population of 1 million, the fifth-largest city should have a population one-fifth as big (that is, 200,000); the hundredth-ranked city should have a population one-hundredth as big (that is, 10,000), and so on. Plotting this relationship on a graph with a logarithmic scale for population sizes would produce a perfectly straight line. The actual rank-size relationships for urban systems at all levels of economic development come close to this. Over time, the slope moves to the right on the graph, reflecting the growth of towns and cities at every level in the urban hierarchy.

In some urban systems, the rank-size distribution is distorted as a result of the disproportionate size of the largest (and sometimes also the second-largest) city. According to the rank-size rule, the largest city should be just twice the size of the second-largest city. In the United Kingdom, London is more than 9 times the size of Birmingham, the second-largest city. In France, Paris is more than 8 times the size of Marseilles, France's second-largest city. In Brazil, both Rio de Janeiro and São Paulo are 5 times the size of Belo Horizonte, the third-largest city. Geographers call this condition **primacy**, occurring when the population of the largest city in an urban system is disproportionately large in relation to the second- and third-largest cities in that system. Cities like London and Buenos Aires are termed *primate* cities.

Primacy is not simply a matter of size. Some of the largest metropolitan areas in the world—Karachi, New York, and Mumbai (Bombay), for example—are not primate. Further, primacy is a condition that is found in both the core and the periphery of the world-system. This suggests that primacy is a result of the roles played by particular cities within their own national urban systems. A relationship does exist between primacy and the world economy, however. Primacy in peripheral countries is usually a consequence of primate cities' early roles as gateway cities. In core countries, it is usually a consequence of primate cities' roles as imperial capitals and centers of administration, politics, and trade for a much wider urban system than their own domestic system.

When a city's economic, political, and cultural function is disproportionate to its population, the condition is known as **centrality**. Centrality refers to the functional dominance of cities within an urban system. Cities that account for a disproportionately high share of economic, political, and cultural activity have a high degree of centrality within their urban system. Very often primate cities exhibit this characteristic, but cities do not necessarily have to be primate in order to be functionally dominant within their urban system. **Figure 10.18** shows some examples of centrality, revealing the overwhelming dominance of some cities within the world-system periphery. Bangkok, for instance, with around 12 percent of the Thai population, accounts for approximately 38 percent of the country's overall gross domestic product (GDP); over 85 percent of the country's GDP in banking, insurance, and real estate; and 75 percent of its manufacturing.

World Cities and the Global Urban System

Recall from Chapter 7 that ever since the evolution of a world-system in the sixteenth century, certain cities known as world cities (sometimes referred to as *global cities*) have played key roles in organizing space beyond their own national boundaries. In the first stages of world-system growth, these key roles involved the organization of trade and the execution of colonial, imperial, and geopolitical strategies. The world cities of the seventeenth century were London, Amsterdam, Antwerp, Genoa, Lisbon, and Venice. In the eighteenth century, Paris, Rome, and Vienna also became world cities, while Antwerp and Genoa became less influential. In the nineteenth century, Berlin, Chicago, Manchester, New York, and St. Petersburg became world cities, while Venice became less influential.

Today, the globalization of the economy has resulted in the creation of a global urban system in which the key roles of world cities are concerned less with the deployment of imperial power and the orchestration of trade and more with transnational corporate organization, international banking and finance (**Figure 10.19**), supranational government, and the work of international agencies (see Box 10.2, "Visualizing Geography: The World City Network"). World cities have become the control centers for the flows of information, cultural products, and finance that collectively sustain the economic and cultural globalization of the world.

A great deal of synergy exists among the various functional dimensions of world cities. A city like New York, for example, attracts transnational corporations because it is a center of culture and communications. It attracts specialized business services because it is a center of corporate headquarters and of global markets, and so on. These interdependencies represent a special case of the geographical *agglomeration effects* that we discussed in Chapter 7. In the case of New York City, corporate headquarters and specialized legal, financial, and business services cluster together because of the mutual cost savings and advantages of being close to one another.

At the same time, different world cities fulfill different roles within the world-system, making for different emphases and combinations (that is, differences in the nature of their world-city functions) as well as for differences in the absolute and relative localization of particular world-city functions (that is, differences in their degree of importance as world cities). For example, Brussels is relatively unimportant as a corporate headquarters location but qualifies as a world city because it is the administrative center of the European Union and has attracted a large number of nongovernmental organizations and advanced

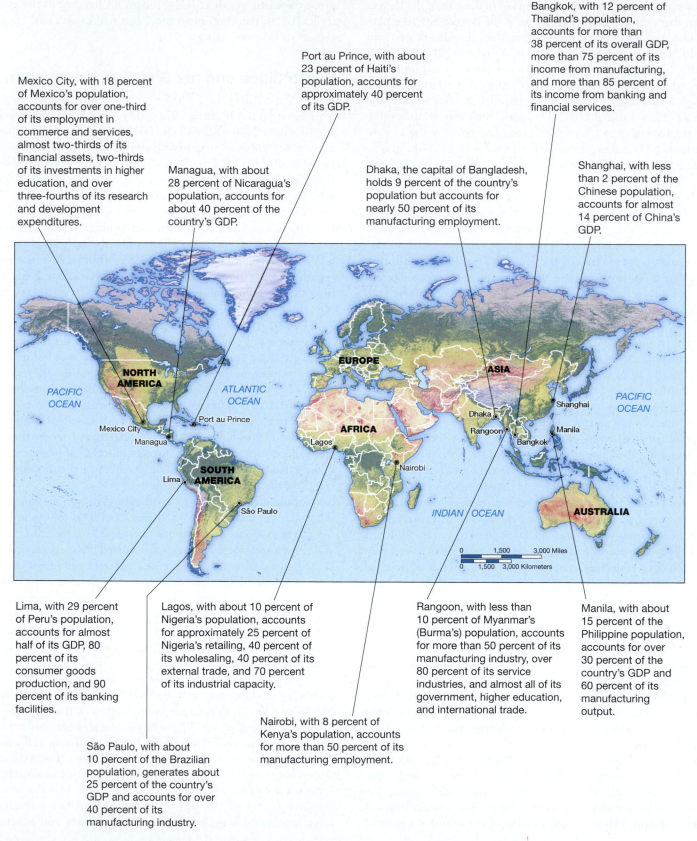

Mexico City, with 18 percent of Mexico's population, accounts for over one-third of its employment in commerce and services, almost two-thirds of its financial assets, two-thirds of its investments in higher education, and over three-fourths of its research and development expenditures.

Managua, with about 28 percent of Nicaragua's population, accounts for about 40 percent of the country's GDP.

Port au Prince, with about 23 percent of Haiti's population, accounts for approximately 40 percent of its GDP.

Dhaka, the capital of Bangladesh, holds 9 percent of the country's population but accounts for nearly 50 percent of its manufacturing employment.

Bangkok, with 12 percent of Thailand's population, accounts for more than 38 percent of its overall GDP, more than 75 percent of its income from manufacturing, and more than 85 percent of its income from banking and financial services.

Shanghai, with less than 2 percent of the Chinese population, accounts for almost 14 percent of China's GDP.

Lima, with 29 percent of Peru's population, accounts for almost half of its GDP, 80 percent of its consumer goods production, and 90 percent of its banking facilities.

Lagos, with about 10 percent of Nigeria's population, accounts for approximately 25 percent of Nigeria's retailing, 40 percent of its wholesaling, 40 percent of its external trade, and 70 percent of its industrial capacity.

Rangoon, with less than 10 percent of Myanmar's (Burma's) population, accounts for more than 50 percent of its manufacturing industry, over 80 percent of its service industries, and almost all of its government, higher education, and international trade.

Manila, with about 15 percent of the Philippine population, accounts for over 30 percent of the country's GDP and 60 percent of its manufacturing output.

São Paulo, with about 10 percent of the Brazilian population, generates about 25 percent of the country's GDP and accounts for over 40 percent of its manufacturing industry.

Nairobi, with 8 percent of Kenya's population, accounts for more than 50 percent of its manufacturing employment.

FIGURE 10.18 Examples of urban centrality The economic, political, and cultural importance of some cities is disproportionate to their population size. This is a reflection of core-periphery differentials within countries and often becomes a political issue because of the economic disparities. The centrality of these cities also leads to localized problems of congestion, land price inflation, and pollution.

World cities also provide an interface between the global and the local. They contain the economic, cultural, and institutional apparatus that channels national and provincial resources into the global economy and that transmits the impulses of globalization back to national and provincial centers. As such, world cities possess several functional characteristics. They are the sites of the following:

■ Most of the leading global markets for commodities, commodity futures, investment capital, foreign exchange, equities, and bonds

■ Clusters of specialized, advanced business services, especially those that are international in scope and that are attached to finance, accounting, insurance, advertising, property development, and law

■ Concentrations of corporate headquarters—not just of transnational corporations but also of major national firms and large foreign firms

■ Concentrations of national and international headquarters of trade and professional associations

■ Most of the leading nongovernmental organizations (NGOs) and intergovernmental organizations (IGOs) that are international in scope (for example, the World Health Organization; United Nations Educational, Scientific, and Cultural Organization (UNESCO); the International Labor Organization, and the International Federation of Agricultural Producers)

■ The most powerful and internationally influential media organizations (including newspapers, magazines, book publishing, and satellite television); news and information services (including news wires and online information services); and culture industries (including art and design, fashion, film, and television)

■ Many terrorist acts because of their importance and visibility

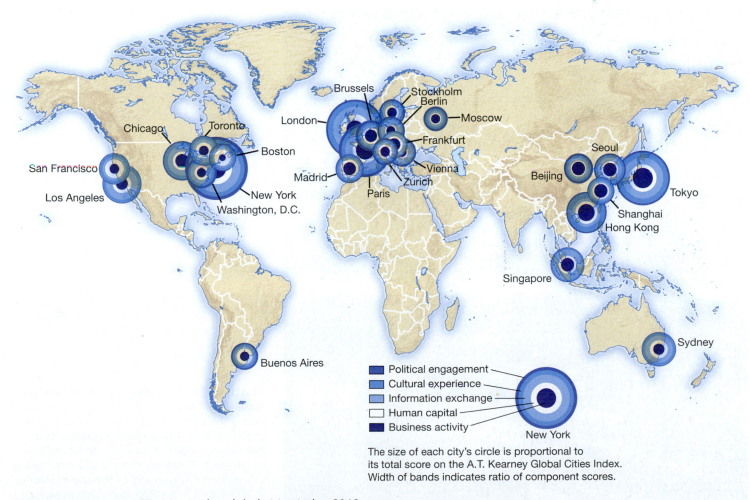

The size of each city's circle is proportional to its total score on the A.T. Kearney Global Cities Index. Width of bands indicates ratio of component scores.

Legend:
- Political engagement
- Cultural experience
- Information exchange
- Human capital
- Business activity

FIGURE 10.D Top 25 cities in the global cities index 2010. (Source: A.T. Kearney, *The Urban Elite: The A.T. Kearney Global Cities Index 2010. Chicago, Illinois, 2010, p. 3.*)

business services that are transnational in scope. Milan is relatively dependent in terms of corporate control and advanced business services but has global status in terms of cultural influence (especially fashion and design) and is an important regional financial center.

Thus there is a geographical complexity to world cities' roles that cannot be reduced to a simple "hierarchy" or ranking. World cities are not simply mini-Londons and little New Yorks. World cities exist in networks of flows among firms and institutions, networks that are complex and multilayered. As a result world cities are connected in different ways and integrated to different degrees in the global urban system. Quantifying these flows and connectivities is difficult, but geographers at the Globalization and World Cities (GaWC) research network have been able to identify breakpoints in cities' aggregate levels of integration within the world city network, resulting in a classification of alpha-, beta- and gamma-level world cities (**Table 10.1**). London and New York are tightly interrelated with one another (so much so that they are sometimes jointly referred to as "NYLON") and are both significantly more integrated into the overall world city network than any other city. Hong Kong is the third most highly integrated city.

Overall, the level of connectivity in the world city network rose steadily between 2000 and 2008, largely as a result of the increased global connectivity of South Asian, Chinese, and Eastern European cities (Shanghai, Beijing, and Moscow, in particular). This reflects the importance of "emerging markets" in globalization. Cities in the United States and sub-Saharan Africa, however, experienced a decline in global connectivity in the period (Los Angeles, San Francisco, and Miami, in particular). U.S. cities have always exhibited lower levels of integration than might be expected because foreign firms find it hard to penetrate the U.S. market while U.S. firms, with a big domestic market, have less reason to gamble on global expansion.

APPLY YOUR KNOWLEDGE Choose two world cities and provide two examples each of how these cities serve to connect global and local regions. Be specific in your answers. For example, you might want to consider how certain cities are centers of transnational corporations or the different ways cities culturally connect different places. ■

FIGURE 10.19 Advanced business services The "square mile" of the City of London is the cornerstone of London as a world city. The district contains the London Stock Exchange, Lloyd's of London, and the Bank of England. There are around 500 banks with offices in the City, many of them specializing in areas such as foreign exchange markets, Eurobonds, and energy futures. The City also accounts for a quarter of the world market for marine insurance and over a third of the market in aviation risks.

TABLE 10.1 Alpha-level world cities in 2008

Alpha ++	Alpha +	Alpha	Alpha −
London	Hong Kong	Milan	Warsaw
New York	Paris	Madrid	Jakarta
	Singapore	Seoul	Sao Paulo
	Sydney	Moscow	Zurich
	Tokyo	Brussels	Mexico City
	Shanghai	Toronto	Dublin
	Beijing	Mumbai	Amsterdam
		Buenos Aires	Bangkok
		Kuala Lumpur	Taipei
			Rome
			Istanbul
			Lisbon
			Chicago
			Frankfurt
			Stockholm
			Vienna
			Budapest
			Athens
			Prague
			Caracas
			Auckland
			Santiago

Source: From Taylor et al. (2009b), "Measuring the World City Network: New Developments and Results," *GaWC Research Bulletin 300*; http://www.lboro.ac.uk/gawc/rb/rb300.html.

WORLD URBANIZATION TODAY

It is difficult to say just how urbanized the world has become. In many areas, urban growth is taking place at such a pace and under such chaotic conditions that experts can provide only informed estimates. The most comprehensive source of statistics is the United Nations, whose data suggest that almost half of the world's population is now urban. These data incorporate the very different definitions of *urban* used by different countries. Some countries (Australia and Canada, for example) count any settlement of 1,000 people or more as urban; others (including Italy and Jordan) use 10,000 as the minimum for an urban settlement, and Japan uses 50,000 as the cutoff. This, by the way, tells us something about the nature of urbanization itself: It is a relative phenomenon. In countries like Peru, where the population is thinly distributed and scattered, a settlement of 2,000 represents a significant center. In countries like Japan, however, with more residents, higher population densities, and a tradition of centralized agricultural settlement, a much larger concentration of people is required to count as "urban."

In order to get around these national differences in definitions of what counts as "urban," the World Bank has developed a uniform definition of what constitutes an urban area based on an "agglomeration index" that identifies an area of 1 square kilometer as urban if its population density exceeds 150 and it has access to a settlement of more than 50,000 inhabitants within 60 minutes by road. According to this measure, the overall level of urbanization in the world in 2010 was 51 percent. North America is the most urbanized continent in the world, with 82 percent of its population living in urban areas. In contrast, Africa is less than 40 percent urban. To put these figures in perspective, only 30 percent of the world's population was urbanized in 1950. In that year, there were only 83 metropolitan areas of a million or more, and only 8 of 5 million or more existed; in 2010, there were 468 metropolitan areas of a million or more people, and 51 with over 5 million.

The World Bank's analyses clearly show how levels of urbanization change with a country's level of economic development. There is a strong positive relationship between levels of urbanization and levels of economic development until a GDP per capita of around $10,000. The early stages of urbanization are associated with a rapid shift in the number of people moving from rural to urban areas. Subsequently, the pace of urbanization slows and density levels off as the proportion of the country's population living in cities surpasses 60 percent and the level of GDP per capita surpasses $10,000.

The Periphery and Semiperiphery: Overurbanization and Megacities

Few peripheral countries are able to handle the urban population crush, which is causing problems on an unprecedented scale with everything from clean water to disease prevention. Ten million people are dying annually in densely populated urban areas from conditions produced by substandard housing and poor sanitation. In 2005, there were 1.6 billion inadequately housed people across the world and an estimated 100 million homeless. According to the World Bank, 2.6 billion people worldwide do not have access to sanitation. About 1.7 million deaths a year—90 percent of which are children—are attributed to unsafe drinking water and poor sanitation and hygiene, mainly through infectious diarrhea.

In contrast to the world's core regions, where urbanization has largely resulted from economic growth, the urbanization of peripheral regions has been a consequence of demographic growth that preceded economic development. Although demographic transition is a fairly recent phenomenon in the peripheral regions of the world (see Chapter 3), it has generated large increases in population well in advance of any significant levels of industrialization or rural economic development.

The result, for the mainly rural populations of peripheral countries, has been more and more of a bad thing. Problems with agricultural development (see Chapter 8) mean an apparently hopeless future of drudgery and poverty for fast-growing rural populations. Emigration has provided one potential safety valve, but as the frontiers of the world-system closed, the more affluent core countries have put up barriers to immigration. The only option for the growing numbers of impoverished rural residents is to move to the larger towns and cities, where at least there is the hope of employment and the prospect of access to schools, health clinics, piped water, and the kinds of public facilities and services that are often unavailable in rural regions. Cities also have the lure of modernization and the appeal of consumer goods—attractions that rural areas are now directly exposed to via satellite TV. Overall, the metropolises of the periphery have absorbed 4 out of 5 of the 1.2 billion city dwellers added to the world's population since 1970.

When natural disasters, environmental degradation, or civil war impacts highly populated rural regions in the periphery, rates of rural-urban migration increase dramatically:

> For several years now, relentless cycles of drought and flooding have wreaked havoc on the tiny country of Malawi, in the heart of southern Africa. In 2002 and 2003, torrential rains caused massive mudslides, washing away bridges and homes and devastating harvests of maize, the main food staple. Unable to eke a living from the ravaged countryside, rural residents have flocked in droves to the country's bourgeoning cities—giving Malawi the dubious distinction of being the world's fastest-urbanizing nation today.[1]

Rural migrants have poured into cities out of desperation and hope rather than being drawn by jobs and opportunities. Because these migration streams have been composed disproportionately of teenagers and young adults, an important additional component of urban growth has followed—exceptionally high rates of natural population increase. In most peripheral countries, the rate of natural increase of the population in cities exceeds that of net in-migration. On average, about 60 percent of urban population growth in peripheral countries is attributable to natural increase.

The consequence of all this urban population growth is **overurbanization**, which occurs when cities grow more rapidly than they can sustain jobs and housing. In such circumstances, urban growth produces instant slums—shacks set on unpaved streets, often with open sewers and no basic utilities. In general, it is the rate, not simply the level, of urbanization that produces slums; higher levels of urbanization tend to be associated with relatively fewer slums. Shacks are constructed out of any material that comes to hand, such as planks, cardboard, tar paper, thatch, mud, and corrugated iron. Many of these instant slums are squatter settlements, built illegally by families who are desperate for shelter. **Squatter settlements** are residential developments on land that is neither owned nor rented by its occupants. Squatter settlements are often, but not always, slums. In Chile, squatter settlements are called *callampas*, "mushroom cities"; in Turkey, they are called *gecekondu*, meaning that they were built after dusk and before dawn. In India, they are called *bustees*; in Tunisia, *gourbevilles;* in Brazil, *favelas*; and in Argentina, simply *villas miserias.* These settlements typically account for well over one-third and sometimes up to three-quarters of the population of major cities (**Figure 10.20**). The most extensive slums, according to United Nations statistics, are in the cities of

FIGURE 10.20 Slum housing in peripheral cities Throughout much of the world, the scale and speed of urbanization, combined with the scarcity of formal employment, have resulted in very high proportions of slum housing, much of it erected by squatters. This photograph shows part of the huge district of Mathare in Nairobi, Kenya.

sub-Saharan Africa, where over 70 percent of the population lives in unfit accommodations.

The United Nations International Children's Fund (UNICEF) has blamed "uncontrollable urbanization" in less developed countries for the widespread creation of "danger zones" in which increasing numbers of children become beggars, prostitutes, and laborers before reaching their teens.[4] Pointing out that urban populations are growing at twice the general population rate, UNICEF has concluded that too many people are being squeezed into cities that do not have enough jobs, shelter, or schools to accommodate them. As a consequence, the family and community structures that support children are being destroyed, with the result that increasing numbers of children have to work (**Figure 10.21**). For hundreds of thousands of street kids in less developed countries, "work" means anything that contributes to survival: shining shoes, guiding cars into parking spaces, chasing other street kids away from patrons at an outdoor café, working as domestic help, making fireworks, selling drugs. In Abidjan, in Côte d'Ivoire, 15-year-old Jean-Pierre Godia, who cannot read or write, spends about 6 hours every day trying to sell 10-roll packets of toilet paper to motorists at a busy intersection. He buys the packets for about $1.20 and sells them for $2. Some days he doesn't sell any. In the same city, 7-year-old Giulio guides cars into parking spaces outside a chic pastry shop. He has been doing this since he was 5, to help his mother and four siblings, who beg on a nearby corner.

Megacities are very large cities characterized by both primacy and a high degree of centrality within their national economy. Their most important common denominator is their sheer size—most of them number 10 million or more in population. This, together with their functional centrality, means that in many ways they have more in common with one another than with the smaller metropolitan areas and cities within their own countries. Megacities include Bangkok, (**Figure 10.22**), Dhaka, Jakarta, Lagos, Manila, Mumbai (**Figure 10.23**), New Delhi, São Paulo, and Teheran. Each has more inhabitants than 100 of the member countries of the United Nations. While most of them do not function as world cities, they do serve important intermediate roles between the upper tiers of the system of world cities and the provincial towns and villages of large regions of the world. They not only link local and provincial economies with the global economy but also provide a point of contact between the traditional and the modern, and between formal and informal economic sectors (see Box 10.3, "Window on the World: The Pearl River Delta: An Extended Metropolis").

The **informal sector** of an economy involves a wide variety of economic activities whose common feature is that they take place beyond official record and are not subject to formalized systems of regulation or remuneration. As we shall see in the next chapter, the slums and squatter settlements in megacities are often associated with severe problems of social disorganization and environmental degradation. Nevertheless, many neighborhoods are able to

[4] *The State of the World's Children 2011.* New York: United Nations International Children's Fund (UNICEF), 2011.

FIGURE 10.21 Child labor A 13-year-old girl makes fireworks at a factory in Sivakasi, India. One in twelve of the world's children (180 million young people aged 5 to 17) are involved in the worst forms of child labor—hazardous work, slavery, forced labor, the military, commercial sexual exploitation, and illicit activities. Of these children, 97 percent are in peripheral and semi-peripheral countries. Globally, an estimated 114 million children of primary school age are not enrolled in school, depriving one in five children of an education. They are exposed to exploitation and abuse and miss out on developing the knowledge and employable skills that could lift them and their own children out of the poverty cycle.

develop self-help networks and organizations that form the basis of community amid dauntingly poor and crowded cities.

APPLY YOUR KNOWLEDGE Conduct an Internet search to find two megacities. List two ways a megacity differs from a world city. ∎

The Core: Mature Metropolises

The high levels of urbanization and relatively slow rates of urban growth within the world's core regions are reflected in relatively stable urban systems. There is constant change, nevertheless,

The Pearl River Delta: An Extended Metropolis

The Pearl River Delta (**Figure 10.E**) is one of the fastest-growing urban regions in the world. Anchored by the major metropolitan centers of Guangzhou, Hong Kong, Macau, Shenzhen, and Zhuhai, it is an extended metropolitan region of nearly 50 million people. It is one of three extended metropolitan regions—Beijing-Tianjin and Shanghai are the others—that have been fostered by the Chinese government to be engines of capitalist growth since liberal economic reforms were introduced in the late 1970s.

Hong Kong (**Figure 10.F**) was a British colony until 1997. It is now a metropolis of 7.4 million with a thriving industrial and commercial base that is recognized as a capitalist economic dynamo by the Chinese government, which has created a Special Administrative District for the metropolis. As a result, Hong Kong's citizens have retained their British-based legal system and its guaranteed rights of property ownership and democracy. Hong Kong is the world's largest container port, the third-largest center for foreign exchange trade, the seventh-largest stock market, and the tenth-largest trading economy.

Hong Kong's financial success encouraged the Chinese government to establish two of its first Special Economic Zones (SEZs) in nearby Shenzhen and Zhuhai. Designed to attract foreign capital, technology, and management practices, these SEZs were established as export-processing zones that offered cheap labor and land, along with tax breaks, to transnational corporations. Investors from Hong Kong and Taiwan responded quickly and enthusiastically. By 1993, more than 15,000 manufacturers from Hong Kong alone had set up businesses in Guangdong Province,

FIGURE 10.E Pearl River Delta One of the fastest-growing regions of the world, the Pearl River Delta is an extended metropolitan region of more than 50 million people.

FIGURE 10.F City of Hong Kong
Although most of its manufacturing has been transferred to neighboring Guangdong Province, where wages are much lower, thousands of companies are located in Hong Kong simply for the purpose of doing business with China. As a result, Hong Kong remains a major world city—a major financial hub with a thriving commercial sector and a population of 7.4 million.

and a similar number had established subcontracting relationships, contracting out assembly-line work to Chinese companies in the Pearl River Delta. Meanwhile, the Chinese government designated the entire delta region an Open Economic Region, where local governments, individual enterprises, and farm households enjoy a high degree of autonomy in economic decision making.

The relaxation of state control over the regional economy allowed the region's dense and growing rural population to migrate to urban areas in search of assembly-line jobs or to stay in rural areas and diversify agricultural production from paddy-rice cultivation to more profitable activities such as market-farming activities, livestock husbandry, and fishery. Economic freedom also facilitated rural industrialization—mostly low-tech, small-scale, labor-intensive, and widely scattered across the countryside. The triangular area between Guangzhou, Hong Kong, and Macau has quickly emerged as an especially important zone because of its relatively cheap land and labor and because of significant levels of investment by regional and local governments in the transport and communications infrastructure. The result is a distinctive "extended metropolis" in which numerous small towns play an increasingly important role in fostering the process of urbanization, with an intense mixture of agricultural and nonagricultural activities and an intimate interaction between urban and rural areas.

The metropolitan cores of the region, aiming to increase their competitiveness and prominence in the globalizing world economy, have invested heavily in infrastructure improvements. The Guangzhou municipal government, for example, invested more than $10 billion between 1998 and 2004 in infrastructure construction—including a metro system and an elevated railway network to link the city's new international airport, railway stations, and port. Throughout the region, enormous investments have been made in showpiece infrastructure projects geared to the needs of local and international capital. These include major airports, high-speed tolled highways, satellite ground stations, port installations, metro and light-rail networks, and new water-management systems. In turn, these projects have attracted business and technology parks, financial centers, and resort complexes in a loose-knit sprawl of urban development.

Today, the Pearl River Delta provides a thriving export-processing platform that has driven double-digit annual economic growth for much of the past two decades. The region's GDP grew from just over US$8 billion in 1980 to nearly US$270 billion in 2010. During that period, the average real rate of GDP growth in the Pearl River Delta Economic Zone exceeded 16 percent, well above the People's Republic of China national figure of 9.8 percent. By 2010 and with only 3.5 percent of the country's population, the region was contributing 10 percent of the country's GDP and 29 percent of its total trade.

Guangzhou is a megacity with a 2010 population of around 10 million (**Figure 10.G**). Shenzhen has grown from a population of just 19,000 in 1975 to 8.1 million in 2010, with an additional 2 million in the surrounding municipalities. The southern border of the Shenzhen Special Economic Zone adjoins Hong Kong, but the northern border is walled off from the rest of China by an electrified fence to prevent smuggling and to keep back the mass of people trying to migrate illegally into Shenzhen and Hong Kong.

FIGURE 10.G Guangzhou, China An ancient Chinese city that became known as Canton by European traders, the city has grown rapidly in recent decades, its modern architecture almost completely replacing the old city.

in patterns and processes of urbanization as the metropolises, cities, and towns adjust to the opportunities of new technologies and new industries and to the constraints of obsolescent urban infrastructure and land-use conflicts. New rounds of urbanization are initiated in the places most suited to new technologies and new industries. Those places least suited are likely to suffer spirals of deindustrialization and urban decline.

Deindustrialization and Agglomeration Diseconomies

Deindustrialization involves a decline in industrial employment in core regions as firms scale back their activities in response to lower levels of profitability (see Chapter 7). Such adversity has particularly affected cities like Pittsburgh and Cleveland (United States), Sheffield and Liverpool (United Kingdom), Lille (France), and Liège (Belgium)—places where heavy manufacturing constituted a key economic sector. Cities like these have suffered substantial reductions in employment since the 1970s and 1980s when better and more flexible transport and communications networks allowed many industries to choose from a broader range of potential locations.

In many instances, deindustrialization has been intensified by the dampening effects of *agglomeration diseconomies* (Chapter 7) on the growth of larger metropolitan areas. Agglomeration diseconomies, the negative effects of urban size and density, include noise, air pollution, increased crime, high commuting costs, inflated land and housing prices, traffic congestion, and crowded port and railroad facilities. They also include higher taxes levied to rebuild decaying infrastructure and to support services and amenities previously considered unnecessary—traffic police, city planners, and homeless shelters, for example.

The result of deindustrialization has been a *decentralization* of jobs and people from larger to smaller cities within the urban systems of core countries, and from metropolitan cores to suburban and ex-urban fringes. In some cases, routine production

FIGURE 10.22 Mexico City Every year another half million or more people are added to the city. In 2010, the population of the agglomeration had reached almost 20 million.

FIGURE 10.23 Mumbai With more than 20 million people, Mumbai is India's most populous city.

activities relocated to smaller metropolitan areas or to rural areas with lower labor costs and more hospitable business climates. In other cases, these activities moved overseas—as part of the new international division of labor (see Chapter 2)—or were eliminated entirely.

Counterurbanization and Reurbanization

The combination of deindustrialization in core manufacturing regions, agglomeration diseconomies in major metropolitan areas, and the improved accessibility of smaller towns and rural areas can give rise to the phenomenon of counterurbanization. **Counterurbanization** occurs when cities experience a net loss of population to smaller towns and rural areas. This process results in the deconcentration of population within an urban system. This is what happened in the United States, Britain, Japan, and many other developed countries in the 1970s and 1980s. Metropolitan growth slowed dramatically, while the growth rates of small and medium-size towns and of some rural areas increased. In these countries, counties that for decades had recorded stable populations grew by 15 or 20 percent. Some of the strongest gains were registered in counties that were within commuting range of metropolitan areas, but some remote counties also registered big population increases.

Counterurbanization was a major reversal of long-standing trends, but it seems to have been a temporary adjustment rather than a permanent change. The globalization of the economy and the growth of postindustrial activities in revamped and expanded metropolitan settings have restored the trend toward the concentration of population within urban systems. Most of the cities that were declining fast in the 1970s and 1980s are now either recovering (New York, London) or bottoming out (Paris, Chicago), while most of those that were growing only slowly (Tokyo, Barcelona) are now expanding more quickly. This trend of **reurbanization** involves the growth of population in metropolitan central cores following a period of absolute or relative decline in population.

In the United States, two principal streams are driving reurbanization. One consists of new migrants, mainly from Latin America and Asia, who have moved into the central districts of major metropolitan areas, especially New York and Los Angeles, and some medium-size metropolitan areas of the West, Southwest, and South. A second, very different migration stream to the central districts of metropolitan areas consists of "baby boomers" (see Chapter 3) electing to pursue urban rather than suburban lifestyles. This stream has been most pronounced in the central districts of metropolitan areas with expanding high-tech and defense-oriented economies, such as Boston and Seattle.

APPLY YOUR KNOWLEDGE Identify a town or city near you that has experienced counterurbanization or reurbanization over the last 10 years. Provide two reasons for this occurrence. ■

Future Geographies

The United Nations Human Settlements Program (UN-Habitat) estimates that by 2030, more than 65 percent of the world's population will be living in urban areas, and there will be around 575 cities with a population of a million or more, including about 50 cities of 5 million or more. The number of megacities—those with a population of 10 million or more—will increase, and the populations of most of them will swell significantly (**Table 10.2**). The single most important aspect of future patterns of world urbanization is the striking difference in trends and projections between the core regions and the semiperipheral and peripheral regions. In 1950, two-thirds of the world's urban population was concentrated in the more developed countries of the core economies. By 2030, around 80 percent of all city dwellers will be in peripheral and semiperipheral countries. In 1950, 21 of the world's largest 30 metropolitan areas were located in core countries—11 of them in Europe and 6 in North America. By 1980, the situation was completely reversed, with 19 of the largest 30 located in peripheral and semiperipheral regions. By 2030, all but two or three of the 30 largest metropolitan areas are expected to be located in peripheral and semiperipheral regions.

Asia provides some of the most dramatic examples of this trend. From a region of villages, Asia is fast becoming a region of cities and towns. Between 1950 and 2005, for example, Asia's urban population rose more than 10-fold, to over 1.5 billion people. By 2020, about two-thirds of its population will be living in urban areas. Nowhere is the trend toward rapid urbanization more pronounced than in China, where for decades the communist government imposed strict controls on where people were allowed to live, fearing the transformative and liberating effects of cities. By tying people's jobs, school admission, and even the right to buy food to the places where people were registered to live, the government made it almost impossible for rural residents to migrate to towns or cities. As a result, more than 70 percent of China's 1 billion people still lived in the countryside in 1985. Now, however, China is rapidly making up for lost time. The Chinese government, having decided that towns and cities can be engines of economic growth within a communist system, has not only relaxed residency laws but also drawn up plans to establish over 430 new cities.

Whatever the current level of urbanization in peripheral countries, almost all are forecast to experience high rates of urbanization, with growth forecasts of unprecedented speed and unmatched size. Karachi, Pakistan, a metropolis of 1.03 million in 1950, reached 8.5 million in 1995 and is expected to reach 16.2 million by 2015. Likewise, Cairo, Egypt, grew from 2.44 million to 9.7 million between 1950 and 1995 and is expected to reach 13.1 million by 2015. Mumbai (India; formerly Bombay), Delhi (India), Mexico City (Mexico), Dhaka (Bangladesh), Jakarta (Indonesia),

TABLE 10.2 Growth and More Urban Growth

The World's Megacities, 2007 and 2025

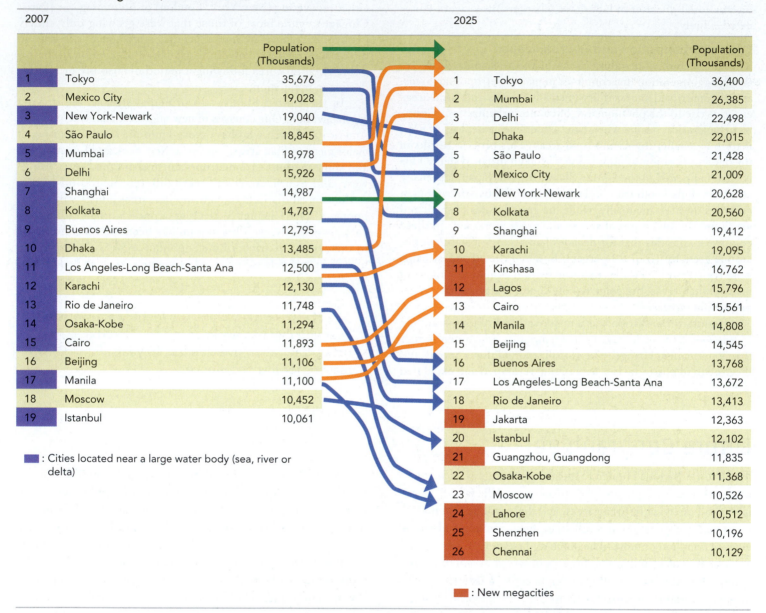

2007		Population (Thousands)		2025		Population (Thousands)
1	Tokyo	35,676		1	Tokyo	36,400
2	Mexico City	19,028		2	Mumbai	26,385
3	New York-Newark	19,040		3	Delhi	22,498
4	São Paulo	18,845		4	Dhaka	22,015
5	Mumbai	18,978		5	São Paulo	21,428
6	Delhi	15,926		6	Mexico City	21,009
7	Shanghai	14,987		7	New York-Newark	20,628
8	Kolkata	14,787		8	Kolkata	20,560
9	Buenos Aires	12,795		9	Shanghai	19,412
10	Dhaka	13,485		10	Karachi	19,095
11	Los Angeles-Long Beach-Santa Ana	12,500		11	Kinshasa	16,762
12	Karachi	12,130		12	Lagos	15,796
13	Rio de Janeiro	11,748		13	Cairo	15,561
14	Osaka-Kobe	11,294		14	Manila	14,808
15	Cairo	11,893		15	Beijing	14,545
16	Beijing	11,106		16	Buenos Aires	13,768
17	Manila	11,100		17	Los Angeles-Long Beach-Santa Ana	13,672
18	Moscow	10,452		18	Rio de Janeiro	13,413
19	Istanbul	10,061		19	Jakarta	12,363
				20	Istanbul	12,102
				21	Guangzhou, Guangdong	11,835
				22	Osaka-Kobe	11,368
				23	Moscow	10,526
				24	Lahore	10,512
				25	Shenzhen	10,196
				26	Chennai	10,129

■ : Cities located near a large water body (sea, river or delta)

■ : New megacities

Source: UN Habitat: State of the World's Cities 2008/2009. Harmonious Cities. Copyright © United Nations Human Settlement Programme, 2008. *Note:* Population figures are for urban agglomeration, not city proper. Megacities are cities with populations of more than 10 million.

Lagos (Nigeria), São Paulo (Brazil), and Shanghai (China) are all projected to have populations in excess of 17 million by 2015. For the most part, this growth will be a consequence of the onset of the demographic transition (see Chapter 3), which has produced fast-growing rural populations in regions that face increasing problems with agricultural development (see Chapter 8). As a response, many people in these regions will continue to migrate to urban areas seeking a better life.

CONCLUSION

Urbanization is one of the most important geographic phenomena. Cities can be seedbeds of economic development and cultural innovation. Cities and groups of cities also organize space—not just the territory immediately around them, but in some cases national and even international space. The causes and consequences of urbanization, however, are very different in different parts of the world. The urban experience of the world's peripheral regions stands in sharp contrast to that of the developed core regions, for example. This contrast is a reflection of some of the demographic, economic, and political factors that we have explored in previous chapters.

Much of the developed world has become almost completely urbanized, with highly organized systems of cities. Today, levels of urbanization are high throughout the world's core countries, while rates of urbanization are relatively low. At the top of the urban hierarchies of the world's core regions are world cities such as London, New York, Tokyo, Paris, and Zürich, which have become control centers for the flows of information, cultural products, and finance that collectively sustain the economic and cultural globalization of the world. In doing so, they help to consolidate the hegemony of the world's core regions.

Few of the metropolises of the periphery, on the other hand, are world cities occupying key roles in the organization of global economics and culture. Rather, they operate as connecting links between provincial towns and villages and the world economy. They have innumerable economic, social, and cultural linkages to their provinces on one side and to major world cities on the other. Almost all peripheral countries, meanwhile, are experiencing high rates of urbanization, with forecasted growth of unprecedented speed and unmatched size. In many peripheral and semiperipheral regions, current rates of urbanization have given rise to unintended metropolises and fears of "uncontrollable urbanization," with urban "danger zones" where "work" means anything that contributes to survival. The result, as we shall see in Chapter 11, is that these unintended metropolises are quite different from the cities of the core as places in which to live and work.

Learning Outcomes Revisited

- **Describe how the earliest towns and cities developed independently in the various hearth areas of the first agricultural revolution.**

  The very first region of independent urbanism, in the Middle East, produced successive generations of urbanized world-empires, including those of Greece, Rome, and Byzantium. By 2500 B.C.E., cities had appeared in the Indus Valley, and by 1800 B.C.E., urban areas were established in northern China. Other areas of independent urbanism include Mesoamerica (from around 100 B.C.E.) and Andean America (from around 800 C.E.). The classical archaeological interpretation emphasizes the availability of an agricultural surplus large enough to allow the emergence of specialized, nonagricultural workers. Some urbanization, however, may have resulted from the pressure of population growth.

- **Scrutinize how the expansion of trade around the world, associated with colonialism and imperialism, established numerous gateway cities.**

  European powers founded or developed literally thousands of towns as they extended their trading networks and established their colonies. The great majority of the towns were ports that served as control centers commanding entrance to, and exit from, their particular country or region. The great majority of them were ports. Protected by fortifications and European naval power, they began as trading posts and colonial administrative centers. Before long, they developed manufacturing of their own to supply the pioneers' needs, along with more extensive commercial and financial services. As colonies developed and trading networks expanded, some of these ports grew rapidly, acting as gateways for colonial expansion into continental interiors.

- **Asssess why and how the Industrial Revolution generated new kinds of cities—and many more of them.**

  Industrial economies required the large pools of labor; the transportation networks; the physical infrastructure of factories, warehouses, stores, and offices; and the consumer markets provided by cities. As industrialization spread throughout Europe in the first half of the nineteenth century and then to other parts of the world, urbanization increased at a faster pace. The higher wages and greater variety of opportunities in urban labor markets attracted migrants from surrounding areas. The countryside began to empty. In Europe, the *demographic transition* caused a rapid growth in population as death rates dropped dramatically. This growth in population provided a massive increase in the labor supply, further boosting the rate of urbanization, not only within Europe itself but also in Australia, Canada, New Zealand, South Africa, and the United States as emigration spread industrialization and urbanization to the frontiers of the world-system.

- **Interpret how a small number of "world cities," most of them located within the core regions of the world-system, have come to occupy key roles in the organization of global economics and culture.**

  At the top of a global urban system, these cities experience growth largely as a result of their role as key nodes in the world economy. World cities are the control centers for the flows of information, cultural products, and finance that collectively sustain the economic and cultural globalization of the world. The globalization of the economy has resulted in a global urban system in which the key roles of world cities are concerned with transnational corporate organization, international banking and finance, supranational government, and the work of international agencies. World cities also provide an interface between the global and the local. They contain the economic, cultural, and institutional apparatus that channels national and provincial resources into the global economy and transmits the impulses of globalization back to national and provincial centers.

■ Compare and contrast the differences in trends and projections between the world's core regions and peripheral regions.

In 1950, two-thirds of the world's urban population was concentrated in the more developed countries of the core economies. Since then, the world's urban population has increased threefold, the bulk of the growth having taken place in the less developed countries of the periphery. The world's core regions are highly urbanized, with slow rates of urban growth. Peripheral regions, although less highly urbanized, have been experiencing exceptionally high rates of urban growth, partly due to rural-urban migration and partly due to natural population increase. Much of the resulting urbanization has taken the form of "megacities" of 10 million people or more. Overurbanization has occurred where cities have grown more rapidly than they have been able to generate jobs or housing.

KEY TERMS

central place *(p. 367)*	gateway city *(p. 360)*	rank-size rule *(p. 367)*	urban ecology *(p. 356)*
central place theory *(p. 367)*	informal sector *(p. 375)*	reurbanization *(p. 379)*	urban form *(p. 356)*
centrality *(p. 369)*	megacity *(p. 375)*	shock city *(p. 364)*	urban system *(p. 356)*
colonial city *(p. 366)*	overurbanization *(p. 374)*	squatter settlements	urbanism *(p. 356)*
counterurbanization *(p. 379)*	primacy *(p. 369)*	*(p. 374)*	

REVIEW AND DISCUSSION

1. Go to the library or conduct an Internet search for pictures of what your campus looked like 5, 10, and 30 years ago. How have the urban forms changed? Note five examples of how the campus has changed between then and now. How have the physical structures changed? How is the land used differently? Is the university organized differently? If so, why do you think this is the case? If not, explain why things are the same.

2. Consider the term *world cities*. Choose two cities that you consider to be world cities. Now read over the functional characteristics of world cities that are bulleted on page 359. Pick two bullet points and describe the characteristics in terms of the cities you have chosen. For example, the second bullet point notes how world cites have "clusters of specialized, advanced business services." Do an Internet search and identify the clusters of specialized business services in the cities you have chosen.

3. Counterurbanization refers to the process of cities losing population to smaller towns or rural areas. In light of the recent economic recession, conduct a library search to determine if any American cities or towns have experienced recent counterurbanization. List three reasons for the counterurbanization. Consider if some cities are experiencing reurbanization. List three reasons for this type of population movement.

UNPLUGGED

1. Figures 10.1 and 10.2 show that the United States, like most core countries, is already highly urbanized and has a relatively low rate of urbanization Nevertheless, some U.S. cities have been growing much faster than others. Which have been the fastest-growing U.S cities in recent times, and what reasons can you suggest for their relatively rapid growth? (*Hint:* The U.S. Bureau of the Census publishes data on population change by urban area, as does the Population Division of the UN Department of Economic and Social Affairs, in a volume entitled *World Urbanization Prospects.*)

2. Go to the U.S. Bureau of the Census Web page and find the population of the town or city you know best. Do the same for every census year, going from 2010 back to 2000, 1990, 1980, and so on, all the way back to 1860. Then plot the populations on a simple graph. List four explanations for the pattern that the graph reveals. Now draw a larger version of the same graph, annotating it to show the landmark events that might have influenced the city's growth (or decline).

3. The following cities all have populations in excess of 2 million. How many of them could you locate on a world map? Their size reflects a certain degree of importance, at least within their regional economy. What can you find out about each? Compile for each a 50-word description that explains its chief industries and a little of its history.

Poona	Ibadan	Recife
Bangalore	Turin	Ankara

4. Figure 10.9 features two colonial gateway cities on the eastern seaboard of the United States—Boston and New York. Two other colonial gateway cities were Charleston, South Carolina, and Savannah, Georgia. What can you find out about the commodities and manufactures that Charleston and Savannah imported and exported in pre-Revolutionary times? Where did their exports go? List three reasons why they went to those places. On the import side, where did imports come from? List three reasons why Charleston and Savannah needed to import these particular goods. Which geographic concepts do you consider to be useful in explaining these facts?

Log in to www.masteringgeography.com for MapMaster™ interactive maps, geography videos, RSS feeds, flash cards, Web links, an eText version of *Human Geography: Places and Regions in Global Context,* and self-study quizzes to enhance your study of urbanization.

MapMaster™ presents 13 Place Name and 13 Layered Thematic interactive maps to help students practice and master their geographic literacy, spatial reasoning, and critical thinking skills.

11 CITY SPACES: URBAN STRUCTURE

China seems to have caught onto sprawl. Yet, what is perhaps most distinctive about many of China's affluent new suburbs is not the sprawl itself, but the way it looks. Many suburbs are directly modeled on the tract homes that have defined American suburban growth in the past 30 years. Most of them carry few, if any, Asian influences in their design, layout, and ornamentation. The first suburbs were built to house expatriates, such as diplomats or executives in the local offices of multinational companies; now, an increasing number of Chinese with money are eager to get out of the crowded and dirty milieu of the city.

Want to live in Australia? Beijing residents can buy a home at Sydney Coast, a subdivision that offers its residents a "seven-day Australian-style villa life." Beijingers who prefer California have been able to settle in the Yosemite subdivision. Now there is also the option of moving to Napa Valley, a new development about 50 kilometers (31 miles) outside Beijing. Napa Valley attempts to replicate a Californian Mediterranean lifestyle of laid-back, al fresco leisure. "Rustic stone is widely used, with rich stucco colors, along with wood shutters and wrought-iron accents, to create an intimate scale and village-like feel," according to Napa Valley's architects and planners, who are based in Palm Springs and Newport Beach, California.

Thames Town, an English-styled development, near Shanghai, China.

For those who hanker for seventeenth-century France, there is Chateau Regalia, located on Beijing's northern outskirts. Here, potential buyers can choose from several different models of homes: the Duke I, the Duke II, the Marquis, the Earl, and the Viscount. In both form and decoration, Chateau Regalia's homes are an eccentric amalgam of French Baroque and neo-classical architecture.

The craze has also caught on in Shanghai, where there are more tract homes built in foreign styles. Thames Town, just outside Shanghai, is one of seven satellite towns built by the municipal government to house 500,000 people (the other six are themed with architectural styles adopted from Italy, Spain, Canada, Sweden, Holland, and Germany). Thames Town includes a gothic church, village green, and mock-Tudor pub selling real ale. Built in a little over 3 years, the suburb encompasses five centuries of British architecture. At its center are half-timbered Tudor-style buildings. By the waterfront, Victorian redbrick warehouses have been "converted" into shops. The residential area includes gabled Edwardian houses bordered by privet hedges, manicured lawns, and leafy roads.[1] ■

[1] Based on D. Elsea, "China's Chichi Suburbs: American-style Sprawl All the Rage in Beijing," *San Francisco Chronicle*, April 24, 2005, p. 23.

URBAN LAND USE AND SPATIAL ORGANIZATION

The internal organization of cities reflects the way cities function, both to bring people and activities together and to sort them out into functional subareas and neighborhoods. While there are many complex processes at work in cities, some broad tendencies go a long way toward shaping patterns of urban land use and spatial organization. In this first section of the chapter, we consider two of them: people's need for accessibility and their sense of territoriality.

Accessibility and Land Use

Most urban land users want to maximize the *utility* they derive from a particular location. The utility of a specific place or location refers to its usefulness to particular persons or groups. The price people are prepared to pay for different locations—the bid-rent—is a reflection of this utility. In general, utility is a function of *accessibility*. Commercial land users want to be accessible to one another, to markets, and to workers; private residents want to be accessible to jobs, amenities, and friends; public institutions want to be accessible to clients. In an idealized city built on an isotropic surface, the point of maximum accessibility is the city center. An **isotropic surface** is a hypothetical, uniform plane: flat, and with no variations in its physical attributes. Under these conditions, accessibility decreases steadily with distance from the city center. Likewise, utility decreases, *but at different rates for different land users.* This idealized pattern of decreasing accessibility and utility is reflected in concentric zones of different mixes of land use (**Figure 11.1**).

One counterintuitive implication of this model is that the poorest households will end up occupying the periphery of the city. Although this is true in some parts of the world, we know that in the core countries this is often not the case. In fact, the farthest suburbs are generally the territory of wealthier households, while the poor usually occupy more accessible locations nearer city centers. To reflect reality, some of the assumptions of this model must be modified. In this case, we must assume that wealthier households

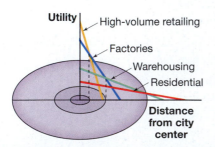

FIGURE 11.1 Accessibility, bid-rent, and urban structure Competition for accessible sites near the city center is an important determinant of land-use patterns. Different users are prepared to pay different amounts—the bid-rents—for locations at various distances from the city center. The result is often a concentric pattern of land uses. (*Source:* Reprinted with permission of Prentice Hall, from P. L. Knox, *Urbanization* © 1994, p. 99.)

trade off the convenience of accessibility for the greater utility of being able to consume larger amounts of (relatively cheap) suburban space. Poorer households, unable to afford the recurrent costs of transportation, trade off living space for accessibility to jobs and end up in high-density areas, at expensive locations, near their low-wage jobs. Because of the presumed trade-off between accessibility and living space, this modified urban land-use model is often referred to as a **trade-off model**.

> **APPLY YOUR KNOWLEDGE** Identify three different trade-offs that you have made in terms of your own living space. List three ways these trade-offs have affected your access to school, transportation, and job. ■

Territoriality, Congregation, and Segregation

In cities, as at other geographic scales, territoriality provides a means of establishing and preserving group membership and identity. The first step in forming group identity is to define "others" in an exclusionary and stereotypical way. **Congregation**—the territorial and residential clustering of specific groups or subgroups of people—enables group identity to be consolidated in relation to people and places outside the group. Congregation is a place-making activity and an important basis for urban structure and land use. It is particularly important in situations where there are one or more distinctive minority groups. **Minority groups** are population subgroups that are seen—or that see themselves—as somehow different from the general population. The defining characteristics of minority groups can be based on race, language, religion, nationality, caste, sexual orientation, or lifestyle.

For minority groups there are several specific advantages of congregation:

- *Congregation provides a means of cultural preservation. It allows religious and cultural practices to be maintained and strengthens group identity through daily involvement in particular routines and ways of life.* Particularly important in this regard is the way that clustering fosters within-group marriage and kinship networks.

- *Congregation helps minimize conflict and provides defense against "outsiders."*

- *Congregation provides a place where mutual support can be established through minority institutions, businesses, social networks, and welfare organizations.*

- *Congregation helps establish a power base in relation to the host society.* This power base can be democratic, organized through local elections, or it can take the form of a territorial heartland for insurrectionary groups.

Congregation is not always voluntary, of course. Host populations are also impelled by territoriality, and they may respond to social and cultural differences by *discrimination* against minority groups. Discrimination can also have a strong territorial basis. It can restrict the territory of minority groups and resist their assimilation into the host society. This resistance can take a variety of forms. Social hostility and the voicing of "keep out" attitudes are probably the most widespread, although other forms of discrimination can have pronounced spatial effects. These forms include exclusion and prejudice in local labor markets, the manipulation of private land and housing markets, the steering of capital investment away from minority areas, and the institutionalization of discrimination through the practices and spatial policies of public agencies.

The combined result of congregation and discrimination is **segregation**, the spatial separation of specific subgroups within a wider population (see Box 11.1, "Visualizing Geography: Racial Segregation"). Segregation varies a great deal in both intensity and form, depending on the relative degree and combination of congregation and discrimination. Geographers have identified three principal situations in terms of the spatial form of segregation.

- In *enclaves*, tendencies toward congregation and discrimination are long-standing but dominated by internal cohesion and identity. The Jewish districts of many of today's cities in Europe and the eastern United States are enclaves.

- *Ghettos* are also long-standing but are more the product of discrimination than congregation. Examples are the segregation of African Americans and Hispanics in American cities.

- *Colonies* may result from relatively weak and short-lasting congregation, discrimination, or both. The persistence of colonies over time depends on the continuing arrival of new minority-group members. For example, U.S. cities in the early twentieth century contained distinctive colonies of German, Scandinavian, Irish, and Italian immigrants. These have all but disappeared. Today there are colonies of Greek and Yugoslav neighborhoods in Australian cities, and colonies of Koreans in Japanese cities.

Economic competition for space and accessibility, along with the tendency toward social and ethnic discrimination, congregation, and segregation, can be traced in many of the world's cities. Patterns are apparent particularly in affluent core regions where economic, social, and cultural forces are broadly similar. Nevertheless, urban structure varies considerably because of the influence of history, culture, and the different roles that cities have played within the world-system. In this chapter, we examine the typical characteristics of North American, European, and Islamic cities, and the unintended metropolises of the periphery.

APPLY YOUR KNOWLEDGE Identify any present or historical enclaves, ghettos, or colonies in your region. List two ways these spatial forms may have resulted from discrimination and two ways residents potentially benefit from congregation in these communities. ■

SPATIAL PATTERNS AND PROCESSES IN NORTH AMERICAN CITIES

In North American cities the very centers have historically been the principal hubs of shops and offices. They have also featured some of the major institutional land uses such as the city hall, libraries, and museums. A city's center, known as the **central business district**, or **CBD**, is the nucleus of commercial land uses. It traditionally contains the densest concentration of shops, offices, and warehouses and the tallest nonresidential buildings (**Figure 11.2**). It usually developed at the nodal point of transportation routes so that it also contains bus stations, railway terminals, and hotels. The CBD usually is surrounded by a zone of mixed land uses: warehouses, small factories and workshops, specialized stores, apartment buildings, public housing projects, and older residential neighborhoods. This zone is often referred to as the **zone in transition** because of its mixture of growth, change, and decline.

Beyond this zone are residential neighborhoods, suburbs of various ages and different social and ethnic composition. Just as different categories of land use attract and repel one another, so do different social and ethnic groups. In North America, where urban population growth has been fueled by streams of migrants and immigrants with very different backgrounds, sociologists have developed an ecological perspective to describe neighborhoods as being structured by the "invasion" of successive waves of migrants and immigrants.

FIGURE 11.2 The central city This photograph of Chicago shows the concentration of high-rise office buildings on the skyline that is typical of central business districts (CBDs) in North American metropolitan areas.

Usually, ethnic neighborhoods on maps are sharply bounded blocks of color at the scale of census tracts that contain hundreds, sometimes thousands, of people. On these maps, one dot equals 25 people, and the dots are color-coded based on race: White is pink, black is blue, Hispanic is orange, and Asian is green. The results, based on U.S. Census data from 2000, show that just as every city is different, every city is segregated (or integrated) in different ways.

Detroit, for example, is highly segregated. Its Eight Mile beltway serves as a precise boundary between the city's black and white populations. In Washington, D.C., there is a sharp east-west divide between white and black. In New York, there are areas of extreme racial concentration, but the sheer number of people in those areas means that boundary regions become intensely rich areas of cross-cultural ferment. Long Beach, California, meanwhile, is almost the opposite: Because no part of the city is particularly dense, there are a number of blended neighborhoods, some of which are more extensive than the racially homogeneous ones.

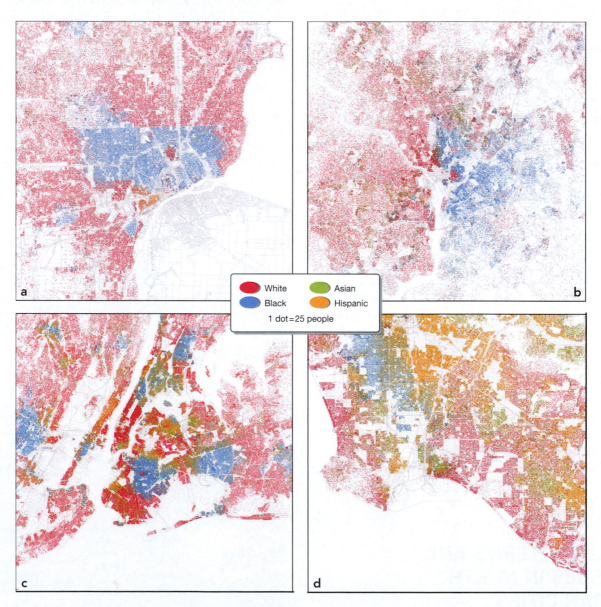

FIGURE 11.A (a) Detroit **(b)** Washington, D.C. **(c)** New York **(d)** Long Beach

Based on infographics by Eric Fischer. See http://www.flickr.com/photos/walkingsf/
sets/72157624812674967/with/4981417821/

Immigration and Neighborhood Change

When immigrants first arrive in a city looking for work and a place to live, they have little choice but to cluster in the cheapest accommodations, found in the zone in transition around the CBD and typically resulting in enclaves, ghettos, or colonies. Chicago in the 1920s and 1930s provides a classic example of this. Immigrants from Scandinavia, Germany, Italy, Ireland, Poland, Bohemia (now part of the Czech Republic), and Lithuania established themselves in Chicago's low-rent areas, the only places they could afford. By congregating in these areas, immigrants accomplished several things—they were able to establish a sense of security; continue speaking their native language; have familiar churches or synagogues, restaurants, bakeries, butcher shops, and taverns; and support their own community newspapers and clubs. African American migrants from the South joined in the city's zone in transition. They also established their own neighborhoods and communities. In Chicago, as in other U.S. cities of the period, the various ethnic groups formed a patchwork or mosaic of communities encircling the CBD.

These ethnic communities lasted from one to three generations, after which they started to break up. Many of the younger, city-born individuals did not feel the need for the security and familiarity of ethnic neighborhoods. Gradually, increasing numbers established themselves in better jobs and moved out into newer, better housing. As the original immigrants and their families left, their place in the transitional zone was taken by a new wave of migrants and immigrants. In this way, Chicago became structured into a series of *concentric zones* of neighborhoods of different ethnicity and social status (**Figure 11.3**).

Throughout this process of invasion and succession, people of the same background tend to stick together—partly because of the advantages of residential clustering and partly because of discrimination. **Invasion and succession** is a process of neighborhood change whereby one social or ethnic group succeeds another in a residential area. The displaced group, in turn, invades other areas, creating over time a rippling process of change throughout the city. The result is that within each concentric zone there exists a mosaic of distinctive neighborhoods. Classic examples include the Chinatowns, Little Italys, Koreatowns, and African American ghettos of big North American cities (**Figure 11.4**). Such neighborhoods can be thought of as *ecological niches* within the overall metropolis—settings where a particular mix of people have come to dominate a particular territory and a particular physical environment, or habitat.

Industry, Class, and Spatial Organization

Households also sort themselves within cities according to differences in class status and affluence and ability to avoid living in or near to industrial districts. The classic study of this tendency resulted in a generalized model of urban land use (**Figure 11.5**). The author of the study, Homer Hoyt, argued that corridors of industry and warehousing will always tend to be surrounded on both sides by sectors of working-class housing, while middle-income housing will always tend to act as a buffer between the

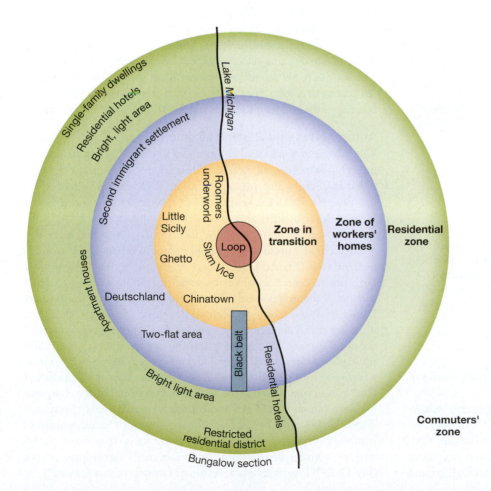

FIGURE 11.3 The ecological model of urban land use: Chicago in the 1920s Competition between members of different migrant and immigrant groups for residential space in the city often results in distinctive neighborhoods that have their own social "ecology." The classic example is Chicago of the 1920s, which had developed a series of concentric zones of distinctive neighborhoods as successive waves of immigrants established themselves. Over time, most immigrant groups made their way from low-rent, inner-city districts surrounding the CBD (known in Chicago as the Loop) to more attractive and expensive districts farther out. (*Source:* After R. E. Park, E. W. Burgess, and R. D. McKenzie, *The City.* Chicago: University of Chicago Press, 1925, p. 53.)

FIGURE 11.4 Chinatown In most larger cities a patchwork of distinctive neighborhoods results from processes of congregation and segregation. Most distinctive are neighborhoods of ethnic minorities, such as the Chinatowns, Little Italys, and Little Koreas of major American cities. Chinatown in New York City is shown here.

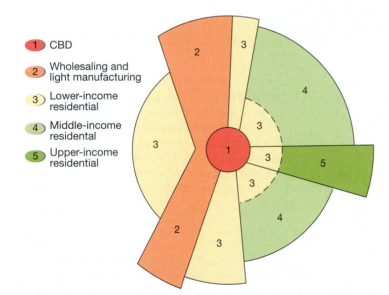

1 CBD

2 Wholesaling and light manufacturing

3 Lower-income residential

4 Middle-income residental

5 Upper-income residential

FIGURE 11.5 Hoyt's model of urban structure Based on his observations of American cities in the first half of the twentieth century, Hoyt saw the dominant pattern in terms of sectors of industry and the relative arrangement of different classes. 1–CBD; 2–wholesaling and light manufacturing; 3–lower-income residential; 4–middle-income residential; 5–upper-income residential.

(*Source*: Based on C.D. Harris and E.L. Ullman, "The Nature of Cities," *Annals*, *American Academy of Political and Social Science*, 1945, Figure 5.)

industrial/working-class half of the city and the city's main sector of elite neighborhoods.

The key to the dynamics that produced these patterns was, as Hoyt observed, the behavior of affluent households. Once the CBD is established and corridors of industrial development laid out, affluent households have first pick of the most desirable sites: away from industry and the press of humanity in the CBD and on high ground

free from the risk of flooding. With urban growth, the high-status area expands along the axes of transportation lines. This happens at first in response to the desire among the most affluent to combine accessibility with suburban or exurban living and, subsequently, to a desire among the almost-as-affluent to live in the same neighborhood as the rich and influential. Middle-class, middle-income housing is established in surrounding sectors by speculative developers, who recognize the desire for "good" addresses among prospective customers. An important consequence of the outward growth of each sector is the banding of the sectors into zones of different age, style, and condition.

Automobiles and Decentralization

The advent of trucks and automobiles allowed both industry and households to become mobile. The tendency for cities to be organized in sectors and zones around a CBD was undermined by this process. In 1945, geographers Chauncy Harris and Edward Ullman described the appearance of new, automobile-based suburban nodes of commercial and industrial activity that were not arranged in any predictable fashion except in relation to surrounding land uses (**Figure 11.6**). These nodes might develop around a government center, a university, a transit stop, or a highway intersection. If the nodes were office and retailing centers, they would attract middle-income residential development, whereas if they were industrial centers, they would attract working-class residential development. Harris and Ullman description was the so-called **multiple-nuclei model**. This model quite accurately describes how American cities began to develop an irregular-shaped patchwork of land uses across which there is a loose functional order. Looking back, we can see that Harris and Ullman were remarkably prescient: The multiple-nuclei city is a uniquely American manifestation of contemporary urbanization—ever-increasing metropolitan sprawl with new nodes of economic and residential development (see Box 11.2, "Geography Matters: Smart Growth versus Sprawl").

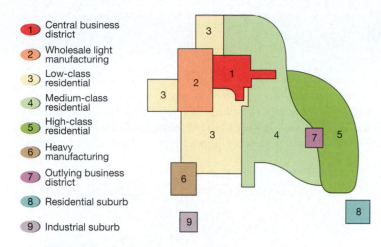

1 Central business district
2 Wholesale light manufacturing
3 Low-class residential
4 Medium-class residential
5 High-class residential
6 Heavy manufacturing
7 Outlying business district
8 Residential suburb
9 Industrial suburb

FIGURE 11.6 Multiple-nuclei model Decentralized nodes of different categories of land use end up in many different configurations, depending on local conditions. (*Source:* Based on C.D. Harris and E.L. Ullman, "The Nature of Cities," *Annals, American Academy of Political and Social Science,* 1945, Figure 5.)

The Polycentric New Metropolis

With continued urban growth, the irregular-shaped, multiple-nuclei patchwork of land uses have scaled up into polycentric metropolitan regions. During the middle decades of the twentieth century, American cities were reshaped by the combination of increased automobility, massive federal outlays on highway construction, and federal mortgage insurance programs that underpinned the growth of home ownership. The resulting spurt of city building produced a dispersed spatial structure and "urban realms," semiautonomous subregions that displaced the traditional core-periphery relationship between city centers and their suburbs. Initially, the shift to an expanded polycentric metropolis was most pronounced in the northeastern United States. Geographer Jean Gottmann captured the moment in 1961 with his conceptualization of "megalopolis"—his term for the highly urbanized region between Boston and Washington, D.C.

In the polycentric new metropolis (**Figure 11.7**), the system of nodes and realms is bound together with ever-expanding four-, six-, and eight-lane highways. It is interspersed with smaller clusters of decentralized employment, studded with micropolitan centers and filled out with booming stand-alone suburbs. Traditional nodal anchors—downtown commercial centers—remain very important, especially as settings for the advanced business services—advertising, banking, insurance, investment management, and logistics services. But in addition there are other nodes. These vary in character and include:

- **edge cities**, decentralized clusters of retailing and office development, often located on an axis with a major airport, sometimes adjacent to a high-speed rail station, and always linked to an urban freeway system;

- newer business centers, often developing in a prestigious residential quarter and serving as a setting for newer services such as corporate headquarters, the media, advertising, public relations, and design;

- outermost complexes of back-office and R&D operations, typically near major transport hubs 20 to 30 miles from the main core; and

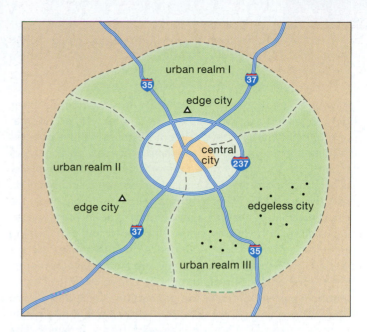

FIGURE 11.7 The twentieth-century metropolis Consisting of a central city, a ring of suburbs, and a series of far-flung urban realms, studded with edge cities (nodal concentrations of office and retail space) and "edgeless cities" of suburban and exurban office parks and shopping malls.

- specialized subcenters, usually for education, entertainment, and sporting complexes and exhibition and convention centers.[2]

Some urbanized regions have been extended and reshaped into polycentric networks of half a dozen or more urban realms and as many as 50 nodal centers of different types and sizes (**Figure 11.8**). These centers are physically separate but functionally networked and they draw enormous economic strength from a new functional division of labor. Bound together through urban freeways, arterial highways, beltways, and interstates, new metropolises coalesce into "megapolitan" regions that dominate national economies. The term **metroburbia** has emerged to capture the way that residential settings in suburban and exurban areas are thoroughly interspersed with office employment and high-end retailing.

Meanwhile, some older inner-urban districts have been redeveloped around mixed-use projects—complexes of shops, offices, and apartments. Some older, centrally located, working-class neighborhoods have also been invaded by higher-income households seeking the character and convenience of centrally located and (for them) less expensive residences—a process known as **gentrification**. Gentrification results in the physical renovation and upgrading of housing (**Figure 11.9**), but it also displaces existing households, which can no longer afford the increased rents that are a consequence of gentrification.

APPLY YOUR KNOWLEDGE Create a map like the ones in Figures 11.7 and 11.8 for the metropolitan area with which you are most familiar. Label urban realms, major highways, edge cities, secondary business centers, and other important features. ■

[2]P. Hall, "Global City-Regions in the Twenty-first Century." In A. J. Scott (ed.), *Global City-Regions: Trends, Theory Policy.* New York: Oxford University Press, 2001, pp. 59–77.

Sprawl is endemic to North American urbanization, and it is widely regarded as problematic. The unplanned, ad hoc nature of most suburban development destroys millions of acres of wildlife habitat and agricultural land every year. Rationalized, standardized, and tightly zoned off-ramp subdivisions are place-less neighborhoods that lack visual, demographic, and social diversity (**Figure 11.E**). The economics of private subdivision—the "sprawl machine"—lead to a lack of public open space, urban infrastructure, and civic amenities. Increased traffic, punishing commutes, and a chronic dependence on automobiles are the result of low-density, single-family suburban developments. The environmental costs of automobile dependency include air pollution—and in particular the generation of millions of tons of greenhouse gases from suburban commuters. Run-off from the roads and parking lots pollutes suburban watersheds. The automobile-dependent lifestyles associated with sprawl, mean-while, lead to increases in rates of asthma, lung cancer, and heart problems. Stress resulting from commuting has adverse effects on marriages and family life. The fragmented and balkanized nature of American local government also intensifies intrametropolitan fiscal disparities: Outlying communities have a larger tax base and fewer social service needs to finance in comparison with central cities. This structured inequality exacerbates the exclusionary nature of suburbia and, from a planning perspective, is simply inefficient.

The counterargument is that sprawl is a logical consequence of economic growth and the democratization of society, providing millions of people with the kinds of mobility, privacy, and choice that were once the prerogatives of the rich and powerful. Sprawl reflects market forces and represents the desires and preferences of the broad mass of people. Joel Kotkin, a blogger and popular speaker on the business circuit, has persistently argued that most people seem to like living in the suburbs: so why all the fuss? Libertarian think tanks like the Reason Foundation and the Heritage Foundation have made efforts to counter antisprawl arguments on the principle of freedom of markets and individual choice. What they overlook, from a geographical perspective, is that the

benefits of sprawl—for example, more housing for less cost with higher eventual appreciation—still tend to accrue to Americans individually, while sprawl's cost in infrastructure building, energy generation, and pollution mitigation tends to be borne by society overall.

FIGURE 11.E Sprawl California farmland in the process of being transformed into suburban sprawl.

FIGURE 11.F Smart growth A Metro Gold Line train passes through Del Mar Station in Pasadena. Transit oriented development there has provided a public plaza, a refurbished train depot, and residential space.

The compromise position is "smart growth." Smart growth is progrowth, but only when it is relatively compact and steered toward strategically designated locales with adequate infrastructure (**Figure 11.F**). Smart growth has been endorsed by the U.S. Environmental Protection Agency (EPA), the Lincoln Institute of Land Policy, and the National Resources Defense Council and has been embraced by an embattled planning profession. A Smart Growth Network has developed a set of ten basic goals (referred to as "principles") for smart growth: mixed land uses; taking advantage of compact (i.e., higher-density) neighborhood design; creating housing opportunities and choices; creating walkable communities; fostering distinctive communities with a strong sense of place; preserving open space, farmland, and critical environmental areas; strengthening and directing development toward existing communities; providing a variety of transportation choices; making development decisions predictable, fair, and cost-effective; and encouraging community and stakeholder collaboration in developer decisions.

In practice, however, smart growth has made little headway against the "sprawl machine." In addition to resistance from landowners, developers, and other real estate interests, the most effective challenges to smart-growth policies, to the dismay of planners, have come from citizens. In classic NIMBY (Not In My Backyard) responses, residential and retail projects around transit stations have been stopped cold or scaled back because of neighborhood opposition. Even projects in formally designated smart-growth districts have run into local opposition from residents contending that they already face crowded roads and schools and need to preserve the remaining open space in the area.

FIGURE 11.8 The new metropolis The largest metropolitan regions are now "megapolitan," with coalescing metropolitan areas merging into disjointed and decentralized urban landscapes with varying-sized urban centers, subcenters, and satellites and unexpected juxtapositions of form and function. The residential settings in suburban and exurban areas are thoroughly interspersed with office employment and high-end retailing, creating "metroburban" landscapes.

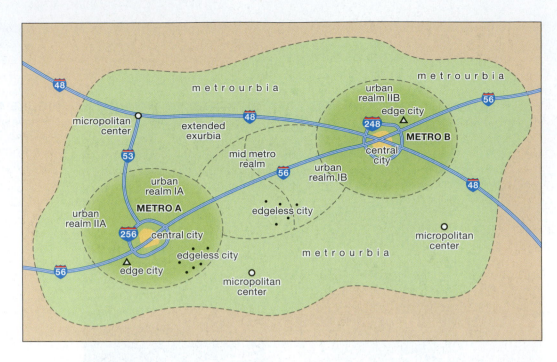

FIGURE 11.9 Gentrification Neighborhoods of row houses in central locations, like this one in Philadelphia, have long been attractive to upwardly-mobile households.

Problems of North American Cities

For all their relative prosperity, the cities of the world's core regions have their share of problems. The most acute problems are localized in central city areas and are interrelated: fiscal problems, infrastructure problems, and localized cycles of poverty and spirals of neighborhood decay. **Central cities** are the original, core jurisdictions of metropolitan areas.

Fiscal Problems

The term *fiscal* refers to government revenue or taxes. Economic restructuring and metropolitan decentralization over the past several decades have left central cities with a chronic "fiscal squeeze." A **fiscal squeeze** occurs when increasing limitations on tax revenues combine with increasing demands for expenditures on urban infrastructure and city services. The revenue-generating potential of most central cities has steadily fallen as metropolitan areas have lost both residential and commercial taxpayers to suburban jurisdictions. Growth industries, white-collar jobs, retailing, and more affluent households have moved out to suburban and exurban jurisdictions, taking their local tax dollars with them.

At the same time, growth in property-tax revenues from older, decaying neighborhoods has slowed as the growth of property values has slowed. At the same time, these older, decaying neighborhoods cost more to maintain and service. The residual populations of these neighborhoods, with high proportions of elderly and indigent households, are increasingly in need of municipal welfare services. Large numbers of low-income migrants and immigrants also bring increased demands for municipal services. Added to all this, central city governments are still responsible for services and amenities used by the entire metropolitan population: municipal galleries and museums, sports facilities, parks, traffic police, and public transport, for example. The net result is that many central city jurisdictions are in a precarious financial position. In a constant drive to develop revenue-generating projects, cities compete fiercely with one another to finance and attract tourist developments, museums, sports franchises, and business and conference centers.

Infrastructure Problems

As fiscal problems have intensified, public spending on the urban infrastructure of roads, bridges, parking spaces, transit systems, communications systems, power lines, gas lines, street lighting, water mains, sewers, and drains has declined. Meanwhile, much of the original infrastructure, put in place 75 or 100 years ago with a design life of 50 or 75 years, is obsolete, worn out, and in some cases perilously near the point of collapse. In 2007, the I-35W bridge in Minneapolis collapsed (**Figure 11.10**), killing 13 and injuring 144; a year later, the American Association of State Highway and Transportation Officials released a report saying at least $140 billion was needed to shore up or replace 152,000 of the nation's 590,000 bridges.[3]

Infrastructure problems are easily overlooked because they build up slowly. Only when a bridge collapses or a water main bursts are infrastructure problems newsworthy. In Boston, three-quarters of the sewer system was built a century or more ago and has now deteriorated to the point where about 20 percent of the system's overall flow is lost to leaks. In Cleveland, the District of Columbia, and Philadelphia, the losses approach 25 percent. Hundreds of kilometers of water mains in New York City have been

identified as being in need of replacement, at an estimated cost of over $5 billion. Almost 50 percent of all wastewater treatment systems in America are operating at 80 percent or more of capacity.[4]

Freshwater supplies are also at risk: Old water systems are unable to cope with the leaching of pollutants into city water. Common pollutants include chlorides, oil, phosphates, and nondegradable toxic chemicals from industrial wastewater; dissolved salts and chemicals from highway de-icing; nitrates and ammonia from fertilizers and sewage; and coliform bacteria from septic tanks and sewage. Many cities still use water-cleaning technology dating to World War I. About one-third of all towns and cities in the United States have contaminated water supplies, and about 8 million people use water that is potentially dangerous.

Poverty and Neighborhood Decay

Inner-city poverty and neighborhood decay have become increasingly pronounced in the past several decades as manufacturing, warehousing, and retailing jobs have moved out to suburban and edge-city locations and as many of the more prosperous households have moved out to be near these jobs. The spiral of neighborhood decay typically begins with substandard housing occupied by low-income households that can afford to rent only a minimal amount of space. The consequent overcrowding not only causes greater wear and tear on the housing itself but also puts pressure on the neighborhood infrastructure of streets, parks, schools, and playgrounds. The need for maintenance and repair increases quickly and is rarely met. Individual households cannot afford it, and landlords have no incentive, because they have a captive market. Public authorities face a fiscal squeeze and are in any case often indifferent to the needs of such neighborhoods because of their relative lack of political power.

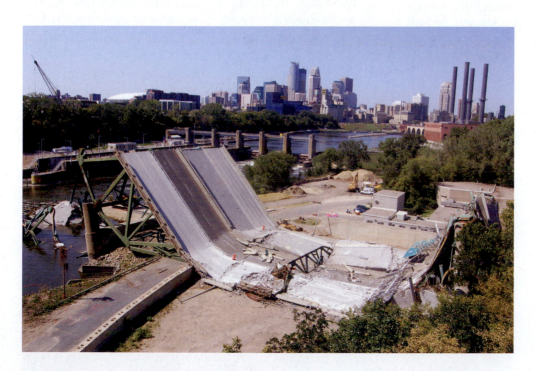

FIGURE 11.10 Decaying infrastructure This freeway bridge on Interstate 35W just outside of downtown Minneapolis, Minnesota, collapsed in 2007. A report issued in 2005 by the American Society of Civil Engineers (ASCE) estimated that it would cost $1.6 trillion to bring the country's infrastructure up to acceptable standards.

[3]American Association of State Highway and Transportation Officials, *Bridging the Gap: Restoring and Rebuilding the Nation's Bridges*, http://www. transportation1.org/BridgeReport/front-page.html, accessed February 4, 2011.

[4]*Renewing America's Infrastructure*. American Society of Civil Engineers, 2005, http://www.asce.org/reportcard/

Shops and privately run services such as restaurants and hair salons are afflicted with the same syndrome of decay. With a low-income clientele, profit margins must be kept low, leaving little to spare for upkeep or improvement. Many small businesses fail, or relocate to more favorable settings, leaving commercial property vacant for long periods. In extreme cases, property is abandoned when the owners are unable to find renters or buyers. Residential buildings may also be left derelict.

Meanwhile, a dismal cycle of poverty intersects with these localized spirals of decay. The **cycle of poverty** involves the transmission of poverty and deprivation from one generation to another through a combination of domestic circumstances and local conditions. This cycle begins with an absence of employment opportunities and, therefore, a concentration of low incomes, poor housing, and overcrowded conditions. Such conditions are unhealthy. Overcrowding makes people vulnerable to poor health, which is compounded by poor diets. This contributes to absenteeism from work, which results in decreased income. Similarly, absenteeism from school because of illness contributes to the cycle of poverty by constraining educational achievement, limiting occupational skills, and leading to low wages. Crowding also produces psychological stress, which contributes to social disorganization and a variety of pathological behaviors, including crime and vandalism. Such conditions not only affect people's educational achievement and employment opportunities but also lead to *labeling* of the neighborhood, whereby all residents may find their employment opportunities affected by the poor image of their community.

One of the most important elements in the cycle of neighborhood poverty is the educational setting. Schools, obsolete and physically deteriorated like their surroundings, are unattractive to teachers—partly because of the physical environment and partly because of the social and disciplinary environment. Fiscal squeeze leaves schools resource-poor, with relatively small budgets for staff, equipment, and materials. Over the long term, poor educational resources translate into poor education, however positive the values of students and their parents. Poor education limits occupational choice and, ultimately, results in lower incomes. Students, faced with evidence all around them of unemployment or low-wage jobs at the end of school careers, find it difficult to be positive about school. The result becomes a self-fulfilling prophecy of failure, and people become trapped in areas of concentrated poverty (**Figure 11.11**).

Many areas of concentrated poverty are also racial ghettos, although not all ghettos are poverty areas. Recall that ethnic and racial congregation can mitigate the effects of poverty. Nevertheless, discrimination is usually the main cause of ghettoization. In the United States, discrimination in housing markets is illegal, but it nevertheless takes place in a variety of ways. One example of housing-market discrimination by banks and other lending institutions is the practice of **redlining**. **Redlining** involves marking off bad-risk neighborhoods on a city map and then using the map to determine lending policy. This practice results in a bias against minorities, female-headed households, and other vulnerable groups who tend to be localized in low-income neighborhoods. Redlining tends to become another self-fulfilling prophecy, since neighborhoods starved of property loans become progressively more run-down and therefore increasingly unattractive to lenders. Discrimination affects education and labor markets as well as housing markets. In the case of ghetto poverty, all three types of discrimination come together, reinforcing the cycle of poverty and intensifying the disadvantages of the minority poor.

APPLY YOUR KNOWLEDGE Conduct an Internet search for data on your city or town. Identify and list two fiscal, industrial, or neighborhood problems. List and explain any cycles of poverty or other patterns you can identify. ∎

FIGURE 11.11 Poverty areas Concentrations of poverty are found not only in decaying inner-city areas but also in newer public housing projects and in older suburbs that have filtered down the housing scale, as in this example in the District of Columbia, a short distance from the Capitol.

EUROPEAN CITIES

European cities, like North American cities, reflect the operation of competitive land markets and social congregation along ethnic lines. They also suffer from similar problems of urban management, infrastructure maintenance, and poverty. What makes most European cities distinctive in comparison with North American cities is that they are the product of several major epochs of urban development.

Features of European Cities

As we saw in Chapter 2, many of today's most important European cities were founded in the Roman period, and it is not uncommon for the outlines of Roman and medieval urban development to be preserved in their street plans. Many distinctive features of European cities derive from their long history. In the historic cores of some older cities, the layout of streets reflects ancient patterns of rural settlement and field boundaries. Beyond these historic cores, narrow, complex streets are the product of the long, slow growth of European cities in the pre-automobile era, when hand-pushed and horse-drawn carts were the principal means of transportation and urban development was piecemeal and small-scale.

Plazas and squares are another important historical legacy in many European cities. Greek, Roman, and medieval cities were all characterized by plazas, central squares, and marketplaces, and those elements are still important nodes of urban activity (**Figure 11.12**). European history also means that its cities bear the accessories and scars of war. The legacy of defensive hilltop and clifftop sites and city walls has limited and shaped the growth of modern European cities, while in more recent times the bombings and shellings of World War II destroyed many city buildings (**Figure 11.13**).

The legacy of a long and varied history includes a rich variety of symbolism. Europeans are reminded of their past not only by large numbers of statues and memorials but also by cathedrals, churches, and monasteries; by guildhalls and city walls; by the palaces of royalty and the mansions of aristocracy; and by city halls

FIGURE 11.12 Vigevano, Italy Widely considered to be one of the finest piazzas in Italy, the Piazza Ducale in Vigevano is a product of early Renaissance town planning, designed by Bramante for Ludovico Maria Sforza in 1492–93 as a noble forecourt to the Castello Sforzesco. Unified by the arcades that completely surround the square, the piazza provides an important social space for the citizens of the town.

FIGURE 11.13 Cologne, Germany, 1945 About 90 percent of Cologne's central area was destroyed or severely damaged by Allied bombing during World War II. One of the few structures to survive intact was the solid, stone-built cathedral, erected between the thirteenth and nineteenth centuries.

and the libraries, museums, sports stadiums, and galleries that are monuments to civic achievement. European cities are also typically compact in form, resulting in high densities of population. A long history of pre-automobile urban development and the constraints of peripheral defensive walls all made urban land expensive and encouraged a tradition of high-density living in tenements and apartment houses.

Other distinctive features of European cities include:

- *Low skylines*—Although the larger European cities have a fair number of high-rise apartment buildings and a sprinkling of office skyscrapers, they all offer a predominantly low skyline. This is partly because much of their growth came before the invention of the elevator and the development of steel-reinforced, concrete building techniques and partly because of master plans and building codes (some written as long ago as the sixteenth century) seeking to preserve the dominance of monumental buildings like palaces and cathedrals.

- *Lively downtowns*—The CBDs of European cities have retained their focal position in residents' shopping and social lives because of the relatively late arrival of the suburbanizing influence of the automobile and strong planning controls directed against urban sprawl.

- *Neighborhood stability*—Europeans change residence, on average, about half as often as Americans. In addition, the physical life cycle of city neighborhoods tends to be longer because of the past use of durable construction materials, such as brick and stone. As a result, European cities provide relatively stable socioeconomic environments.

- *Municipal socialism*—For decades, European welfare states have provided a broad range of municipal services and amenities, from clinics to public transit systems. Perhaps the most important to urban structure is social housing (public housing), which accounts for 20 to 40 percent of all housing in most larger English, French, and German cities. In recent years, neoliberal policies have resulted in a reduction in public services, especially in France and Great Britain.

The richness of European history and the diversity of Europe's geography mean that there are important regional variations: The industrial cities of northern England, northeastern France, and the Ruhr district of Germany, for example, are quite different in character from the cities of Mediterranean Europe. One of the most interesting regional variations is in Eastern Europe, where the legacy of an interlude of 44 years of socialism (1945–1989) was grafted onto cities that had already developed mature patterns of land use and social differentiation. Major examples include Belgrade, Budapest, Katowice, Kraków, Leipzig, Prague, and Warsaw. State control of land and housing meant that huge public housing estates and industrial zones were created in outlying districts. The structure of the older cities was little altered, however, apart from the addition of socialist monuments and the renaming of streets.

Urban Design and Planning

European city planning and design have a long history. Most Greek and Roman settlements were laid out on grid systems, within which the siting of key buildings and the relationship of neighborhoods to one another were carefully considered.

The roots of modern Western urban planning and design can be traced to the Renaissance and Baroque periods (between the fifteenth and seventeenth centuries) in Europe. Artists and intellectuals dreamed of ideal cities, and rich and powerful regimes used urban design to produce extravagant symbolizations of wealth, power, and destiny. Inspired by the classical art forms of ancient Greece and Rome, Renaissance urban design sought to recast cities to show off the power and the glory of the state and the Church. Spreading slowly from its origins in Italy at the beginning of the fifteenth century, Renaissance design had diffused to most of the larger cities of Europe by the end of the eighteenth century. Dramatic advances in military ordnance (cannon and artillery) brought a surge of planned redevelopment that featured impressive fortifications; geometric-shaped redoubts, or strongholds; and an extensive *glacis militaire*—a sloping, clear zone of fire. Inside new walls, cities were recast according to a new aesthetic of grand design (**Figure 11.14**)—fancy palaces and geometrical plans,

FIGURE 11.14 Sabbioneta, Italy Sabbioneta was built by Vespasiano Gonzaga in the mid-sixteenth century as an ideal town, with a central piazza, a ducal palace (shown here), churches, gardens, theater, and residences all encompassed within a star-shaped plan, bounded by thick walls bearing the Gonzaga family crest.

streetscapes, and gardens that emphasized views of dramatic perspectives. These developments were often of such a scale that they effectively fixed the layout of cities well into the eighteenth and even into the nineteenth century, when walls and/or glacis eventually made way for urban redevelopment in the form of parks, railway lines, or beltways.

As societies and economies became more complex with the transition to industrial capitalism, national rulers and city leaders in Europe looked to urban design to impose order, safety, and efficiency, as well as to symbolize the new seats of power and authority. An important early precedent was set in Paris by Napoleon III, who presided over a comprehensive program of urban redevelopment and monumental urban design. The work was carried out by Baron Georges Haussmann between 1853 and 1870. Haussmann demolished large sections of old Paris to make way for broad, new, tree-lined avenues (**Figure 11.15**) and numerous public open spaces and monuments. In doing so, he made the city not only more efficient (wide boulevards meant better flows of traffic) and a better place to live (parks and gardens allowed more fresh air and sunlight into a crowded city and were held to be a "civilizing" influence) but also safer from revolutionary politics (wide boulevards were hard to barricade; monuments and statues instilled a sense of pride and civic identity).

The preferred architectural style for these new designs was the **Beaux Arts** style, which takes its name from L'École des Beaux Arts in Paris. In this school, architects were trained to draw on Classical, Renaissance, and Baroque styles to synthesize new designs for the Industrial Age. The idea was that the new buildings would blend artfully with the older palaces, cathedrals, and civic buildings that dominated European city centers. Haussmann's ideas were widely influential and extensively copied.

Early in the twentieth century there emerged a different intellectual and artistic reaction to the pressures of industrialization and urbanization. The **Modern movement** was based on the idea that buildings and cities should be designed and run like machines. Equally important to the Modernists was the idea that urban design should not simply reflect dominant social and cultural values, but, rather, help to create a new moral and social order. The movement's best-known advocate was Le Corbusier, a Paris-based Swiss who provided the inspiration for technocratic urban design. Modernist buildings sought to dramatize technology, exploit industrial production techniques, and use modern materials and unembellished, functional design. Le Corbusier's ideal city (*La Ville Radieuse*) featured linear clusters of high-density, medium-rise apartment blocks, elevated on stilts and segregated from

FIGURE 11.15 Boulevard des Italiens, Paris Central Paris owes much of its character to the *grandes boulevards* that were key to the urban renewal schemes of Baron Georges-Eugène Haussmann.

industrial districts; high-rise tower office blocks; and transportation routes—all separated by broad expanses of public open space.

After World War II, the International Style became pervasive in urban design. The boxlike steel-frame buildings with concrete and glass façades were avant-garde yet respectable and, above all, comparatively inexpensive to build. This tradition of urban design, more than anything else, has imposed a measure of uniformity on cities around the world. Due to globalization International Style buildings have appeared in big cities in every part of the world. Furthermore, the International Style has often been the preferred style for large-scale urban design projects around the world. One of the best examples is Brasilia (**Figure 11.16**), the capital of Brazil, founded in 1956 in an attempt to shift the country's political, economic, and psychological focus away from the past; differentiate it from the former colonial cities on the coast; and orient the country toward the future and the interior.

Modern urban design has had many critics, mainly on the grounds that it tends to take away the natural life and vitality of cities, replacing varied and human-scale environments with monotonous and austere settings. In response, historic preservation has become an important element of urban planning in every city that can afford it.

APPLY YOUR KNOWLEDGE Search the Internet to find six images that best capture the typical features of European cities, such as narrow, complex streets, plazas and market squares, symbolism, low skylines, lively downtowns, neighborhood stability, and municipal socialism. ■

ISLAMIC CITIES

Islamic cities provide good examples of how social and cultural values and people's responses to their environment are translated into spatial terms through the built environment. Because of similarities in cityscapes, layout, and design, geographers are able to talk about the Islamic city as a meaningful category. It is a category that includes thousands of towns and cities, not only in the Arabian Peninsula and the Middle East—the heart of the Islamic Empire under the prophet Muhammad (570–632 c.e.)—but also in regions into which Islam spread later: North Africa, coastal East Africa, South-Central Asia, and Indonesia. Most cities in North Africa and South-Central Asia are Islamic, and many elements of the classic Islamic city can be found in towns and cities as far away as Seville, Granada, and Córdoba in southern Spain (the western extent of Islam), Kano in northern Nigeria and Dar es Salaam in Tanzania (the southern extent), and Davao in the Philippines (the eastern extent).

The fundamentals of the layout and design of the traditional Islamic city are so closely attached to Islamic cultural values that they are referenced in the Qur'an, the holy book of Islam. Although urban growth in the Islamic world does not have to conform to any overall master plan or layout, certain basic regulations and principles are intended to support Islam's emphasis on personal privacy and virtue, on communal well-being, and on the inner essence of things rather than on their outward appearance.

The most dominant feature of the traditional Islamic city is the *Jami*—the principal mosque (**Figure 11.17**). Located centrally, the mosque complex is a center not only of worship but also of education and the hub of a broad range of welfare functions. As cities grow, new, smaller mosques are built toward the edge of the city, each out

FIGURE 11.16 Brasilia The Brazilian National Congress buildings, Brasilia.

of earshot of the call to prayer from the Jami and from one another. The traditional Islamic city was walled for defense, with several lookout towers and a *Kasbah,* or citadel (fortress), containing palace buildings, baths, barracks, and its own small mosque and shops.

Traditionally, gates controlled access to the city, allowing careful scrutiny of strangers and permitting the imposition of taxes on merchants. The major streets led from these gates to the main covered bazaars or street markets (*suqs,* **Figure 11.18**). The suqs nearest the Jami typically specialize in the cleanest and most prestigious goods, such as books, perfumes, prayer mats, and modern consumer goods. Those nearer the gates feature bulkier and less valuable goods such as basic foodstuffs, building materials, textiles, leather goods, and pots and pans. Within the suqs, every profession and line of business had its own alley, and the residential districts around the suqs are organized into distinctive quarters, or *ahya',* according to occupation (or sometimes ethnicity, tribal affiliation, or religious sect).

Privacy is central to the construction of the Islamic city. Above all, women must be protected, according to Islamic values, from the gaze of unrelated men. Traditionally, doors must not face each other across a minor street, and windows must be small, narrow, and above eye level. Cul-de-sacs (dead-end streets) are used where possible to restrict the number of persons approaching the home. Angled entrances prevent intrusive glances. Larger homes are built around courtyards, which provide an interior and private focus for domestic life.

FIGURE 11.17 Mosque The dominant feature of traditional Islamic cities is the *Jami,* or main mosque. This photograph shows the main prayer hall and court of Badshahi Mosque, Lahore, Pakistan.

FIGURE 11.18 The suq The *suq,* a covered bazaar or open street market, is one of the most important distinguishing features of a traditional Islamic city. Typically, a suq consists of small stalls located in numerous passageways. Many important suqs are covered with vaults or domes. This photograph is of the Bazaar-i Vakil suq in Shiraz, Iran.

The rights of others are also emphasized in Islamic urban design. The Qur'an specifies an obligation to neighborly cooperation and consideration—traditionally this consideration is interpreted as applying to a minimum radius of 40 houses. Roofs, in traditional designs, are surrounded by parapets to preclude views of neighbors' homes, and drainage channels are steered away from neighbors' houses. Refuse and wastewater are carefully recycled. Public thoroughfares were originally designed to be wide enough to allow two fully laden camels to pass each other and high enough to accommodate a camel and rider. The overall result is a compact, cellular urban structure within which it is possible to maintain a high degree of privacy (**Figure 11.19**).

Because most Islamic cities are located in hot, dry climates, these basic principles of urban design have evolved in conjunction with certain practical solutions to intense heat and sunlight. Twisting streets, as narrow as permissible, help to maximize shade, as does latticework on windows and cellular residential courtyard design. In some regions, local architectural styles include air ducts and roof funnels with adjustable shutters that can be used to create dust-free drafts (**Figure 11.20**).

While all these features are still characteristic of Islamic cities, they are most evident in the old cores, or *medinas*. Like cities everywhere, however, Islamic cities also bear the imprint of globalization. Although Islamic culture is self-consciously opposed to many aspects of globalization, it has been unable to resist altogether the penetration of the world economy and the infusion of the Western-based culture of global metropolitanism. The result can be seen in international hotels, skyscrapers and office blocks, modern factories, highways, airports, and stores. Indeed, the leading cities of some oil-rich states have become the "shock cities" of the early twenty-first century, with phenomenal rates of growth characterized by breathtakingly ambitious architectural and urban design projects (see Box 11.3, "Window on the World: Dubai, United Arab Emirates"). Meanwhile, Islamic culture and urban design principles have not always been able to cope with the pressures of contemporary rates of urbanization, so the larger Islamic

FIGURE 11.19 Housing in Kalaa Sghira, Tunisia Seen from above, the traditional Islamic city is a compact mass of residences with walled courtyards.

FIGURE 11.20 Islamic architecture In Islamic societies elaborate precautions are taken through architecture and urban design to ensure the privacy of individuals, especially women. Entrances are L-shaped and staggered across the street from one another. Windows are often placed above pedestrian access. Architectural details also reflect climatic influences: Window screens and narrow, twisting streets maximize shade, while air ducts and roof funnels create dust-free drafts, as in this photograph of traditional wind towers in the city of Yazd, Iran.

cities in less-affluent states—cities such as Algiers, Cairo, Karachi, and Teheran—now share with other peripheral cities the common denominators of unmanageable size, shanty and squatter development, and low-income mass housing. In the next section of the chapter, we examine the problems of large and rapidly growing metropolises throughout the periphery.

> **APPLY YOUR KNOWLEDGE** Research an Islamic city that is similar in size to the community in which you live and list two similarities and two differences between your own town or city and the Islamic city you have researched. ■

CITIES OF THE PERIPHERY: UNINTENDED METROPOLISES

The cities of the world-system periphery, often still referred to as Third World cities, are numerous and varied. What they have in common is the experience of unprecedented rates of growth driven by rural "push"—overpopulation and the lack of employment opportunities in rural areas—rather than the "pull" of prospective jobs in towns and cities. Faced with poverty in overpopulated rural areas, many people regard moving to a city much like playing a lottery: You buy a ticket (in other words, go to the city) in the hope of hitting the jackpot (in other words, landing a good job). As with all lotteries, most people lose, and the net result is widespread underemployment. **Underemployment** occurs when people work less than full-time even though they would prefer to work more hours. Underemployment is difficult to measure with any degree of accuracy, but estimates commonly range from 30 to 50 percent of the employed workforce in peripheral cities.

Because of their rapid growth and high underemployment, the peripheral metropolises of the world, the shock cities of the late twentieth century—Mexico City (Mexico), São Paulo (Brazil), Lagos (Nigeria, **Figure 11.21**), Mumbai (India; formerly Bombay),

Dhaka (Pakistan), Jakarta (Indonesia), Karachi (Pakistan), and Manila (the Philippines)—embodied the most remarkable and unprecedented changes in economic, social, and cultural life. Socioeconomic conditions in these cities are still shocking, in the sense of being deplorably bad, but they are no longer shocking in the sense of being unprecedented. Rather, they have become all too familiar.

Recall from Chapter 10 that peripheral metropolises play a key role in international economic flows, linking provincial regions with the hierarchy of world cities and, thus, with the global economy. Within peripheral metropolises, this role results in a pronounced **dualism**, or juxtaposition in geographic space of the formal and informal sectors of the economy. This dualism is evidenced by the contrast between high-rise modern office and apartment towers and luxurious homes and the slums and shantytowns (**Figure 11.22**).

The Informal Economy

In many peripheral cities, more than one-third of the population is engaged in the informal sector of the urban economy, and in some—for example Chennai, India; Colombo, Sri Lanka; Delhi, India; Guyaquil, Ecuador; and Lahore, Pakistan—the figure is more than one-half. People who cannot find regularly paid work must resort to various ways of gleaning a living. Some of these ways are imaginative, some desperate, some pathetic. Examples range from street vending, shoe shining, craft work, and street-corner repairs to scavenging in garbage dumps (**Figure 11.23**). The informal sector consists of a broad range of activities that represent an important coping mechanism (**Figure 11.24**). For too many, however, coping means resorting to begging, crime, or prostitution. Occupations such as selling souvenirs, driving pedicabs, making home-brewed beer, writing letters for others, and dressmaking may seem marginal from the point of view of the global economy, but more than a billion people around the world must feed, clothe, and house themselves entirely from such occupations. Across Africa, the International Labor Office estimates, informal-sector employment is growing 10 times faster than formal-sector employment.

FIGURE 11.21 Lagos, Nigeria Lagos, like most metropolises in the world's periphery, grew relatively slowly until quite recently. The combination of the demographic transition, political independence, and an economic boom stimulated by the discovery of oil reserves in southeastern Nigeria triggered an explosive growth in population. Because of its difficult site on sand spits and lagoons, this growth has resulted in an irregular sprawl and, in the central area, a density of population higher than that of Manhattan Island in New York.

Dubai is one of seven territories that comprise the United Arab Emirates (UAE). Since the discovery of oil in the UAE more than 30 years ago, the country has undergone a profound transformation, from an impoverished region of small desert principalities to a modern state with a high standard of living. Before the discovery of oil in the 1950s, the UAE's economic base was mainly fishing and pearling. The supply of oil is expected to last in the region only between 20 and 30 years, so there has been a major diversification program with investment in national infrastructure, health care, and education.

Dubai has become a major financial and trading center and an important tourist destination for the wealthy. Until the global financial meltdown of 2008, rates of growth exceeded 10 percent a year, fueling a phenomenal construction boom that depended on more than 500,000 low-skilled, poorly paid workers—immigrant labor from Iran, Bangladesh, Pakistan, India, the Philippines, and other Asian nations, all of them living in substandard conditions, with few rights. The city's skyline quickly came to reflect its emerging role as a globally important business hub, with scores of towers, many of them designed by famed architects from around the world. The Burj Khalifa tower, at 2,313 ft (818 meters), claimed the title of the tallest building in the world on completion in 2010—nearly 40 percent taller than the second-highest building in the world (Taipei 101, in Taipei, Taiwan).

Dubai's "shock city" status derives as much from spectacular affluence as its phenomenal rate of growth. It has established a reputation for luxury shopping, with the duty-free stores at Dubai airport exploiting the city's role as a stopping-off point between Europe and Asia. The over-the-top Mall of the Emirates includes Ski Dubai, an enormous indoor complex that provides five different runs with fresh snow for skiing, snowboarding, and tobogganing. Two artificial islands built in the shape of date palm trees extend from the Dubai City waterfront and support more than 60 luxury hotels, 4,000 exclusive residential villas, 1,000 water homes, 5,000 shoreline apartments, marinas, water theme parks, restaurants, shopping malls, sports facilities, health spas, cinemas, and dive sites.

In the wake of the financial meltdown of 2008, Dubai's real estate boom came to an abrupt halt, and most of the immigrant workforce was immediately sent home. Thousands of skilled white-collar workers from Europe, the United States, and other parts of the Arab world also began to leave. Many of them, having lived the high life during the boom, found themselves unable to meet the payments for expensive real estate or luxury cars. In the months

FIGURE 11.G Dubai cityscape Awash—for a while—with petrodollars, Dubai was able to offer generous tax breaks to companies willing to relocate their activities to the city. The city did so in an effort to create a major business hub to provide an economic base when the oil and gas run out.

following October 2008, more than 3,000 cars were abandoned outside the city's international airport, most of them with keys left in the ignition. Some also had used-to-the-limit credit cards in the glove compartment. Others had notes of apology attached to the windshield; their owners had fled rather than risk jail for defaulting on loans.[5] A year after it was inaugurated with a massive water-and-fireworks display, more than 800 of the Burj Khalifa tower's 900 ultraluxury apartments remained unoccupied, while scores of other towers remained completely unoccupied. Some areas, such as Business Bay, have been left as entire districts of unfinished buildings.

FIGURE 11.H Dubai luxury development Aerial view of villas on the Palm Jumeirah, an artificial island in Dubai.

FIGURE 11.I Dubai luxury shopping The main foyer in The Mall of the Emirates in Dubai.

FIGURE 11.J Dubai real estate bust Development stalled in Dubai in 2010 as a result of the global financial meltdown of October 2008.

[5]Verman, S. "Driven Down by Debt, Dubai Expats Give New Meaning to Long-stay Car Park," *The Times*, February 5, 2009, http://business.timesonline.co.uk/tol/business/markets/the_gulf/article5663618.ece

FIGURE 11.22 Rio de Janeiro This photograph, looking toward the buildings of wealthier Rio on the horizon, shows very clearly the dualism of peripheral metropolises, with a *favela* in the foreground.

In most peripheral countries, the informal labor force includes children. In environments of extreme poverty, every family member must contribute something, and so children are expected to do their share. Industries in the formal sector often take advantage of this situation. Many firms farm out their production under subcontracting schemes that are based not in factories but in home settings that use child workers. In these settings, labor standards are nearly impossible to enforce. In the Philippines, for example, batches of rural children are ferried by syndicates to work in garment-manufacturing sweatshops in urban areas.

Despite this side of the picture, the informal sector has a few positive aspects. Pedicabs, for example, provide an affordable, nonpolluting means of transportation in crowded metropolitan settings. Garbage picking provides an important means of recycling paper, steel, glass, and plastic products. One study of Mexico City estimated that as much as 25 percent of the municipal waste ends up being recycled by the 15,000 or so scavengers who work over the city's official dump sites. This positive contribution to the economy, though, scarcely justifies the lives of poverty and degradation experienced by the scavengers.

FIGURE 11.23 Garbage picking Scavengers picking out recyclable garbage at the Nong Khaem dump, Bangkok, Thailand.

FIGURE 11.24 Informal economic activities In cities where jobs are scarce, people cope through the informal sector of the economy, which includes a very broad variety of activities, including agriculture (backyard hens), manufacturing (craft work), and retailing (street vending). This photograph shows the floating market in Bangkok, Thailand.

Urban geographers also recognize that the informal sector represents an important resource to the formal sector of peripheral economies. The informal sector provides a vast range of cheap goods and services that reduce the cost of living for employees in the formal sector, thus enabling employers to keep wages low. Although this network does not contribute to urban economic growth or help alleviate poverty, it does keep companies competitive within the context of the global economic system. For export-oriented companies, in particular, the informal sector provides a considerable indirect subsidy to production. This subsidy is often passed on to consumers in the core regions in the form of lower prices for goods and consumer products made in the periphery.

Consider, for example, the paper industry in Cali, Colombia. This industry is dominated by one company, Cartón de Colombia, which was established in 1944 with North American capital and subsequently acquired by the Mobil Oil Company. Most of the company's lower-quality paper products are made from recycled waste paper. Sixty percent of this waste paper is gathered by the 1,200 to 1,500 garbage pickers in Cali. Some work the city's municipal waste dump, some work the alleys and yards of shopping and industrial areas, and some work the routes of municipal garbage trucks, intercepting trash cans before the truck arrives. They are part of Cali's informal economy, for they are not employed by Cartón de Colombia nor do they have any sort of contract with the company or its representatives. They simply show up each day to sell their pickings. In this way the company avoids paying both wages and benefits and is able to dictate the price it will pay for various grades of waste paper. The company can operate profitably while keeping the price of its products down—the arrangement is a microcosm of core-periphery relationships.

APPLY YOUR KNOWLEDGE Identify an informal economic function in your town or city. List three positive and three negative effects of the activities involved in this type of economy. ■

Slums of Hope, Slums of Despair

The informal labor market is directly paralleled in informal shantytowns and squatter housing: Because so few jobs with regular wages exist in the cities of the periphery, few families can afford rent or house payments for sound housing. Unemployment, underemployment, and poverty mean overcrowding. In situations where urban growth has swamped the available stock of cheap housing and outstripped the capacity of builders to create affordable new housing, the inevitable outcome is makeshift shanty housing. Such housing has to be constructed on the cheapest and least desirable sites. Often this means building on bare rock, over ravines, on derelict land, on swamps, or on steep slopes. Nearly always it means building without any basic infrastructure of streets or utilities. Sometimes it means adapting to the most extreme ecological niches, as in Lima, Peru, where garbage pickers actually live on the waste dumps, or in Cairo, Egypt, where for generations the poor have adapted catacombs and cemeteries into living spaces. In many cities, more than half of the housing is substandard. The United Nations estimated in 2010 that more than 1.75 billion people worldwide live in inadequate housing in urban areas.

Globalization and the accompanying trend toward neoliberal economic policies have intensified problems of poverty and slum housing in many cities. The United Nations Human Settlement Programme reported in 2004 that

the reduction of fiscal deficits has partly entailed reduction of public expenditure through downsizing of the civil service and privatization of state enterprises, resulting in the laying-off of large numbers of public-sector employees in many countries. Trade liberalization has often resulted in the closure of some industries that have been unable to compete against cheap imports, again leading to massive retrenchment and higher unemployment levels. Rising urban unemployment and increasing poverty have forced large numbers of the urban poor into the informal sector. Underpaid formal-sector employees have also entered the informal sector as a survival strategy. This, in turn, leads to the erosion of the tax base and decreasing ability of national and local governments to assist the poor through social and basic services. The removal of price controls on subsistence goods, and increased utility charges through privatization . . . have resulted in rising inequalities and increasing poverty. . . . As a result, cities end up with their prime resources largely appropriated by the affluent.[6]

Faced with the growth of slums, the first response of many governments has been to eradicate them. Encouraged by Western development economists and housing experts, many cities sought to stamp out unintended urbanization through large-scale eviction and clearance programs. In Caracas (Venezuela), Lagos (Nigeria), Bangkok (Thailand), Kolkata (India; formerly Calcutta), Manila (the Philippines), and scores of other cities in the periphery, hundreds of thousands of shanty dwellers were ordered out on short notice and their homes bulldozed to make way for public works, land speculation, luxury housing, and urban renewal and, on occasion, to improve the appearance of cities for special visitors. Seoul, South Korea, has probably had the most forced evictions of any city in the world. Since 1966, as part of sustained government clean-up campaigns millions of people in Seoul have been forced out of accommodations that they owned or rented. The Beijing Olympics in 2008 displaced 1.5 million people, according to the Geneva-based Centre on Housing Rights and Evictions (COHRE). Its report[7] blames the Chinese government for widespread forced evictions along with other human rights violations during preparations for the Olympics. Demolitions and evictions were often undertaken without due process, without the provision of adequate compensation sufficient to attain alternative accommodation, and without access to legal recourse. In some cases, tenants were given little or no notice of their eviction and did not receive the promised compensation. The city also subjected alleged unlicensed taxi operators, street vendors, vagrants, and beggars to "Re-education through Labour" (a form of imprisonment without charge).

---

[6]United Nations Human Settlement Programme, *The State of the World's Cities 2004/2005. Globalization and Urban Culture.* Sterling, VA: Earthscan, 2004, p. 102.

[7]Centre on Housing Rights and Evictions (COHRE), 2008, "One World, Whose Dream?: Housing Rights Violations and the Beijing Olympic Games," http://www.cohre.org/store/attachments/One_World_Whose_Dream_July08.pdf, accessed February 4, 2010.

Yet Seoul and Beijing, more than most other cities, could afford to build new low-income housing to replace the demolished neighborhoods. Most peripheral cities cannot do so, which means that displaced slum dwellers have no option but to create new squatter and shanty settlements elsewhere in the city. Most cities, in fact, cannot evict and demolish fast enough to keep pace with the growth of slums caused by in-migration. The futility of slum clearance has led to a widespread reevaluation of the wisdom of such policies. The thinking now is that informal-sector housing should be seen as a rational response to poverty. Shanty and squatter neighborhoods not only provide affordable shelter but also function as important reception areas for migrants to the city, with supportive communal organizations and informal employment opportunities that help them to adjust to city life. They can, in other words, be "slums of hope." City authorities, recognizing the positive functions of informal housing, are now increasingly disposed to be tolerant and even helpful to squatters rather than sending in police and municipal workers with bulldozers.

In fact, many informal settlements are the product of careful planning. In parts of Latin America, for example, it is common for community activists to draw up plans for invading unused land and then quickly build shanty housing before landowners can react. The activists' strategy is to organize a critical mass of people large enough to be able to negotiate with the authorities to resist eviction. It is also common for activists to plan their invasions for public holidays so that the risk of early detection is minimized. As the risk of eviction diminishes over time, some residents of informal housing are able to gradually improve their dwellings through self-help (**Figure 11.25**).

Nevertheless, there are many shanty and squatter neighborhoods where self-help and community organization do not emerge. Instead, grim and desperately miserable conditions prevail. These are "slums of despair," where overcrowding, lack of adequate sanitation, and lack of maintenance lead to shockingly high levels of ill health and infant mortality and where social pathologies are at their worst. Consider, for example, the squatter settlement of Dharavi, a "mega-slum" in Mumbai, India.[8] Dharavi is about 1 square mile in size and home to over a million people. This means that as many as 18,000 people crowd into each single acre of land. In a city where house rents are among the highest in the world, Dharavi provides a cheap and affordable option to those who move to Mumbai to earn their living. Rents can be as low as 185 rupees ($4) per month. Recycling is one of the slum's biggest industries. In Dharavi nothing is considered garbage. Thousands of tons of scrap plastic, metals, paper, cotton, soap, and glass revolve through Dharavi each day. Ruined plastic toys are tossed into massive grinders, chopped into tiny pieces, and melted down into multicolored pellets, ready to be refashioned into knockoff Barbie dolls. Dharavi also houses about 15,000 hutment factories, each typically stuffed with children as well as adults sewing cotton, melting plastic, hammering iron, molding clay, or producing embroidered garments, export-quality leather goods, pottery, and plastic.

The aggregate annual turnover of these businesses is estimated to be more than $650 million a year, yet conditions for residents are miserable. In Dharavi, there is only one toilet per 1,440 people—and during the monsoon rains, flooded lanes run with human excrement. Most people have food intakes of less than the recommended minimum of 1,500 calories a day; 90 percent of all infants and children under 4 have less than the minimum calories needed for a healthy diet. More than half of the children and almost half the adults have intestinal worm infections. Infant and child mortality is high—though nobody knows just how high—with malaria, tetanus, diarrhea, dysentery, and cholera as the principal causes of death among under-fives.

FIGURE 11.25 Self-help as a solution to housing problems Self-help is often the only solution to housing problems because wages are so low and so scarce that builders cannot construct even the most inexpensive new housing and make a profit and because municipalities cannot afford to build sufficient quantities of subsidized housing. One of the most successful ways of encouraging self-help housing is for municipal authorities to create the preconditions by clearing sites, putting in the footings for small dwellings, and installing a basic framework of water and sewage utilities. This "sites-and-services" approach has become the mainstay of urban housing policies in many peripheral countries. This photograph shows self-help housing in Ndola, Zambia.

[8]See *The Economist*, 2007, "A Flourishing Slum," http://www.economist.com/world/asia/displaystory.cfm?story_id=10311293, accessed August 5, 2008; M. Jacobson, "Mumbai Slum," *National Geographic, 2007*, http://ngm.nationalgeographic.com/ngm/0705/feature3/index.html, accessed August 5, 2008; C. W. Dugger, "Toilets Underused to Fight Disease, U.N. Study Finds," *New York Times*, 2006, http://www.nytimes.com/2006/11/10/world/10toilet.html?_r=1&ex=1189828800&en=905358c57769b677&ei=5070&oref=slogin, accessed February 4, 2010.

Transport and Infrastructure Problems

The informal labor market is also reflected in cities' inability to provide a basic infrastructure of highways, transportation, schools, utilities, and emergency services. Because the informal sector yields no tax revenues, municipal funds are insufficient to provide an adequate infrastructure or to maintain a safe and sanitary environment. Even though the governments of peripheral cities typically spend nearly all of their budgets on transport and infrastructure in an effort to keep up with population growth, conditions are bad and getting rapidly worse. Peripheral cities have always been congested, but in recent years the modernizing influence of formal-sector activities has turned the congestion into near gridlock. Sharp increases have occurred in the availability and use of automobiles. India has become one of the world's fastest-growing automobile markets. More than 2.5 million cars were sold there in 2010. Yet India's transportation infrastructure has not caught up with this sudden growth, nor have its drivers. The auto accident rate is the highest in the world. India has about 1 percent of the world's cars yet still manages to kill over 100,000 people in traffic accidents each year. This amounts to 10 percent of the entire world's traffic fatalities (the United States, with more than 40 percent of the world's cars, creates just 43,000 fatalities).

Not only are there more people and more traffic but the changing spatial organization of peripheral cities has increased the need for transportation. Traditional patterns of land use have been superseded by the agglomerating tendencies inherent in modern industry and the segregating tendencies inherent in modernizing societies. The greatest single change has been the separation of home from work, however, which has meant a significant increase in commuting.

In spite of innovative responses to urban transportation needs, road transportation in many cities is breaking down, with poorly maintained roads, traffic jams, long delays at intersections, and frequent accidents. Governments have invested in expensive new freeways and street-widening schemes, but because they tend to focus on city centers (which are still the settings for most jobs and most services and amenities), they ultimately fail. New freeways disgorge vehicles into a congested and chaotic mixture of motorized traffic, bicycles, animal-drawn vehicles, and hand-drawn carts. Some of the worst traffic tales come from Mexico City—where traffic backups total more than 90 kilometers (60 miles) each day, on average—and Bangkok, where the 24-kilometer (15-mile) trip into town from Don Muang Airport can take 3 hours. In São Paulo, Brazil, gridlock can span 160 kilometers (100 miles), rush-hour traffic jams average 85 kilometers (53 miles) in length, and 15-hour traffic jams are not unusual. The costs of these traffic backups are enormous. The annual costs of traffic delays in Singapore have been estimated at $305 million; in Bangkok, Thailand, they have been estimated at $272 million—the equivalent of around 1 percent of Thailand's gross national product.

Water supplies and sewerage also present acute problems for many cities (**Figure 11.26**). Definitions of what constitutes an adequate amount of safe drinking water and sanitation vary from country to country. Although many governments classify the existence of a water tap within 100 meters (328 feet) of a house as "adequate," such a tap does not guarantee that the household will be able to secure enough water for good health. Communal taps often function only a few hours each day, so residents must wait in long lines to fill even one bucket. In Rajkot, India, a city of 600,000 people, piped water routinely runs for only 20 minutes each day.

FIGURE 11.26 Infrastructure problems When city governments are unable to raise sufficient revenues through local taxes, infrastructure is neglected. Putting water and sewage lines into neighborhoods that were built without these basic utilities is arduous and expensive, yet without them, public health is seriously threatened. This photograph shows water lines being installed in a low-income neighborhood of Cartagena, Colombia.

The World Bank estimates that worldwide around 65 percent of urban residents in less developed countries have access to a satisfactory water source, and only about 40 percent are connected to sewers (90 percent of which discharge their waste untreated into a river, a lake, or the sea). Hundreds of millions of urban dwellers have no alternative but to use contaminated water or water whose quality is not guaranteed. A small minority, usually the residents of the most affluent neighborhoods, have water piped into their homes, while the majority have piped water nearby. Those not served are obliged to carry water in small quantities over long distances or to use water from streams or other surface sources (**Figure 11.27**). In Colombo, Sri Lanka, about one-third of all houses have indoor piped water, and another one-fourth have piped water outside. In Dar es Salaam (Tanzania), Kinshasa (Democratic Republic of the Congo), and many other peripheral cities, almost half the population has no access to piped water, either indoors or outdoors.

In many cities, including Bangkok, Bogotá, Dar es Salaam, Jakarta, Karachi, and São Paulo, only one-fourth to one-third of all garbage and solid waste is collected and removed—the rest is partially recycled informally, tipped into gullies, canals, or rivers, or simply left to rot. Sewage services are just as bad. In Latin America, for example, only about 2 percent of collected sewage receives any

FIGURE 11.27 Water-supply problems
Many peripheral cities have grown so quickly and under such difficult conditions that large sections of the population do not have access to supplies of clean water, even in new housing developments, as in this slum-clearance project in a suburb of Bangalore, India.

treatment. In Mexico, more than 90 percent of wastewater treatment plants are nonfunctional, and in cities like Bogotá, Buenos Aires, Mexico City, and Santiago, some 50 to 60 million cubic meters of mostly untreated sewage is discharged every day into nearby bodies of water. São Paulo has over 1,600 kilometers (1,000 miles) of open sewers, and raw sewage from the city's slums drains into the Billings reservoir, a major source of the city's drinking water. In Bangkok, less than 5 percent of the population is connected to a sewer system. Jakarta has no waterborne sewage system at all. Septic tanks serve about one-quarter of the city's population; others must use pit latrines, cesspools, and ditches along the roadside. A survey of over 3,000 towns and cities in India found that only 8 had full sewage-treatment facilities, and another 209 had partial treatment facilities. Along one river, the Ganga (better known to English speakers as the Ganges), 114 towns and cities dump untreated sewage into the river every day, along with waste from DDT factories, tanneries, paper and pulp mills, petrochemical and fertilizer complexes, and other industrial pollutants. Each day the Yamuna River picks up 200 million liters of untreated sewage and 20 million liters of industrial effluents as it passes through Delhi. In China, inadequate sewerage and meager municipal wastewater plants have resulted in widespread water-quality deterioration. Shanghai has had to move its water supply intake 40 kilometers (25 miles) upstream at a cost of US$300 million because of degradation of river water quality around the city.

These problems can provide opportunities for the informal sector, however. Street vendors, who get their water from private tanker and borehole operators, sell water from 2- or 4-gallon cans. Vendors typically charge 5 to 10 times the local rate set by public water utilities; in some cities, they charge 60 to 100 times as much. Similarly, many cities have evolved informal-sector mechanisms for sewage disposal. In many Asian cities, for example, human waste is removed overnight by handcart operators. Unfortunately, it is rarely disposed of properly and often ends up polluting the rivers or lakes from which the urban poor draw their water.

Environmental Degradation

With pressing problems of poverty, slum housing, and inadequate infrastructure, it is not surprising that peripheral cities are unable to devote many resources to environmental problems. Because

of the speed of population growth, these problems are escalating rapidly. Industrial and human wastes pile up in lakes and lagoons and pollute long stretches of rivers, estuaries, and coastal zones. Groundwater is polluted through the leaching of chemicals from uncontrolled dump sites, and the forests around many cities are denuded by the demand of cities for timber and domestic fuels. This environmental degradation is, of course, directly linked to human health. People living in such environments have much higher rates of respiratory infections, tuberculosis, and diarrhea, and much shorter life expectancies than people living in surrounding rural communities. Children in squatter settlements may be 50 times as likely to die before the age of 5 as those born in affluent core countries.

Air pollution has escalated to very harmful levels in many cities. With the development of a modern industrial sector and the growth of automobile ownership, but without enforceable regulations on pollution and vehicle emissions, tons of lead, sulfur oxides, fluorides, carbon monoxide, nitrogen oxides, petrochemical oxidants, and other toxic chemicals are pumped into the atmosphere every day in large cities. The burning of charcoal, wood, and kerosene for fuel and cooking in low-income neighborhoods also contributes significantly to dirty air. In cities where sewerage systems are deficient, the problem is compounded by the presence of airborne dried fecal matter. Worldwide, according to UN data, more than 1.1 billion people live in urban areas where air pollution exceeds healthful levels.

A UN study of 20 megacities found that every one of them had at least one major pollutant at levels exceeding World Health Organization (WHO) guidelines. Fourteen of the 20 had *two* major pollutants exceeding WHO guidelines, and 7 had *three*. Such pollution is not only unpleasant but dangerous. The World Bank has cited China as having 16 of the 20 most air-polluted cities on Earth. China's spectacular economic growth has brought with it air pollution levels that are blamed for more than 400,000 premature deaths a year.

The European Space Agency, utilizing satellite imagery, determined that Beijing and its surrounding areas have the world's highest levels of nitrogen dioxide, a substance poisonous to the lungs. In Manila, the Philippines, the Asian Development Bank found levels of suspended particulate matter in the air to be 200–400 percent above guideline levels. In Mexico City, where sulfur dioxide and lead concentrations are 2 to 4 times higher than the WHO guidelines, and where national ozone levels are exceeded on more than half of the days throughout the year, 7 in 10 newborns have dangerously high levels of lead in their bloodstream.

WHO studies demonstrate that it is unhealthy for human beings to breathe air with more than 100 to 120 parts per billion (ppb) of ozone contaminants for more than one day a year. Yet Mexico City residents breathe this level, or more, for over 300 days a year. In Bangkok, Thailand, where air pollution is almost as severe as in Mexico City, research has shown that lead-bearing air pollutants reduce children's IQ by an average of 3.5 points per year until they are 7 years old. It has also been estimated that Bangkok's pall of dust and smoke causes more than 1,400 deaths annually and $3.1 billion each year in lost productivity resulting from traffic- and pollution-linked illnesses. Proximity to industrial facilities, often the result of the need and desire of the poor to live near places of employment, poses another set of risks. A notorious accident at the Union Carbide factory in Bhopal, India, in 1984 caused 2,988 deaths and more than 100,000 injuries, mostly among residents of the shantytowns near the chemical factory.

APPLY YOUR KNOWLEDGE Research a specific megacity (as categorized by the United Nations) and examine it in terms of the presence of slum housing, environmental degradation, and infrastructure concerns. List and explain three causes of any one of these problems. ■

Future Geographies

The future of urban structure in 10–20 years seems relatively easy to predict. In spite of many ideas about urban planning and the prospect of new and improved transportation, the patterns and problems described in this chapter seem set to continue. The reasons are straightforward. Demographic trends are established and accessibility, agglomeration, territoriality, congregation, segregation, and sprawl will continue to shape people's behavior. In North American cities, continuing fiscal constraints dictate that infrastructure problems will continue and likely intensify. The political and financial interests behind the "sprawl machine" ensure a continuation of polycentric, metroburban development. The past 200 years of economic, social, and urban history provide no reason at all to expect any mitigation of urban poverty or neighborhood decay. Similarly, the basic templates of urban structure are well established in both European and Islamic cities. Change will likely be marginal, even as economic and demographic modifications take place and as the impact of new technologies—electric cars, perhaps—alter the locational behavior of people and businesses.

We can apply a similar logic to the unintended metropolises of the periphery. Faced with continuing streams of migration as well as high rates of natural increase among their relatively youthful populations, the megacities of the periphery will continue to expand, virtually unchecked. Informal economic activities will continue to have an important role, and cityscapes will continue to be dominated by stark contrasts between the towers of international business and elite residences and slums and informal housing (**Figure 11.28**). Sheer pressure of numbers will ensure continuing problems of congestion, water supply, sanitation, and environmental degradation. What is less predictable is how individual cities might change as a result of the introduction of progressive planning policies or new transportation systems or—from a more pessimistic perspective—as a result of localized economic problems, political unrest, or environmental disasters.

FIGURE 11.28 The future is already here Demographic trends mean that the megacities of the periphery will continue to expand, virtually unchecked, for the foreseeable future. As a result, cityscapes in much of the world will continue to be dominated by stark contrasts between rich and poor, as in these slums of the Worli village, Mumbai, India.

CONCLUSION

Patterns of land use and the functional organization of economic and social subareas in cities are partly a product of economic, political, and technological conditions at the time of the city's growth, partly a product of regional cultural values, and partly a product of processes of globalization. Geographers can draw on several perspectives in looking at patterns of land use within cities, including an economic perspective that emphasizes competition for space and a sociocultural perspective that emphasizes ethnic congregation and segregation. Nevertheless, urban structure varies considerably because of the influence of history, culture, and the different roles that different cities have played within the world-system.

The evolution of the unintended metropolis of the periphery has been very different from the evolution of metropolitan areas in the world's core regions. Similarly, the problems they have faced are very different. In the core regions, the consequences of an economic transformation to a postindustrial economy have dominated urban change. Traditional manufacturing and related activities have been moved out of central cities, leaving decaying neighborhoods and a residual population of elderly and marginalized people. New, postindustrial activities have begun to cluster in redeveloped CBDs and in edge cities around metropolitan fringes. In a few cases, metropolitan growth has become so complex and extensive that 100-mile cities have begun to emerge, with half a dozen or more major commercial and industrial centers forming the nuclei of a series of interdependent urban realms.

In other parts of the world, traditional patterns of land use and the functional organization of economic and social subareas have been quite different, reflecting different historical legacies and different environmental and cultural influences. A basic trend affecting the cities of the world's periphery is demographic—the phenomenal rates of natural increase and in-migration. An ever-growing informal sector of the economy, in which people seek economic survival, is reflected in extensive areas of shanty housing. High rates of unemployment, underemployment, and poverty generate acute social problems, which are overwhelming for city governments that are understaffed and underfunded. If present trends continue, such problems are likely to characterize increasing numbers of the world's largest settlements. Meanwhile, globalization processes are recasting metropolitan structure and intensifying social and economic inequalities.

Learning Outcomes Revisited

- Assess how the internal structure of cities is shaped by competition for territory and location.

  In general, all categories of land users—commercial and industrial, as well as residential—compete for the most convenient and accessible locations within the city. An important exception is that wealthier households tend to trade off the convenience of accessibility for the greater utility of being able to consume larger amounts of (relatively cheap) suburban space. Poorer households, unable to afford the recurrent costs of transportation, trade off living space for accessibility to jobs.

- Appraise the ways in which social patterns in cities are influenced by human territoriality.

  Territoriality provides a means of establishing and preserving group membership and identity. Processes of congregation and discrimination often result in segregation, the spatial separation of specific subgroups within a wider population. Segregation varies a great deal in both intensity and form.

- Describe the spatial structure of the typical North American city.

  Most larger cities are structured around a central business district (CBD); a transitional zone; suburbs; secondary business districts and commercial strips; and industrial districts. In larger metropolitan areas, a polycentric structure is typical, with "edge cities," new business centers, and specialized subcenters. The internal organization of cities reflects the way that they function, both to bring certain people and activities together and to sort them out into neighborhoods and functional subareas. During the middle decades of the twentieth century, North American cities were reshaped by the combination of increased automobility, federal outlays on highway construction, and federal mortgage insurance programs that promoted home ownership. The resulting spurt of city building produced a dispersed spatial structure and the emergence of a polycentric metropolitan structure.

- Compare and contrast urban structures in different regions of the world.

  Urban structure varies considerably because of the influence of history, culture, and the different roles that cities have played within the world-system. European cities have evolved under circumstances very different from those in American cities and consequently exhibit some distinctive characteristics that reflect their history. Islamic cities provide examples of how social and cultural values and people's responses to their environment are translated into spatial terms through urban form and the design of the built environment. The new cities of the world's peripheral regions are characterized and shaped by explosive growth.

- Explain the nature and causes of the problems associated with urbanization in various world regions.

  The most acute problems of the cities of the world's core regions are localized in the central city areas that have borne the brunt of restructuring from an industrial to a postindustrial economy, while the problems of the cities of the periphery stem from the way in which their demographic growth has out-stripped their economic growth. The central city districts of cities in the world's core regions typically experience several interrelated problems: fiscal problems, infrastructure problems, and localized spirals of neighborhood decay and cycles of poverty. In peripheral cities high rates of long-term unemployment and underemployment, low and unreliable wages of informal-sector jobs, chronic poverty, and slum housing are common.

KEY TERMS

Beaux Arts *(p. 399)*
central business district
(CBD) *(p. 387)*
central cities *(p. 394)*
colonies *(p. 387)*
congregation *(p. 386)*

cycle of poverty *(p. 396)*
dualism *(p. 403)*
edge cities *(p. 391)*
enclaves *(p. 387)*
fiscal squeeze *(p. 394)*
gentrification *(p. 391)*

ghettos *(p. 387)*
invasion and succession *(p. 389)*
isotropic surface *(p. 386)*
metroburbia *(p. 391)*
minority groups *(p. 386)*
Modern movement *(p. 399)*

multiple-nuclei model *(p. 390)*
redlining *(p. 396)*
segregation *(p. 387)*
trade-off model *(p. 386)*
underemployment *(p. 403)*
zone in transition *(p. 387)*

REVIEW AND DISCUSSION

1. Research the different ethnic, class, and gender makeup of your city or town. Examine the data through the lens of congregation and the territorial and residential clustering of specific groups or subgroups of people. Identify where these different groups congregated. Does a pattern of discrimination or segregation emerge? Explain your observations in a paragraph.

2. Conduct research on the different ethnic groups in your city over the past 20 years. Map out and list three ways the congregation of these groups changed. Would you identify the change as "invasion and

succession?" Has the infrastructure around these groups changed? Explain your observations in a paragraph.

3. Create a list of the various modes of transportation that are commonly available in your community. List three ways that the transportation shapes your city or town. Identify two major problems with the current transportation configuration of your city (*Hint:* Compare how long it takes a person to get from point A to B using various modes of transportation.) On a piece of paper map out how these problems could be corrected.

UNPLUGGED

1. Collect a week's worth of local newspapers and review the coverage of urban problems. What kinds of problems in what kinds of communities are covered? Compile a list of the different categories of problems, and then carefully analyze the content of the week's coverage, calculating the number of column inches devoted to each category.

2. On a tracing-paper overlay of a street map of your town or city, plot the distribution of houses and apartments for sale or rent in different price brackets. (You can obtain the information from the real estate pages of your city's local newspaper; in smaller cities you may have

to gather data from several issues of the paper—your local library will likely have back issues.) Explain the spatial distributions that you observe and monitor.

3. Most cities consist of "ordinary" cityscapes that are strongly evocative because they are widely understood as being a particular kind of place. Write a brief essay (500 words, or two double-spaced, typed pages) describing an "ordinary" cityscape with which you are familiar. What are its principal features, and how might it be considered typical of a particular kind of place?

Mastering GEOGRAPHY™

Log in to www.masteringgeography.com for MapMaster™ interactive maps, geography videos, RSS feeds, flash cards, Web links, an eText version of *Human Geography: Places and Regions in Global Context,* and self-study quizzes to enhance your study of city spaces urban structure.

MapMaster™ presents 13 Place Name and 13 Layered Thematic interactive maps to help students practice and master their geographic literacy, spatial reasoning, and critical thinking skills.

GLOSSARY

A

accessibility: the opportunity for contact or interaction from a given point or location in relation to other locations.

acid rain: the wet deposition of acids upon Earth created by the natural cleansing properties of the atmosphere.

actor-network theory: an orientation that views the world as composed of "heterogeneous things," including humans and nonhumans and objects.

affect: emotions that are embodied reactions to the social and physical environment.

age-sex pyramid: a representation of the population based on its composition according to age and sex.

agglomeration diseconomies: the negative economic effects of urbanization and the local concentration of industry.

agglomeration effects: interdependencies associated with various kinds of economic linkages, including the cost advantages that accrue to individual firms because of their location among functionally related activities.

agrarian: referring to the culture of agricultural communities and the type of tenure system that determines access to land and the kind of cultivation practices employed there.

agribusiness: a set of economic and political relationships that organizes agro-food production from the development of seeds to the retailing and consumption of the agricultural product.

agricultural density: ratio between the number of agriculturists per unit of arable land and a specific area.

agricultural industrialization: process whereby the farm has moved from being the centerpiece of agricultural production to becoming one part of an integrated string of vertically organized industrial processes including production, storage, processing, distribution, marketing, and retailing.

agriculture: a science, art, and business directed at the cultivation of crops and the raising of livestock for sustenance and profit.

ancillary industries: industries that manufacture parts and components to be used by larger industries.

anthropocene: the modern geological era during which humans have dramatically affected the environment.

aquaculture: the cultivation of fish and shellfish under controlled conditions, usually in coastal lagoons.

arithmetic density (crude density): total number of people divided by the total land area.

autarky: an economic policy or situation in which a nation is independent of international trade and not reliant upon imported goods.

azimuthal projection: a map projection on which compass directions are correct only from one central point.

B

baby boom: population of individuals born between the years 1946 and 1964.

backwash effects: the negative impacts on a region (or regions) of the economic growth of some other region.

Beaux Arts: a style of urban design that sought to combine the best elements of all of the classic architectural styles.

biofuels: renewable fuels derived from biological materials that can be regenerated.

biomass: fuels made from biological material from living or recently living organisms that include wood, waste, gas, and alcohol fuels.

biometric census: a census in which the government photographs and fingerprints individuals to create a national database.

biopharming: an application of biotechnology in which genes from other life forms (plant, animal, fungal, bacterial, or human) are inserted into a host plant.

bioprospecting: the search for plant and animal species that may yield medicinal drugs and other commercially valuable compounds.

Biorevolution: the genetic engineering of plants and animals with the potential to exceed the productivity of the Green Revolution.

biotechnology: technique that uses living organisms (or parts of organisms) to make or modify products, to improve plants and animals, or to develop microorganisms for specific uses.

bioterrorism: deliberate use of microorganisms or toxins from living organisms to induce death or disease.

Blue Revolution: the introduction of new production techniques, processing technology, infrastructure, and larger, motorized boats as well as the application of transgenics into peripheral country fisheries.

Borlaug hypothesis: restricting crop usage to traditional low-yield methods (such as organic farming) in the face of rising global food demand would require either the world population to decrease or the further conversion of forest land into cropland.

C

capitalism: a form of economic and social organization characterized by the profit motive and the control of the means of production, distribution, and the exchange of goods by private ownership.

carrying capacity: the maximum number of users that can be sustained, over the long term, by a given set of natural resources.

cartography: the body of practical and theoretical knowledge about making distinctive visual representations of Earth's surface in the form of maps.

census: count of the number of people in a country, region, or city.

central business district (CBD): the central nucleus of commercial land uses in a city.

central cities: the original, core jurisdictions of metropolitan areas.

central place: a settlement in which certain products and services are available to consumers.

central place theory: a theory that seeks to explain the relative size and spacing of towns and cities as a function of people's shopping behavior.

centrality: the functional dominance of cities within an urban system.

chemical farming: application of synthetic fertilizers to the soil—and herbicides, fungicides, and pesticides to crops—in order to enhance yields.

children's rights: the fundamental right of children to life, liberty, education, and health care codified in 1989 by the United Nations Convention on the Rights of the Child.

citizenship: a category of belonging to a nation-state that includes civil, political, and social rights.

climate change: any significant change in measures of climate (such as temperature, precipitation, or wind) lasting for an extended period (decades or longer).

cognitive distance: the distance that people perceive to exist in a given situation.

cognitive images (mental maps): psychological representations of locations that are made up from people's individual ideas and impressions of these locations.

cognitive space: space defined and measured in terms of the nature and degree of people's values, feelings, beliefs, and perceptions about locations, districts, and regions.

cohort: a group of individuals who share a common temporal demographic experience.

colonial city: a city that was deliberately established or developed as an administrative or commercial center by colonial or imperial powers.

colonialism: the establishment and maintenance of political and legal domination by a state over a separate and alien society.

colonization: the physical settlement of a new territory of people from a colonizing state.

Columbian Exchange: interaction between the Old World—originating with the voyages of Columbus—and the New World.

commercial agriculture: farming primarily for sale, not direct consumption.

commodity chain: network of labor and production processes beginning with the extraction or production of raw materials and ending with the delivery of a finished commodity.

comparative advantage: principle whereby places and regions specialize in activities for which they have the greatest advantage in productivity relative to other regions—or for which they have the least disadvantage.

confederation: a group of states united for a common purpose.

conformal projection: a map projection on which compass bearings are rendered accurately.

conglomerate corporations: companies that have diversified into various economic activities, usually through a process of mergers and acquisitions.

congregation: the territorial and residential clustering of specific groups or subgroups of people.

conservation: the view that natural resources should be used wisely and that society's effects on the natural world should represent stewardship and not exploitation.

contract farming: an agreement between farmers and processing and/or marketing firms for the production, supply, and purchase of agricultural products—from beef, cotton, and flowers to milk, poultry, and vegetables.

conventional farming: approach that uses chemicals in the form of plant protectants and fertilizers, or intensive, hormone-based practices in breeding and raising animals.

core regions: regions that dominate trade, control the most advanced technologies, and have high levels of productivity within diversified economies.

cosmopolitanism: an intellectual and aesthetic openness toward divergent experiences, images, and products from different cultures.

cost/price squeeze: the simultaneous decrease in selling prices and rise in production costs that reduce a business's profit margin.

counterurbanization: the net loss of population from cities to smaller towns and rural areas.

creative destruction: the withdrawal of investments from activities (and regions) that yield low rates of profit in order to reinvest in new activities (and new places).

crop rotation: method of maintaining soil fertility in which the fields under cultivation remain the same but the crop being planted is changed.

crude birthrate (CBR): ratio of the number of live births in a single year for every thousand people in the population.

crude death rate (CDR): the number of deaths in a single year for every thousand people in the population.

crude density (arithmetic density): total number of people divided by the total land area.

cultural complex: combination of traits characteristic of a particular group.

cultural ecology: study of the relationship between a cultural group and its natural environment.

cultural geography: how space, place, and landscape shape culture at the same time that culture shapes space, place, and landscape.

cultural hearths: the geographic origins or sources of innovations, ideas, or ideologies.

cultural landscape: a characteristic and tangible outcome of the complex interactions between a human group and a natural environment.

cultural nationalism: an effort to protect regional and national cultures from the homogenizing impacts of globalization, especially from the penetrating influence of U.S. culture.

cultural region: the areas within which a particular cultural system prevails.

cultural system: a collection of interacting elements that taken together shape a group's collective identity.

cultural trait: a single aspect of the complex of routine practices that constitute a particular cultural group.

culture: a shared set of meanings that are lived through the material and symbolic practices of everyday life.

cumulative causation: a spiral buildup of advantages that occurs in specific geographic settings as a result of the development of external economies, agglomeration effects, and localization economies.

cycle of poverty: the transmission of poverty and deprivation from one generation to another through a combination of domestic circumstances and local, neighborhood conditions.

D

decolonization: the acquisition by colonized peoples of control over their own territory.

deep ecology: approach to nature revolving around two key components: self-realization and biospherical egalitarianism.

deforestation: the removal of trees from a forested area without adequate replanting.

deindustrialization: a relative decline in industrial employment in core regions.

democratic rule: a system in which public policies and officials are directly chosen by popular vote.

demographic collapse: phenomenon of near genocide of native populations.

demographic transition: replacement of high birth and death rates by low birth and death rates.

demography: the study of the characteristics of human populations.

dependency: high level of reliance by a country on foreign enterprises, investment, or technology.

dependency ratio: measure of the economic impact of the young and old on the more economically productive members of the population.

derelict landscapes: landscapes that have experienced abandonment, misuse, disinvestment, or vandalism.

desertification: the degradation of land cover and damage to the soil and water in grasslands and arid and semiarid lands.

dialects: regional variations in standard languages.

diaspora: spatial dispersion of a previously homogeneous group.

digital divide: inequality of access to telecommunications and information technology, particularly the Internet.

discourse: institutionalized ways of constituting knowledge.

distance-decay function: the rate at which a particular activity or process diminishes with increasing distance.

division of labor: the specialization of different people, regions, or countries in particular kinds of economic activities.

domino theory: the theory that if one country in a region chooses or is forced to accept a communist political and economic system, then neighboring countries would be irresistibly susceptible to communism.

double cropping: practice used in the milder climates whereby intensive subsistence fields are planted and harvested more than once a year.

doubling time: measure of how long it will take the population of an area to grow to twice its current size.

dualism: the juxtaposition in geographic space of the formal and informal sectors of the economy.

E

East/West divide: communist and noncommunist countries, respectively.

ecofeminism: view that patriarchal ideology is at the center of our present environmental malaise.

ecological footprint: measure of the human pressures on the natural environment from the consumption of renewable resources and the production of pollution indicating how much space a population needs compared to what is available.

ecological imperialism: introduction of exotic plants and animals into new ecosystems.

eco-migration: population movement caused by the degradation of land and essential natural resources.

economies of scale: cost advantages to manufacturers that accrue from high-volume production, since the average cost of production falls with increasing output.

ecosystem: community of different species interacting with each other and with the larger physical environment that surrounds it.

ecotheology: the view that it is necessary to address the current environmental crisis through belief systems that will overcome the inadequacies of humanly created institutions.

edge cities: nodal concentrations of shopping and office space situated on the outer fringes of metropolitan areas, typically near major highway intersections.

elasticity of demand: degree to which levels of demand for a product or service change in response to changes in price.

emigration: move from a particular location.

environmental determinism: doctrine holding that human activities are controlled by the environment.

environmental ethics: philosophical perspective on nature that prescribes moral principles as guidance for our treatment of it.

environmental justice: movement reflecting a growing political consciousness, largely among the world's poor, that their immediate environs are far more toxic than those in wealthier neighborhoods.

equal-area (equivalent) projection: map projection that portrays areas on Earth's surface in their true proportions.

equidistant projection: map projection that allows distance to be represented as accurately as possible.

ethnicity: socially created system of rules about who belongs and who does not belong to a particular group based upon actual or perceived commonality.

ethnocentrism: attitude that one's own race and culture are superior to others'.

ethology: scientific study of the formation and evolution of human customs and beliefs.

export-processing zones (EPZs): small areas within which especially favorable investment and trading conditions are created by governments in order to attract export-oriented industries.

external arena: regions of the world not yet absorbed into the modern world system.

external economies: cost savings that result from circumstances beyond a firm's own organization and methods of production.

F

famine: acute starvation associated with a sharp increase in mortality.

fast food: edibles that can be prepared and served very quickly, sold in a restaurant, and served to customers in packaged form.

federal state: form of government in which power is allocated to units of local government within the country.

fiscal squeeze: increasing limitations on city revenues, combined with increasing demands for expenditure.

flexible production systems: ability of manufacturers to shift quickly and efficiently from one level of output to another, or from one product configuration to another.

folk culture: traditional practices of small groups, especially rural people with a simple lifestyle who are seen to be homogeneous in their belief systems and practices.

food chain: five central and connected sectors (inputs, production, product processing, distribution, and consumption) with four contextual elements acting as external mediating forces (the state, international trade, the physical environment, and credit and finance).

food manufacturing: adding value to agricultural products through a range of treatments—such as processing, canning, refining, packing, and packaging—that occur off the farm and before products reach the market.

food regime: specific set of links that exists among food production and consumption and capital investment and accumulation opportunities.

food security: assured access by a person, household, or even a country to enough food at all times to ensure active and healthy lives.

food sovereignty: right of peoples, communities, and countries to define their own agricultural, labor, fishing, food, and land policies that are ecologically, socially, economically, and culturally appropriate to their unique circumstances.

forced migration: movement of an individual against his or her will.

Fordism: principles for mass production based on assembly-line techniques, scientific management, mass consumption based on higher wages, and sophisticated advertising techniques.

foreign direct investment: total of overseas business investments made by private companies.

friction of distance: deterrent or inhibiting effect of distance on human activity.

functional regions: regions with some variability in certain attributes but with an overall coherence to the structure and dynamics of economic, political, and social organization.

G

gateway city: serves as a link between one country or region and others because of its physical situation.

gender: social differences between men and women rather than the anatomical differences that are related to sex.

genetically modified organism (GMO): any organism that has had its DNA modified in a laboratory rather than through cross-pollination or other forms of evolution.

genre de vie: functionally organized way of life that is seen to be characteristic of a particular cultural group.

gentrification: invasion of older, centrally located, working-class neighborhoods by higher-income households seeking the character and convenience of less expensive and well-located residences.

geodemographic analysis: practice of assessing the location and composition of particular populations.

geodemographic research: study of census data and commercial data (such as sales data and property records) about the populations of small districts to create profiles of those populations for market research.

geographic information system (GIS): organized collection of computer hardware, software, and geographic data that is designed to capture, store, update, manipulate, and display geographically referenced information.

geographical imagination: capacity to understand changing patterns, changing processes, and changing relationships among people, places, and regions.

geographical path dependence: historical relationship between the present activities associated with a place and the past experiences of that place.

geopolitics: state's power to control space or territory and shape the foreign policy of individual states and international political relations.

gerrymandering: practice of redistricting for partisan purposes.

ghetto: an area of a city inhabited by a minority group, sometimes by choice but more often as a result of social, legal, or economic discrimination.

global change: combination of political, economic, social, historical, and environmental problems at the world scale.

global civil society: set of institutions, organizations, and behaviors situated between the state, business world, and family, including voluntary and nonprofit organizations, philanthropic institutions, and social and political movements.

Global Positioning System (GPS): system of satellites that orbit Earth on precisely predictable paths, broadcasting highly accurate time and locational information.

globalization: increasing interconnectedness of different parts of the world through common processes of economic, environmental, political, and cultural change.

globalized agriculture: system of food production increasingly dependent upon an economy and set of regulatory practices that are global in scope and organization.

Green Revolution: export of a technological package of fertilizers and high-yielding seeds from the core to the periphery to increase global agricultural productivity.

greenhouse gases (GHG): any gas that absorbs infrared radiation in the atmosphere, including, but not limited to, water vapor, carbon dioxide (CO_2), methane (CH_4), and nitrous oxide (N_2O).

greening: adding biomass, including grasses and trees, through rainfall to an area that was formerly a desert.

gross domestic product (GDP): estimate of the total value of all materials, foodstuffs, goods, and services produced by a country in a particular year.

gross migration: total number of migrants moving into and out of a place, region, or country.

gross national income (GNI): similar to GDP, but also includes the value of income from abroad.

growth poles: economic activities that are deliberately organized around one or more high-growth industries.

guest workers: individuals who migrate temporarily to take up jobs in other countries.

H

hajj: religious pilgrimage.

hearth areas: geographic settings where new practices have developed and from which they have subsequently spread.

hegemony: domination over the world economy exercised by one national state in a particular historical epoch through a combination of economic, military, financial, and cultural means.

hinterland: sphere of economic influence of a town or city.

historical geography: geography of the past.

human geography: study of the spatial organization of human activity and of people's relationships with their environments.

human rights: people's individual rights to justice, freedom, and equality, considered by most societies to belong automatically to all people.

humanistic approach: point of view that places the individual—especially individual values, meaning systems, intentions, and conscious acts—at the center of analysis.

hunting and gathering: activities whereby people feed themselves through killing wild animals and fish and gathering fruits, roots, nuts, and other edible plants to sustain themselves.

hybridity: a mixing of different types; in cultural geography, hybridity is most often associated with movements across a binary of, for instance, the racial categories of black and white such that identities are more multiple and ambivalent.

I

identity: sense that people make of themselves through their subjective feelings based on their everyday experiences and wider social relations.

immigration: move to another location.

imperialism: extension of the power of a nation through direct or indirect control of the economic and political life of other territories.

import substitution: process by which domestic producers provide goods or services that formerly were bought from foreign producers.

infant mortality rate: annual number of deaths of infants under 1 year of age compared to the total number of live births for that same year.

inflation: increased supply of printed currency that leads to higher prices and international financial differentials.

informal sector: economic activities that take place beyond official record, not subject to formalized systems of regulation or remuneration.

infrastructure (or fixed social capital): underlying framework of services and amenities needed to facilitate productive activity.

initial advantage: critical importance of an early start in economic development; a special case of external economies.

intensive subsistence agriculture: practice that involves the effective and efficient use—usually through a considerable expenditure of human labor and application of fertilizer—of a small parcel of land in order to maximize crop yield.

internal migration: move within a particular country or region.

internally displaced persons (IDPs): individuals who are uprooted within the boundaries of their own country because of conflict or human rights abuse.

international division of labor: specialization, by countries, in particular products for export.

international migration: move from one country to another.

international organization: group that includes two or more states seeking political and/or economic cooperation with each other.

international regime: orientation of contemporary politics around the international arena instead of the national.

intersubjectivity: shared meanings among people, derived from their lived experience of everyday practice.

intertillage: practice of mixing different seeds and seedlings in the same swidden.

invasion and succession: process of neighborhood change whereby one social or ethnic group succeeds another.

irredentism: assertion by the government of a country that a minority living outside its formal borders belongs to it historically and culturally.

Islam: Arabic term that means submission to God's will.

Islamism: anticolonial, anti-imperial, and generally anticore political movement.

isotropic surface: hypothetical, uniform plain that is flat and has no variations in its physical attributes.

J

jihad: sacred struggle.

just-in-time production: manufacturing process in which daily or hourly delivery schedules of materials allow for minimal or zero inventories.

K

kinship: relationship based on blood, marriage, or adoption.

L

land reform: redistribution of land by the state with a goal of increasing productivity and reducing social unrest.

landscape as text: idea that landscapes can be read and written by groups and individuals.

language: communicating ideas or feelings by means of a conventionalized system of signs, gestures, marks, or articulate vocal sounds.

language branch: collection of languages that possess a definite common origin but have split into individual languages.

language family: collection of individual languages believed to be related in their prehistorical origin.

language group: collection of several individual languages that are part of a language branch, share a common origin, and have similar grammar and vocabulary.

latitude: angular distance of a point on Earth's surface, measured north or south from the equator, which is 0°.

law of diminishing returns: tendency for productivity to decline, after a certain point, with the continued application of capital and/or labor to a given resource base.

leadership cycles: periods of international power established by individual states through economic, political, and military competition.

life expectancy: average number of years a newborn infant can expect to live.

lifeworld: taken-for-granted pattern and context for everyday living through which people conduct their lives.

local food: food that is organically grown and produced within a fairly limited distance from where it is consumed.

localization economies: cost savings that accrue to particular industries as a result of clustering together at a specific location.

longitude: angular distance of a point on Earth's surface, measured east or west from the prime meridian (the line that passes through both poles and through Greenwich, England, and that has the value of 0°).

M

malnutrition: the condition that develops when the body does not get the right amount of the vitamins, minerals, and other nutrients it needs to maintain healthy tissues and organ function.

map projection: systematic rendering on a flat surface of the geographic coordinates of the features found on Earth's surface.

materialism: a theory which emphasizes that the material world—its objects and nonhuman entities—is at least partly separate from humans and possesses the power to affect humans; materialism attempts to understand the ways that specific properties of material things affect the interactions between humans and nonhuman entities.

mechanization: replacement of human farm labor with machines.

medical geography: subarea of the discipline that specializes in understanding the spatial aspects of health and illness.

megacity: very large city characterized by both primacy and high centrality within its national economy.

metroburbia: suburban and exurban areas where residential settings are thoroughly interspersed with office employment and high-end retailing.

middle cohort: members of the population 15 to 64 years of age who are considered economically active and productive.

migration: move beyond the same political jurisdiction, involving a change of residence—either as emigration or as immigration.

minisystem: society with a single cultural base and a reciprocal social economy.

minority groups: population subgroups that are seen—or that see themselves—as somehow different from the general population.

mobility: ability to move, either permanently or temporarily.

Modern movement: architectural movement based on the idea that buildings and cities should be designed and run like machines.

modernity: forward-looking view of the world that emphasizes reason, scientific rationality, creativity, novelty, and progress.

multiple-nucleii model: model of urbanization proposed by Chauncy Harris and Edward Ullman in which decentralized nodes of different categories of land use end up in many different configurations, depending on local conditions.

Muslim: member of the Islamic community of believers whose duty is obedience and submission to the will of God.

N

nation: group of people often sharing common elements of culture, such as religion or language or a history or political identity.

nationalism: feeling of belonging to a nation as well as the belief that a nation has a natural right to determine its own affairs.

nation-state: ideal form consisting of a homogeneous group of people governed by their own state.

natural decrease: difference between CDR and CBR, which is the deficit of births relative to deaths.

natural increase: difference between the CBR and CDR, which is the surplus of births relative to deaths.

nature: social creation as well as the physical universe that includes human beings.

neocolonialism: economic and political strategies by which powerful states in core economies indirectly maintain or extend their influence over other areas or people.

neo-Fordism: economic principles in which the logic of mass production coupled with mass consumption is modified by the addition of more flexible production, distribution, and marketing systems.

neoliberal policies: economic policies that are predicated on a minimalist role for the state, assuming the desirability of free markets as the ideal condition not only for economic organization but also for political and social life.

neoliberalism: reduction in the role and budget of government, including reduced subsidies and the privatization of formerly publicly owned and operated concerns, such as utilities.

net migration: gain or loss in the total population of a particular area as a result of migration.

new world order: triumph of capitalism over communism, wherein the United States becomes the world's only superpower and therefore its policing force.

newly industrializing countries (NICs): countries formerly peripheral within the world system that have acquired a significant industrial sector, usually through foreign direct investment.

non-representational theory: an approach to human (and non-human) practices that explores how they are performed and what are their effects such as how music produces in humans both remembering and forgetting.

nontraditional agricultural exports (NTAEs): new export crops that contrast with traditional exports.

North/South divide: differentiation made between the colonizing states of the Northern Hemisphere and the formerly colonized states of the Southern Hemisphere.

nutritional density: ratio between the total population and the amount of land under cultivation in a given unit of area.

O

offshore financial centers: islands or microstates that have become a specialized node in the geography of worldwide financial flows.

old-age cohort: members of the population 65 years of age and older who are considered beyond their economically active and productive years.

ordinary landscapes (vernacular landscapes): everyday landscapes that people create in the course of their lives.

organic farming: farming or animal husbandry done without commercial fertilizers, synthetic pesticides, or growth hormones.

Orientalism: discourse that positions the West as culturally superior to the East.

overurbanization: condition in which cities grow more rapidly than the jobs and housing they can sustain.

P

pandemic: an epidemic that spreads rapidly around the world with high rates of illness and death.

pastoralism: subsistence activity that involves the breeding and herding of animals to satisfy the human needs of food, shelter, and clothing.

peripheral regions: regions with undeveloped or narrowly specialized economies with low levels of productivity.

physical geography: subarea of the discipline that studies Earth's natural processes and their outcomes.

place: specific geographic setting with distinctive physical, social, and cultural attributes.

plantation: large landholding that usually specializes in the production of one particular crop for market.

political ecology: approach to cultural geography that studies humans in their environment through the relationships of patterns of resource use to political and economic forces.

popular culture: practices and meaning systems produced by large groups of people whose norms and tastes are often heterogeneous and change frequently, often in response to commercial products.

population policy: official government policy designed to effect any or all of several objectives, including the size, composition, and distribution of population.

postmodernity: view of the world that emphasizes an openness to a range of perspectives in social inquiry, artistic expression, and political empowerment.

preservation: approach to nature advocating that certain habitats, species, and resources should remain off-limits to human use, regardless of whether the use maintains or depletes the resource in question.

primacy: condition in which the population of the largest city in an urban system is disproportionately large in relation to the second- and third-largest cities.

primary activities: economic activities that are concerned directly with natural resources of any kind.

producer services: services that enhance the productivity or efficiency of other firms' activities or that enable them to maintain specialized roles.

proxemics: study of the social and cultural meanings that people give to personal space.

pull factors: forces of attraction that influence migrants to move to a particular location.

purchasing power parity (PPP): measures how much of a common "market basket" of goods and services each currency can purchase locally, including goods and services that are not traded internationally.

push factors: events and conditions that impel an individual to move from a location.

Q

quaternary activities: economic activities that deal with the handling and processing of knowledge and information.

R

race: problematic classification of human beings based on skin color and other physical characteristics.

racialization: practice of categorizing people according to race or of imposing a racial character or context.

rank-size rule: statistical regularity in size distributions of cities and regions.

reapportionment: process of allocating electoral seats to geographical areas.

redistricting: defining and redefining of territorial district boundaries.

redlining: practice whereby lending institutions delimit "bad-risk" neighborhoods on a city map and then use the map as the basis for determining loans.

refugee: individual who crosses national boundaries to seek safety and asylum.

region: larger-sized territory that encompasses many places, all or most of which share similar attributes in comparison with the attributes of places elsewhere.

regional geography: study of the ways unique combinations of environmental and human factors produce territories with distinctive landscapes and cultural attributes.

regionalism: feeling of collective identity based on a population's politico-territorial identification within a state or across state boundaries.

regionalization: classification of individual places or areal units.

religion: belief system and set of practices that recognize the existence of a power higher than humans.

remote sensing: collection of information about parts of Earth's surface by means of aerial photography or satellite imagery designed to record data on visible, infrared, and microwave sensor systems.

reurbanization: growth of population in metropolitan central cores, following a period of absolute or relative decline in population.

risk society: contemporary societies in which politics is increasingly about avoiding hazards.

rites of passage: ceremonial acts, customs, practices, or procedures that recognize key transitions in human life, such as birth, menstruation, and other markers of adulthood such as marriage.

romanticism: philosophy that emphasizes interdependence and relatedness between humans and nature.

S

sacred space: area recognized by individuals or groups as worthy of special attention as a site of special religious experiences or events.

secondary activities: economic activities that process, transform, fabricate, or assemble the raw materials derived from primary activities or that reassemble, refinish, or package manufactured goods.

sectionalism: extreme devotion to local interests and customs.

segregation: spatial separation of specific population subgroups within a wider population.

self-determination: right of a group with a distinctive politico-territorial identity to determine its own destiny, at least in part, through the control of its own territory.

semiotics: practice of writing and reading signs.

semiperipheral regions: regions that are able to exploit peripheral regions but are themselves exploited and dominated by core regions.

sense of place: feelings evoked among people as a result of the experiences and memories that they associate with a place and the symbolism that they attach to it.

sexuality: set of practices and identities that a given culture considers related to each other and to those things it considers sexual acts and desires.

shifting cultivation: system in which farmers aim to maintain soil fertility by rotating the fields within which cultivation occurs.

shock city: city that is seen as the embodiment of surprising and disturbing changes in economic, social, and cultural life.

site: physical attributes of a location—its terrain, its soil, vegetation, and water sources, for example.

situation: location of a place relative to other places and human activities.

slash-and-burn (swidden): system of cultivation in which plants are cropped close to the ground, left to dry for a period, and then ignited.

slow food: attempt to resist fast food by preserving the cultural cuisine and the associated food and farming of an ecoregion.

society: sum of the inventions, institutions, and relationships created and reproduced by human beings across particular places and times.

sovereignty: exercise of state power over people and territory, recognized by other states and codified by international law.

spatial analysis: study of geographic phenomena in terms of their arrangement as points, lines, areas, or surfaces on a map.

spatial diffusion: way that things spread through space and over time.

spatial interaction: movement and flows involving human activity.

spatial justice: fairness of the distribution of society's burdens and benefits, taking into account spatial variations in people's needs and in their contribution to the production of wealth and social well-being.

spread effects: positive impacts on a region (or regions) of the economic growth of some other region.

squatter settlements: residential developments that occur on land that is neither owned nor rented by its occupants.

state: an independent political units with territorial boundaries that are internationally recognized by other states.

strategic alliances: commercial agreements between transnational corporations, usually involving shared technologies, marketing networks, market research, or product development.

subsistence agriculture: farming for direct consumption by the producers; not for sale.

suburbanization: growth of population along the fringes of large metropolitan areas.

supranational organizations: collections of individual states with a common goal that may be economic and/or political in nature.

sustainable development: vision of development that seeks a balance among economic growth, environmental impacts, and social equity.

swidden: land that is cleared using the slash-and-burn process and is ready for cultivation.

symbolic landscapes: representations of particular values or aspirations that the builders and financiers of those landscapes want to impart to a larger public.

T

technology: physical objects or artifacts, activities or processes, and knowledge or know-how.

technology systems: clusters of interrelated energy, transportation, and production technologies that dominate economic activity for several decades at a time.

terms of trade: ratio of prices at which exports and imports are exchanged.

territorial organization: system of government formally structured by area, not by social groups.

territoriality: specific attachment of individuals or peoples to a specific location or territory.

territory: delimited area over which a state exercises control and which is recognized by other states.

terrorism: threat or use of force to bring about political change.

tertiary activities: economic activities involving the sale and exchange of goods and services.

time-space convergence: rate at which places move closer together in travel or communication time or costs.

topological space: connections between, or connectivity of, particular points in space.

topophilia: emotions and meanings associated with particular places that have become significant to individuals.

total fertility rate (TFR): average number of children a woman will have throughout the years that demographers have identified as her childbearing years, approximately ages 15 through 49.

trade-off model: a modified urban land-use model that describes how poorer households, unable to afford the recurrent costs of transportation, trade off living space for accessibility to jobs and end up in high-density areas, at expensive locations, near their low-wage jobs.

trading blocs: groups of countries with formalized systems of trading agreements.

transcendentalism: philosophy in which a person attempts to rise above nature and the limitations of the body to the point where the spirit dominates the flesh.

transhumance: movement of herds according to seasonal rhythms: warmer, lowland areas in the winter; cooler, highland areas in the summer.

transnational corporations: companies with investments and activities that span international boundaries and with subsidiary companies, factories, offices, or facilities in several countries.

transnational migrant: migrants who set up homes and/or work in more than one nation-state.

tribe: form of social identity created by groups who share a set of ideas about collective loyalty and political action.

U

underemployment: situation in which people work less than full-time even though they would prefer to work more hours.

undernutrition: inadequate intake of one or more nutrients and/or calories.

unitary state: form of government in which power is concentrated in the central government.

urban agriculture (peri-urban agriculture): establishment or performance of agricultural practices in or near an urban or citylike setting.

urban ecology: social and demographic composition of city districts and neighborhoods.

urban form: physical structure and organization of cities.

urban system: interdependent set of urban settlements within a specified region.

urbanism: way of life, attitudes, values, and patterns of behavior fostered by urban settings.

urbanization: increasing concentration of population into growing metropolitan areas.

utility: usefulness of a specific place or location to a particular person or group.

V

vertical disintegration: evolution from large, functionally integrated firms within a given industry toward networks of specialized firms, subcontractors, and suppliers.

virgin soil epidemics: conditions in which the population at risk has no natural immunity or previous exposure to a disease within the lifetime of the oldest member of the group.

virtual water: water embedded in the production of the food and other things we consume.

vital records: information about births, deaths, marriages, divorces, and the incidence of certain infectious diseases.

voluntary migration: movement by an individual based on choice.

W

world city: city in which a disproportionate part of the world's most important business is conducted.

world-empire: minisystems that have been absorbed into a common political system while retaining their fundamental cultural differences.

world music: musical genre defined largely in response to the sudden increase of non-English-language recordings released in the United Kingdom and the United States in the 1980s.

world-system: interdependent system of countries linked by economic and political competition.

Y

youth cohort: members of the population who are less than 15 years of age and generally considered to be too young to be fully active in the labor force.

Z

zone in transition: area of mixed commercial and residential land uses surrounding the CBD.

PHOTO CREDITS

INDEX

A

Abdullah (king), 313
al-Abidine Ben Ali, Zine, 312
Abidjan, Côte d'Ivoire, 375
Abraham, 157
Absolute distance, 19
Accessibility, 386
    spatial analysis and, 21–22
Accra, Ghana, 360
Acid rain, 128–129
    global emissions, 134
Acquired immunodeficiency syndrome (AIDS).
    See HIV/AIDS
Actor-network theory (ANT), 174
Adidas, 238, 249
Administrative centers, 357
Advanced Very High Resolution Radiometer
    (AVHRR), 15
Advertising, 197, 208
    Wal-Mart, 249
Aerial photography, 9
Affect, 175
Affluence
    birth and death rates, 84
    in China, 384–385
    environmental impact, 108–109
    international culture, 239
    patterns of American cities, 390
Afghanistan, 328–330
    ecological footprint, 222
    educating girls in, 98
    England and, 328
    fertility rate, 80
    IDPs in, 89
    infant mortality rate, 81
    Islamic population, 164
    migration in, 86
    Soviet Union and, 328–329
    war, 240
Africa. *See also* Specific countries
    Arab Spring, 312–315
    birthrate, 79
    boundaries, 305–306
    cocoa in, 229
    colonization of, 43, 311, 316
    cultivable land, 220
    debt problem, 230
    decolonization of, 320
    desertification of, 138
    energy consumption of, 220
    food production, 3, 34
    food production in, 137
    fuelwood depletion, 127

    geopolitics, 302
    HIV/AIDS in, 81–82
    IDPs in, 89
    internal displacement, 88
    kinship social organization, 163
    labor force, 223
    land purchased in, 137
    language in, 163
    movie production, 179
    natural resources, 221
    pastoralism in, 265
    population, 69, 71, 96
    precipitation map, 10
    shifting cultivation, 261
    tourism spending, 251
    urbanization in, 350–351, 372–375
    world-empires in, 39
African Americans
    cognitive image of Los Angeles, 192
    hip-hop culture, 148
    HIV/AIDS, 81
    identity, 70
    infant mortality rates, 80–81
    internal migration of, 91
    life expectancy, 81
Agency, 174
Agent orange, 325
Age-sex pyramid, 73–75
    for Germany, 74
Agglomeration diseconomies,
    235, 378–379
Agglomeration effects, 233
Agha-Soltan, Neda, 312
Agrarian, 260
Agribusiness, TNCs and, 281
Agricultural density, 71
Agricultural revolutions
    environment and, 120
    first, 266
    industrialization and, 270, 274
    second, 266–268
    third, 268–269
Agriculture
    alternative food movements, 283–285
    biotechnology and, 274, 278–279
    contract farming, 320
    in Corsica, 152
    cultivable land and, 137, 220, 221
    definition of, 260
    diversity of foodstuffs, 139–140
    environmental impact, 290–292
    fast-food, 289–290

    food production and, 107, 258–295
    food regimes, 282–283
    gender division of labor, 264
    globalization and, 261, 279–281
    global warming, 132
    industrialization of, 270, 274
    intensive subsistence, 264–265
    organization of agro-food system, 281
    pastoralism, 265–266
    shifting cultivation, 261–264
    subsistence, 36, 187, 261
    in Sudan, 187
    urban, 286–288
Ahmadinejad, Mahmoud, 312
Aigues-Mortes, France, 359
Air pollution
    in Asia, 125
    peripheral cities, 410
Air travel
    disruption by Icelandic ash
        plume, 34–35
    hubs for, 21–22
Alaska, 94
Albania, 164
Albany, N.Y., 22
Alexander the Great, 328
Alfred P. Murrah Federal Office Building,
    in Oklahoma City, 326, 327
Algeria, 312
    Arab Spring, 313
    energy resources, 217
    IDPs in, 89
    Islamic population, 164
al-Qaeda, 328–329, 331
Alternative energy, 133–134, 294
Alternative food movements, 283–285
Althusser, Louis, 310–311
Altria, 241
Amazon tribes, 37
American Dreams group, 16
American Express, 247
American Girl Place, 253
Americanization, 60
    in France, 164
    globalization and, 175, 179
American Society of Civil Engineers
    (ASCE), 395
Americas. *See also* North America
    hearth areas in, 37
Amin, Ash, 354
Amsterdam, Netherlands, 360
Analysis, 7
Anan, Kofi, 334

Water Use

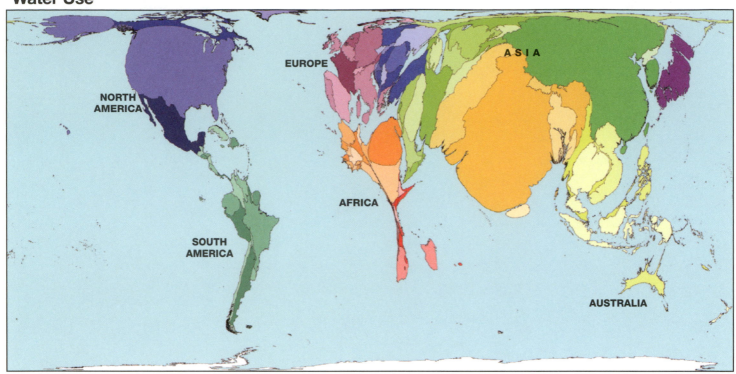

Wealth 2015

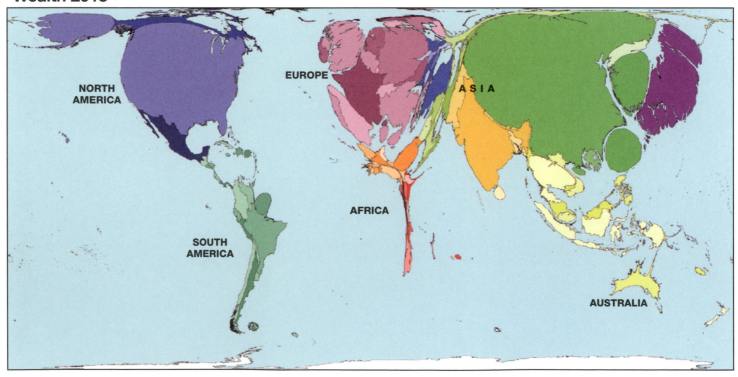

Human Development

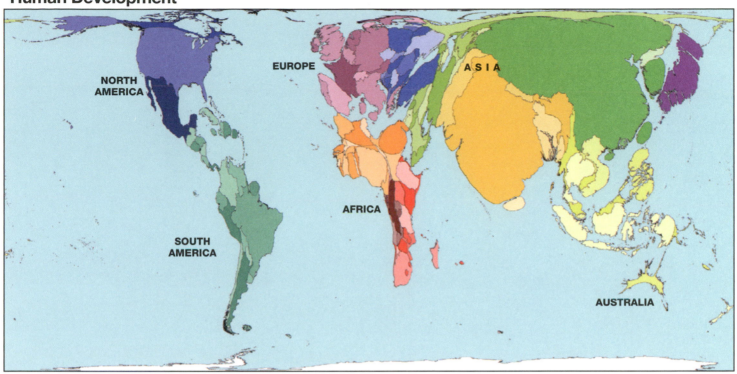

NORTH AMERICA

EUROPE

ASIA

AFRICA

SOUTH AMERICA

AUSTRALIA

Infant Mortality

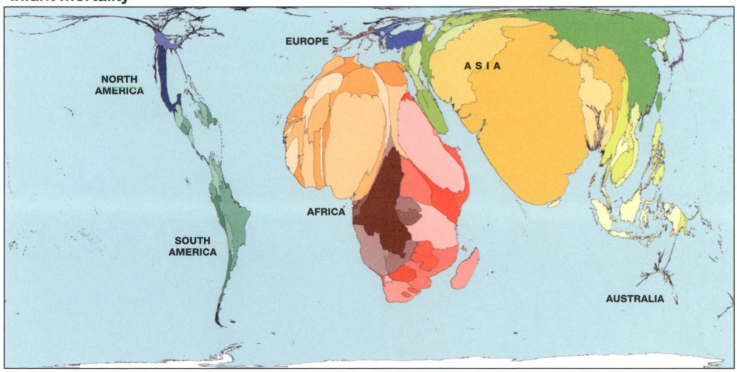

NORTH AMERICA

EUROPE

ASIA

AFRICA

SOUTH AMERICA

AUSTRALIA